Nervous Tissue Cells in Health and Disease

Nervous Tissue Cells in Health and Disease

Editor: Emma Hixon

MURPHY & MOORE
www.murphy-moorepublishing.com

www.murphy-moorepublishing.com

MURPHY & MOORE

Cataloging-in-Publication Data

Nervous tissue cells in health and disease / edited by Emma Hixon.
p. cm.
Includes bibliographical references and index.
ISBN 978-1-63987-792-8
1. Nerve tissue. 2. Cells. 3. Nervous system--Health aspects.
4. Nervous system--Diseases. I. Hixon, Emma.
RC346 .N47 2023
616.8--dc23

Murphy & Moore Publishing
1 Rockefeller Plaza,
New York City,
NY 10020, USA

ISBN 978-1-63987-792-8

Contents

Preface

The main aim of this book is to educate learners and enhance their research focus by presenting diverse topics covering this vast field. This is an advanced book which compiles significant studies by distinguished experts in the area of analysis. This book addresses successive solutions to the challenges arising in the area of application, along with it; the book provides scope for future developments.

Nervous tissue is referred to as the primary tissue component of the nervous system. It can be found in the nerves, brain and spinal cord. Glial cells and neurons are the two types of cells that make up nervous tissue. Neurons are responsible for the communication and computation provided by the nervous system. Glial cells play a supporting role for nervous tissues and are considerably smaller than neurons. They enhance signal conduction within neurons and defend neurons from pathogens. Many body activities are controlled and coordinated by the nervous tissue. It develops environmental awareness, promotes muscle contraction and plays an important role in thinking, emotions and memory. The function of the nervous tissue is to create the communication network of the nervous system by conducting electric signals through the tissue. This book provides comprehensive insights on the role of nervous tissue cells in health and disease. It presents researches and studies performed by experts across the globe. The readers would gain knowledge that would broaden their perspective in this area of study.

It was a great honour to edit this book, though there were challenges, as it involved a lot of communication and networking between me and the editorial team. However, the end result was this all-inclusive book covering diverse themes in the field.

Finally, it is important to acknowledge the efforts of the contributors for their excellent chapters, through which a wide variety of issues have been addressed. I would also like to thank my colleagues for their valuable feedback during the making of this book.

Editor

1

Microglial Lectins in Health and Neurological Diseases

Jian Jing Siew [1,2] *and Yijuang Chern* [1,2]*

[1] *Taiwan International Graduate Program in Molecular Medicine, National Yang-Ming University and Academia Sinica, Taipei, Taiwan,* [2] *Institute of Biomedical Sciences, Academia Sinica, Taipei, Taiwan*

***Correspondence:**
Yijuang Chern
bmychern@ibms.sinica.edu.tw

Microglia are the innate sentinels of the central nervous system (CNS) and are responsible for the homeostasis and immune defense of the CNS. Under the influence of the local environment and cell-cell interaction, microglia exhibit a multidimensional and context-dependent phenotypes that can be cytotoxic and neuroprotective. Recent studies suggest that microglia express multitudinous types of lectins, including galectins, Siglecs, mannose-binding lectins (MBLs) and other glycan binding proteins. Because most studies that examine lectins focus on the peripheral system, the functions of lectins have not been critically investigated in the CNS. In addition, the types of brain cells that contribute to the altered levels of lectins present in diseases are often unclear. In this review, we will discuss how galectins, Siglecs, selectins and MBLs contribute to the dynamic functions of microglia. The interacting ligands of these lectins are complex glycoconjugates, which consist of glycoproteins and glycolipids that are expressed on microglia or surrounding cells. The current understanding of the heterogeneity and functions of glycans in the brain is limited. Galectins are a group of pleotropic proteins that recognize both β-galactoside-containing glycans and non- β-galactoside-containing proteins. The function and regulation of galectins have been implicated in immunomodulation, neuroinflammation, apoptosis, phagocytosis and oxidative bursts. Most Siglecs are expressed at a low level on the plasma membrane and bind to sialic acid residues for immunosurveillance and cell-cell communication. Siglecs are classified based on their inhibitory and activatory downstream signaling properties. Inhibitory Siglecs negatively regulate microglia activation upon recognizing the intact sialic acid patterns and vice versa. MBLs are expressed upon infection in cytoplasm and can be secreted in order to recognize molecules containing terminal mannose as an innate immune defense machinery. Most importantly, multiple studies have reported dysregulation of lectins in neurological disorders. Here, we reviewed recent studies on microglial lectins and their functions in CNS health and disease, and suggest that these lectin families are novel, potent therapeutic targets for neurological diseases.

Keywords: microglia, galectin, Siglec, selectin, mannose-binding lectin, neurological disease, neurodegeneration

INTRODUCTION

The central nervous system (CNS) consists of neurons and glial cells such as microglia, astrocytes and oligodendrocytes. Microglia are the CNS counterpart of macrophages and differ from other types of glial cells in terms of their origin, morphology and functions (Kettenmann et al., 2011). They are widely distributed throughout the brain and spinal cord, and constitute 5%–10% of total brain cells depending on the region (Lawson et al., 1990).

Microglia play multiple roles in healthy and diseased brains. During development, microglia regulate the number of neuronal cells by triggering programmed cell death and actively phagocytose the apoptotic neurons (de la Rosa and de Pablo, 2000; Bessis et al., 2007). In addition, microglia are capable of promoting neurogenesis by secreting trophic factors (such as nerve growth factor (NGF) and insulin-like growth factor 1 (IGF-1; Frielingsdorf et al., 2007; Thored et al., 2009). In addition to regulating the number of neurons, microglia also strengthen and prune excessive synapses to ensure that neuronal circuits are connected correctly (Paolicelli et al., 2011). Hence, microglia have been implicated in synaptic plasticity.

As the immune sentinels of the CNS, microglia are responsible for detecting and swiftly responding to infection, injury, molecules released by damaged cells and misfolded proteins (Davalos et al., 2005; Nimmerjahn et al., 2005). In this context, microglia are believed to play exacerbating roles in the inflammatory response of neurological diseases, including multiple sclerosis (MS; Politis et al., 2012), ischemic stroke (Patel et al., 2013), Alzheimer's disease (AD; Fuhrmann et al., 2010), Parkinson's disease (PD; Subramaniam and Federoff, 2017), amyotrophic lateral sclerosis (ALS; Brites and Vaz, 2014), Huntington's disease (HD; Hsiao et al., 2014) and virus infection in the brain (Cenker et al., 2017). Hence, elucidating the machinery that regulates the multidimensional phenotypes of microglia is critical for understanding neurological diseases and developing effective therapeutic interventions.

Microglia express various types of lectins that functionally regulate their dynamic phenotypes. These lectins include galectins, Siglecs, C-type lectins (such as mannose-binding lectins, MBLs) and other glycan binding proteins. Some of these lectins are expressed constitutively, while others are expressed upon stimulation. For instance, microglia generally do not express galectin-3, but its expression can be up-regulated by inflammatory mediators (Burguillos et al., 2015). Galectins have multiple cellular localizations and different functions. For instance, galectin-1, -3 and -9 have been reported to exist in the nucleus, cytoplasm, plasma membrane and extracellular matrix. The cell types in the CNS that express galectins and their cellular localization and functions are summarized in **Table 1**. Contrary to galectins, all CNS Siglecs (except Siglec-4) are localized on the plasma membrane of microglia. Details about the CNS Siglecs are summarized in **Table 2**. Similar to Siglecs, selectins are localized on the plasma membrane of endothelial cells. In contrast, MBL is found in the cytoplasm of all brain cells in disease conditions and can be secreted into the extracellular matrix. Details on these C-type lectins are summarized in **Table 3**. The cell surface contains a complex layer of glycosylated molecules, such as glycoproteins and glycolipids that are recognized by these lectins and store important information for cell-cell communication and response (Brandley and Schnaar, 1986; Groves, 2013; Letschert et al., 2014). Classification of lectins is based on the homology of amino acid sequences and specificity of glycan structures. Their functions have been reviewed in detail elsewhere (Varki et al., 2009; Fujimoto et al., 2014).

TABLE 1 | Summary of galectins in central nervous system (CNS).

Type	Ligands	Cell type/Subcellular location		Functions	References
Galectin-1	β-Galactosides	Microglia Astrocytes Neurons	• Nucleus • Cytoplasm • Plasma membrane • Extracellular	• Suppression of microglial activation • Regulation of neurogenesis • Regulation of oligodendrocyte proliferation	Sakaguchi et al. (2006, 2007) and Starossom et al. (2012)
Galectin-3	β-Galactosides	Microglia	• Nucleus • Cytoplasm • Plasma membrane • Extracellular	• Microglial activation • Promotion of neuroinflammation • Suppression of neuroinflammation • Regulation of oligodendrocyte differentiation	Jiang et al. (2009), Lerman et al. (2012) and James et al. (2016)
Galectin-4	β-Galactosides	Neurons	• Plasma membrane	• Axonal growth • Axonal transport • Suppression of axonal myelination	Cao and Guo (2016)
Galectin-8	β-Galactosides	Microglia Others	• High levels in hippocampus and choroid plexus • Other brain regions • CSF	• Immunosuppression	Stancic et al. (2011), John and Mishra (2016) and Pardo et al. (2017)
Galectin-9	β-Galactosides	Microglia Astrocytes	• Nucleus • Cytoplasm • Plasma membrane • Extracellular	• Modulation of neuroinflammation	Stancic et al. (2011), Lerman et al. (2012) and Steelman et al. (2013)

TABLE 2 | Summary of Siglecs in CNS.

Type	Ligands	Cell type/ Subcellular location		Functions	References
Siglec-1 (sialoadhesin, CD169)	α2–3, α2–6 sialic acid residues	Microglia	• Plasma membrane	• Cell-cell interaction	Macauley et al. (2014) and Groh et al. (2016)
Siglec-3 (CD33)	α2–6, α2–3 sialic acid residues	Microglia	• Plasma membrane	• Modulation of microglial phagocytosis activity	Griciuc et al. (2013) and Macauley et al. (2014)
Siglec-4 (MAG)	α2–3, α2–6 sialic acid residues	On myelin sheath, produced by oligodendrocytes	• Inner-most, myelin periaxonal membrane	• Myelin-axon interaction, maintaining long-term axon stability	Sun et al. (2004), Huang et al. (2012) and Macauley et al. (2014)
Siglec-E	α2–3, α2–6, α2–8 sialic acid residues	Microglia	• Plasma membrane	• Inhibition of inflammation, ROS production and phagocytosis	Claude et al. (2013) and Macauley et al. (2014)
Siglec-F	α2–3 sialic acid residues	Microglia	• Plasma membrane	• Inhibition of phagocytosis	Wielgat and Braszko (2012) and Macauley et al. (2014)
Siglec-H	Abnormal sialic acid residues in glioma	Microglia	• Plasma membrane	• Promotion of phagocytosis	Kopatz et al. (2013) and Macauley et al. (2014)
Siglec-11	α2–8 sialic acid residues	Microglia	• Plasma membrane	• Inhibition of inflammation and phagocytosis	Wang and Neumann (2010), Wang X. et al. (2012) and Macauley et al. (2014)
Siglec-16	α2–8 sialic acid residues	Microglia	• Plasma membrane	• No functional data	Wang X. et al. (2012) and Macauley et al. (2014)

TABLE 3 | Summary of C-type lectins in CNS.

Type	Ligands	Cell type/Subcellular location		Functions	References
E-Selectin	Sialyl-Lewis X terminal	Endothelial cells	• Plasma membrane	• Facilitate the adhesion of neutrophils	Jin et al. (2010)
P-Selectin	Sialyl-Lewis X terminal	Endothelial cells	• Plasma membrane	• Facilitate the adhesion of neutrophils	Atkinson et al. (2006) and Jin et al. (2010)
Mannose-binding lectin	Terminal mannose, fucose, GlcNAc	Microglia Astrocytes Oligodendrocytes Neuron	• Cytoplasm • Extracellular	• Innate immune defense	Singh et al. (2011)

In this review, we focus on the role of lectins and how they affect the functions of microglia in the CNS. **Tables 1–3** summarizes the major lectins that have been reported in the CNS so far.

GALECTINS

Galectins represent a family of lectins characterized by conserved carbohydrate-recognition domains (CRDs) and bind to β-galactosides with different specificities and affinities (Liu and Rabinovich, 2005). Through the conserved CRDs, 15 galectins have been identified; they can be categorized into three groups: (1) proto, or one CRD; (2) tandem repeats, or two distinct CRDs in tandem; and (3) chimera, consisting of CRD with unusual tandem repeat of glycine- and proline-rich short stretches (Kasai and Hirabayashi, 1996). The prototype galectins include galectins-1, -2, -5, -7, -10, -11, -13, -14 and -15, while the tandem repeats type consists of galectins -4, -6, -8, -9 and -12; and galectin-3 is the only galectin classified as chimera type. Among these galectins, only galectins -1, -2, -3, -4, -7, -8, -9, -10, -12 and -13 are expressed in humans (Chang et al., 2017). Furthermore, many of these galectins have differentially spliced forms that may be expressed in different tissues. For instance, galectins-8 and -9 have seven and three isoforms, respectively (Zhang et al., 2009; Troncoso et al., 2014). In particular, the galectin-9M and 9S isoforms positively promote the expression of E-selectin, while the galectin-9L isoform suppresses the level of E-selectin (Zhang et al., 2009). These isoforms are critical in regulating important cellular activity. The mechanisms involved in coordinating the expression of these isoforms and the functions they serve are still in need of extensive exploration. To date, no specific surface receptor for galectins has been reported. Nonetheless, galectins can bind to and interact with various glycoproteins and cell membrane glycoconjugates via their carbohydrate moieties and trigger a cascade of transmembrane signaling events (Liu and Rabinovich, 2005; Wan and Liu, 2016). In addition, galectins have non-carbohydrate binding partners and function intracellularly to regulate gene transcription, mRNA splicing, cell growth, apoptosis and immune responses (Liu et al., 2002; Wang J. L. et al., 2012). For instance, intracellular galectin-1 and galectin-3 interact with Gemin4 to facilitate the assembly of spliceosomes (Park et al., 2001). The functions of galectins in the peripheral system have been extensively studied. To date, only galectins-1, -3, -4, -8 and -9 have been found in the CNS. **Table 4** summarizes the role of galectins in neurological disorders.

Galectins and Myelination

During the maturation of neurons, axons are myelinated in a specific manner that leaves certain parts of the segments

TABLE 4 | Summary of the role of galectins in neurological disease.

Neurological conditions	Type of galectin	Disease-associated features	References
Multiple sclerosis	Galectin-1	• Astrocytes produce galectin-1 to deactivate microglia through the p38, CREB and NF-κB pathways. • The function of galectin-1 produced by MS microglia is unknown.	Starossom et al. (2012)
	Galectin-3	• Modulation of immune cell infiltration into the CNS • Up-regulation of inflammatory cytokines and down-regulation of anti-inflammatory cytokines in dendritic and T cells. • Inhibition of neural cell proliferation	Jiang et al. (2009) and James et al. (2016)
	Galectin-8	• Up-regulated in microglia. • Immunosuppression in the periphery, unknown effects in the brain.	Stancic et al. (2011) and Pardo et al. (2017)
	Galectin-9	• Up-regulated in CSF of secondary progressive MS patients and microglia, its function is not clear. • Located in the cytosol of microglia in inactive lesions. • Located in the nuclei and cytosol of microglia in active lesions.	Stancic et al. (2011) and Burman and Svenningsson (2016)
Stroke and ischemia	Galectin-3	• Up-regulated in the microglia, promotes proliferation. • Promotes the production of IGF-1. • Suppression of IL6 inflammatory cytokine. • Acts as the ligand of TLR4 to promote inflammatory responses.	Lalancette-Hébert et al. (2007, 2012) and Burguillos et al. (2015)
Traumatic brain injury	Galectin-3	• Up-regulated in the corpus callosum and is associated with the production of NGF. • Up-regulated in the cortex and hippocampus and promotes inflammation.	Venkatesan et al. (2010) and Yip et al. (2017)
Amyotrophic Lateral Sclerosis (ALS)	Galectin-1	• Up-regulated in the spinal cord at the end stage.	Lerman et al. (2012)
	Galectin-3	• Up-regulated in the microglia and suppresses inflammation and oxidative damage in the spinal cord.	Lerman et al. (2012)
	Galectin-9	• Up-regulated in the spinal cord at the symptomatic stage	Lerman et al. (2012)
Parkinson's disease	Galectin-3	• Up-regulated in the microglia to promote phagocytosis of α-synuclein and secretion of inflammatory cytokines.	Boza-Serrano et al. (2014)
Prion disease	Galectin-3	• Up-regulated in the medulla and pons. • Up-regulation associated with impairment of lysosomal function and autophagy.	Riemer et al. (2004) and Mok et al. (2007)

unmyelinated; these unmyelinated nodes are marked by contactin-1 (Çolakoğlu et al., 2014; **Figure 1**). Beginning in development, the expression of galectin-4 is different in each cortical layer and brain region. Galectin-4 is absent in the striatum and cortical layer VI, which are heavily myelinated, while its level remains high in cortical layers II to III, which contain neurons with less myelination at the mature stage. In particular, galectin-4 is found to cover these unmyelinated segments, colocalize with contactin-1 and inhibit oligodendrocytes from depositing myelin on these segments (Díez-Revuelta et al., 2017). Additionally, galectin-4 is secreted by neurons and inhibits the maturation of oligodendrocytes (Stancic et al., 2012). In contrast, astrocytes and microglia are involved in regulating and supporting myelination in different manners. In particular, astrocytes promote the proliferation of oligodendrocytes, while microglia induce the differentiation of oligodendrocytes, as demonstrated by culturing oligodendrocytes with conditioned medium from the two cell types (Pang et al., 2013). In addition, galectin-1, which is secreted mainly by astrocytes, shift microglia from an inflammatory to an anti-inflammatory status, which subsequently drives oligodendrocytes toward the myelination process (Starossom et al., 2012; Miron et al., 2013; Sirko et al., 2015). In parallel with these findings, galectin-3 controls the integrity of myelin *in vivo*, as demonstrated by the loosely wrapped myelin structures in galectin-3-depleted mice (Pasquini et al., 2011). Furthermore, in the cuprizone-induced demyelination mouse model, the level of galectin-3 is increased in microglia, which promotes the differentiation of oligodendrocytes for remyelination. The abovementioned machinery is impaired in galectin-3-depleted mice (Hoyos et al., 2014, 2016).

Galectins in Multiple Sclerosis (MS)

MS is a chronic inflammatory and autoimmune neurological disease marked by damage to the myelin that surrounds the axons in the CNS (Goldenberg, 2012). The pathogenesis of MS can be divided into two processes: (1) the overactivated immune system damages the myelin sheath and axon; and (2) remyelination is impaired (Dendrou et al., 2015). Both microglia and macrophages are chronically activated throughout the progression of MS and contribute to the inflammatory and demyelination processes (Fischer et al., 2012). Importantly, galectins-1, -3, -8 and -9 are reported to be up-regulated in human MS white matter samples. Immunohistochemical staining reveals that galectin-1 is up-regulated in microglia and macrophages, while its quantity is decreased in the astrocytes, at the lesioned area. The reduction of astrocytic galectin-1 may be due to the enhanced release of galectin-1 by astrocytes and may play an important role in MS progression. Conversely, the levels of galectins-8 and -9 are increased in microglia and macrophages located in active MS lesions. Interestingly,

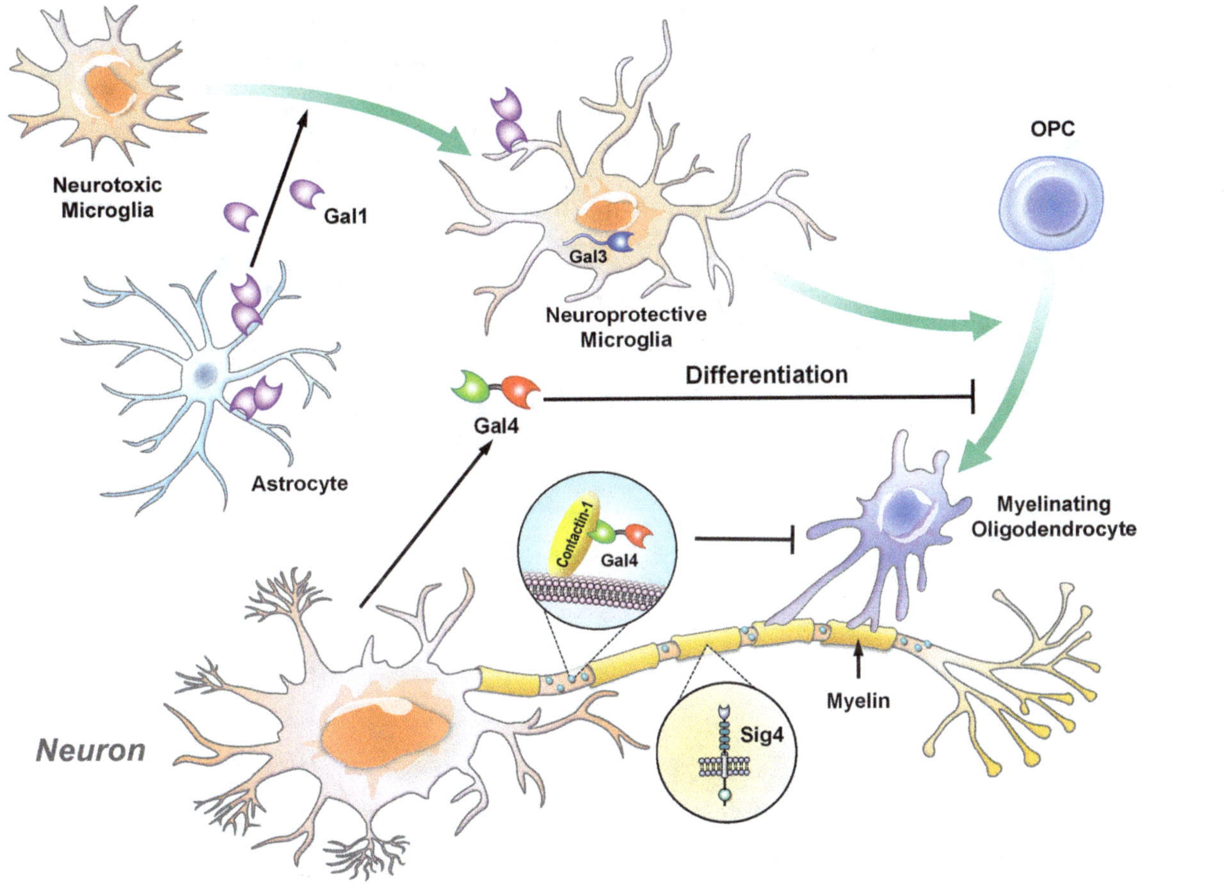

FIGURE 1 | Galectins and Siglec-4 contribute to the glia-mediated regulation of neuronal myelination. Oligodendrocytes are the cells responsible for myelination in the central nervous system (CNS). Under normal conditions, galectin-4 is delivered to the unmyelinated axonal regions and co-localizes with contactin-1 (a marker for nodes of Ranvier). The presence of galectin-4 inhibits oligodendrocytes from myelinating the region (Çolakoğlu et al., 2014). In addition, soluble galectin-4 inhibits the maturation of oligodendrocytes (Stancic et al., 2012). Siglec-4 is produced by oligodendrocytes and is expressed in the myelin sheath (Sun et al., 2004; Huang et al., 2012). Astrocytes and microglia promote the proliferation and differentiation, respectively, of oligodendrocytes. In particular, astrocytes secrete galectin-1 to drive microglia toward neuroprotective phenotype, while these microglia trigger the differentiation of oligodendrocytes in a galectin-3-dependent manner (Hoyos et al., 2014, 2016). Gal, galectin; Sig, Siglec.

galectin-9 alters cellular localization in MS microglia upon activation. In inactive lesion, galectin-9 is located in the cytoplasmic region of microglia, while it is present in both the nuclei and the cytosol of microglia in active lesions (Stancic et al., 2011). The functional relevance of this nuclear localization of galectin-9 remains elusive and requires further investigation.

To date, the most widely used animal model for the study of MS pathogenesis has been the experimental autoimmune encephalomyelitis (EAE) model (Steinman and Zamvil, 2006). In the EAE model, the expression of galectin-1 is up-regulated, with the highest level at the peak of the disease. Further analysis shows that galectin-1 is highly expressed in astrocytes (GFAP+ cells) and moderately expressed in a subset of T cells (CD4+ cells) and microglia/macrophages (CD11b+ cells) during the acute phase. Interestingly, astrocytes, but not other cell types, continue to express galectin-1 during the chronic phase. One of the specific binding targets of the secreted galectin-1 consists of the core 2 O-glycans of CD45 on the cell surface of microglia. Treatment of the pro-inflammatory microglia with galectin-1 inhibits the activation of the p38, CREB and NF-κB pathways (**Figure 2**). In the absence of galectin-1, MS progresses much more rapidly. More microglia with inflammatory phenotype and much more severe demyelination can be observed in the galectin-1 depleted EAE model (Starossom et al., 2012). These studies suggest a beneficial role of galectin-1 in the EAE model.

The level of galectin-3 is also up-regulated in the phagocytosing microglia in areas of demyelination in the EAE model (Reichert and Rotshenker, 1999; Stancic et al., 2011). Further studies suggested that the onset and severity of EAE were markedly reduced in galectin-3 KO mice (Jiang et al., 2009). In a demyelination experimental model of MS, generated by infection with Theiler's murine encephalomyelitis virus (TMEV), the level of galectin-3 is also up-regulated in microglia residing in the subventricular zone

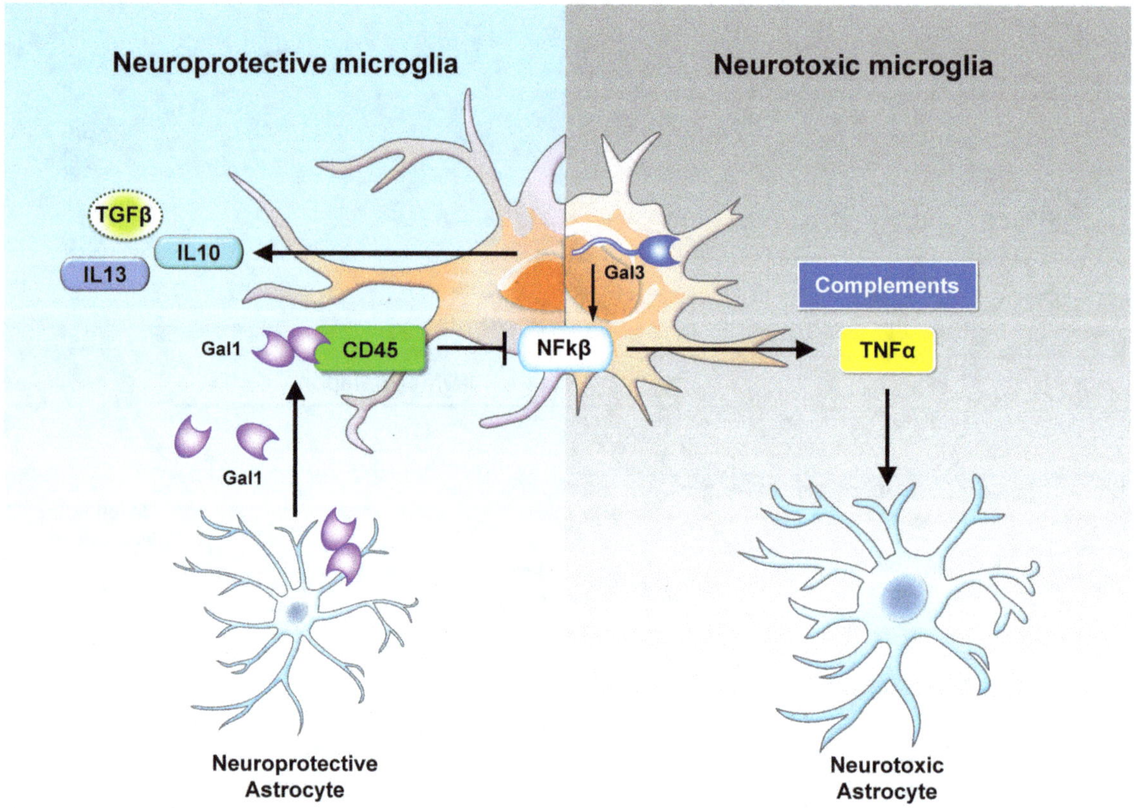

FIGURE 2 | Cross-regulation between astrocytes and microglia by galectin-1 and galectin-3. Neuroprotective astrocytes secrete galectin-1 to inhibit the neurotoxic activation of microglia. The secreted galectin-1 binds with CD45 on the microglia surface, suppresses NF-κB activation and promotes the polarization of neuroprotective microglia (Starossom et al., 2012). Neurotoxic microglia express various types of lectins. In particular, the up-regulated galectin-3 promotes inflammation response through NF-κB activation (Boza-Serrano et al., 2014). These microglial cells secrete inflammatory cytokines (TNFα) and complements that trigger the activation of neurotoxic astrocytes (Liddelow et al., 2017). Gal, galectin; Sig, Siglec.

(the neurogenetic niche that is markedly inflamed in MS and its murine models) and the cerebral cortex. In this model, genetic removal of galectin-3 greatly normalizes the immune response observed in the SVZ, reduces the migration of immune cells into the SVZ, and increases the number of progenitors in the corpus callosum (James et al., 2016). These lines of evidence suggest that galectin-3 may play an essential role in MS.

Galectin-8 can be detected in the cerebrospinal fluid (CSF) of MS patients. Genetic removal of galectin-8 hastens the onset of the disease and worsens the progression. Daily treatment with galectin-8 significantly delays the disease progression of EAE. In particular, galectin-8 inhibits the activity of Th17-positive cells to induce inflammation by influencing cells toward apoptosis (Pardo et al., 2017). Similarly, galectin-9 can also be detected in the CSF of secondary progressive MS patients (Burman and Svenningsson, 2016). Compared to the peripheral nervous system, the roles of galectin-8 and galectin-9 in the CNS (such as their expression profiles and their functions in different brain cells) have not been extensively explored and are worth further investigation.

Galectins in Stroke and Ischemia

Galectin-3 is up-regulated in microglia after stroke and induces the expression of IGF-1 (Lalancette-Hébert et al., 2007). Importantly, galectin-3 interacts directly with IGF-1 to promote post-ischemic microglial proliferation. These microglia are suggested to serve as the reservoir of neurotrophic factors, such as IGF-1. Genetic depletion of galectin-3 significantly reduces microglial activation and proliferation; these effects are associated with larger ischemic lesion size, decreased IGF-1 levels and neuronal death. Moreover, there is a twofold increase in IL6 in the ischemic brains of galectin-3-depleted mice (Lalancette-Hébert et al., 2012). These studies suggest that the presence of galectin-3 promotes alternative activation of microglia toward an anti-inflammatory status after stroke. Nonetheless, in a separate study, secreted galectin-3 was shown to act as a ligand for the toll-like receptor 4 (TLR4) and promote inflammatory response in a murine neuroinflammatory LPS model and in the brain of stroke patients. Compared to wild-type controls, galectin-3-knockout (KO) mice have elevated neuronal survival in the hippocampus after ischemia, suggesting that galectin-3 prolongs microglia at the pro-inflammatory status (Burguillos et al., 2015). Consistent with a detrimental role, galectin-3 is found to be up-regulated in the CA1 and CA2 hippocampal regions following transient ischemia. The expression of galectin-3 in these regions is associated with neuronal death and can be prevented by intra-ischemic hypothermia treatment (Satoh et al., 2011; Hisamatsu et al., 2016). Moreover, in the permanent middle cerebral artery occlusion (pMCAO) model, the lack of galectin-3 rescues significant neuronal loss. Treatment with sera collected from pMCAO-treated WT mice also induces increased death of enteric neurons compared with sera collected from pMCAO-treated galectin-3 KO mice. The death of these neurons can be prevented by inhibition of TAK1 and AMPK (transforming growth factor and beta-activated kinase 1 and AMP activated kinase, respectively; Cheng et al., 2016). Collectively, the inflammatory roles of galectin-3 are diverse and may depend on the following factors: the type of ischemic insult, the stage of disease/trauma and the subcellular localization.

Galectins in Traumatic Brain Injury (TBI)

Traumatic Brain Injury (TBI) is clinically categorized into three groups, namely, mild, moderate and severe. Its effects can be further divided into primary and secondary injuries. Primary injuries (such as direct damage to the brain parenchyma that may shear axons) are usually irreversible. On the other hand, secondary injuries are the detrimental effects resulting from the primary injury (such as inflammation and ischemia; Blennow et al., 2016). In response to TBI, in order to restore the normal brain environment, microglia rapidly migrate to the site of injury to remove damaged and dead cells. In a mouse closed-skull model of TBI, the level of galectin-3 increased in the corpus callosum for about 1 month following injury, with the peak at 24 h after injury. The galectin-3- positive microglia engulf damaged axons and produce NGF that facilitates the repair of damaged neurons (Venkatesan et al., 2010). In another study using the controlled cortical impact (CCI) model of head injury, galectin-3 is also up-regulated in the microglia but binds with TLR4 to promote inflammation in the cortex and hippocampus. Treatment with a galectin-3-neutralizing antibody reduces the expression of pro-inflammatory genes *IL1β*, *IL6* and *NOS2* while promoting the expression of *Ym1*, *Arg1* and *TGFβ*. Consistent with a damaging role of galectin-3, galectin-3-deficient mice show less neuronal death and microglial activation in a CCI model (Yip et al., 2017). These data suggest that the role of microglial galectin-3 in TBI can be complex depending on the severity and type of trauma.

Galectins in Amyotrophic Lateral Sclerosis (ALS)

ALS is a rare, progressive and fatal neurodegenerative disease. Multiple genes have been identified as being associated with familial ALS, but those genes explain only approximately 10% of cases. Nearly 90% of ALS cases are sporadic (Abhinav et al., 2007). Up-regulation of plasma galectin-3 is detected in ALS patients with limb onset or disease duration longer than 12 months. The level of plasma galectin-3 is correlated with the duration of disease. Female patients exhibit higher increase of galectin-3 in the plasma than male patients, although the reason is still unknown (Yan et al., 2016). A study using the B6SJL SOD1(G93A) transgenic mouse model of familial ALS (Smittkamp et al., 2008) revealed that microglia undergo a transition stage from the protective phenotype in early-stage ALS to the detrimental phenotype in end-stage ALS. Furthermore, early-stage microglia promote the survival of motor neurons in a co-culture system, while end-stage microglia trigger neuronal death (Liao et al., 2012).

In particular, galectin-1 is increased in the spinal cord of SOD1 mice, while galectins-3 and -9 are increased in SOD1 mice and sporadic ALS patients. The alterations in these galectins during ALS progression are different. Specifically, the amount of galectin-3 can be detected in microglia of spinal cord at the presymptomatic stage, and its level continues to increase until the end stage. Conversely, the expression of galectin-9 is increased at the symptomatic stage, while the levels of galectin-1 is up-regulated only at the end stage (Lerman et al., 2012). The types of cells that contribute to the higher levels of galectin-1 and -9 in the spinal cord are currently unknown.

Nonetheless, it is of great interest to observe that treatment of SOD1 mice with recombinant galectin-1 significantly enhances the survival of motor neurons, delays the onset of disease, improves motor performance, and prolongs the lifespan of ALS mice. Because galectin-1 treatment has no specific effect on isolated primary neurons, the therapeutic effect of galectin-1 may be mediated by non-neuronal cells (Chang-Hong et al., 2005). In particular, microglia and macrophages are among the target cells of galectin-1. For example, treatment with galectin-1 induces macrophages to produce an axonal regeneration-promoting factor that is critical to the survival of degenerating motor neurons (Horie et al., 2004). Overall, although the underlying mechanism remain to be clarified, galectin-1 is a potential therapeutic target for ALS.

Another interesting drug target for ALS is galectin-3. Genetic depletion of galectin-3 in SOD1 mice accelerates disease progression (i.e., a shorter life span and faster motor function impairment) with a significant increase in microglial activation, along with higher inflammatory response and oxidative damage (Lerman et al., 2012). In the CNS of healthy rats, the basal expression of galectin-3 is relatively higher in the spinal cord than in the cortex. Flow cytometry analysis detects at least two major populations of microglia (namely, galectin-3- positive and negative) in the spinal cord of SOD1(G93A) rats after disease onset. Compared to microglia in wild-type rats, microglia in the spinal cord of SOD1(G93A) rats express significantly higher levels of galectin-3 from early to late stage of the disease. The levels of both neurotoxic and neuroprotective molecules (i.e., TNFα, IL6, arginase-1, IL10 and BDNF) in the late stage of the disease are lower, compared with in the early stage. Intriguingly, these alterations only appeared in the spinal cord but not in the cortex, suggesting that the function of galectin-3 and the features of microglia are different in the spinal cord and cortex (Nikodemova et al., 2014). Collectively, these findings suggest that microglial galectin-3 plays a protective role in ALS, which appears different from most of the neurodegenerative diseases where the role of galectin-3 has been investigated so far; further investigation is required.

Galectin-3 in Parkinson's Disease (PD)

The pathology of PD is characterized by the degeneration of dopaminergic neurons in the substantia nigra and widespread Lewy bodies that contain α-synuclein (α-Syn) aggregates. Accumulated evidence suggest that neurons can release α-Syn to neighboring astrocytes and microglia (Kim et al., 2013). Microglia would take up α-Syn either by phagocytosis or endocytosis. Treatment of a BV2 microglial cell line with different forms of α-Syn (monomer and aggregates) promotes microglial activation as assessed by the levels of iNOS and pro-inflammatory cytokines (TNFα, IL2 and IL12). Importantly, knockdown of galectin-3 or treatment with a galectin-3 inhibitor abolishes the effect of α-Syn on the release of iNOS and pro-inflammatory cytokines. Inhibition of galectin-3 reduces the phagocytic capability of microglia to take up exogenous α-Syn, while treatment with recombinant galectin-3 promotes phagocytosis. Injection of oligomeric forms of α-Syn into the olfactory bulb also significantly enhances the expression of galectin-3 and the phagocytotic activity of microglia. These findings suggest that exposure of microglia to α-Syn causes up-regulation of galectin-3 and subsequently enhances phagocytosis (Boza-Serrano et al., 2014).

The glycan binding targets of galectin (i.e., β-galactosides) are located in the plasma membrane outer leaflet and the luminal side of the intracellular vesicles. Cytosolic galectins (e.g., galectin-3) therefore have no access to β-galactosides. In a neuronal model (SY5Y cells), treatments with α-Syn trigger the formation of galectin-3 puncta because the endocytosed α-Syn cannot be effectively degraded and triggers cellular stress (e.g., elevated ROS), leading to vesicle rupture (Flavin et al., 2017). Exposure of surface glycans from the ruptured vesicular membranes to the cytoplasm recruits galectin-3, which forms galectin-3 puncta (Freeman et al., 2013; Flavin et al., 2017). Such endosomal and lysosomal ruptures not only expose the internal compartment of the vesicles but also release into the cytoplasm substances (e.g., cathepsin B) that need to be confined inside the vesicle, potentially inducing the mitochondrial dysfunction and inflammation that are the common features of PD. It is likely that cell-to-cell transmission of cytosolic protein aggregates (e.g., α-Syn) may induce cellular stress that causes vesicle rupture and the formation of galectin-3 puncta as illustrated in **Figure 3**. Similar to galectin-8 (Thurston et al., 2012), galectin-3 may serve as a cellular danger receptor that senses detrimental conditions

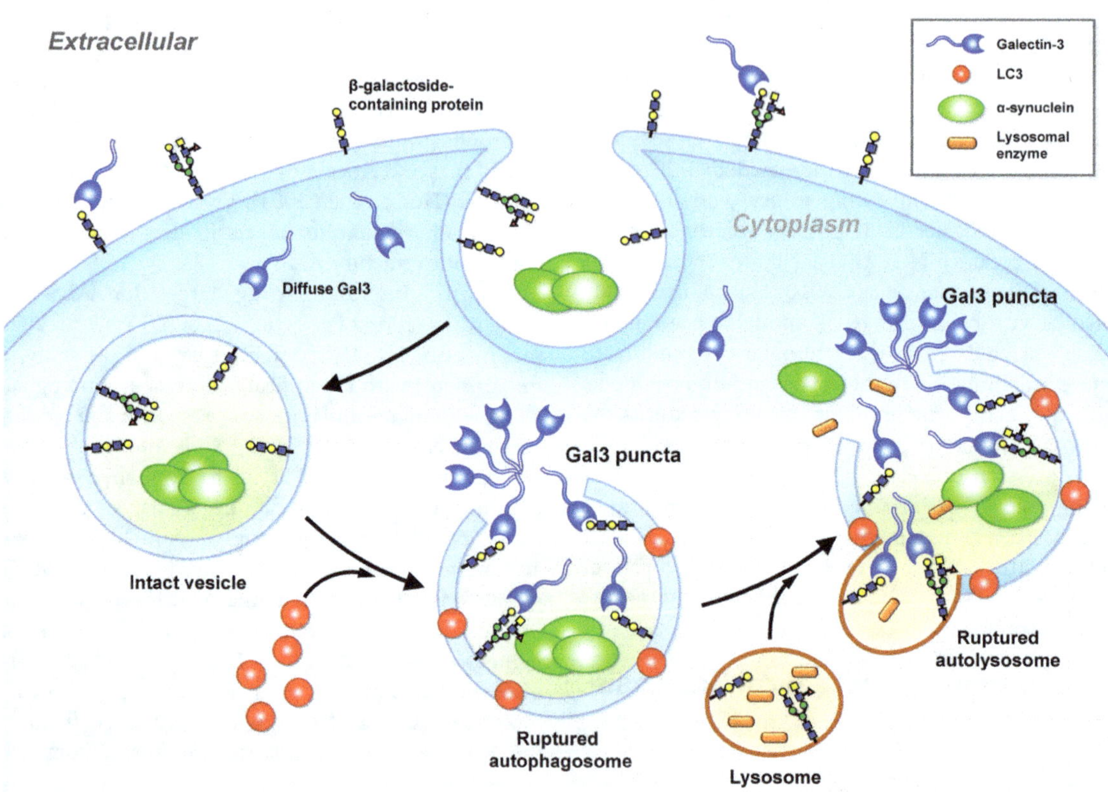

FIGURE 3 | Galectin-3 is a sensor of ruptured intracellular vesicles. Galectin-3 recognizes and binds to β-galactosides. β-galactosides containing proteins are located in the plasma membrane outer leaflet and the luminal side of intact intracellular vesicles (e.g., lysosome, autophagosome and autolysosome). Once these vesicles leak or rupture under stress, galectin-3 gains access and binds to β-galactosides located inside of ruptured vesicles and forms puncta (Freeman et al., 2013; Flavin et al., 2017). Gal3, galectin-3; α-Syn, α-synuclein.

during neuronal degeneration. Further investigation is required to determine whether the formation of galectin-3 puncta at these ruptured vesicles affects cellular functions.

Galectin-3 in Prion Diseases

Prion diseases consist of a group of neurodegenerative disorders that are caused by misfolded proteins called prions. Prions are infectious and the disease can occur in a sporadic, genetic or acquired manner. Genetic prion disease is caused by the mutation of prion-related protein gene (*PRPN*; Geschwind, 2015). In the scrapie-infected mouse model of prion disease, gene array analysis of the brain reveals 114 altered mRNA levels, and many of the identified genes were related to immune and stress responses. Among the genes with altered expression, the mRNA of galectin-3 is significantly up-regulated in the medulla and pons (Riemer et al., 2004). In a separate study, the protein level of galectin-3 was also found to be increased in scrapie-infected brains. Because galectin-3 KO mice with scrapie infection survive longer than the scrapie-infected wild-type mice and because the levels of a lysosomal activation marker (LAMP-2) and autophagy-related proteins (i.e., Beclin-1 and Atg5) are significantly decreased in the scrapie-infected galectin-3 KO mice, it is possible that galectin-3 may play a detrimental role by damaging the integrity of lysosomes and suppressing autophagy in prion disease (Mok et al., 2007).

SIGLECS

Siglecs (sialic acid-binding immunoglobulin-like lectins) represent the best-characterized subgroup of I-type lectins, which contain the immunoglobulin domain that recognizes structural diversity of carbohydrates (Crocker et al., 2007). In general, Siglecs are expressed on plasma membrane, containing an amino terminal V-set immunoglobulin domain that binds sialic acid and a number of C2-set immunoglobulin domains. As their full name suggests, Siglecs recognize sialic acid residues (best described as N-acetyl-neuraminic acids, Neu5Ac) that are presented on glycoproteins and glycolipids with distinct yet overlapping specificities. The presence of C-2 α-hydroxyl group at the terminal of sialic acid enables covalent linkage to other hydroxyl groups on the C-3 and C-6 positions to form the α2–3 and α2–6 linkages, respectively. Enzymes that are responsible for generating such linkages are named sialyltransferases. There are multiples types of sialyltransferases that produce a variety of linkages that can be recognized by Siglecs with different specificities.

Currently, up to 16 Siglecs have been identified and can be categorized based on their sequence similarity, conservation among murine and human and the regulatory properties (i.e., inhibitory or activatory; for review, see Pillai et al., 2012; Macauley et al., 2014). In brief, Siglec-1 (sialoadhesin), Siglec-2 (CD22), Siglec-4 (myelin-associated glycoprotein, MAG) and Siglec-15 are conserved across mammals. Conversely, a variety of Siglecs-3 (CD33)- related Siglecs have been found across mammals. The human CD33-related Siglecs include Siglecs-3, -5, -6, -7, -8, -9, -10, -11, -12, -14 and -16, while the murine CD33-related Siglecs consist of Siglecs-3, -E, -F, -G and -H (Macauley et al., 2014). Most Siglecs (excluding Siglec-1 and MAG) contain immunoreceptor tyrosine-based inhibitory motifs (ITIMs) at the cytoplasmic regions to recruit SH2 domain-containing protein tyrosine phosphatase SHP1 and SHP2; these Siglecs are classified as inhibitory Siglecs. On the other hand, Siglecs-14, -15 and -16 contain the immunoreceptor tyrosine-based activation motifs (ITAMs) on the adaptor of DAP12 to recruit spleen tyrosine kinase (SYK); these Siglecs are classified as activatory Siglecs.

Siglecs are expressed specifically on the cells from hematopoietic lineage, with the exceptions of MAG on oligodendrocytes and Siglec-6 on placental trophoblasts (Macauley et al., 2014). Knowledge of Siglecs is collected mainly from the peripheral system, while Siglecs that are reported to exist in the CNS include sialoadhesin, CD33, MAG, Siglec-E, Siglec-F, Siglec-H, Siglec-11 and Siglec-16. Siglecs are generally absent or scarce under normal physiological condition and are induced to express by various stimuli that activate microglia. These Siglecs, with exception of MAG that is located in oligodendrocytes, are expressed on microglia in the brain. **Table 5** summarizes the role of Siglecs in neurological disorders.

Siglec-1 (Sialoadhesin, CD169) During the Breakdown of the BBB

Siglec-1 does not have the ITIM or ITAM at its cytoplasmic regions but contains a long immunoglobulin domain that allows the binding domain to reach a distal site from the membrane. Therefore, Siglec-1 is thought to be important for cell-cell adhesion and cell-pathogen recognition (Crocker et al., 1994). Siglec-1 is present in peripheral blood mononuclear phagocytes. Nevertheless, its expression in microglia is rather controversial and may be due to the exposure to plasma proteins upon the breakdown of the blood-brain barrier (BBB) during brain trauma (e.g., TBI model, Perry et al., 1992). Similarly, during retinal degeneration, activated microglia may express Siglec-1 after the

TABLE 5 | Summary of the role of Siglecs in neurological disease.

Neurological conditions	Type of Siglec	Disease-associated features	References
Traumatic brain injury	Siglec-1	• Up-regulated in the microglia upon exposure to plasma proteins, unclear function.	Perry et al. (1992)
Ceriod lipofuscinoses	Siglec-1	• Up-regulated in the microglia of the optic nerve. • Serves as an APC and interacts with CD8+ T cells to promote inflammation.	Groh et al. (2016)
Alzheimer's disease	CD33	• Up-regulated in the microglia to suppress phagocytosis of Aβ.	Bradshaw et al. (2013) and Griciuc et al. (2013)
Brain tumor	Siglec-H	• Phagocytosis of glioma cells	Kopatz et al. (2013)
Amyotrophic lateral sclerosis	Siglec-H	• Up-regulated over time in the spinal cord of SOD1 mice, uncharacterized function.	Chiu et al. (2013)

breakdown of the blood-retinal barrier (Hughes et al., 2003). Whether these Siglec-1-expressing microglia are originated in the brain or blood-derived cells (i.e., macrophages) that migrate into the brain and their functions are not clear at this time.

In ceroid lipofuscinosis (CLN), a disease that is characterized by lysosomal storage dysfunction, neurodegeneration and early death, the expression of Siglec-1 is significantly up-regulated. These Siglec-1-expressing microglia serve as antigen-presenting cells (APCs) to interact with CD8+ T cells and promote the pathology of the disease. In Siglec-1-deficient CLN mice, microglia express significantly lower inflammatory cytokines (e.g., IL1β and TNFα) and higher level of anti-inflammatory cytokines (e.g., TGFβ). In addition, Siglec-1-deficient CLN mice exhibit reduced axonal degeneration and longer lifespan, supporting a harmful role of Siglec-1 (Groh et al., 2016). Hence, antagonizing the function of Siglec-1 appears as a potential therapeutic approach during the breakdown of the BBB from disease or injury.

Siglec-3 (CD33) in AD

AD is a chronic neurodegenerative disorder that affects up to 30% of the population over 65 years old (Masters et al., 2015). The pathology of the disease is characterized by intracellular neurofibrillary tangles and failure to clear extracellular amyloid-β (Aβ) peptides from the brain. Most patients with AD have the sporadic form, while a small portion of patients carry genes that cause ineffective clearing of Aβ. Recent genome-wide association studies have identified CD33 as a genetic risk factor of AD (Hollingworth et al., 2011; Naj et al., 2011). The risk allele *rs3865444* is associated with higher expression of CD33 on the cell surface without affecting the number of microglial cells. Moreover, these CD33-immunoreactive microglia exhibit a positive correlation with Aβ aggregates and plaque burden in the brain of AD patients (Bradshaw et al., 2013; Griciuc et al., 2013). Further *in vitro* studies confirmed the presence of CD33 with the risk allele *rs3865444*, which significantly suppresses the phagocytosis capability of microglia. Interestingly, the levels of CD33 with the minor allele of *rs3865444* single nucleotide polymorphism (SNP) are markedly reduced. Similarly, primary microglia derived from CD33 KO mice showed greater uptake of Aβ than the control cells, while the rates of Aβ degradation are unaffected. Moreover, the KO of CD33 in the APP/PS1 mouse model significantly reduces Aβ plaques in the cortex and hippocampus (Griciuc et al., 2013). Therefore, tuning the capability of microglia to take up and degrade Aβ through inhibition of CD33 activity may provide a novel therapeutic intervention for AD. The regulation of microglial phagocytosis by CD33 is summarized in **Figure 4**.

Siglec-H in Brain Tumors

Siglec-H is expressed only in murine microglia and plasmacytoid dendritic cells (pDCs), while they are absent in human monocytes (Zhang et al., 2006; Kopatz et al., 2013). In contrast to most Siglecs, Siglec-H is an activatory Siglec with complex functions. Siglec-H suppresses the downstream signaling of TLR9 to produce interferon-α in pDCs (Takagi et al., 2011; Puttur et al., 2013). Although Siglec-H is known to be expressed in microglia, its ligands and functions are only now being discovered, as it does not bind to the typical sialic acid-containing glycoproteins or glycolipids of healthy cells (Varki and Angata, 2006; Varki, 2009). Treatment of a microglial cell line and primary microglia with interferon-γ significantly enhances inflammatory phenotypes (e.g., up-regulation of iNOS, CXCL10 and TNFα) and up-regulation of Siglec-H. These Siglec-H-enriched microglia exhibit high phagocytosis activity that can be suppressed by down-regulation of Siglec-H. Importantly,

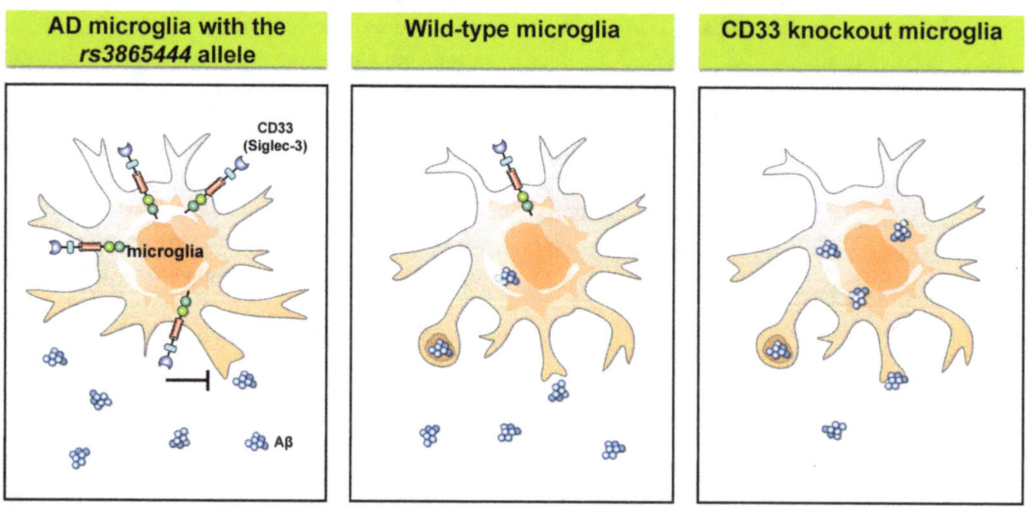

FIGURE 4 | Siglecs regulate microglial phagocytosis. Wild-type microglia express a low level of CD33/Siglec 3, a member of the Siglec family. The risk allele *rs3865444* was originally identified in patients with Alzheimer's disease (AD) and is associated with higher levels of CD33 that suppress the microglial phagocytosis. Conversely, knockdown of CD33 significantly improves the phagocytosis capability of microglia (Hollingworth et al., 2011; Naj et al., 2011; Griciuc et al., 2013). Aβ, amyloid β.

Siglec-H binds to and triggers the engulfment of glioma cell lines (namely, GL261 and SMA560) but not normal astrocytes, splenocytes, or fibroblasts. It is possible that microglial Siglec-H may function to monitor the surface sialylation profiles present on tumor cells in the brain (Kopatz et al., 2013). These studies identified the functions of Siglec-H in the brain. However, whether enhancing Siglec-H activity is a potent approach to target brain tumor required further investigation.

Siglec-H in ALS

Microglia share many markers with peripheral monocytes and macrophages. Transcriptome analysis reveals that Siglec-H is the third highest gene (out of 29 identified genes) that can distinguish microglia from monocytes and macrophages. The level of Siglec-H is also up-regulated in the spinal cord of an ALS mouse model (SOD1; Chiu et al., 2013). The function of Siglec-H in ALS remains unclear.

C-TYPE LECTINS

C-type lectins are a group of proteins that bind to glycan in a calcium-dependent manner. The proteins contain a C-type lectin domain (CTLD), which is also present in many proteins that do not recognize or bind to any glycan. C-type lectins and protein with CTLD are found in all organism and can be categorized to at least 17 groups based on their domain architecture, such as selectins, collectins, etc. The classification of C-type lectins and their roles in the peripheral immunity have been reviewed extensively elsewhere (Sancho and Reis e Sousa, 2012; Dambuza and Brown, 2015; Yan et al., 2015). In this review, we focus on a few C-type lectins (**Table 6**) that have been implicated in neurological disorder.

Selectins

Selectins are single-chain transmembrane glycoproteins of the cell adhesion molecule family. They are highly homologous proteins, which contain a CTLD, an epidermal growth factor-like domain, a few consensus repeats composed of sushi domain, a transmembrane domain and an intracellular cytoplasmic tail. The CTLD of selectins recognizes sialyl Lewis X antigens (sialylated and fucosylated glycoprotein and glycolipid structures; Ley, 2003; Angiari and Constantin, 2013). Three types of selectins (i.e., E-, L- and P-Selectins) have being identified so far. In the brain, selectins (particularly the E- and P-type) are located in endothelial cells and are responsible for controlling leukocyte infiltration and the BBB permeability during trauma or acute inflammation (Tang et al., 1996; Carvalho-Tavares et al., 2000; Bernardes-Silva et al., 2001). This is an important aspect because accumulation and adhesion of leukocytes to the endothelial cells are the critical steps that cause cell death at the ischemic regions (Kataoka et al., 2004; Langer and Chavakis, 2009).

In the animal model of focal cerebral ischemia and reperfusion injury, E-Selectins are induced and expressed on the endothelial cells (Wang et al., 1995; Zhang et al., 1996). Because E-Selectins bind sialyl Lewis X glycans enriched in neutrophils (Lasky, 1992) and the administration of sialyl Lewis X significantly decreases the recruitment of neutrophils to the ischemic region and decrease the infarct volume (Zhang et al., 1996), E-Selectins appear to be an important component of ischemia pathogenesis. Another intriguing observation is that the polysialylated E-Selectin Ligand-1 (ESL1, a ligand for E-Selectins) was found to be enriched in the Golgi compartments of microglia. Brain injuries translocate ESL1 to the cell surfaces and therefore trigger the binding of E-Selectins with microglia (Werneburg et al., 2016). Binding of E-Selectins to myeloid cells (i.e., neutrophils) has been demonstrated to regulate cell homing to sites of inflammation (Mondal et al., 2016), while the effect of binding of E-Selectins to microglia remains to be determined.

The expression of P-Selectin in endothelial cells is also up-regulated in a complement C3-dependent manner in an ischemia mouse model (Atkinson et al., 2006). Both P-Selectin deficiency and P-Selectin inhibition protect mice from the cerebral injury following ischemia (Zhou et al., 2000; Atkinson et al., 2006). Up-regulation of P-Selectin also contributes to the transient breakdown of BBB following ischemia (Jin et al., 2010). It is interesting to note that the level of P-Selectin is significantly up-regulated in the experimental subarachnoid hemorrhage (eSAH) model as well. The presence of P-Selectins on endothelial cells recruit neutrophils and microglia to the site of injury. It is likely that these microglia are activated by the inflammatory cytokines released by the neutrophils promoted

TABLE 6 | Summary of the role of C-type lectins in neurological disease.

Neurological conditions	Type of C-type lectin	Disease-associated features	References
Stroke and ischemia	E-Selectin	• Facilitates the adhesion of neutrophils to endothelial cells and promotes cell damage.	Zhang et al. (1996) and Jin et al. (2010)
	P-Selectin	• Facilitates the adhesion of neutrophils to endothelial cells and promotes the activation of complement. • Contributes to the breakdown of the BBB.	Atkinson et al. (2006) and Jin et al. (2010)
	Mannose-binding lectin	• Up-regulated and contributes to cerebral infarction. • Contributes to complement activation and promotes inflammation.	Cervera et al. (2010), Morrison et al. (2011), Orsini et al. (2012) and De Blasio et al. (2017)
Subarachnoid hemorrhage	P-Selectin	• Recruits neutrophils and subsequently activates microglia through intravascular inflammation.	Atangana et al. (2017)
HIV infection	Mannose-binding lectin	• Up-regulated in microglia, astrocytes, oligodendrocytes and neurons. • Expression is associated with MCP-1, which may trigger inflammation.	Singh et al. (2011)

by intravascular inflammation (Schneider et al., 2015; Atangana et al., 2017). In addition, induction of eSAH in mice lacking the P-Selectin glycoprotein ligand-1 (PSGL1; a P-Selectin ligand) results in significant reductions in the endothelial-neutrophil interactions, the number of activated microglia and the extent of neuronal death (Atangana et al., 2017). Collectively, these studies suggest that P-Selectin is likely to play a detrimental role in neurological traumas by facilitating the microglia-mediated neuroinflammation. **Figure 5** summarizes the roles of E-selectins and P-selectins in brain ischemia.

Mannose-Binding Lectins (MBLs)

MBLs, also called mannan-binding lectins, are from the collectin class of the C-type lectin family and are important components of the innate immune system. These proteins contain a cysteine-rich N-terminal region, a collagenous domain, a short α-helical coiled-coil domain and a CRD at the C-terminal. It is important to note that CRDs of MBL recognize glycans on the surface of pathogens as well as cryptic self-antigens that are modified in disease conditions or injuries (Ip et al., 2009). They exist in oligomeric forms (i.e., from dimer to hexamer) and recognizes mannose, fucose and N-acetyl-glucosamine (GlcNAc), but not sialic acid and galactose (Drickamer, 1992; Weis et al., 1992; Yokota et al., 1995). MBLs participate in a range of biological functions including the recognition of pathogen invasion, activation of complement system, phagocytosis and mediation of inflammatory signaling (Ip et al., 2009; Auriti et al., 2017).

Activation of complement system is one of the main causes for brain damage during ischemia and stroke. MBL has been implicated to participate in the activation of complement system via the lectin pathway (Morrison et al., 2011). MBL-KO mice are protected from transient and permanent ischemic injuries with the reduction of infarct volume and sensorimotor impairments. Blocking MBL using an anti-MBL-A antibody also produce similar results (Orsini et al., 2012; Longhi et al., 2014). Consistently, ischemic stroke patients, who have MBL-low genotypes and lower MBL in their serum, are associated with a favorable stroke outcome and decreased levels of proinflammatory cytokines (e.g., TNFα and IL6). These patients with lower MBL levels (<100 ng/ml) are likely to recover from movement disabilities in less than 3 months (Cervera et al., 2010; Osthoff et al., 2011). Collectively, effective inhibition of MBL may reduce the damage caused by ischemic stroke.

In patients infected by HIV, viruses can enter the CNS and continue to replicate within the brain. Infection of HIV in the

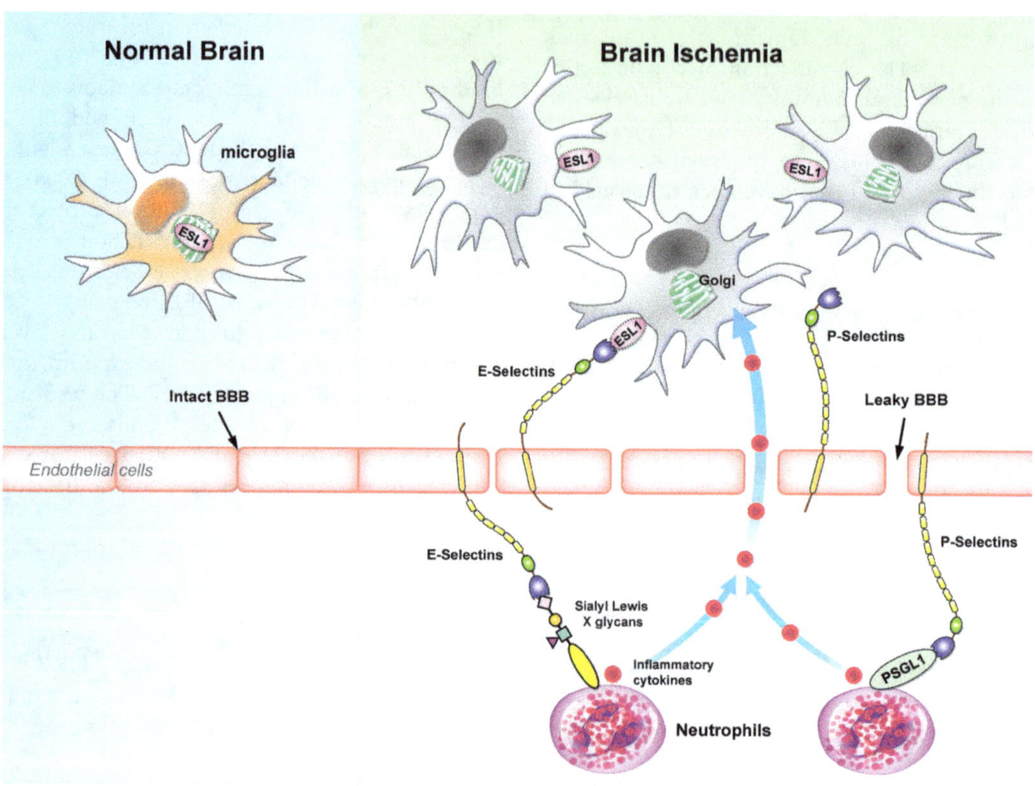

FIGURE 5 | Selectins regulates microglial activation and homing to sites of inflammation during brain ischemia. E-selectins and P-selectins that bind to Sialyl Lewis X glycans are up-regulated in endothelial cells during brain ischemia. Binding of selectins serves as a homing mechanism to retain Sialyl Lewis X glycans-containing cells close to endothelial cells. In particular, ESL1 located in the Golgi compartments of microglia is translocated to the plasma membrane and binds to E-selectins on the surface of endothelial cells (Werneburg et al., 2016). In contrast, P-selectins cause a transient leakage of BBB (Jin et al., 2010). P-selectins bind to PSGL1 on neutrophils, which secrete inflammatory cytokines and further activate microglia in the CNS (Schneider et al., 2015; Atangana et al., 2017). BBB, blood-brain barrier; ESL1, E-selectin ligand-1; PSGL1, P-selectin glycoprotein ligand-1.

brain therefore causes a chronic inflammatory response and a spectrum of neurological dysfunction (Saylor et al., 2016). By studying post-mortem brain tissues, it was shown that the expression of MBL is up-regulated in the HIV- infected brains. Interestingly, the expression of MBL is detected in all brain cells, including microglia, astrocytes, oligodendrocytes and neurons. In particular, axons in the HIV-infected brain show high levels of MBL, while the level of MBL is associated with the level of monocytes chemoattractant protein-1 (MCP-1 or CCL2; Singh et al., 2011). MCP-1 is the key chemokine that triggers the migration of microglia or macrophages (Deshmane et al., 2009; Hinojosa et al., 2011). Hence, the expression of MBL is suggested to be associated with the activation of complement and neuroinflammation, which subsequently cause neuronal damage in the HIV-infected brain.

CONCLUSION AND FUTURE PERSPECTIVES

Accumulated evidence suggest that dysregulation of lectins is a critical pathogenetic factor in neurological diseases. The altered expressions of lectins on microglia affect the phenotypes of microglia (Yip et al., 2017). Importantly, the aberrant expression of lectins on the surrounding cells can also trigger the activation of microglia (Atangana et al., 2017). These may independently or synergistically lead to the chronic pathogenic inflammatory response that causes neuronal dysfunctions or death. Likewise, these activated microglia may also secrete inflammatory cytokines and complement that will influence the activity of astrocytes to produce neurotoxic factors (Liddelow et al., 2017).

One major area that still requires clarification is the specific type of cells that express these lectins. Many lectins are not expressed in the healthy CNS, nevertheless are induced to express or up-regulated in disease condition. It is crucial to identify the specific types of cells that show aberrant expression of lectins and their functional effect on the cells themselves or the surrounding cells. For instance, galectin-3 is expressed virtually on activated microglia in various neurological disorders *in vivo*, while it is expressed in various cell lines derived from brain glial cells *in vitro* (Pasquini et al., 2011). These studies raise the questions of why cell lines express galectin-3 even under normal conditions. There is a possibility that galectin-3 is involved in the immortalization and proliferation of cell lines, as it does in tumor cells (i.e., pancreatic cancer cells, thyroid carcinoma cells; Song et al., 2012; Cardoso et al., 2016). At the same time, it is also of interest to investigate whether other glial cells (astrocytes and oligodendrocytes) in abnormal conditions also express galectin-3 *in vivo*. Meanwhile, Galectin-4 is expressed in non-myelinated region of axons and Siglec-4 is expressed specifically on oligodendrocytes to regulate myelination (**Figure 1**). Contrary to Galectins that can be expressed in various types of cells, all CD33-related Siglecs are expressed specifically on myeloid- origin cells, namely, microglia in the CNS. Selectins are mainly on endothelial cells and their expressions directly affects the integrity of BBB and the activity of nearby microglia. MBL has been reported to increase in stroke and ischemia, while the sources of MBL are elusive. It is unclear whether MBLs are produced by brain cells or migrate from the peripheral nervous system due to the leakage of the BBB. Interestingly, in HIV-infected brain, all brain cells show positive signals for MBL, indicating brain cells are capable of producing MBL upon stimulation. Further understandings on how lectins affect microglia and its interaction with other brain cells certainly would broaden our view on neurodegenerative diseases and raise new questions as well. Are these lectins druggable targets for neurological disease? Would pharmacological inhibition or activation of these lectins be effective in treating neurological diseases? Notably, these inhibitors/activators also need to have the capacity to cross the BBB to treat brain disorders.

The concept of lectin-based therapeutics deploys the specific targets of lectins or the delivery of drugs that interacts with lectins to the target site. For example, Galectin-3 has been implicated in numerous neurological disorders and diseases in the peripheral nervous system. Pharmacological suppression of galectin-3 appears to be a promising approach. Specifically, TD139 that binds to galectin-3 and inhibit its functions have passed the clinical phase Ib/IIa trials for treating idiopathic pulmonary fibrosis (http://ClinicalTrials.gov identifier, NCT02257177). Although this development is encouraging, the ability of TD139 to cross the BBB is poor and therefore limits its application to neurological diseases with abnormal up-regulation of Gal3. Further modification of TD139 to increase its permeability across the BBB may pave the way for the development of therapeutic intervention for brain diseases.

On the other hand, lectin-based therapeutics may enhance the exposure of drugs to the targeted cells/sites, and reduce the dosage and side effects. Moreover, internalization of drug also increases the uptake of therapeutic agents with low cellular permeability. The decoration of drug enveloped with surface sialic acid ligands can directly activate Siglecs downstream signaling or deploy their internalization properties. For example, nanoparticles coated with di(α2–8) N-acetylneuraminic acid effectively target Siglec-7 and Siglec-9 to exert anti-inflammatory effects of macrophages in a human *ex vivo* lung perfusion model (Spence et al., 2015). Furthermore, liposomal nanoparticles that are coated with high-affinity CD22 ligands result in substantial uptake of the nanoparticles by B cell lymphomas that express the endocytic receptor CD22 (Chen et al., 2012).

Collectively, emerging studies have shown that modulating the lectins activity via direct interference in the lectin itself or the receptor-ligand interaction in CNS is a promising area for further exploration. Currently, treatment of neurological disorders using a lectin-based approach has not been envisaged, likely owing to an incomplete understanding of the roles of lectins in the CNS and the limited ability of drugs to cross the BBB. A greater understanding of microglial lectins may facilitate the development of novel therapeutic interventions for patients with neurological disorders.

AUTHOR CONTRIBUTIONS

JJS designed and wrote the manuscript. YC designed and edited the manuscript.

ACKNOWLEDGMENTS

We are grateful to the Medical Art Room of Institute of Biomedical Sciences at Academia Sinica for the assistance in preparing the illustration.

REFERENCES

Abhinav, K., Stanton, B., Johnston, C., Hardstaff, J., Orrell, R. W., Howard, R., et al. (2007). Amyotrophic lateral sclerosis in South-East England: a population-based study. The South-East England register for amyotrophic lateral sclerosis (SEALS Registry). *Neuroepidemiology* 29, 44–48. doi: 10.1159/000108917

Angiari, S., and Constantin, G. (2013). Selectins and their ligands as potential immunotherapeutic targets in neurological diseases. *Immunotherapy* 5, 1207–1220. doi: 10.2217/imt.13.122

Atangana, E., Schneider, U. C., Blecharz, K., Magrini, S., Wagner, J., Nieminen-Kelha, M., et al. (2017). Intravascular inflammation triggers intracerebral activated microglia and contributes to secondary brain injury after experimental subarachnoid hemorrhage (eSAH). *Transl. Stroke Res.* 8, 144–156. doi: 10.1007/s12975-016-0485-3

Atkinson, C., Zhu, H., Qiao, F., Varela, J. C., Yu, J., Song, H., et al. (2006). Complement-dependent P-selectin expression and injury following ischemic stroke. *J. Immunol.* 177, 7266–7274. doi: 10.4049/jimmunol.177.10.7266

Auriti, C., Prencipe, G., Moriondo, M., Bersani, I., Bertaina, C., Mondì, V., et al. (2017). Mannose-binding lectin: biologic characteristics and role in the susceptibility to infections and ischemia-reperfusion related injury in critically Ill neonates. *J. Immunol. Res.* 2017:7045630. doi: 10.1155/2017/7045630

Bernardes-Silva, M., Anthony, D. C., Issekutz, A. C., and Perry, V. H. (2001). Recruitment of neutrophils across the blood-brain barrier: the role of E- and P-selectins. *J. Cereb. Blood Flow Metab.* 21, 1115–1124. doi: 10.1097/00004647-200109000-00009

Bessis, A., Béchade, C., Bernard, D., and Roumier, A. (2007). Microglial control of neuronal death and synaptic properties. *Glia* 55, 233–238. doi: 10.1002/glia.20459

Blennow, K., Brody, D. L., Kochanek, P. M., Levin, H., McKee, A., Ribbers, G. M., et al. (2016). Traumatic brain injuries. *Nat. Rev. Dis. Primers* 2:16084. doi: 10.1038/nrdp.2016.84

Boza-Serrano, A., Reyes, J. F., Rey, N. L., Leffler, H., Bousset, L., Nilsson, U., et al. (2014). The role of Galectin-3 in α-synuclein-induced microglial activation. *Acta Neuropathol. Commun.* 2:156. doi: 10.1186/s40478-014-0156-0

Bradshaw, E. M., Chibnik, L. B., Keenan, B. T., Ottoboni, L., Raj, T., Tang, A., et al. (2013). CD33 Alzheimer's disease locus: altered monocyte function and amyloid biology. *Nat. Neurosci.* 16, 848–850. doi: 10.1038/nn.3435

Brandley, B. K., and Schnaar, R. L. (1986). Cell-surface carbohydrates in cell recognition and response. *J. Leukoc. Biol.* 40, 97–111. doi: 10.1002/jlb.40.1.97

Brites, D., and Vaz, A. R. (2014). Microglia centered pathogenesis in ALS: insights in cell interconnectivity. *Front. Cell. Neurosci.* 8:117. doi: 10.3389/fncel.2014.00117

Burguillos, M. A., Svensson, M., Schulte, T., Boza-Serrano, A., Garcia-Quintanilla, A., Kavanagh, E., et al. (2015). Microglia-secreted galectin-3 acts as a toll-like receptor 4 ligand and contributes to microglial activation. *Cell Rep.* 10, 1626–1638. doi: 10.1016/j.celrep.2015.02.012

Burman, J., and Svenningsson, A. (2016). Cerebrospinal fluid concentration of Galectin-9 is increased in secondary progressive multiple sclerosis. *J. Neuroimmunol.* 292, 40–44. doi: 10.1016/j.jneuroim.2016.01.008

Cao, Z. Q., and Guo, X. L. (2016). The role of galectin-4 in physiology and diseases. *Protein Cell* 7, 314–324. doi: 10.1007/s13238-016-0262-9

Cardoso, A. C., Andrade, L. N., Bustos, S. O., and Chammas, R. (2016). Galectin-3 determines tumor cell adaptive strategies in stressed tumor microenvironments. *Front. Oncol.* 6:127. doi: 10.3389/fonc.2016.00127

Carvalho-Tavares, J., Hickey, M. J., Hutchison, J., Michaud, J., Sutcliffe, I. T., and Kubes, P. (2000). A role for platelets and endothelial selectins in tumor necrosis factor-α-induced leukocyte recruitment in the brain microvasculature. *Circ. Res.* 87, 1141–1148. doi: 10.1161/01.res.87.12.1141

Cenker, J. J., Stultz, R. D., and McDonald, D. (2017). Brain microglial cells are highly susceptible to HIV-1 infection and spread. *AIDS Res. Hum. Retroviruses* 33, 1155–1165. doi: 10.1089/aid.2017.0004

Cervera, A., Planas, A. M., Justicia, C., Urra, X., Jensenius, J. C., Torres, F., et al. (2010). Genetically-defined deficiency of mannose-binding lectin is associated with protection after experimental stroke in mice and outcome in human stroke. *PLoS One* 5:e8433. doi: 10.1371/journal.pone.0008433

Chang, W. A., Tsai, M. J., Kuo, P. L., and Hung, J. Y. (2017). Role of galectins in lung cancer. *Oncol. Lett.* 14, 5077–5084. doi: 10.3892/ol.2017.6882

Chang-Hong, R., Wada, M., Koyama, S., Kimura, H., Arawaka, S., Kawanami, T., et al. (2005). Neuroprotective effect of oxidized galectin-1 in a transgenic mouse model of amyotrophic lateral sclerosis. *Exp. Neurol.* 194, 203–211. doi: 10.1016/j.expneurol.2005.02.011

Chen, W. C., Sigal, D. S., Saven, A., and Paulson, J. C. (2012). Targeting B lymphoma with nanoparticles bearing glycan ligands of CD22. *Leuk. Lymphoma* 53, 208–210. doi: 10.3109/10428194.2011.604755

Cheng, X., Boza-Serrano, A., Turesson, M. F., Deierborg, T., Ekblad, E., and Voss, U. (2016). Galectin-3 causes enteric neuronal loss in mice after left sided permanent middle cerebral artery occlusion, a model of stroke. *Sci. Rep.* 6:32893. doi: 10.1038/srep32893

Chiu, I. M., Morimoto, E. T., Goodarzi, H., Liao, J. T., O'Keeffe, S., Phatnani, H. P., et al. (2013). A neurodegeneration-specific gene-expression signature of acutely isolated microglia from an amyotrophic lateral sclerosis mouse model. *Cell Rep.* 4, 385–401. doi: 10.1016/j.celrep.2013.06.018

Claude, J., Linnartz-Gerlach, B., Kudin, A. P., Kunz, W. S., and Neumann, H. (2013). Microglial CD33-related Siglec-E inhibits neurotoxicity by preventing the phagocytosis-associated oxidative burst. *J. Neurosci.* 33, 18270–18276. doi: 10.1523/jneurosci.2211-13.2013

Çolakoğlu, G., Bergstrom-Tyrberg, U., Berglund, E. O., and Ranscht, B. (2014). Contactin-1 regulates myelination and nodal/paranodal domain organization in the central nervous system. *Proc. Natl. Acad. Sci. U S A* 111, E394–E403. doi: 10.1073/pnas.1313769110

Crocker, P. R., Mucklow, S., Bouckson, V., McWilliam, A., Willis, A. C., Gordon, S., et al. (1994). Sialoadhesin, a macrophage sialic acid binding receptor for haemopoietic cells with 17 immunoglobulin-like domains. *EMBO J.* 13, 4490–4503.

Crocker, P. R., Paulson, J. C., and Varki, A. (2007). Siglecs and their roles in the immune system. *Nat. Rev. Immunol.* 7, 255–266. doi: 10.1038/nri2056

Dambuza, I. M., and Brown, G. D. (2015). C-type lectins in immunity: recent developments. *Curr. Opin. Immunol.* 32, 21–27. doi: 10.1016/j.coi.2014.12.002

Davalos, D., Grutzendler, J., Yang, G., Kim, J. V., Zuo, Y., Jung, S., et al. (2005). ATP mediates rapid microglial response to local brain injury *in vivo*. *Nat. Neurosci.* 8, 752–758. doi: 10.1038/nn1472

De Blasio, D., Fumagalli, S., Longhi, L., Orsini, F., Palmioli, A., Stravalaci, M., et al. (2017). Pharmacological inhibition of mannose-binding lectin ameliorates neurobehavioral dysfunction following experimental traumatic brain injury. *J. Cereb. Blood Flow Metab.* 37, 938–950. doi: 10.1177/0271678x16647397

de la Rosa, E. J., and de Pablo, F. (2000). Cell death in early neural development: beyond the neurotrophic theory. *Trends Neurosci.* 23, 454–458. doi: 10.1016/s0166-2236(00)01628-3

Dendrou, C. A., Fugger, L., and Friese, M. A. (2015). Immunopathology of multiple sclerosis. *Nat. Rev. Immunol.* 15, 545–558. doi: 10.1038/nri3871

Deshmane, S. L., Kremlev, S., Amini, S., and Sawaya, B. E. (2009). Monocyte chemoattractant protein-1 (MCP-1): an overview. *J. Interferon Cytokine Res.* 29, 313–326. doi: 10.1089/jir.2008.0027

Díez-Revuelta, N., Higuero, A. M., Velasco, S., Peñas-de-la-Iglesia, M., Gabius, H. J., and Abad-Rodríguez, J. (2017). Neurons define non-myelinated axon segments by the regulation of galectin-4-containing axon membrane domains. *Sci. Rep.* 7:12246. doi: 10.1101/115758

Drickamer, K. (1992). Engineering galactose-binding activity into a C-type mannose-binding protein. *Nature* 360, 183–186. doi: 10.1038/360183a0

Fischer, M. T., Sharma, R., Lim, J. L., Haider, L., Frischer, J. M., Drexhage, J., et al. (2012). NADPH oxidase expression in active multiple sclerosis lesions in relation to oxidative tissue damage and mitochondrial injury. *Brain* 135, 886–899. doi: 10.1093/brain/aws012

Flavin, W. P., Bousset, L., Green, Z. C., Chu, Y., Skarpathiotis, S., Chaney, M. J., et al. (2017). Endocytic vesicle rupture is a conserved mechanism of cellular invasion by amyloid proteins. *Acta Neuropathol.* 134, 629–653. doi: 10.1007/s00401-017-1722-x

Freeman, D., Cedillos, R., Choyke, S., Lukic, Z., McGuire, K., Marvin, S., et al. (2013). α-synuclein induces lysosomal rupture and cathepsin dependent reactive oxygen species following endocytosis. *PLoS One* 8:e62143. doi: 10.1371/journal.pone.0062143

Frielingsdorf, H., Simpson, D. R., Thal, L. J., and Pizzo, D. P. (2007). Nerve growth factor promotes survival of new neurons in the adult hippocampus. *Neurobiol. Dis.* 26, 47–55. doi: 10.1016/j.nbd.2006.11.015

Fuhrmann, M., Bittner, T., Jung, C. K., Burgold, S., Page, R. M., Mitteregger, G., et al. (2010). Microglial Cx3cr1 knockout prevents neuron loss in a mouse model of Alzheimer's disease. *Nat. Neurosci.* 13, 411–413. doi: 10.1038/nn.2511

Fujimoto, Z., Tateno, H., and Hirabayashi, J. (2014). Lectin structures: classification based on the 3-D structures. *Methods Mol. Biol.* 1200, 579–606. doi: 10.1007/978-1-4939-1292-6_46

Geschwind, M. D. (2015). Prion diseases. *Continuum* 21, 1612–1638. doi: 10.1212/CON.0000000000000251

Goldenberg, M. M. (2012). Multiple sclerosis review. *P T* 37, 175–184.

Griciuc, A., Serrano-Pozo, A., Parrado, A. R., Lesinski, A. N., Asselin, C. N., Mullin, K., et al. (2013). Alzheimer's disease risk gene CD33 inhibits microglial uptake of amyloid β. *Neuron* 78, 631–643. doi: 10.1016/j.neuron.2013.04.014

Groh, J., Ribechini, E., Stadler, D., Schilling, T., Lutz, M. B., and Martini, R. (2016). Sialoadhesin promotes neuroinflammation-related disease progression in two mouse models of CLN disease. *Glia* 64, 792–809. doi: 10.1002/glia.22962

Groves, J. T. (2013). Cell membranes: glycans' imprints. *Nat. Mater.* 12, 96–97. doi: 10.1038/nmat3555

Hinojosa, A. E., Garcia-Bueno, B., Leza, J. C., and Madrigal, J. L. (2011). CCL2/MCP-1 modulation of microglial activation and proliferation. *J. Neuroinflammation* 8:77. doi: 10.1186/1742-2094-8-77

Hisamatsu, K., Niwa, M., Kobayashi, K., Miyazaki, T., Hirata, A., Hatano, Y., et al. (2016). Galectin-3 expression in hippocampal CA2 following transient forebrain ischemia and its inhibition by hypothermia or antiapoptotic agents. *Neuroreport* 27, 311–317. doi: 10.1097/wnr.0000000000000538

Hollingworth, P., Harold, D., Sims, R., Gerrish, A., Lambert, J. C., Carrasquillo, M. M., et al. (2011). Common variants at ABCA7, MS4A6A/MS4A4E, EPHA1, CD33 and CD2AP are associated with Alzheimer's disease. *Nat. Genet.* 43, 429–435. doi: 10.1038/ng.803

Horie, H., Kadoya, T., Hikawa, N., Sango, K., Inoue, H., Takeshita, K., et al. (2004). Oxidized galectin-1 stimulates macrophages to promote axonal regeneration in peripheral nerves after axotomy. *J. Neurosci.* 24, 1873–1880. doi: 10.1523/jneurosci.4483-03.2004

Hoyos, H. C., Marder, M., Ulrich, R., Gudi, V., Stangel, M., Rabinovich, G. A., et al. (2016). The role of galectin-3: from oligodendroglial differentiation and myelination to demyelination and remyelination processes in a cuprizone-induced demyelination model. *Adv. Exp. Med. Biol.* 949, 311–332. doi: 10.1007/978-3-319-40764-7_15

Hoyos, H. C., Rinaldi, M., Mendez-Huergo, S. P., Marder, M., Rabinovich, G. A., Pasquini, J. M., et al. (2014). Galectin-3 controls the response of microglial cells to limit cuprizone-induced demyelination. *Neurobiol. Dis.* 62, 441–455. doi: 10.1016/j.nbd.2013.10.023

Hsiao, H. Y., Chiu, F. L., Chen, C. M., Wu, Y. R., Chen, H. M., Chen, Y. C., et al. (2014). Inhibition of soluble tumor necrosis factor is therapeutic in Huntington's disease. *Hum. Mol. Genet.* 23, 4328–4344. doi: 10.1093/hmg/ddu151

Huang, J. Y., Wang, Y. X., Gu, W. L., Fu, S. L., Li, Y., Huang, L. D., et al. (2012). Expression and function of myelin-associated proteins and their common receptor NgR on oligodendrocyte progenitor cells. *Brain Res.* 1437, 1–15. doi: 10.1016/j.brainres.2011.12.008

Hughes, E. H., Schlichtenbrede, F. C., Murphy, C. C., Sarra, G. M., Luthert, P. J., Ali, R. R., et al. (2003). Generation of activated sialoadhesin-positive microglia during retinal degeneration. *Invest. Ophthalmol. Vis. Sci.* 44, 2229–2234. doi: 10.1167/iovs.02-0824

Ip, W. K., Takahashi, K., Ezekowitz, R. A., and Stuart, L. M. (2009). Mannose-binding lectin and innate immunity. *Immunol. Rev.* 230, 9–21. doi: 10.1111/j.1600-065X.2009.00789.x

James, R. E., Hillis, J., Adorján, I., Gration, B., Mundim, M. V., Iqbal, A. J., et al. (2016). Loss of galectin-3 decreases the number of immune cells in the subventricular zone and restores proliferation in a viral model of multiple sclerosis. *Glia* 64, 105–121. doi: 10.1002/glia.22906

Jiang, H. R., Al Rasebi, Z., Mensah-Brown, E., Shahin, A., Xu, D., Goodyear, C. S., et al. (2009). Galectin-3 deficiency reduces the severity of experimental autoimmune encephalomyelitis. *J. Immunol.* 182, 1167–1173. doi: 10.4049/jimmunol.182.2.1167

Jin, A. Y., Tuor, U. I., Rushforth, D., Kaur, J., Muller, R. N., Petterson, J. L., et al. (2010). Reduced blood brain barrier breakdown in P-selectin deficient mice following transient ischemic stroke: a future therapeutic target for treatment of stroke. *BMC Neurosci.* 11:12. doi: 10.1186/1471-2202-11-12

John, S., and Mishra, R. (2016). mRNA transcriptomics of galectins unveils heterogeneous organization in mouse and human brain. *Front. Mol. Neurosci.* 9:139. doi: 10.3389/fnmol.2016.00139

Kasai, K., and Hirabayashi, J. (1996). Galectins: a family of animal lectins that decipher glycocodes. *J. Biochem.* 119, 1–8. doi: 10.1093/oxfordjournals.jbchem.a021192

Kataoka, H., Kim, S. W., and Plesnila, N. (2004). Leukocyte-endothelium interactions during permanent focal cerebral ischemia in mice. *J. Cereb. Blood Flow Metab.* 24, 668–676. doi: 10.1097/01.wcb.0000117812.35136.5b

Kettenmann, H., Hanisch, U. K., Noda, M., and Verkhratsky, A. (2011). Physiology of microglia. *Physiol. Rev.* 91, 461–553. doi: 10.1152/physrev.00011.2010

Kim, C., Ho, D. H., Suk, J. E., You, S., Michael, S., Kang, J., et al. (2013). Neuron-released oligomeric α-synuclein is an endogenous agonist of TLR2 for paracrine activation of microglia. *Nat. Commun.* 4:1562. doi: 10.1038/ncomms2534

Kopatz, J., Beutner, C., Welle, K., Bodea, L. G., Reinhardt, J., Claude, J., et al. (2013). Siglec-h on activated microglia for recognition and engulfment of glioma cells. *Glia* 61, 1122–1133. doi: 10.1002/glia.22501

Lalancette-Hébert, M., Gowing, G., Simard, A., Weng, Y. C., and Kriz, J. (2007). Selective ablation of proliferating microglial cells exacerbates ischemic injury in the brain. *J. Neurosci.* 27, 2596–2605. doi: 10.1523/JNEUROSCI.5360-06.2007

Lalancette-Hébert, M., Swarup, V., Beaulieu, J. M., Bohacek, I., Abdelhamid, E., Weng, Y. C., et al. (2012). Galectin-3 is required for resident microglia activation and proliferation in response to ischemic injury. *J. Neurosci.* 32, 10383–10395. doi: 10.1523/JNEUROSCI.1498-12.2012

Langer, H. F., and Chavakis, T. (2009). Leukocyte-endothelial interactions in inflammation. *J. Cell. Mol. Med.* 13, 1211–1220. doi: 10.1111/j.1582-4934.2009.00811.x

Lasky, L. A. (1992). Selectins: interpreters of cell-specific carbohydrate information during inflammation. *Science* 258, 964–969. doi: 10.1126/science.1439808

Lawson, L. J., Perry, V. H., Dri, P., and Gordon, S. (1990). Heterogeneity in the distribution and morphology of microglia in the normal adult mouse brain. *Neuroscience* 39, 151–170. doi: 10.1016/0306-4522(90)90229-w

Lerman, B. J., Hoffman, E. P., Sutherland, M. L., Bouri, K., Hsu, D. K., Liu, F. T., et al. (2012). Deletion of galectin-3 exacerbates microglial activation and accelerates disease progression and demise in a SOD1(G93A) mouse model of amyotrophic lateral sclerosis. *Brain Behav.* 2, 563–575. doi: 10.1002/brb3.75

Letschert, S., Gohler, A., Franke, C., Bertleff-Zieschang, N., Memmel, E., Doose, S., et al. (2014). Super-resolution imaging of plasma membrane glycans. *Angew. Chem. Int. Ed. Engl.* 53, 10921–10924. doi: 10.1002/anie.201406045

Ley, K. (2003). The role of selectins in inflammation and disease. *Trends Mol. Med.* 9, 263–268. doi: 10.1016/s1471-4914(03)00071-6

Liao, B., Zhao, W., Beers, D. R., Henkel, J. S., and Appel, S. H. (2012). Transformation from a neuroprotective to a neurotoxic microglial phenotype in a mouse model of ALS. *Exp. Neurol.* 237, 147–152. doi: 10.1016/j.expneurol.2012.06.011

Liddelow, S. A., Guttenplan, K. A., Clarke, L. E., Bennett, F. C., Bohlen, C. J., Schirmer, L., et al. (2017). Neurotoxic reactive astrocytes are induced by activated microglia. *Nature* 541, 481–487. doi: 10.1038/nature21029

Liu, F. T., Patterson, R. J., and Wang, J. L. (2002). Intracellular functions of galectins. *Biochim. Biophys. Acta* 1572, 263–273. doi: 10.1016/s0304-4165(02)00313-6

Liu, F. T., and Rabinovich, G. A. (2005). Galectins as modulators of tumour progression. *Nat. Rev. Cancer* 5, 29–41. doi: 10.1038/nrc1527

Longhi, L., Orsini, F., De Blasio, D., Fumagalli, S., Ortolano, F., Locatelli, M., et al. (2014). Mannose-binding lectin is expressed after clinical and experimental traumatic brain injury and its deletion is protective. *Crit. Care Med.* 42, 1910–1918. doi: 10.1097/CCM.0000000000000399

Macauley, M. S., Crocker, P. R., and Paulson, J. C. (2014). Siglec-mediated regulation of immune cell function in disease. *Nat. Rev. Immunol.* 14, 653–666. doi: 10.1038/nri3737

Masters, C. L., Bateman, R., Blennow, K., Rowe, C. C., Sperling, R. A., and Cummings, J. L. (2015). Alzheimer's disease. *Nat. Rev. Dis. Primers* 1:15056. doi: 10.1038/nrdp.2015.56

Miron, V. E., Boyd, A., Zhao, J. W., Yuen, T. J., Ruckh, J. M., Shadrach, J. L., et al. (2013). M2 microglia and macrophages drive oligodendrocyte differentiation during CNS remyelination. *Nat. Neurosci.* 16, 1211–1218. doi: 10.1038/nn.3469

Mok, S. W., Riemer, C., Madela, K., Hsu, D. K., Liu, F. T., Gultner, S., et al. (2007). Role of galectin-3 in prion infections of the CNS. *Biochem. Biophys. Res. Commun.* 359, 672–678. doi: 10.1016/j.bbrc.2007.05.163

Mondal, N., Stolfa, G., Antonopoulos, A., Zhu, Y., Wang, S. S., Buffone, A. Jr., et al. (2016). Glycosphingolipids on human myeloid cells stabilize E-selectin-dependent rolling in the multistep leukocyte adhesion cascade. *Arterioscler. Thromb. Vasc. Biol.* 36, 718–727. doi: 10.1161/ATVBAHA.115.306748

Morrison, H., Frye, J., Davis-Gorman, G., Funk, J., McDonagh, P., Stahl, G., et al. (2011). The contribution of mannose binding lectin to reperfusion injury after ischemic stroke. *Curr. Neurovasc. Res.* 8, 52–63. doi: 10.2174/156720211794520260

Naj, A. C., Jun, G., Beecham, G. W., Wang, L. S., Vardarajan, B. N., Buros, J., et al. (2011). Common variants at MS4A4/MS4A6E, CD2AP, CD33 and EPHA1 are associated with late-onset Alzheimer's disease. *Nat. Genet.* 43, 436–441. doi: 10.1038/ng.801

Nikodemova, M., Small, A. L., Smith, S. M., Mitchell, G. S., and Watters, J. J. (2014). Spinal but not cortical microglia acquire an atypical phenotype with high VEGF, galectin-3 and osteopontin, and blunted inflammatory responses in ALS rats. *Neurobiol. Dis.* 69, 43–53. doi: 10.1016/j.nbd.2013.11.009

Nimmerjahn, A., Kirchhoff, F., and Helmchen, F. (2005). Resting microglial cells are highly dynamic surveillants of brain parenchyma *in vivo*. *Science* 308, 1314–1318. doi: 10.1126/science.1110647

Orsini, F., Villa, P., Parrella, S., Zangari, R., Zanier, E. R., Gesuete, R., et al. (2012). Targeting mannose-binding lectin confers long-lasting protection with a surprisingly wide therapeutic window in cerebral ischemia. *Circulation* 126, 1484–1494. doi: 10.1161/CIRCULATIONAHA.112.103051

Osthoff, M., Katan, M., Fluri, F., Schuetz, P., Bingisser, R., Kappos, L., et al. (2011). Mannose-binding lectin deficiency is associated with smaller infarction size and favorable outcome in ischemic stroke patients. *PLoS One* 6:e21338. doi: 10.1371/journal.pone.0021338

Pang, Y., Fan, L. W., Tien, L. T., Dai, X., Zheng, B., Cai, Z., et al. (2013). Differential roles of astrocyte and microglia in supporting oligodendrocyte development and myelination *in vitro*. *Brain Behav.* 3, 503–514. doi: 10.1002/brb3.152

Paolicelli, R. C., Bolasco, G., Pagani, F., Maggi, L., Scianni, M., Panzanelli, P., et al. (2011). Synaptic pruning by microglia is necessary for normal brain development. *Science* 333, 1456–1458. doi: 10.1126/science.1202529

Pardo, E., Cárcamo, C., Uribe-San Martin, R., Ciampi, E., Segovia-Miranda, F., Curkovic-Pena, C., et al. (2017). Galectin-8 as an immunosuppressor in experimental autoimmune encephalomyelitis and a target of human early prognostic antibodies in multiple sclerosis. *PLoS One* 12:e0177472. doi: 10.1371/journal.pone.0177472

Park, J. W., Voss, P. G., Grabski, S., Wang, J. L., and Patterson, R. J. (2001). Association of galectin-1 and galectin-3 with Gemin4 in complexes containing the SMN protein. *Nucleic Acids Res.* 29, 3595–3602. doi: 10.1093/nar/29.17.3595

Pasquini, L. A., Millet, V., Hoyos, H. C., Giannoni, J. P., Croci, D. O., Marder, M., et al. (2011). Galectin-3 drives oligodendrocyte differentiation to control myelin integrity and function. *Cell Death Differ.* 18, 1746–1756. doi: 10.1038/cdd.2011.40

Patel, A. R., Ritzel, R., McCullough, L. D., and Liu, F. (2013). Microglia and ischemic stroke: a double-edged sword. *Int. J. Physiol. Pathophysiol. Pharmacol.* 5, 73–90.

Perry, V. H., Crocker, P. R., and Gordon, S. (1992). The blood-brain barrier regulates the expression of a macrophage sialic acid-binding receptor on microglia. *J. Cell Sci.* 101, 201–207.

Pillai, S., Netravali, I. A., Cariappa, A., and Mattoo, H. (2012). Siglecs and immune regulation. *Annu. Rev. Immunol.* 30, 357–392. doi: 10.1146/annurev-immunol-020711-075018

Politis, M., Giannetti, P., Su, P., Turkheimer, F., Keihaninejad, S., Wu, K., et al. (2012). Increased PK11195 PET binding in the cortex of patients with MS correlates with disability. *Neurology* 79, 523–530. doi: 10.1212/WNL.0b013e3182635645

Puttur, F., Arnold-Schrauf, C., Lahl, K., Solmaz, G., Lindenberg, M., Mayer, C. T., et al. (2013). Absence of Siglec-H in MCMV infection elevates interferon α production but does not enhance viral clearance. *PLoS Pathog.* 9:e1003648. doi: 10.1371/journal.ppat.1003648

Reichert, F., and Rotshenker, S. (1999). Galectin-3/MAC-2 in experimental allergic encephalomyelitis. *Exp. Neurol.* 160, 508–514. doi: 10.1006/exnr.1999.7229

Riemer, C., Neidhold, S., Burwinkel, M., Schwarz, A., Schultz, J., Krätzschmar, J., et al. (2004). Gene expression profiling of scrapie-infected brain tissue. *Biochem. Biophys. Res. Commun.* 323, 556–564. doi: 10.1016/j.bbrc.2004.08.124

Sakaguchi, M., Imaizumi, Y., and Okano, H. (2007). Expression and function of galectin-1 in adult neural stem cells. *Cell. Mol. Life Sci.* 64, 1254–1258. doi: 10.1007/s00018-007-6476-5

Sakaguchi, M., Shingo, T., Shimazaki, T., Okano, H. J., Shiwa, M., Ishibashi, S., et al. (2006). A carbohydrate-binding protein, Galectin-1, promotes proliferation of adult neural stem cells. *Proc. Natl. Acad. Sci. U S A* 103, 7112–7117. doi: 10.1073/pnas.0508793103

Sancho, D., and Reis e Sousa, C. (2012). Signaling by myeloid C-type lectin receptors in immunity and homeostasis. *Annu. Rev. Immunol.* 30, 491–529. doi: 10.1146/annurev-immunol-031210-101352

Satoh, K., Niwa, M., Goda, W., Binh, N. H., Nakashima, M., Takamatsu, M., et al. (2011). Galectin-3 expression in delayed neuronal death of hippocampal CA1 following transient forebrain ischemia and its inhibition by hypothermia. *Brain Res.* 1382, 266–274. doi: 10.1016/j.brainres.2011.01.049

Saylor, D., Dickens, A. M., Sacktor, N., Haughey, N., Slusher, B., Pletnikov, M., et al. (2016). HIV-associated neurocognitive disorder—pathogenesis and prospects for treatment. *Nat. Rev. Neurol.* 12:309. doi: 10.1038/nrneurol.2016.53

Schneider, U. C., Davids, A. M., Brandenburg, S., Müller, A., Elke, A., Magrini, S., et al. (2015). Microglia inflict delayed brain injury after subarachnoid hemorrhage. *Acta Neuropathol.* 130, 215–231. doi: 10.1007/s00401-015-1440-1

Singh, K. K., Nathamu, S., Adame, A., Alire, T. U., Dumaop, W., Gouaux, B., et al. (2011). Expression of mannose binding lectin in HIV-1-infected brain: implications for HIV-related neuronal damage and neuroAIDS. *Neurobehav. HIV Med.* 3, 41–52. doi: 10.2147/nbhiv.s19969

Sirko, S., Irmler, M., Gascon, S., Bek, S., Schneider, S., Dimou, L., et al. (2015). Astrocyte reactivity after brain injury-: the role of galectins 1 and 3. *Glia* 63, 2340–2361. doi: 10.1002/glia.22898

Smittkamp, S. E., Brown, J. W., and Stanford, J. A. (2008). Time-course and characterization of orolingual motor deficits in B6SJL-Tg(SOD1-G93A)1Gur/J mice. *Neuroscience* 151, 613–621. doi: 10.1016/j.neuroscience.2007.10.017

Song, S., Ji, B., Ramachandran, V., Wang, H., Hafley, M., Logsdon, C., et al. (2012). Overexpressed galectin-3 in pancreatic cancer induces cell proliferation and invasion by binding Ras and activating Ras signaling. *PLoS One* 7:e42699. doi: 10.1371/journal.pone.0042699

Spence, S., Greene, M. K., Fay, F., Hams, E., Saunders, S. P., Hamid, U., et al. (2015). Targeting Siglecs with a sialic acid-decorated nanoparticle abrogates inflammation. *Sci. Transl. Med.* 7:303ra140. doi: 10.1126/scitranslmed.aab3459

Stancic, M., Slijepcevic, D., Nomden, A., Vos, M. J., de Jonge, J. C., Sikkema, A. H., et al. (2012). Galectin-4, a novel neuronal regulator of myelination. *Glia* 60, 919–935. doi: 10.1002/glia.22324

Stancic, M., van Horssen, J., Thijssen, V. L., Gabius, H. J., van der Valk, P., Hoekstra, D., et al. (2011). Increased expression of distinct galectins in multiple sclerosis lesions. *Neuropathol. Appl. Neurobiol.* 37, 654–671. doi: 10.1111/j.1365-2990.2011.01184.x

Starossom, S. C., Mascanfroni, I. D., Imitola, J., Cao, L., Raddassi, K., Hernandez, S. F., et al. (2012). Galectin-1 deactivates classically activated microglia and protects from inflammation-induced neurodegeneration. *Immunity* 37, 249–263. doi: 10.1016/j.immuni.2012.05.023

Steelman, A. J., Smith, R. III., Welsh, C. J., and Li, J. (2013). Galectin-9 protein is up-regulated in astrocytes by tumor necrosis factor and promotes encephalitogenic T-cell apoptosis. *J. Biol. Chem.* 288, 23776–23787. doi: 10.1074/jbc.M113.451658

Steinman, L., and Zamvil, S. S. (2006). How to successfully apply animal studies in experimental allergic encephalomyelitis to research on multiple sclerosis. *Ann. Neurol.* 60, 12–21. doi: 10.1002/ana.20913

Subramaniam, S. R., and Federoff, H. J. (2017). Targeting microglial activation states as a therapeutic avenue in Parkinson's disease. *Front. Aging Neurosci.* 9:176. doi: 10.3389/fnagi.2017.00176

Sun, J., Shaper, N. L., Itonori, S., Heffer-Lauc, M., Sheikh, K. A., and Schnaar, R. L. (2004). Myelin-associated glycoprotein (Siglec-4) expression is progressively and selectively decreased in the brains of mice lacking complex gangliosides. *Glycobiology* 14, 851–857. doi: 10.1093/glycob/cwh107

Takagi, H., Fukaya, T., Eizumi, K., Sato, Y., Sato, K., Shibazaki, A., et al. (2011). Plasmacytoid dendritic cells are crucial for the initiation of inflammation and T cell immunity *in vivo*. *Immunity* 35, 958–971. doi: 10.1016/j.immuni.2011.10.014

Tang, T., Frenette, P. S., Hynes, R. O., Wagner, D. D., and Mayadas, T. N. (1996). Cytokine-induced meningitis is dramatically attenuated in mice deficient in endothelial selectins. *J. Clin. Invest.* 97, 2485–2490. doi: 10.1172/jci118695

Thored, P., Heldmann, U., Gomes-Leal, W., Gisler, R., Darsalia, V., Taneera, J., et al. (2009). Long-term accumulation of microglia with proneurogenic phenotype concomitant with persistent neurogenesis in adult subventricular zone after stroke. *Glia* 57, 835–849. doi: 10.1002/glia.20810

Thurston, T. L., Wandel, M. P., von Muhlinen, N., Foeglein, A., and Randow, F. (2012). Galectin 8 targets damaged vesicles for autophagy to defend cells against bacterial invasion. *Nature* 482, 414–418. doi: 10.1038/nature10744

Troncoso, M. F., Ferragut, F., Bacigalupo, M. L., Cárdenas Delgado, V. M., Nugnes, L. G., Gentilini, L., et al. (2014). Galectin-8: a matricellular lectin with key roles in angiogenesis. *Glycobiology* 24, 907–914. doi: 10.1093/glycob/cwu054

Varki, A. (2009). Natural ligands for CD33-related Siglecs? *Glycobiology* 19, 810–812. doi: 10.1093/glycob/cwp063

Varki, A., and Angata, T. (2006). Siglecs—the major subfamily of I-type lectins. *Glycobiology* 16, 1R–27R. doi: 10.1093/glycob/cwj008

Varki, A., Etzler, M. E., Cummings, R. D., and Esko, J. D. (2009). "Discovery and classification of glycan-binding proteins," in *Essentials of Glycobiology*, 2nd Edn. eds A. Varki, R. D. Cummings, J. D. Esko, H. H. Freeze, P. Stanley, C. R. Bertozzi, et al. (New York, NY: Cold Spring Harbor), 1–22.

Venkatesan, C., Chrzaszcz, M., Choi, N., and Wainwright, M. S. (2010). Chronic upregulation of activated microglia immunoreactive for galectin-3/Mac-2 and nerve growth factor following diffuse axonal injury. *J. Neuroinflammation* 7:32. doi: 10.1186/1742-2094-7-32

Wan, L., and Liu, F.-T. (2016). Galectin-3 and inflammation. *Glycobiol. Insights* 6, 1–9. doi: 10.4137/gbi.s13921

Wang, J. L., Haudek, K. C., Voss, P. G., Patterson, R. J., and Arnoys, E. J. (2012). "Intracellular galectins: platforms for assembly of macromolecular complexes," in *Galectins and Disease Implications for Targeted Therapeutics*, eds A. A. Klyosov and P. G. Traber (Washington, DC: ACS Symposium Series), 81–93. doi: 10.1021/bk-2012-1115.ch004

Wang, X., Mitra, N., Cruz, P., Deng, L., NISC Comparative Sequencing Program, Varki, N., et al. (2012). Evolution of siglec-11 and siglec-16 genes in hominins. *Mol. Biol. Evol.* 29, 2073–2086. doi: 10.1093/molbev/mss077

Wang, Y., and Neumann, H. (2010). Alleviation of neurotoxicity by microglial human Siglec-11. *J. Neurosci.* 30, 3482–3488. doi: 10.1523/JNEUROSCI.3940-09.2010

Wang, X., Yue, T. L., Barone, F. C., and Feuerstein, G. Z. (1995). Demonstration of increased endothelial-leukocyte adhesion molecule-1 mRNA expression in rat ischemic cortex. *Stroke* 26, 1665–1668; discussion 1668–1669. doi: 10.1161/01.str.26.9.1665

Weis, W. I., Drickamer, K., and Hendrickson, W. A. (1992). Structure of a C-type mannose-binding protein complexed with an oligosaccharide. *Nature* 360, 127–134. doi: 10.1038/360127a0

Werneburg, S., Buettner, F. F., Erben, L., Mathews, M., Neumann, H., Muhlenhoff, M., et al. (2016). Polysialylation and lipopolysaccharide-induced shedding of E-selectin ligand-1 and neuropilin-2 by microglia and THP-1 macrophages. *Glia* 64, 1314–1330. doi: 10.1002/glia.23004

Wielgat, P., and Braszko, J. J. (2012). The participation of sialic acids in microglia-neuron interactions. *Cell. Immunol.* 273, 17–22. doi: 10.1016/j.cellimm.2011.12.002

Yan, H., Kamiya, T., Suabjakyong, P., and Tsuji, N. M. (2015). Targeting C-type lectin receptors for cancer immunity. *Front. Immunol.* 6:408. doi: 10.3389/fimmu.2015.00408

Yan, J., Xu, Y., Zhang, L., Zhao, H., Jin, L., Liu, W. G., et al. (2016). Increased expressions of plasma galectin-3 in patients with amyotrophic lateral sclerosis. *Chin. Med. J.* 129, 2797–2803. doi: 10.4103/0366-6999.194656

Yip, P. K., Carrillo-Jimenez, A., King, P., Vilalta, A., Nomura, K., Chau, C. C., et al. (2017). Galectin-3 released in response to traumatic brain injury acts as an alarmin orchestrating brain immune response and promoting neurodegeneration. *Sci. Rep.* 7:41689. doi: 10.1038/srep41689

Yokota, Y., Arai, T., and Kawasaki, T. (1995). Oligomeric structures required for complement activation of serum mannan-binding proteins. *J. Biochem.* 117, 414–419. doi: 10.1093/jb/117.2.414

Zhang, R. L., Chopp, M., Zhang, Z. G., Phillips, M. L., Rosenbloom, C. L., Cruz, R., et al. (1996). E-selectin in focal cerebral ischemia and reperfusion in the rat. *J. Cereb. Blood Flow Metab.* 16, 1126–1136. doi: 10.1097/00004647-199611000-00006

Zhang, J., Raper, A., Sugita, N., Hingorani, R., Salio, M., Palmowski, M. J., et al. (2006). Characterization of Siglec-H as a novel endocytic receptor expressed on murine plasmacytoid dendritic cell precursors. *Blood* 107, 3600–3608. doi: 10.1182/blood-2005-09-3842

Zhang, F., Zheng, M., Qu, Y., Li, J., Ji, J., Feng, B., et al. (2009). Different roles of galectin-9 isoforms in modulating E-selectin expression and adhesion function in LoVo colon carcinoma cells. *Mol. Biol Rep.* 36, 823–830. doi: 10.1007/s11033-008-9251-2

Zhou, T., Li, X., Wu, P., Zhang, D., Zhang, M., Chen, N., et al. (2000). Effect of anti-P-selectin monoclonal antibody on renal ischemia/reperfusion injury in rats. *Chin. Med. J.* 113, 790–793.

2

Proteomic and Metabolomic Analyses of Vanishing White Matter Mouse Astrocytes Reveal Deregulation of ER Functions

*Lisanne E. Wisse[1‡], Renske Penning[2‡], Esther A. Zaal[2], Carola G. M. van Berkel[1], Timo J. ter Braak[1], Emiel Polder[1], Justin W. Kenney[3†], Christopher G. Proud[3†], Celia R. Berkers[2], Maarten A. F. Altelaar[2], Dave Speijer[4], Marjo S. van der Knaap[1] and Truus E. M. Abbink[1]**

[1] Pediatrics, VU University Medical Center, Amsterdam, Netherlands, [2] Biomolecular Mass Spectrometry and Proteomics Group, Utrecht Institute for Pharmaceutical Sciences, Bijvoet Center for Biomolecular Research, Utrecht University, Utrecht, Netherlands, [3] Centre for Biological Sciences, University of Southampton, Southampton, United Kingdom, [4] Medical Biochemistry, Academic Medical Center, Amsterdam, Netherlands

***Correspondence:**
Truus E. M. Abbink
g.abbink@vumc.nl

†Present address:
Justin W. Kenney,
The Hospital for Sick Children,
Toronto, ON, Canada
Christopher G. Proud,
South Australian Health and Medical Research Institute, School of Biological Sciences, University of Adelaide, Adelaide, SA, Australia

‡These authors have contributed equally to this work.

Vanishing white matter (VWM) is a leukodystrophy with predominantly early-childhood onset. Affected children display various neurological signs, including ataxia and spasticity, and die early. VWM patients have bi-allelic mutations in any of the five genes encoding the subunits of the eukaryotic translation factor 2B (eIF2B). eIF2B regulates protein synthesis rates under basal and cellular stress conditions. The underlying molecular mechanism of how mutations in eIF2B result in VWM is unknown. Previous studies suggest that brain white matter astrocytes are primarily affected in VWM. We hypothesized that the translation rate of certain astrocytic mRNAs is affected by the mutations, resulting in astrocytic dysfunction. Here we subjected primary astrocyte cultures of wild type (wt) and VWM ($2b5^{ho}$) mice to pulsed labeling proteomics based on stable isotope labeling with amino acids in cell culture (SILAC) with an L-azidohomoalanine (AHA) pulse to select newly synthesized proteins. AHA was incorporated into newly synthesized proteins in wt and $2b5^{ho}$ astrocytes with similar efficiency, without affecting cell viability. We quantified proteins synthesized in astrocytes of wt and $2b5^{ho}$ mice. This proteomic profiling identified a total of 80 proteins that were regulated by the eIF2B mutation. We confirmed increased expression of PROS1 in $2b5^{ho}$ astrocytes and brain. A DAVID enrichment analysis showed that approximately 50% of the eIF2B-regulated proteins used the secretory pathway. A small-scale metabolic screen further highlighted a significant change in the metabolite 6-phospho-gluconate, indicative of an altered flux through the pentose phosphate pathway (PPP). Some of the proteins migrating through the secretory pathway undergo oxidative folding reactions in the endoplasmic reticulum (ER), which produces reactive oxygen species (ROS). The PPP produces NADPH to remove ROS. The proteomic and metabolomics data together suggest a deregulation of ER function in $2b5^{ho}$ mouse astrocytes.

Keywords: vanishing white matter, eIF2B, AHA, SILAC, astrocytes, metabolomics, endoplasmic reticulum

INTRODUCTION

Vanishing white matter (VWM) is one of the more prevalent leukodystrophies (van der Knaap et al., 1999). Patients with VWM display chronic neurological deterioration and additionally episodes of stress-provoked rapid and severe deterioration. Neuropathology of post-mortem brain shows cystic degeneration of the cerebral white matter with lack of appropriate astrogliotic scar formation, profound lack of myelin, increased numbers of oligodendrocyte precursor cells and immature astrocytes. The morphology of especially astrocytes in cerebral white matter is abnormal. They look coarse and have fewer, thicker processes than normal (van der Knaap et al., 2006; Bugiani et al., 2011). A small proportion of the oligodendrocytes look foamy (Wong et al., 2000).

Recessive mutations in the eukaryotic translation factor 2B (eIF2B) cause VWM (Leegwater et al., 2001; van der Knaap et al., 2002). eIF2B is an enzyme composed of five different subunits (eIF2Bα-ε), encoded by five genes (*EIF2B1-5*). Mutations reduce the activity of eIF2B (van Kollenburg et al., 2006; Horzinski et al., 2009; Liu et al., 2011).

The eIF2B complex functions as a guanine nucleotide exchange factor (GEF), mediating the exchange of GDP for GTP on eIF2. eIF2-GTP binds to methionine-charged initiator tRNA (Met-tRNA$^{Met}{}_i$), thereby forming the ternary complex eIF2.GTP.Met-tRNA$^{Met}{}_i$. The ternary complex together with the small ribosomal subunit binds the 5′-end of the mRNA and scans the 5′-untranslated region (5′-UTR) until it encounters a start codon in a suitable context, whereupon translation of the open reading frame (ORF) starts. Simultaneously, GTP on eIF2 is hydrolyzed to GDP which makes the complex inactive (Konieczny and Safer, 1983; Kleijn et al., 1998; Proud, 2001). GEF activity is needed to recharge eIF2 with GTP.

Translation initiation is a complex process, involving multiple eukaryotic initiation factors (eIFs) (Voorma et al., 1994; Sonenberg and Hinnebusch, 2009). Translation initiation efficiency is profoundly influenced by the nucleotides flanking the start codon (usually AUG), the Kozak consensus sequence (Kozak, 1987). Purines at the −3 and +4 position relative to the AUG start codon are most important in determining translation initiation efficiency (Kozak, 1987; Noderer et al., 2014). The architecture of the 5′-UTR with regard to various sequences, structural motifs and length also determines the translation efficiency of an mRNA (Hinnebusch et al., 2016). eIF2B mutations are expected to reduce ternary complex levels and thus overall mRNA translation. However, this reduced activity can actually enhance translation of certain mRNAs with 5′-upstream open reading frames (uORFs) in their 5′-UTR. uORFs can inhibit translation of the main ORF (mORF) and translation of these mORFs depends on translation reinitiation, which is regulated by ternary complex levels (Meijer and Thomas, 2002; Hinnebusch et al., 2016).

A VWM mouse model, homozygous for the Arg191His mutation in the eIF2Bε subunit has been developed, representative of the human disease (*2b5^{ho}* mice) (Dooves et al., 2016). The Arg191His mutation corresponds to the Arg195His mutation in patients. This mutation reduces eIF2B activity *in vitro* and gives a severe VWM phenotype (Fogli et al., 2004; Li et al., 2004). Astrocytes in brains of *2b5^{ho}* mice are positive for the immaturity marker nestin. *In vitro* experiments show that *2b5^{ho}* astrocytes inhibit maturation of wild type (wt) oligodendrocytes (Dooves et al., 2016).

The mechanism by which mutations in eIF2B lead to astrocytic dysfunction and disease remains unclear. Here we aim to improve understanding of the molecular mechanism underlying VWM. We expected that mutant eIF2B would *not* have a general effect on mRNA translation but rather affects translation of a small number of specific mRNAs, leading to disruption of cellular balances and dysfunction of astrocytes in particular. To identify these translational differences, we subjected adult mouse astrocytes to high-resolution quantitative proteomics after a pulse labeling with AHA (L-azidohomoalanine) combined with SILAC (stable isotope-labeling by amino acids in cell culture) (Dieterich et al., 2006; Eichelbaum et al., 2012; Kenney et al., 2016).

MATERIALS AND METHODS

Mice

All experiments were carried out under the Dutch and European law with approval of the local Institutional Animal Care and Use Committee (IACUC) of the VU University (Amsterdam, Netherlands). Wt and *2b5^{ho}* animals were used. All animals were weaned at P21 and had *ad libitum* access to food and water. The mice were housed with a 12 h light and dark cycle.

Astrocyte Culture

Four-month-old mice were sacrificed by cervical dislocation. Brains were taken out and the olfactory bulbs, cerebella and forebrains were removed. Astrocytes from the remaining structures (gray and white matter structures, including striatum, hippocampi and basal nuclei) were isolated. Brain tissue was minced with a scalpel in Hank's balanced salt solution (HBSS) without magnesium and calcium (Gibco) at 4°C. The tissue was dissociated with a papain solution containing 20 mM PIPES (pH 7.4), 120 mM NaCl, 5 mM KCl, 1.1 mM EDTA, 5.5 mM L-cysteine-HCl, 40 U/ml DNase and 20 U/ml papain for 30 min at 37°C. Cells were plated in poly-L-ornithine (PLO)-coated flasks and cultured in DMEM/F12 with 15% fetal bovine serum (FBS) (Hyclone), 1% sodium pyruvate (Gibco), 100U penicillin, 100 μg/ml streptomycin (Gibco) and 10 μg/ml gentamicin (Gibco). The cells were passaged twice before they were used for experiments. All chemicals were purchased from Sigma–Aldrich unless otherwise stated. Every experiment was replicated in independent cultures derived from different mice (number of experiments is indicated in figure legends as, e.g., $n = 3$).

Rate of Protein Synthesis Assay

Astrocytes were plated in 6 cm dishes (~250,000 cells/dish) and cultured until 80–90% confluent. Cells were starved for methionine for 15 min in AHA medium [DMEM/F12 without methionine, lysine, arginine and phenol red (Thermo Fisher Scientific)] supplemented with 91.3 mg/L L-lysine, 147.5 mg/L

L-arginine, dialysed FBS and phenol red, after which AHA (Bachem) was added to AHA medium for the indicated time in a final concentration of 2 mM. Cells were harvested in lysis buffer composed of 50 mM Tris pH 7.5, 100 mM NaCl, 0.5 mM EDTA and 0.5% sodium dodecyl sulfate (SDS).

The newly synthesized proteins were labeled using the Click-iT® Protein buffer kit according to the manufacturer's protocol (Invitrogen). In short, the AHA molecules in the cell lysate were coupled to biotin. Proteins were precipitated by subsequent addition of six volumes of methanol, 1.5 volume of chloroform and 4 volumes of water. The protein pellet was dissolved in 50 mM Tris-HCl pH 7.5 with 1% SDS before subjection to Western blot analysis. Samples were loaded on 12% Criterion™ TGX Stain-Free™ Protein Gels (Bio-Rad), allowing detection of total protein load. Biotinylated AHA-labeled proteins were stained with 400 ng/ml streptavidin (680 nm) and visualized at 700 nm in an Odyssey system (Odyssey® Fc, LI-COR). The amount of AHA-staining was corrected for the total protein load determined by Gel Doc™ EZ System (Bio-Rad).

Cell Viability Assay

Astrocytes were plated in ½ area 96 well plates (~5000 cells/well) and cultured for 2 days in AHA-SILAC medium (Thermo Fisher Scientific, custom-made DMEM/F12 without methionine, arginine, lysine and phenol red), with addition of 15% dialyzed serum, 0.005% phenol red, SILAC amino acids (94.2 mg/L [$^{13}C_6$] L-lysine 152 mg/L [$^{13}C_6$] L-arginine or 96 mg/L [$^{13}C_6$, $^{15}N_4$] L-lysine and 154.3 mg/L [$^{13}C_6$, $^{15}N_4$] L-arginine) and 17.2 mg/L L-methionine. The cells were starved for methionine from 15 min and cultured with 2 mM AHA (Bachem) or methionine overnight in half well area plates. Cell viability (ATP levels, CellTiter-Glo) was measured according to manufacturer's instructions (Promega). In short, cells were kept at room temperature for 30 min. CellTiter-Glo was added in the same volume as the culture medium. The plate was shaken for 2 min and left standing for 10 min before measuring luminescence with a Victor2 plate reader (PerkinElmer Life Sciences).

AHA Enrichment and On-Bead Digestion

Astrocytes were plated in 10 cm dishes (~750,000 cells/dish) and cultured until 80% confluent. The culture medium was replaced with AHA-SILAC medium. Cells were grown for 4 days. SILAC labels were reversed in biological duplicates. Cells were subsequently starved for methionine for 15 min and AHA was added as described in the previous section. Astrocytes were further cultured for 2 h or otherwise indicated. The cells were subsequently washed with cold PBS and lysed in urea lysis buffer (supplied with the Click-IT Protein enrichment Kit, Invitrogen).

For the protein analysis of secreted proteins conditioned medium samples from wt and *2b5^ho^* astrocytes (2 h) were concentrated using Amicon® Ultra-15 centrifuge filter tubes (3 kDa, Merck). Samples were diluted with urea lysis buffer to a volume of 400 μl per sample.

The AHA-labeled proteins from cell lysates as well as from the conditioned medium were enriched using the Click-iT® Protein Enrichment Kit according to the manufacturer's protocol (Invitrogen) with some minor modifications. In short, the AHA-labeled proteins were bound to the resin (16 h), following an iodoacetamide treatment and several washing steps. The AHA-labeled proteins bound to the resin were dissolved in 50 mM ammonium bicarbonate with 3 M urea. Digestion was performed at 37°C by adding 0.1 μg Lys-C for 4 h followed by addition of 1 μg trypsin overnight. The peptides were separated from the resin by briefly centrifuging them through a 0.8 ml spin columns (Thermo Fisher Scientific). The flow through contains the peptides. Peptide samples were stored at −80°C until further use. See also **Figure 1** for an overview of the enrichment procedure.

LC-MS/MS Analysis

Peptides were analyzed using a Q Exactive™ Hybrid Quadrupole-Orbitrap™ Mass Spectrometer (Thermo Fisher Scientific), which was connected to a 1290 Infinity II LC System (Agilent). The trap column was made of C18 (Dr. Maisch Reprosil) material and the analytical column was a 50 cm, 50 μm inner diameter Poroshell C18 (Agilent) column. Both the trap and analytical columns were packed in-house. Solvent A consisted of 0.1% formic acid (Merck) in deionized water (Merck) and Solvent B of 0.1% formic acid in 80% acetonitrile (Biosolve). Peptides were first trapped at 50 μl/min with solvent A and then eluted with solvent B in a 120 min gradient at 100 nl/min: 0–10 min, 100% solvent A; 10.1–105 min, 13–40% solvent B; 105–108 min, 40–100% solvent B; 108–109 min, 100% solvent B; 109–110 min, 0–100% solvent A; 110–120 min, 100% solvent A. The Orbitrap was operated in a data-dependent manner, with the following settings: ESI voltage, 1700 V; inlet capillary temperature 320°C; full-scan automatic gain control (AGC) target, 3×10^6 ions at 35000 resolution; scan range, 350–1500 *m/z*; Orbitrap full-scan maximum injection time, 250 ms; MS2 scan AGC target, 5×10^4 ions at 17500 resolution; maximum injection, 120 ms; normalized collision energy, 25; dynamic exclusion time, 30; isolation window 1.5 *m/z*; 10 MS2 scans per full scan.

Data Processing

The LC-MS/MS data were processed with MaxQuant (v1.5.2.8) and MS2 spectra were searched with the Andromeda search engine against the mouse proteome in UniProt (17155 entries, downloaded on 2015-10-27). Enzyme specificity was set to Trypsin/P and two missed cleavages were allowed. SILAC labels (K6/8, R6/10), methionine for AHA substitution and methionine oxidation were set as variable modifications and cysteine carbamidomethylation was set as fixed modification. Minimum peptide length was 7 amino acids. Mass tolerance was set to 20 ppm for peptide masses and 0.6 Da for fragmentation masses. The false discovery rate (FDR) threshold was set to 1% for identifications. Minimal ratio count was set to 2 for protein quantification and the functions "match between runs" and "requantify" were enabled. Data were further analyzed in Perseus (v1.5.0.0). Protein groups were kept for further analysis if they were detected in at least three of the biological replicates.

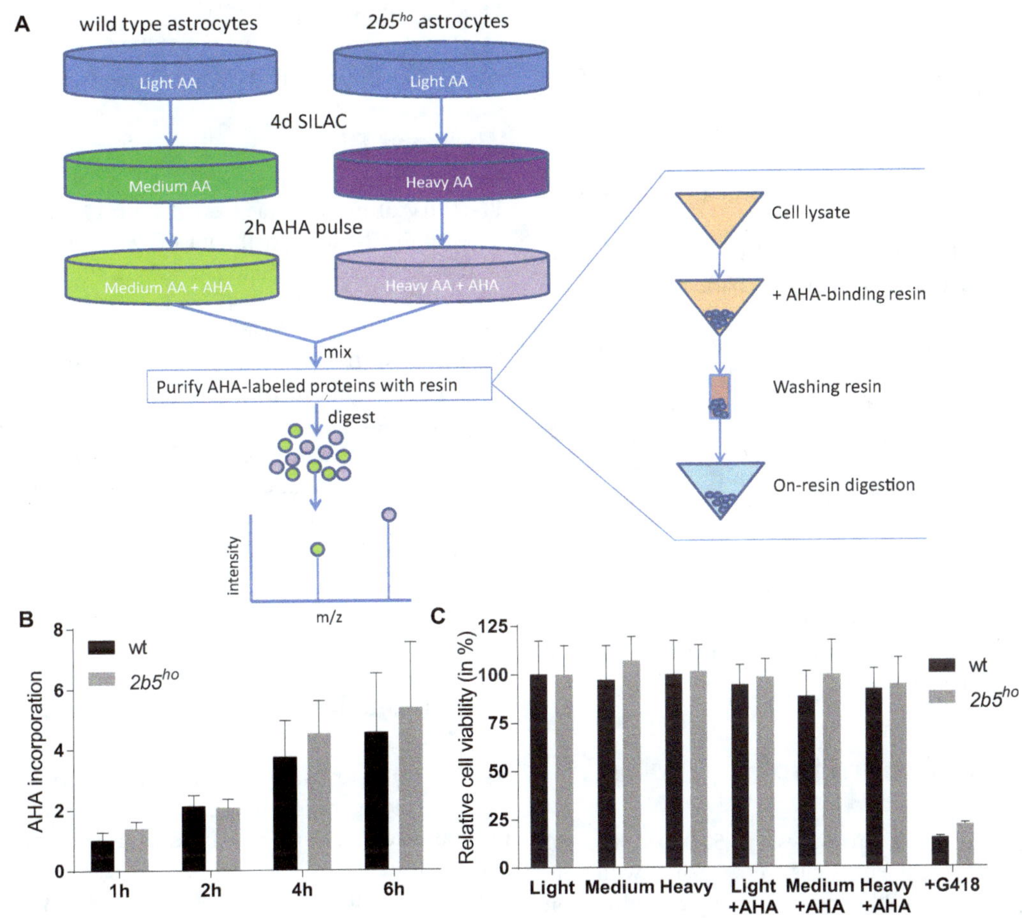

FIGURE 1 | L-azidohomoalanine (AHA) incorporation rates are similar between wt and *2b5^ho^* astrocytes and assay conditions do not affect wt and *2b5^ho^* cell viability. **(A)** Overview of the enrichment protocol. Heavy refers to labeling with [$^{13}C_6$, $^{15}N_4$] L-lysine and [$^{13}C_6$, $^{15}N_4$] L-arginine, while medium refers to labeling with [$^{13}C_6$] L-lysine and [$^{13}C_6$] L-arginine. **(B)** Astrocytes were treated with AHA for 1–6 h to determine a suitable AHA labeling time to measure newly synthesized proteins as well as confirm a similar AHA incorporation between wt and *2b5^ho^* astrocytes. The graph shows the mean + SD ($n = 3$) and the 1-h-incorporation in wt cells was set to 1. **(C)** Astrocytes were grown in normal (light) or SILAC culture medium (medium or heavy) for 2 days with or without a 16-h AHA pulse. Cell viability was determined by measuring intracellular ATP levels (CellTiter-glo, Promega). As a positive control for the assay, cells were grown in the presence of G418. The graph shows the mean ± SD ($n = 3$).

A *t*-test was performed and protein groups were considered significant if the *p*-value was <0.05.

Signal Peptide Analysis

To assess the presence of an N-terminal signal peptide in the proteins, we subjected the proteins from the proteomic screen to the SignalP 4.1 server (Bendtsen et al., 2004; Petersen et al., 2011). For all UniProt IDs the FASTA files were downloaded from the UniProt website and all Fasta files are automatically subjected to the SignalP 4.1 server. The presence or absence of a signal peptide for each UniProt ID was predicted with the SignalP 4.1 algorithms and the results were saved to Excel. The UniProt IDs linked back to the list of all identified protein names. Biomart was used to identify proteins containing a transmembrane domain. The number of proteins with a signal peptide was calculated for the proteins, the synthesis of which was affected by eIF2B$\varepsilon^{Arg191His}$ as well as for the other proteins that are not regulated by the same mutation. Next the number of regulated proteins with a signal peptide were subdivided in 'upregulated' and 'downregulated'.

In Silico Analysis of mRNA Features of eIF2B$\varepsilon^{Arg191His}$-Regulated Proteins

The sequences of the mRNA variants encoding the proteins identified in the AHA-SILAC proteomes were downloaded from the NCBI database including the start and stop codon location. The 5′-UTR features "uORFs, %GC and thermodynamic stability (ΔG in kcal/mol)" were determined for all mRNAs (Babendure et al., 2006). We compared each mRNA characteristic for all proteins found in the AHA-SILAC proteomic screen and compared the characteristics for the proteins regulated by eIF2B$\varepsilon^{Arg191His}$ and non-regulated proteins. The 5′-UTR length,

%GC and ΔG from mRNAs encoding the proteins found in the AHA-SILAC proteomes were analyzed per protein. When a protein is potentially translated from more than 1 mRNA variant (taken from the NCBI database), we counted and analyzed all features of all mRNA variants per protein and plotted all as a proportional representation per protein so that the analysis was not skewed by proteins with multiple variants. 5′-UTR lengths of less than 13 nucleotides were omitted from analyses as AUG codons less than 13 nt from the cap are inefficient to initiate translation (Hinnebusch, 2011). The % of GC in the 5′-UTRs was calculated. To calculate the thermodynamic stability, the free energy (ΔG) of 5′-UTRs was determined using UNAFold[1], which predicts thermodynamic stability. The uORF was defined as upstream AUG/CUG/GUG/UUG/ACG with a purine (A/G) at the -3 position and a guanine at the +4 position (Ingolia et al., 2011; Ferreira et al., 2013; Noderer et al., 2014). Scripts used for the analyses are found at: https://github.com/LisanneWisse/UORF or https://github.com/LisanneWisse/RefSeq.

DAVID Analysis

Overrepresentation of specific pathways was analyzed with DAVID [based on gene ontology (GO)-terms] between the proteins that are affected by the eIF2Bε^{Arg191His} mutation and all the proteins found in the AHA-SILAC proteome (Huang et al., 2009a,b).

Cell Lysates for Western Blot Experiments

Astrocytes were plated in 10 cm dishes (~750,000 cells/dish) and cultured until 80% confluent. Cells were washed with cold PBS, collected by scraping in PBS and pelleted by centrifugation (5 min, 1000 × *g*, 4°C). The method is based on the protocol described[2]. Cells were lysed in harvesting buffer containing 10 mM HEPES pH 7.9, 50 mM NaCl, 0.5 M sucrose, 0.1 mM EDTA, 0.5% Triton, 1 mM DTT, phosphatase inhibitors (0.5 mM activated $NaVO_3$, 25 mM β-glycerophosphate, 50 mM NaF) and protease inhibitors (Roche). Lysates were centrifuged (10 min 500 × *g*, 4°C) and the supernatant was used for Western blot analyses.

Whole Brain Lysates for qPCR and Western Blot Experiments

Mice were sacrificed by cervical dislocation at 4 months of age. Brains were removed, snap-frozen in liquid nitrogen and stored at −80°C until further use. Lysates were prepared by grinding the brain samples with a pestle and mortar under liquid nitrogen. The powder was lysed in cytoplasmic lysis buffer (20 mM Tris pH 7.4, 100 mM KAc, 3 mM $MgAc_2$, 2 mM DTT, 1.5% IGEPAL, 1.5% sodium deoxycholate (SODC), 1x HALT (protease and phosphatase inhibitory cocktail, Thermo Fisher Scientific]. The samples were homogenized with a pestle followed by trituration through a 23G needle. The samples were centrifuged (10 min, 10.000 × *g*, 4°C) and the supernatant was aliquoted into pre-chilled microfuge tubes. Protein concentrations were determined using a Quick Start™ Bradford Protein Assay (Bio-Rad). For RNA isolation, TRIzol™ Reagent (Invitrogen) was added to 50 μl of the supernatant and total RNA was isolated as described under RNA isolation and cDNA synthesis.

Western Blot

For cell lysates approximately 10 μg and for brain lysates 50–60 μg of protein was loaded on a 12% SDS-polyacrylamide gel with 2,2,2-trichloroethanol (TCE) which allow detection of total protein load (Ladner et al., 2004). Proteins were transferred onto a PVDF membrane (Bio-Rad). Membranes were blocked in 5% (w/v) milk powder and stained with the antibody against PROS1 (16910-1-AP, Proteintech) or SLC3A2 (LS-C334231, LSBio) overnight (16 h) at 4°C. Membranes were washed with TBS-Tween20 (0.1%) and incubated with an HRP-conjugated secondary goat anti-rabbit IgG antibody (Dako, P0448) for 2 h at room temperature. The membranes were washed three times with TBS-Tween (0.1%) and once with TBS, incubated with SuperSignal™ West Femto (Thermo Fisher Scientific) and imaged (Odyssey® Fc, LI-COR). Protein expression was corrected for total amount of protein determined by Gel Doc™ EZ System (Bio-Rad).

RNA Isolation, cDNA Synthesis and qPCR

Astrocytes were washed twice with PBS and collected in TRIzol™ Reagent (Invitrogen). RNA was isolated according to the manufacturer's protocol. In short, 1/5 volume chloroform was added to the TRIzol™ Reagent. The samples were centrifuged (10 min, 12000 × *g*, 4°C) and the water layer was transferred to a fresh tube. Half a volume of isopropanol and 1/250 volume of linear acrylamide (Ambion) was added and the samples were centrifuged for 20 min at 4°C and 12000 × *g*. The pellet was washed twice with 70% ethanol and resuspended in non-DEPC treated water (Ambion). RNA was precipitated with 175 mM sodium acetate (pH 5.2) and 70% ethanol and incubated for 30 min at −20°C. The samples were centrifuged (30 min, 4°C, 12000 × *g*) and the pellets were washed twice with 70% ethanol and resuspended in non-DEPC-treated water.

RNA quality and quantity were determined by measuring the A260 and A280 (NanoDrop 2000, Thermo Fisher Scientific). cDNA was synthesized in a 20 μl reverse transcription reaction: 1x first strand buffer (Invitrogen), random hexamers (0.02 μg/μl; Qiagen), oligoDT (0.02 μg/μl; Qiagen), dNTPs (1 mM each; Roche), DTT (1 mM; Invitrogen), RNaseOUT (0.25 U/μl; Invitrogen), Superscript III (5 U/μl; Invitrogen) and 1.5 μg total RNA were incubated for 2 h at 50°C. RNAseH (62.5 U/μl; Invitrogen) was added and incubated for 30 min at 37°C followed by 15 min at 70°C.

mRNA levels were determined with qPCR using a LightCycler® 480 II Instrument (Roche). For each 10 μl sample a mixture of LightCycler® 480 SYBR Green I Master (Roche), primers (1 pmol/μl) and cDNA (0.1 μl) was used. Used primers are listed in **Table 1**. *Gapdh* mRNA was used as reference.

[1] http://unafold.rna.albany.edu//?q=DINAMelt/Quickfold

[2] http://ron.cimr.cam.ac.uk/protocols/NucCyto.html

TABLE 1 | Primers used for qPCR.

Gene name	Forward (5′->3′)	Reverse (5′->3′)
Gapdh	GTGCTGAGTATGTCGTGGAG	TCGTGGTTCACACCCATCAC
Akt	AAGAAGGAGGTCATCGTCGC	GGTCGTGGGTCTGGAATGAG
Gas6	CTAAAACTATCCCCAGACAT	GGTACAAGGACTTCACGCTCT
Nrxn1	CCCCACAAAGGAACCCATCA	GTTGGCTAACCCACCTGAGC
Sorcs1	GGGACATCAGCCGAGTCATC	AACACCGCCACCAGGATATG
Fndc3c1	AGAGCGAGGCTTTTGGAGAA	TGGCACTGTTGGAGGTATTTCA
Pros1	TCCCTGGAGGCTACTCTTGTT	AGGTCCAAGGAAAGGCACAC
Scrn1	TGTCTGTCTTGCCTCAGAACA	TTTTGGCTGGGTCATCGTCA
Fgfr1	GCCAGACAACTTGCCGTATG	TCCGATAGAGTTACCCGCCA
Txnip	GTCTCAGCAGTGCAAACAGAC	CCTTCACCCAGTAGTCTACGC
Ddr1	CTGCTGCTTCTCATCATCGC	GTCAGCTCCTCCTCCAACAC
Slc3a2	GAAAGCTGATGAATGCACCC	CAATTTGCTGCAGGTCAGAG
Sec31A	GAGAGTCCTGCTGCTGAAGAG	AGACCGTCAATATCCCCACT

Constructs

Construct to Determine Secretion

The pNL1.3 plasmid (Promega) expresses the nanoluciferase protein with an *N*-terminal signal peptide from interleukin-6 (IL-6; Secluc). The signal peptide promotes secretion of the nanoluciferase into the culture medium. The pNL1.3-*Gapdh* construct contains the promoter including 5′-UTR of *Gapdh*, which was inserted in the pNL1.3 vector using an infusion reaction, according to the manufacturer's protocol (In-Fusion® HD Cloning Kit, Clontech). The PCR-amplified sequence of the constructs was confirmed by sequence analysis.

Constructs to Determine Translation

The pNL1.1 plasmids express the nanoluciferase protein (Nluc). Promoter and 5′-UTR sequences of candidate genes were taken from NCBI and Ensembl databases (**Table 2**). Infusion primers were designed to amplify promoter (approximately 2000 bps upstream of the transcription start site) and 5′-UTR-encoding sequences using the primer design tool on the Clontech website. Internal primers were designed to merge the *Gapdh* promoter region and the 5′-UTR of the candidate.

The promoter and 5′-UTR sequences were amplified using the infusion primers and the overlapping internal primers on gDNA from mouse liver or cDNA from cultured mouse astrocytes (**Table 2**) according to the manufacturer's protocol (ClonAmp, In-Fusion HD, Clontech). For the chimeric constructs comprising the murine *Gapdh* core promoter and 5′-UTR of the candidate mRNA a triple infusion reaction was performed with the pNL1.1 vector, the amplified promoter product and the amplified 5′-UTR product (**Table 2**). All promoter and 5′-UTR sequences were inserted into the pNL1.1 vector (Promega) using an infusion reaction according to the manufacturer's protocol (In-Fusion® HD Cloning Kit, Clontech). The resulting plasmids are listed in **Table 2**. PCR-amplified sequences of all constructs were confirmed by sequence analysis.

Construct Used As Internal Control

The pGL3 plasmid expresses the firefly luciferase protein (Fluc). The murine *Gapdh* promoter and 5′-UTR were digested from the pNL1.1 vector with KpnI and NcoI, purified from agarose gel (High Pure PCR Cleanup, Roche) and inserted into an empty pGL3 vector digested with the same restriction enzyme using a T4 DNA ligase (Promega). The resulting plasmid pGL3-*Gapdh* was used as internal standard in transfection studies. PCR-amplified sequence of the constructs was confirmed by sequence analysis.

Measurement of Secretion and Secretory Pathway Flux

Astrocytes were plated in half-area 96 well plates (~3000 cells/dish, CELLSTRAR) to 80% confluency. pNL1.3-Gapdh (80 ng) was transfected into the cells using 0.24 μl FuGENE® 6 according to the manufacturer's protocol (Promega). Luciferase activity in cells and culture medium was measured using the protocol of Promega and a Wallac 1420 Victor2 Microplate Reader (Perkin Elmer). The ratio of extracellular/intracellular luciferase activity determines the relative secretion.

Transfections

Astrocytes were plated in half-area 96 well plates (~3000 cells/dish, CELLSTAR). Each pNL1.1 promoter/5′-UTR construct (40 ng) was co-transfected with internal control pGL3-*Gapdh* (20 ng) using FuGENE® 6 Transfection Reagent (0.24 μl) according to the manufacturer's protocol. Luciferase activity was measured using the protocol of Promega and a GloMax® Discover System (Promega). The ratio Nluc:Fluc determines the expression of Nluc corrected for well-by-well differences.

Metabolomics

Astrocytes were cultured in 6 cm dishes (~250,000 cells/dish) until 70% confluent. Medium was replaced 72 and 24 h before harvesting. At the time of harvesting culture medium was removed from the cells and stored at −80°C until analysis. Cells were washed with ice-cold PBS and metabolites were extracted from cells in 0.5 ml lysis buffer containing methanol/acetonitrile/dH_2O (2:2:1). Samples were spun at 16.000 × *g* for 15 min at 4°C. Supernatants were collected for LC-MS analysis. 10 μl of conditioned medium was added to 1 mL of lysis buffer containing methanol/acetonitrile/H_2O (2:2:1) and prepared as above.

LC-MS analysis was performed on an Exactive mass spectrometer (Thermo Fisher Scientific) coupled to a Dionex Ultimate 3000 autosampler and pump (Thermo Fisher Scientific). The MS operated in polarity-switching mode with spray voltages of 4.5 and −3.5 kV. Metabolites were separated using a SeQuant® ZIC®-pHILIC HPLC Columns (2.1 mm × 150 mm, 5 μm, guard column 2.1 mm × 20 mm, 5 μm; Merck) using a linear gradient of acetonitrile and eluent A [20 mM $(NH_4)_2CO_3$, 0.1% NH_4OH in ULC/MS grade water (Biosolve)]. Flow rate was set at 150 μl/min. Metabolites were identified and quantified using LCQUAN™ Quantitative Software (Thermo Fisher Scientific) on the basis of exact mass within 5 ppm and further validated by concordance with retention times of standards. Metabolites were quantified using LCQUAN^{TM} Quantitative Software (Thermo Fisher Scientific). Peak intensities were normalized based on median peak intensity.

Statistical Analyses

The program Factor was used to correct for differences between experiments (qPCR, Western blot, cell viability) but not between other conditions (genotype, treatments) (Ruijter et al., 2006). Statistical analysis of qPCR, Western blot and cell viability experiments were performed with a two way ANOVA followed by a Sidaks multiple comparison test, using GraphPad Prism software. Statistical analysis of the transfection data was performed using a *T*-test per construct.

A Chi-square analysis was used to measure significant differences in the presence of signal peptides, transmembrane regions and SP-targeting proteins as well as for the analysis on the 5′-UTR features using GraphPad Prism software.

The secretion assay was statistically tested with a multilevel analysis in SPSS. Differences were significant when $p < 0.05$.

TABLE 2 | Primer sequences used for cloning.

Gene number	Construct name	Primer	Sequence
NM_008084	pNL1.3-*Gapdh*	Infusion FW primer	TGGCCTAACTGGCCGGTACCTGCTGTGTCACTACCGAAGAACAACGAGGAGAAGAT
		Infusion RV primer	GGCTAGCGAGCTCAGGTACCTTTGTCTACGGGACGAGGCTGGCACTGCACAAGAAG
NM_008084	pNL1.1-*Gapdh*	Infusion FW primer	TGGCCTAACTGGCCGGTACCTGCTGTGTCACTACCGAAGAACAACGAGGAGAAGAT
		Infusion RV primer	AGTGTGAAGACCATGGTTTGTCTACGGGACGAGGCTGGCACTGCACAAGAAG
NM_011173	pNL1.1-*Pros1*	Infusion FW primer	TGGCCTAACTGGCCGGTACCAACTGGCTTCTTTGTGGTG
		Infusion RV primer	AGTGTGAAGACCATGGCTGAGAGGATGGCCGGG
NM_019521	pNL1.1-*Gas6*	Infusion FW primer	TGGCCTAACTGGCCGGTACCGGACAGGCACTCTTTGGA
		Infusion RV primer	AGTGTGAAGACCATGGCGAGGCCGGTGCCGGG
NM_027268	pNL1.1-*Scrn1*	Infusion FW primer	TGGCCTAACTGGCCGGTACCCTTTTTTGATTCTGGAAA
		Infusion RV primer	AGTGTGAAGACCATGCTGCCAAGCAGCCGGCT
NM_010206.3	pNL1.1-*Fgfr1*	Infusion FW primer	TGGCCTAACTGGCCGGTACCCAGGGCAAGGATATTGCTA
		Infusion RV primer	AGTGTGAAGACCATGCCAGTTCTGCGGTTAGAG
		Internal primer FW	ACCGCAGCGCCAAGTGAG
		Internal primer RV	CTCACTTGGCGCTGCGGT
NM_011173	pNL1.1-*Pros1* 5′UTR	Infusion FW primer	TGGCCTAACTGGCCGGTACCTGCTGTGTCACTACCGAAGAACAACGAGGAGAAGAT
		Infusion RV primer	AGTGTGAAGACCATGGCTGAGAGGATGGCCGG
		Internal primer FW	GGGTCCAAAGAGAGGGAGGAGCTCGGGCTGGGCCGCGGCAG
		Internal primer RV	TCCTCCCTCTCTTTGGACCCGCCTCATTTT
NM_019521	pNL1.1-*Gas6* 5′UTR	Infusion FW primer	TGGCCTAACTGGCCGGTACCTGCTGTGTCACTACCGAAGAACAACGAGGAGAAGAT
		Infusion RV primer	AGTGTGAAGACCATGGCGAGGCCGGTGCCGGGG
		Internal primer FW	GGGTCCAAAGAGAGGGAGGAACCCGCTGCCTCCTTCACGGC
		Internal primer RV	TCCTCCCTCTCTTTGGACCCGCCTCATTTT
NM_010206.3	pNL1.1-*Fgfr1* 5′UTR	Infusion FW primer	TGGCCTAACTGGCCGGTACCTGCTGTGTCACTACCGAAGAACAACGAGGAGAAGAT
		Infusion RV primer	AGTGTGAAGACCATGCCAGTTCTGCGGTTAGA
		Internal primer FW	GGGTCCAAAGAGAGGGAGGAGCACAGCGCTCGGAGCGCTCC
		Internal primer RV	GGAGCGCTCCGAGCGCTGTGCTCCTCCCTCTCTTTGGACCC
NM_020252.3	pNL1.1-*Nrxn1* 5′UTR	Infusion FW primer	TGGCCTAACTGGCCGGTACCTGCTGTGTCACTACCGAAGAACAACGAGGAGAAGAT
		Infusion RV primer	AGTGTGAAGACCATGCTCGGGGCTGGGGTGCG
		Internal primer FW	GGGTCCAAAGAGAGGGAGGACCTTTTTCCCTCTCCTCCTCC
		Internal primer RV	GGAGGAGGAGAGGGAAAAAGGTCCTCCCTCTCTTTGGACCC
NM_021377	pNL1.1-*Sorcs1* 5′UTR	Infusion FW primer	TGGCCTAACTGGCCGGTACCTGCTGTGTCACTACCGAAGAACAACGAGGAGAAGAT
		Infusion RV primer	AGTGTGAAGACCATGTCTGGAGCGTAGAGAAG
		Internal primer FW	GGGTCCAAAGAGAGGGAGGAAGCCTGGGCGAGCGGCAGGCA
		Internal primer RV	TCCTCCCTCTCTTTGGACCCGCCTCATTTT
NM_001198833	pNL1.1-*Ddr1* 5′UTR	Infusion FW primer	TGGCCTAACTGGCCGGTACCTGCTGTGTCACTACCGAAGAACAACGAGGAGAAGAT
		Infusion RV primer	AGTGTGAAGACCATGGCTCTCCGGGGCGGACC
		Internal primer FW	GGGTCCAAAGAGAGGGAGGATGGCTCCTCTCCCCGGAACAG
		Internal primer RV	TCCTCCCTCTCTTTGGACCCGCCTCATTTT
NM_011173	pNL1.1-*Pros1* promoter	Infusion FW primer	TGGCCTAACTGGCCGGTACCAACTGGCTTCTTTGTGGTG
		Infusion RV primer	AGTGTGAAGACCATGGTTTGTCTACGGGACGAGGCTGGCACTGCACAAGAAG
		Internal primer FW	TGGCTGCTCCGCCCGCCCGCGGGGAAATGAGAGAGGCCCA
		Internal primer RV	GCGGGCGGGCGGAGCAGCCA

RESULTS

AHA-SILAC Incorporation Is Not Affected by Mutations in eIF2B and Does Not Affect Astrocyte Viability

L-azidohomoalanine incorporation was measured in wt and *2b5^{ho}* astrocytes to investigate if the general protein synthesis rate is affected by the homozygous Arg191His mutation in eIF2Bε (**Figure 1**). Label incorporation increased linearly for a period of 4 h (**Figure 1B**). The incorporation occurred with similar kinetics in *2b5^{ho}* and wt astrocytes (**Figure 1B**), indicating that the eIF2Bε Arg191His mutation did not significantly affect protein synthesis rate. The viability of both wt and *2b5^{ho}* astrocytes was not influenced by AHA or SILAC treatment (**Figure 1C**). Expression of stress-related mRNAs was unaffected by the AHA-labeling protocol (Supplementary Figure 1). These results allow further proteomic analyses with an AHA incorporation pulse of 2 h.

Pulsed Labeling Proteomics of Astrocytes Reveals 80 Proteins Regulated by the eIF2Bε Arg191His Mutation

The AHA proteomic labeling approach was performed to identify and quantify proteins that are differentially translated in *2b5^{ho}* astrocytes. SILAC-labeled wt and *2b5^{ho}* astrocytes were subjected to AHA labeling for 2 h. After AHA labeling cells were harvested and subjected to bead-based enrichment using a Click-iT chemistry approach. Bound proteins were digested and resulting peptides were analyzed by LC-MS/MS. In four biological replicates, we identified a total of 2888 proteins across both wt and *2b5^{ho}* astrocytes. 1240 proteins were detected in at least 3 out of 4 biological replicates for both wt and *2b5^{ho}* astrocytes and used for further analysis. Accumulation of 72 proteins was increased and of eight proteins decreased in the *2b5^{ho}* AHA-proteome ($p < 0.05$, student's T-test) (**Figure 2A** and Supplementary File 1).

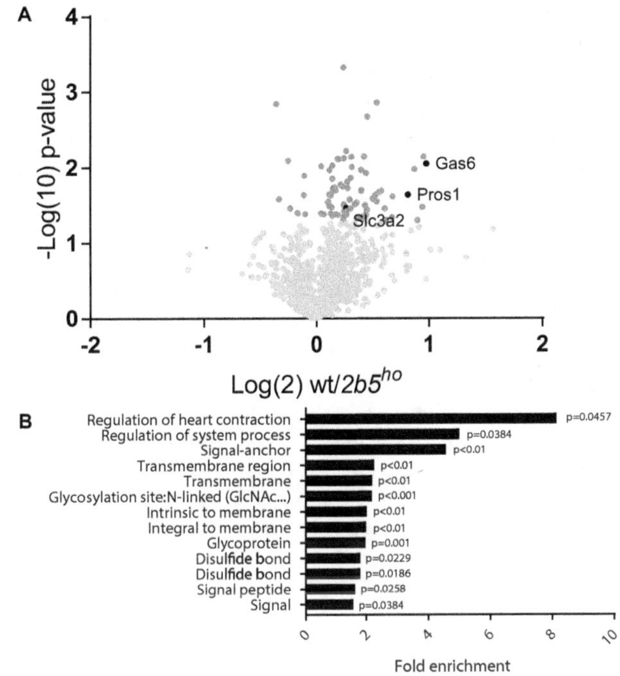

FIGURE 2 | Volcano plot of all 1240 proteins from the proteomic screen and DAVID analysis of the 80 proteins regulated by eIF2Bε^{Arg191His} identifies a difference in proteins that migrate through the secretory pathway. **(A)** In the volcano plot the proteins in darker gray have a p-value < 0.05 and are considered significantly differently synthesized between wt and *2b5^{ho}* astrocyte cultures. The p-values are plotted as Log(10) and $p < 0.05$ corresponds to -Log10 > 1.12. The fold change between wt and *2b5^{ho}* is plotted as Log(2) and a fold change of >2 corresponds to Log(2) > 1. 80 proteins are significantly different between wt and *2b5^{ho}* astrocytes of which 72 are increased and 8 are decreased in *2b5^{ho}* astrocytes. The proteins SLC3A2, PROS1, and GAS6 are highlighted in black. **(B)** The figure shows the overrepresentation analysis (p-value < 0.05) between the significantly different proteins with the proteins found in the proteome as a background. The pathway analysis reveals an overrepresentation of proteins with a signal peptide, transmembrane domain, N-linked glycosylation site(s) and/or disulfide bond(s). These terms together pinpoint targeting of proteins to the secretory pathway, some of which are post-translationally modified, as being affected by this mutation.

eIF2Bε^{Arg191His}-Regulated Proteins Localize Predominantly to the Plasma Membrane or Extracellular Space

We performed a DAVID pathway analysis to investigate if the eIF2Bε^{Arg191His}-regulated proteins share common functions, localizations or other features. This analysis showed enrichment for N-terminally glycosylated proteins, membrane proteins, proteins with disulfide bonds as well as proteins that contain a signal peptide among the set of eIF2Bε^{Arg191His}-regulated

TABLE 3 | Analysis of the proteins that are synthesized and migrate in the secretory pathway (SP).

AHA-SILAC data set*)	% of proteins with a signal peptide (absolute #)	% of proteins with a transmembrane domain (absolute #)	% of SP-targeting proteins (absolute #)
All proteins (1240)	19.1% (237)	14.3% (177)	25.6% (317)
Regulated (80)	33.8% (27)	30.0% (24)	47.5% (38)
Upregulated (72)	37.5% (27)	33.3% (24)	52.7% (38)
Downregulated (8)	0% (0)	0% (0)	0% (0)

**All proteins, proteins measured in wt and 2b5^{ho} astrocytes; regulated, proteins that differ significantly in intracellular accumulation between wt and 2b5^{ho} astrocytes (divided in upregulated and downregulated).*

proteins (**Figure 2B**). This outcome suggests that many regulated proteins follow the secretory pathway. Since signal peptides and transmembrane domains are associated with targeting to the secretory pathway (von Heijne, 1990; Petersen et al., 2011; Borgese, 2016), we tested all 1240 identified proteins for signal peptides or transmembrane domains. Signal peptides were found in 237 proteins and transmembrane domains in 177. The number of proteins predicted to pass through the secretory pathway was significantly enriched in the *2b5^{ho}* proteome (**Table 3**). 27 out of the 80 regulated proteins harbor a signal peptide and 24 have at least one transmembrane domain; 13 of the proteins harbor both a signal peptide and a transmembrane domain (in total 38 proteins out of 80). Intriguingly, the proteins annotated as migrating through the secretory pathway were all upregulated in *2b5^{ho}* astrocytes. It is possible that these proteins accumulate more in *2b5^{ho}* astrocytes during the 2-h labeling pulse due to a secretory pathway flux difference and not to an actual increase in translation rate. To investigate this, we analyzed the AHA-enriched secretome from wt and *2b5^{ho}* cultures after a 2-h AHA labeling. We detected 22 labeled proteins in the secretome, of which three (APOE, CST3 and POSTN) were significantly reduced in the *2b5^{ho}* astrocyte-conditioned medium. Intracellularly, APOE and POSTN were not changed by eIF2Bε^{Arg191His}. CST3 was significantly increased intracellularly in *2b5^{ho}* astrocytes. No other protein from the original 27 eIF2Bε^{Arg191His}-regulated proteins with a signal peptide was detected in the secretomes of wt and *2b5^{ho}* astrocytes. To further address the hypothesis that *2b5^{ho}* astrocytes have an altered secretory pathway flux, we evaluated the flux of proteins migrating through the secretory pathway with a reporter assay. The assay showed robust secretion of the Secluc reporter (Supplementary Figure 2) and secretion differences between wt and *2b5^{ho}* astrocytes were not detected (**Figure 3**).

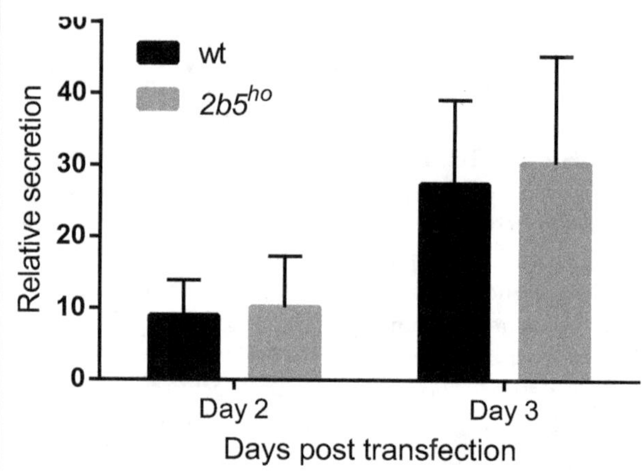

FIGURE 3 | The secretion of Secluc nanoluciferase via the secretory pathway is not affected in *2b5^{ho}* astrocyte cultures. Astrocytes were transfected with a luciferase reporter construct that expresses nanoluciferase fused to the IL-6 signal peptide. Luciferase activity was measured 2 and 3 days post transfection in cell lysates and cell culture medium. The ratio of extracellular:intracellular luciferase activity was determined as a measure for secretion through the secretory pathway. In the absence of the signal peptide, the extracellular luciferase activity was similar to the background signal (Supplementary Figure 2). The graph shows the mean ± SD (three independent cultures, six wells per culture).

Validation of eIF2Bε^{Arg191His}-Regulated Proteins

To investigate if the proteins that were found in the screen are indeed differentially expressed at the translational level, we selected eleven candidates that are either increased or decreased in wt vs. *2b5^{ho}* cultures. We quantified their mRNA levels to discriminate whether differences in protein amounts are regulated at the transcriptional or the translational level. Only one of the selected candidates differed in mRNA expression between wt and *2b5^{ho}* astrocytes (**Figure 4A**), suggesting that the increased accumulation of the other 10 proteins was not due to increased transcription or mRNA stability. Of the 11 candidates two were further investigated at the protein level by Western blot. We found that PROS1 and SLC3A2 were indeed significantly increased at the total protein level (**Figure 4B**). This result confirms the data from the proteomic screen and also demonstrates increased accumulation of these candidate proteins.

We next investigated several targets in mouse brain. At the mRNA level seven out of nine targets were similar between wt and *2b5^{ho}* brains (Supplementary Figure 3A). Two proteins were further assessed using Western blot (Supplementary Figure 3B). SLC3A2 protein abundance was also increased in brain lysates; however, this correlated with increased *Slc3a2* mRNA levels, suggesting transcriptional regulation in brain. PROS1 protein abundance was also increased in brain lysates, but *Pros1* mRNA abundance was not. This finding suggests translational upregulation of PROS1 both in cell culture and brain.

In Silico Analysis of mRNAs

eIF2B plays an essential role in the regulation of protein synthesis. For this reason, we investigated if the mRNAs of the eIF2Bε^{Arg191His}-regulated proteins share specific features that could explain the translational regulation by eIF2B. We investigated the 5′-UTRs for length, structural stability (%GC and ΔG) and number of uORFs with a Kozak sequence. The 5′-UTR length was significantly increased for mRNAs encoding eIF2Bε^{Arg191His}-regulated proteins (**Figure 5**). Of note, the mRNAs for the eight proteins *decreased* in the eIF2Bε^{Arg191His}-regulated proteome have relatively short 5′-UTRs (29-343 bases, median 133). This analysis showed that the eIF2Bε^{Arg191His} mutation differentially influences expression of proteins translated from mRNAs with *short* 5′-UTRs (<150 bases), which were underrepresented, compared to those with lengthy 5′-UTRs (>550 nucleotides), which were overrepresented; those translated from mRNAs with short 5′-UTRs, if regulated, tend to go down in expression in the mutant cells. The %GC in the 5′-UTRs did not significantly differ between eIF2Bε^{Arg191His}-regulated proteins and other proteins (**Figure 5**). The thermodynamic stability (ΔG in kcal/mole) of the 5′-UTR seemed to be higher for mRNAs encoding the eIF2Bε^{Arg191His}-regulated proteins, although not significantly (**Figure 5**). The

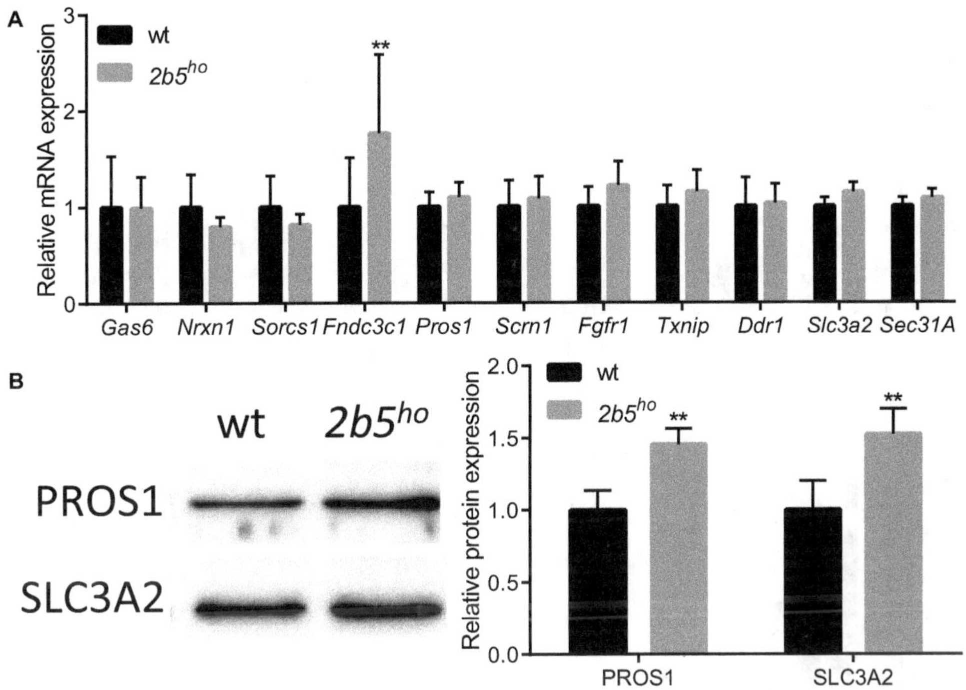

FIGURE 4 | mRNA and protein levels of candidate proteins in astrocyte cultures. **(A)** mRNA levels of the investigated candidate were investigated with qPCR. The mRNA levels were similar between wt and $2b5^{ho}$ astrocytes ($n \geq 5$), with the exception of the *Fndc3c1* mRNA, which was increased in $2b5^{ho}$ astrocytes. **(B)** The total protein levels of PROS1 and SLC3A2 were determined by Western blot. The graph shows the mean ± SD. The protein levels were increased in $2b5^{ho}$ cultures compared to wt cultures ($n \geq 3$). $^{**}p < 0.01$. Post-translational modifications were overall not affected in $2b5^{ho}$ astrocytes. Western blot analyses did not show whether PROS1 in astrocytes was post-translationally modified or not. Staining of all proteins (loading control) is shown in Supplementary Figure 4.

overall number of uORFs did not significantly differ for mRNA encoding the eIF2Bε$^{\text{Arg191His}}$-regulated proteins compared to the non-regulated proteins (**Figure 5**). However, mRNAs with high numbers of uORFs (8 or more) were clearly overrepresented in the group of significantly altered proteins.

Translation Efficiency Was Assessed in Reporter Assays

We investigated the potential translational regulation of some eIF2Bε$^{\text{Arg191His}}$-regulated proteins using reporter constructs encoding nanoluciferase driven by promoter sequences (including the 5′-UTR) of the candidates GAS6, PROS1, SCRN1 and FGFR1. A *Gapdh* promoter construct was included as non-regulated control. Transfection of each construct yielded similar luciferase expression in wt and $2b5^{ho}$ cultures (**Figure 6**). Transfection of the pNL1.1-*Fgfr1* promoter construct did not yield a reliable level of nanoluciferase (approximately two–threefold over the background signal in non-transfected cells) and was therefore omitted from analysis. The levels of nanoluciferase activity expressed with the other constructs were still low in comparison to the *Gapdh*-promoter-driven expression. Because expression could be increased by replacing the candidate promoter with the *Gapdh* promoter, we therefore constructed *Gapdh*-candidate chimeras. With these chimeras we tested the 5′-UTR efficiency of candidates GAS6, SORCS1, PROS1, FGFR1, NRXN1, and DDR1 in wt and $2b5^{ho}$ cultures. Neither the pNL1.1-*Fgfr1*-5′-UTR nor the pNL1.1-*Nrxn1*-5′-UTR constructs yielded quantifiable nanoluciferase expression. These constructs were omitted from further analyses. The expression from the chimeric promoter-5′-UTR constructs of *Gas6* and *Pros1* was increased by the *Gapdh* core promoter in both wt and $2b5^{ho}$ cells. Still, none of these constructs yielded an increased nanoluciferase expression in $2b5^{ho}$ compared to wt astrocytes (**Figure 6**).

Metabolic Screen of Astrocytes Conditioned Medium and Lysates

Because the proteomic data only gave small differences, the VWM phenotype remained difficult to explain. Tiny differences in protein levels and post-translational modifications occasionally lead to metabolic shifts. Thus we checked whether they affected cellular metabolism, using a mass spectrometry-based metabolic screen. We measured intra- and extracellular metabolites 24 and 72 h after replacing the culture medium. We looked at general energy consumption by measuring the uptake and secretion of metabolites from the culture medium (**Figure 7A**). As expected, both cultures had taken up glutamate (Martin-Jimenez et al., 2017) and aspartate (which use the same transporters) (Bender et al., 1997), reflecting specific astrocyte function. The uptake of glutamate and aspartate did not influence their intracellular levels (Supplementary File 2). Also, both wt and $2b5^{ho}$ cultures showed uptake of pyruvate. Astrocytes convert

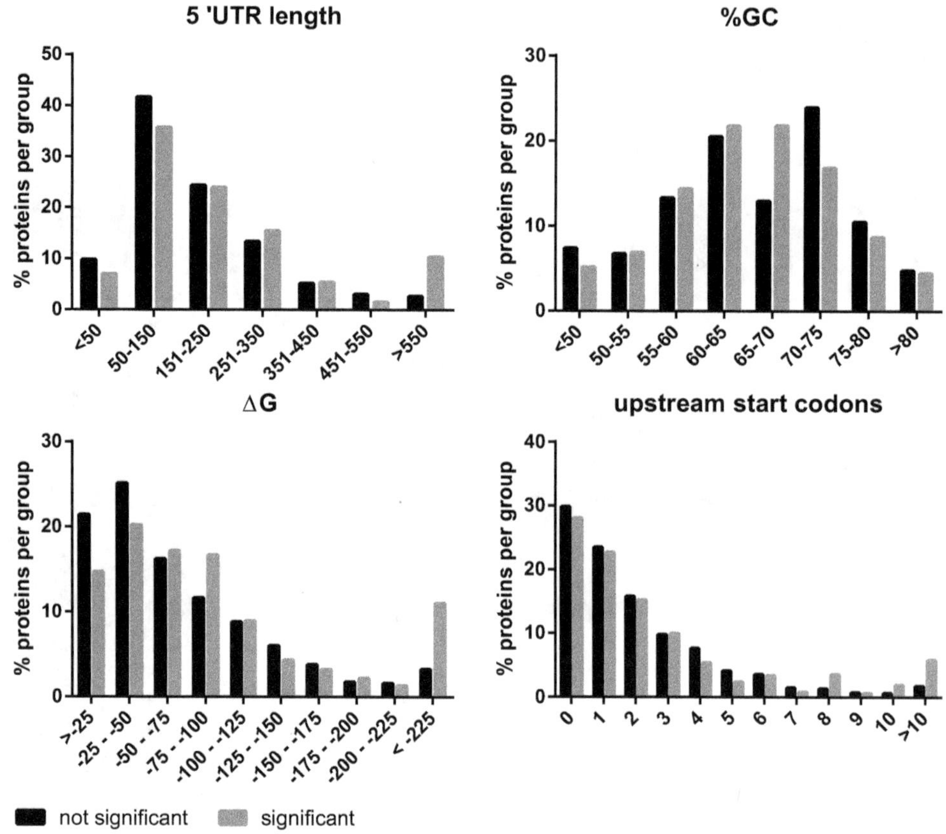

FIGURE 5 | *In silico* 5′UTR analysis of the eIF2B^{Arg191His}-regulated proteins. The 5′UTRs of the mRNAs encoding the proteins regulated by eIF2Bε^{Arg191His} mRNAs are significantly longer than of those encoding non-regluated proteins. They have an overall similar percentage GC content to the non regulated proteins. However, they seem to tend to higher ΔG's. These findings combined are indicative of relatively structured 5′-UTRs in the mRNAs of eIF2Bε^{Arg191His}-regulated proteins. The number of uORFs is similar between the significant and not significant proteins, though mRNAs with very high numbers (8 or more) of predicted uORFs are overrepresented in the group of significantly altered proteins.

pyruvate into lactate and glutamate/aspartate into glutamine, *cis*-aconitate and α-ketoglutarate (Martin-Jimenez et al., 2017). Indeed, the wt as well as the *2b5^{ho}* cultures secreted lactate, glutamine, *cis*-aconitate and α-ketoglutarate (**Figure 7A**). In addition, we investigated intracellular metabolites (**Figure 7B**). 6P-gluconate, an intermediate of the pentose phosphate pathway (PPP), was detected in *2b5^{ho}* but not in wt cells at both time points (**Figure 7B**). A fluctuation over time in NADPH levels was observed. The remaining metabolites tested did not show consistent differences over time between wt and *2b5^{ho}* astrocytes (**Figure 7B** and Supplementary File 2).

DISCUSSION

Here we aimed to identify whether specific proteins are translationally regulated by eIF2Bε^{Arg191His} in astrocytes. We used primary astrocyte cultures of the *2b5^{ho}* mouse model. Both wt and *2b5^{ho}* cells display an astrocyte morphology, synthesize CD44, nestin and vimentin (Supplementary Figure 5 and File 1), supporting astrocyte identity. We first established that cultures of primary astrocytes are suitable for sensitive labeling of newly synthesized proteins. Consistent with previous studies (Howden et al., 2013; Genheden et al., 2015), we found that AHA-labeling did not affect murine astrocyte viability for at least 16 h. No differences in AHA-labeling efficiency were observed between wt and *2b5^{ho}* astrocytes. These results confirm that the pulsed AHA-SILAC labeling protocol is suitable to study differences in proteins synthesized in wt and *2b5^{ho}* astrocytes, enabling us to investigate whether the Arg191His mutation in eIF2Bε affects translation of specific mRNAs.

We identified 80 proteins that accumulate differentially in wt vs. *2b5^{ho}* astrocyte cultures upon a 2 h AHA pulse. The majority (72 out of 80) is upregulated in *2b5^{ho}* astrocytes. We investigated the proteomic results by qPCR (11 candidates) and Western blot (4 candidates) (**Figure 4**). The tested antibodies detecting PROS1 and SLC3A2 showed consistent differences between wt and *2b5^{ho}* while TXNIP gave variable results and SCRN1 was not detected. Changes observed at the proteomic level were not due to altered mRNA levels, indicating that differences arose at the translational level. PROS1 and SLC3A2 protein levels were also increased in *2b5^{ho}* mouse brain lysates, supporting the findings in astrocyte cultures. Of these, *Pros1* was not increased at the mRNA level.

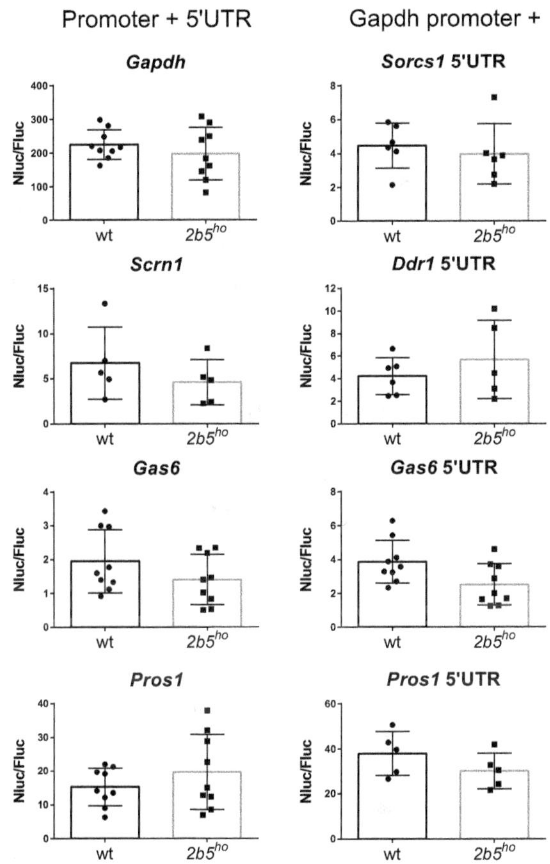

FIGURE 6 | *In vitro* analyses of candidate gene expression regulation do not reveal difference in promoter or 5′-UTR efficiency in wt and $2b5^{ho}$ astrocytes. pNL1.1- *Scrn1*, -*Gas6* and -*Pros1* constructs and pNL1.1-*Sorcs1*, -*Ddr1*, -*Gas6* and -*Pros1* 5′UTR constructs were transfected into wt and $2b5^{ho}$ astrocytes. Nanoluciferase activity was measured 3 days post transfection and corrected for firefly luciferase activity expressed from the pGL3-*Gapdh* internal standard. Every culture is shown as a single dot and bars represent the mean ± SD (n = 5–9). Replacement of the authentic *Pros1* or *Gas6* promoter by the *Gapdh* promoter increased the nanoluciferase expression by approximately twofold. pNL1.1-*Gapdh* was used as reference non-regulated promoter.

Our analyses indicate significant enrichment for proteins containing a signal peptide and motifs for *N*-glycosylation and disulfide bond formation in $2b5^{ho}$ astrocytes (**Figure 2B**). These motifs highlight involvement of the endoplasmic reticulum (ER), the first organelle of the secretory pathway (Farhan and Rabouille, 2011). Approximately 50% of the eIF2Bε^{Arg191His}-regulated proteins were predicted to use the secretory pathway, suggesting an increased flux in $2b5^{ho}$ astrocytes. The analysis of the secretome after 2 h AHA labeling as well as a standardized secretion assay did not indicate a difference in general secretion between wt and $2b5^{ho}$ astrocytes. The increased amount of proteins using the secretory pathway observed in the proteomic screen may still indicate a deregulation of ER function. The secreted luciferase reporter used in the standardized secretion assay was not modified in the ER. Protein modifications could be further investigated by expressing a reporter protein that is, e.g., *N*-glycosylated, actively folded by chaperones and/or subjected to extensive disulfide bond formation in ER and Golgi. However, transfections with such reporter constructs probably do not yield protein levels sufficient to quantify the small differences in these processes. Overexpression of such reporters will overburden the ER and induce an unfolded protein response, compromising ER function, preventing reliable conclusions.

Next we investigated if the mRNAs for eIF2Bε^{Arg191His}-regulated proteins contain uORFs as reduced eIF2B activity increases translation of mRNAs with one or more uORFs in the 5′-UTR (Meijer and Thomas, 2002). *In silico* analyses of the 5′-UTR sequences of the candidates did not reveal obvious overall differences in uORF numbers. Nonetheless, mRNAs with high numbers of uORFs (8 or more) were clearly overrepresented in the group of significantly altered proteins (**Figure 5**). Thus, when eIF2B activity is reduced, protein synthesis of those mRNAs may be enhanced due to inefficient initiation on inhibitory uORFs, stimulating initiation on the "normal" start codon of the mORF (Meijer and Thomas, 2002). Our analyses highlighted some overrepresentation for mRNAs with (especially) long 5′-UTRs and probably higher ΔG in $2b5^{ho}$ astrocytes (**Figure 5**). These length and ΔG observations suggest greater dependence on RNA helicase activity during translation initiation of eIF2Bε^{Arg191His}-regulated proteins (Sen et al., 2016). To further characterize the 5′-UTR efficiency in wt and $2b5^{ho}$ astrocytes, we performed transient transfection assays for several candidates, some of which were longer than 700 nucleotides (*Fgfr1* and *Nrxn1*). Unfortunately, transfection of the *Fgfr1* and *Nrxn1* constructs did not yield reliable expression, which precluded testing whether they are translated more efficiently in $2b5^{ho}$ than in wt astrocytes. The transfection assay is not robust enough to detect differences that were picked up with the pulsed AHA-SILAC proteomics and *in silico* 5′UTR analyses.

PROS1 was shown to be upregulated in both $2b5^{ho}$ astrocytes and brain without increased mRNA level. In brain, PROS1 is expressed by astrocytes and microglia (Zhang et al., 2014). Interestingly, PROS1 and GAS6 (which was also found to be upregulated) bind the same class of receptors (Hafizi and Dahlback, 2006). PROS1 and GAS6 enhance myelination and support oligodendrocyte survival in mice *in vitro* and *in vivo* (Tsiperson et al., 2010; Gruber et al., 2014; Goudarzi et al., 2016; Akkermann et al., 2017). Recently, PROS1 was described to function in neural stem cells (NSCs) as a regulator for NSC quiescence, proliferation and NSC development into neurons or astrocytes (Zelentsova et al., 2016).

Other proteins enriched in the eIF2Bε^{Arg191His}-regulated proteome are implicated in astrocyte development. For example NRXN1 (Zeng et al., 2013) and SORCS1, that transports NRXN1, can induce differentiation toward astrocytes (Savas et al., 2015). Moreover TGFβ2 (Park et al., 2012) and its binding partners, ERBIN and latent TGFβ-binding protein 3 (LTBP3) are increased during astrocyte differentiation (Robertson et al., 2015). All these results suggest that $2b5^{ho}$ astrocytes have a slightly altered differentiation state as shown before in $2b5^{ho}$ mice and VWM patients (Bugiani et al., 2011; Dooves et al., 2016).

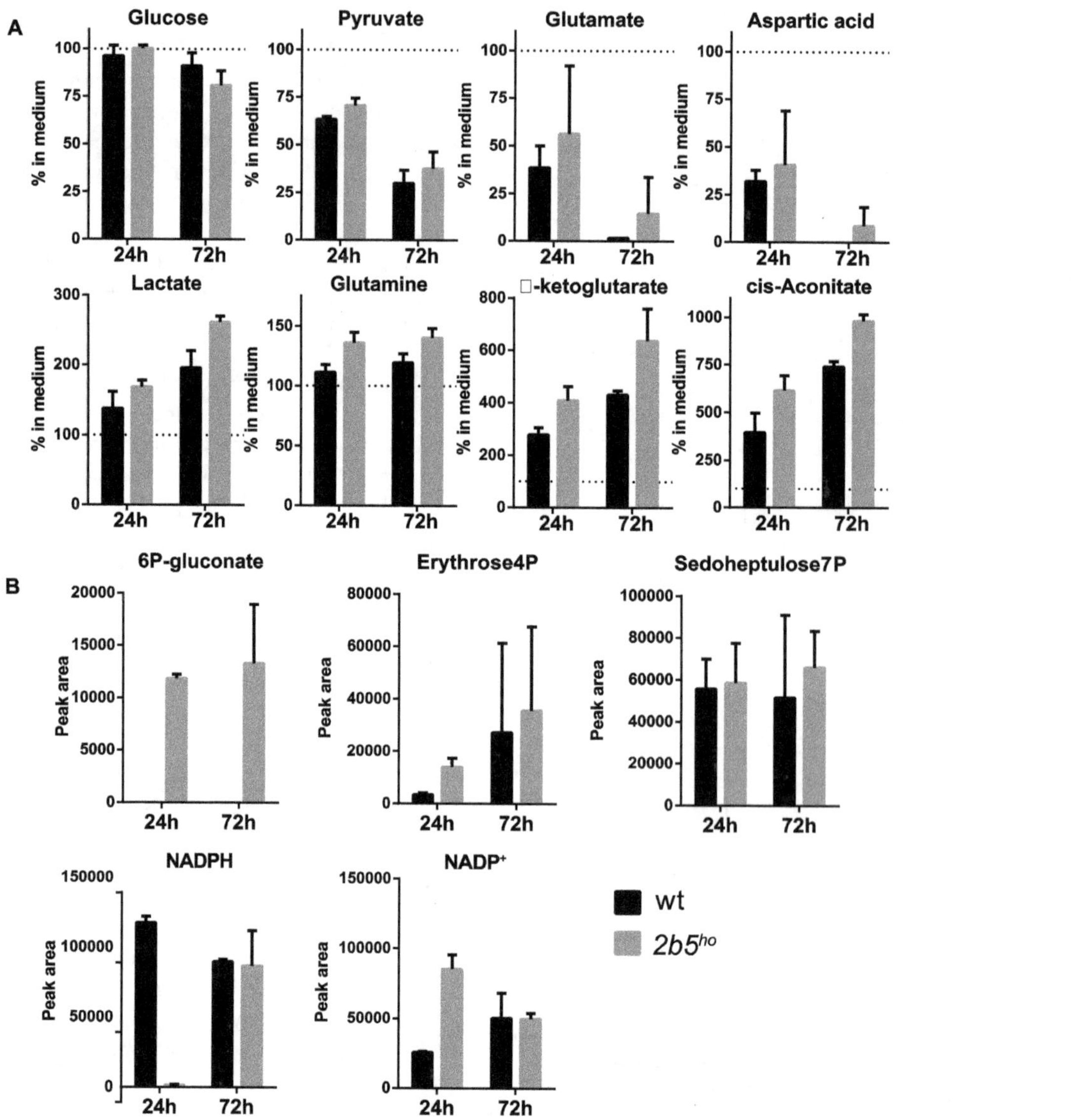

FIGURE 7 | Intracellular and extracellular metabolic analysis after 24 and 72 h of culture showed increased levels of intracellular levels of a pentose phosphate pathway intermediate (6P-gluconate). The astrocytes were cultured until 70% confluent. Media of the cultures were replaced with fresh medium. The cells as well as the media were collected after 24 and 72 h. The graphs show the mean ± SD. **(A)** The metabolite concentrations in the culture media of wt and *2b5*^{ho} astrocytes were compared to unconditioned media (n = 2). Glucose levels and pyruvate levels in conditioned medium of wt and *2b5*^{ho} astrocytes were decreased as well as glutamate and aspartate levels. Lactate, glutamine, *cis*-aconitate and α-ketoglutarate were increased in the conditioned medium of both cultures. **(B)** 6P-gluconate levels were consistently detected in *2b5*^{ho} cultures but not in wt cultures (24 and 72 h after replacing the culture medium). Four other metabolites of the pentose phosphate pathway were detected in the metabolic screen (erythrose-4P and sedoheptulose-7P, NADPH and NADP^{+}), but these were not consistently altered between wt and *2b5*^{ho} cultures.

To investigate other functional changes, we compared the metabolomes of wt and *2b5*^{ho} cells. We detected glutamate uptake in both cultures, which is a typical function of astrocytes, (Martin-Jimenez et al., 2017). This observation further confirms the astrocytic identity of the cultured cells. The most consistent difference observed was an increase in 6P-gluconate, an intermediate of the PPP. The PPP shunt is an alternative route for the metabolism of glucose which can supply ribose for nucleotide production in the non-oxidative part and NADPH in the oxidative part of the pathway. The reducing equivalents provided by NADPH are used both for biosynthesis and repair of oxidative damage (Stincone et al., 2015). The oxidative intermediate 6P-gluconate was increased in *2b5*^{ho} astrocytes while the non-oxidative intermediates sedoheptulose 7-phosphate and erythrose 4-phosphate were not consistent over time. These results suggest an increased activity of the PPP, especially at earlier time points. Interestingly, the observed PPP signature has been found in lung cancer cells in response to

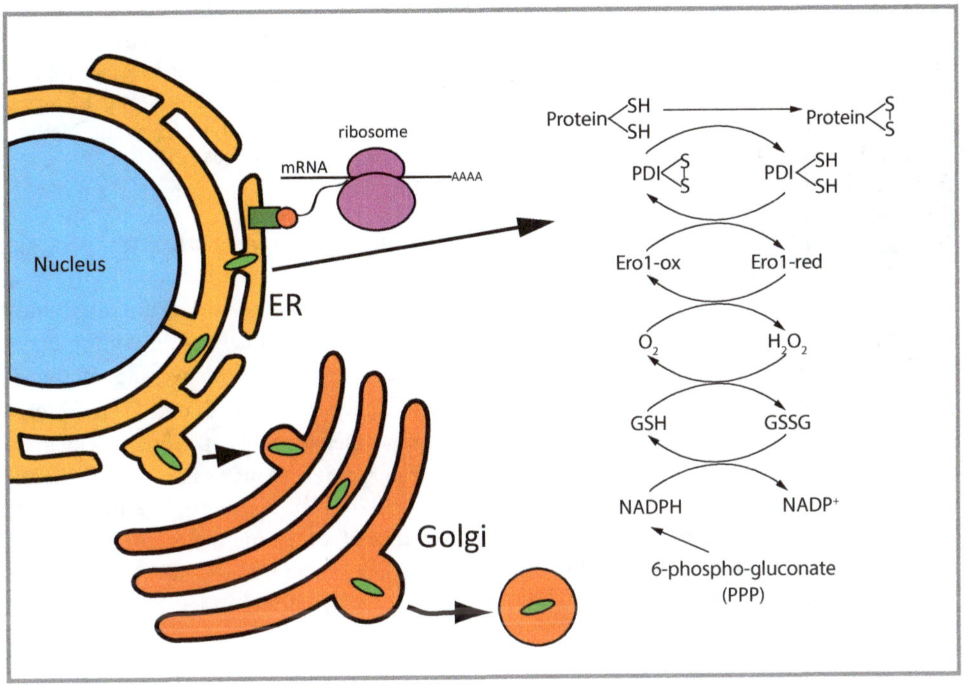

FIGURE 8 | Overview of differentially regulated functions in *2b5^ho^* astrocytes. Proteins with a signal peptide or transmembrane domain enter the ER to be folded. Proteins going through the secretory pathway that are subjected to *N*-glycosylation and disulfide bond formation are enriched in the set of eIF2Bε^{Arg191His}-regulated proteins. Disulfide bridge formation through oxidation of these proteins can induce H_2O_2. 6-Phospho-gluconate is involved in the production of NADPH which can be used to remove H_2O_2. PDI, protein disulfide isomerase; GSH, reduced glutathione; GSSG, oxidized glutathione; Ero1-ox, oxidized Ero1; Ero1-red, reduced Ero1; PPP, pentose phosphate pathway.

reduced 6P-gluconate dehydrogenase (6PGDH) activity leading to reactive oxygen species (ROS) production (Patra and Hay, 2014).

Could there be a link between the two main observations, i.e., the deregulated transport of some proteins in the ER and the increased 6P-gluconate concentrations? We think this is likely; although PPP is cytosolic, there are links to the ER. Part of the pathway appears to be associated with the ER (Stockwell et al., 2012) and NADPH produced by the PPP might be needed to combat ROS formation, which can be produced in the ER by processes related to oxidative protein folding. On the basis of the combined proteomic and metabolomic findings, we hypothesize that *2b5^{ho}* astrocytes may accumulate ROS as a result of an increased flux of proteins (e.g., requiring correct disulfide bond formation) through the secretory pathway (**Figures 2, 8**). Correct disulfide bond formation is dependent on protein disulfide isomerases (PDIs) and Ero1 in the ER (Bechtel and Weerapana, 2017). This process can result in H_2O_2 as by-product and increased PPP flux might compensate by producing NADPH (Espinosa-Diez et al., 2015). 6P-gluconate is the only metabolite in the metabolic screen that was changed in *2b5^{ho}* astrocytes and it is not completely clear whether this reflects increased production or decreased breakdown or a combination of both. There are indications that G6PDH is the only NADPH-producing enzyme activated upon oxidative stress and therefore it has been called "guardian of the cell redox potential" (Filosa et al., 2003). Lack of consistent changes over time in NADPH in *2b5^{ho}* astrocytes suggest that the balance between H_2O_2 and NADPH in steady state is "successfully" compensated (**Figure 8**). On the basis of our results we cannot discriminate between a compensated system or a deregulated system, the latter leading to pathology (Bechtel and Weerapana, 2017).

CONCLUSION

The 80 proteins differentially detected in the AHA-SILAC proteome are most likely translationally regulated as their mRNA levels are similar. Regulation thus must occur via increased synthesis or accumulation in *2b5^{ho}* astrocytes. We observed that the function of some of these differentially affected proteins is linked to astrocyte differentiation, which is known to be disturbed in VWM. As we would expect from deregulated protein synthesis in *2b5^{ho}* cells, the regulated proteins differ from their non-regulated counterparts in aspects of their 5′-UTRs. Altered synthesis may lead to an increase of proteins in the ER and secretory pathway without an overt effect on overall secretion. Possibly, increased demands of protein folding in the ER may result in generation of H_2O_2 or other ROS by-products (Espinosa-Diez et al., 2015). Thus, the increased synthesis of proteins that undergo disulfide formation and elevated 6P-gluconate might point to increased ROS in astrocytes, which may affect their function (**Figure 8**).

AUTHOR CONTRIBUTIONS

Substantial contributions to the conception or design of the work: LW, RP, CP, MvdK, and TA. The acquisition, analysis: LW, RP, EZ, CvB, TtB, and EP. Interpretation of data for the work: LW, RP, EZ, JK, CP, CB, MA, DS, and TA. Drafting the work or revising it critically for important intellectual content: LW, RP, EZ, CvB, TtB, EP, JK, CP, CB, MA, DS, MvdK, and TA. Final approval of the version to be published: LW, RP, EZ, CvB, TtB, EP, JK, CP, CB, MA, DS, MvdK, and TA. Agreement to be accountable for all aspects of the work in ensuring that questions related to the accuracy or integrity of any part of the work are appropriately investigated and resolved: LW, RP, EZ, CvB, TtB, EP, JK, CP, CB, MA, DS, MvdK, and TA.

FUNDING

LW, TA, TtB, EP, and MvdK are supported by ZonMw (TOP grant 91211005), Fonds NutsOHRA (1204-032), Hersenstichting Project grant (BGWS2014(1)-04), the Phelps Foundation (grant 2011.040). MA and RP are supported by Netherlands Organization for Scientific Research (NWO) as part of the National Roadmap Large-scale Research Facilities of the Netherlands, Proteins@Work (project number 184.032.201) and a VIDI grant (723.012.102) CB and EZ are supported by VENI grant (project 722.013.009) from the Netherlands Organization for Scientific Research (NWO).

ACKNOWLEDGMENTS

The authors thank Adri Thomas (University of Utrecht, Netherlands) for critical reading of the manuscript. They thank Anton den Hoed for the writing of the scripts used for the 5′UTR analysis. They thank I. G. Metgod and C. M. T. Beertsen (Department of Clinical Chemistry, VU University Medical Center, Amsterdam, Netherlands) for use of the Wallac 1420 Victor2 Microplate Reader. They thank R. Groen (Department of Hematology, VU University Medical Center, Amsterdam, Netherlands) for use of the GloMax® Discover System. They thank Sander Bobeldijk for facilitating the ΔG calculations.

REFERENCES

Akkermann, R., Aprico, A., Perera, A. A., Bujalka, H., Cole, A. E., Xiao, J., et al. (2017). The TAM receptor Tyro3 regulates myelination in the central nervous system. *Glia* 65, 581–591. doi: 10.1002/glia.23113

Babendure, J. R., Babendure, J. L., Ding, J. H., and Tsien, R. Y. (2006). Control of mammalian translation by mRNA structure near caps. *RNA* 12, 851–861. doi: 10.1261/rna.2309906

Bechtel, T. J., and Weerapana, E. (2017). From structure to redox: the diverse functional roles of disulfides and implications in disease. *Proteomics* 17:1600391. doi: 10.1002/pmic.201600391

Bender, A. S., Woodbury, D. M., and White, H. S. (1997). The rapid L- and D-aspartate uptake in cultured astrocytes. *Neurochem. Res.* 22, 721–726. doi: 10.1023/A:1027358211472

Bendtsen, J. D., Jensen, L. J., Blom, N., Von Heijne, G., and Brunak, S. (2004). Feature-based prediction of non-classical and leaderless protein secretion. *Protein Eng. Des. Sel.* 17, 349–356. doi: 10.1093/protein/gzh037

Borgese, N. (2016). Getting membrane proteins on and off the shuttle bus between the endoplasmic reticulum and the Golgi complex. *J. Cell Sci.* 129, 1537–1545. doi: 10.1242/jcs.183335

Bugiani, M., Boor, I., van Kollenburg, B., Postma, N., Polder, E., van Berkel, C., et al. (2011). Defective glial maturation in vanishing white matter disease. *J. Neuropathol. Exp. Neurol.* 70, 69–82. doi: 10.1097/NEN.0b013e318203ae74

Dieterich, D. C., Link, A. J., Graumann, J., Tirrell, D. A., and Schuman, E. M. (2006). Selective identification of newly synthesized proteins in mammalian cells using bioorthogonal noncanonical amino acid tagging (BONCAT). *Proc. Natl. Acad. Sci. U.S.A.* 103, 9482–9487. doi: 10.1073/pnas.0601637103

Dooves, S., Bugiani, M., Postma, N. L., Polder, E., Land, N., Horan, S. T., et al. (2016). Astrocytes are central in the pathomechanisms of vanishing white matter. *J. Clin. Invest.* 126, 1512–1524. doi: 10.1172/JCI83908

Eichelbaum, K., Winter, M., Berriel Diaz, M., Herzig, S., and Krijgsveld, J. (2012). Selective enrichment of newly synthesized proteins for quantitative secretome analysis. *Nat. Biotechnol.* 30, 984–990. doi: 10.1038/nbt.2356

Espinosa-Diez, C., Miguel, V., Mennerich, D., Kietzmann, T., Sanchez-Perez, P., Cadenas, S., et al. (2015). Antioxidant responses and cellular adjustments to oxidative stress. *Redox Biol.* 6, 183–197. doi: 10.1016/j.redox.2015.07.008

Farhan, H., and Rabouille, C. (2011). Signalling to and from the secretory pathway. *J. Cell Sci.* 124(Pt 2), 171–180. doi: 10.1242/jcs.076455

Ferreira, J. P., Overton, K. W., and Wang, C. L. (2013). Tuning gene expression with synthetic upstream open reading frames. *Proc. Natl. Acad. Sci. U.S.A.* 110, 11284–11289. doi: 10.1073/pnas.1305590110

Filosa, S., Fico, A., Paglialunga, F., Balestrieri, M., Crooke, A., Verde, P., et al. (2003). Failure to increase glucose consumption through the pentose-phosphate pathway results in the death of glucose-6-phosphate dehydrogenase gene-deleted mouse embryonic stem cells subjected to oxidative stress. *Biochem. J.* 370, 935–943. doi: 10.1042/bj20021614

Fogli, A., Schiffmann, R., Bertini, E., Ughetto, S., Combes, P., Eymard-Pierre, E., et al. (2004). The effect of genotype on the natural history of eIF2B-related leukodystrophies. *Neurology* 62, 1509–1517. doi: 10.1212/01.WNL.0000123259.67815.DB

Genheden, M., Kenney, J. W., Johnston, H. E., Manousopoulou, A., Garbis, S. D., and Proud, C. G. (2015). BDNF stimulation of protein synthesis in cortical neurons requires the MAP kinase-interacting kinase MNK1. *J. Neurosci.* 35, 972–984. doi: 10.1523/JNEUROSCI.2641-14.2015

Goudarzi, S., Rivera, A., Butt, A. M., and Hafizi, S. (2016). Gas6 promotes oligodendrogenesis and myelination in the adult central nervous system and after lysolecithin-induced demyelination. *ASN Neuro* 8:1759091416668430. doi: 10.1177/1759091416668430

Gruber, R. C., Ray, A. K., Johndrow, C. T., Guzik, H., Burek, D., de Frutos, P. G., et al. (2014). Targeted GAS6 delivery to the CNS protects axons from damage during experimental autoimmune encephalomyelitis. *J. Neurosci.* 34, 16320–16335. doi: 10.1523/JNEUROSCI.2449-14.2014

Hafizi, S., and Dahlback, B. (2006). Gas6 and protein S. Vitamin K-dependent ligands for the Axl receptor tyrosine kinase subfamily. *FEBS J.* 273, 5231–5244. doi: 10.1111/j.1742-4658.2006.05529.x

Hinnebusch, A. G. (2011). Molecular mechanism of scanning and start codon selection in eukaryotes. *Microbiol. Mol. Biol. Rev.* 75, 434–467. doi: 10.1128/MMBR.00008-11

Hinnebusch, A. G., Ivanov, I. P., and Sonenberg, N. (2016). Translational control by 5′-untranslated regions of eukaryotic mRNAs. *Science* 352, 1413–1416. doi: 10.1126/science.aad9868

Horzinski, L., Huyghe, A., Cardoso, M. C., Gonthier, C., Ouchchane, L., Schiffmann, R., et al. (2009). Eukaryotic initiation factor 2B (eIF2B) GEF activity as a diagnostic tool for EIF2B-related disorders. *PLOS ONE* 4:e8318. doi: 10.1371/journal.pone.0008318

Howden, A. J., Geoghegan, V., Katsch, K., Efstathiou, G., Bhushan, B., Boutureira, O., et al. (2013). QuaNCAT: quantitating proteome dynamics in primary cells. *Nat. Methods* 10, 343–346. doi: 10.1038/nmeth.2401

Huang, D. W., Sherman, B. T., and Lempicki, R. A. (2009a). Bioinformatics enrichment tools: paths toward the comprehensive functional analysis of large gene lists. *Nucleic Acids Res.* 37, 1–13. doi: 10.1093/nar/gkn923

Huang, D. W., Sherman, B. T., and Lempicki, R. A. (2009b). Systematic and integrative analysis of large gene lists using DAVID bioinformatics resources. *Nat. Protoc.* 4, 44–57. doi: 10.1038/nprot.2008.211

Ingolia, N. T., Lareau, L. F., and Weissman, J. S. (2011). Ribosome profiling of mouse embryonic stem cells reveals the complexity and dynamics of mammalian proteomes. *Cell* 147, 789–802. doi: 10.1016/j.cell.2011.10.002

Kenney, J. W., Genheden, M., Moon, K. M., Wang, X., Foster, L. J., and Proud, C. G. (2016). Eukaryotic elongation factor 2 kinase regulates the synthesis of microtubule-related proteins in neurons. *J. Neurochem.* 136, 276–284. doi: 10.1111/jnc.13407

Kleijn, M., Scheper, G. C., Voorma, H. O., and Thomas, A. A. (1998). Regulation of translation initiation factors by signal transduction. *Eur. J. Biochem.* 253, 531–544. doi: 10.1046/j.1432-1327.1998.2530531.x

Konieczny, A., and Safer, B. (1983). Purification of the eukaryotic initiation factor 2-eukaryotic initiation factor 2B complex and characterization of its guanine nucleotide exchange activity during protein synthesis initiation. *J. Biol. Chem.* 258, 3402–3408.

Kozak, M. (1987). At least six nucleotides preceding the AUG initiator codon enhance translation in mammalian cells. *J. Mol. Biol.* 196, 947–950. doi: 10.1016/0022-2836(87)90418-9

Ladner, C. L., Yang, J., Turner, R. J., and Edwards, R. A. (2004). Visible fluorescent detection of proteins in polyacrylamide gels without staining. *Anal. Biochem.* 326, 13–20. doi: 10.1016/j.ab.2003.10.047

Leegwater, P. A., Vermeulen, G., Konst, A. A., Naidu, S., Mulders, J., Visser, A., et al. (2001). Subunits of the translation initiation factor eIF2B are mutant in leukoencephalopathy with vanishing white matter. *Nat. Genet.* 29, 383–388. doi: 10.1038/ng764

Li, W., Wang, X., Van Der Knaap, M. S., and Proud, C. G. (2004). Mutations linked to leukoencephalopathy with vanishing white matter impair the function of the eukaryotic initiation factor 2B complex in diverse ways. *Mol. Cell. Biol.* 24, 3295–3306. doi: 10.1128/MCB.24.8.3295-3306.2004

Liu, R., van der Lei, H. D., Wang, X., Wortham, N. C., Tang, H., van Berkel, C. G., et al. (2011). Severity of vanishing white matter disease does not correlate with deficits in eIF2B activity or the integrity of eIF2B complexes. *Hum. Mutat.* 32, 1036–1045. doi: 10.1002/humu.21535

Martin-Jimenez, C. A., Salazar-Barreto, D., Barreto, G. E., and Gonzalez, J. (2017). Genome-scale reconstruction of the human astrocyte metabolic network. *Front. Aging Neurosci.* 9:23. doi: 10.3389/fnagi.2017.00023

Meijer, H. A., and Thomas, A. A. (2002). Control of eukaryotic protein synthesis by upstream open reading frames in the 5'-untranslated region of an mRNA. *Biochem. J.* 367(Pt 1), 1–11. doi: 10.1042/bj20011706

Noderer, W. L., Flockhart, R. J., Bhaduri, A., Diaz de Arce, A. J., Zhang, J., Khavari, P. A., et al. (2014). Quantitative analysis of mammalian translation initiation sites by FACS-seq. *Mol. Syst. Biol.* 10:748. doi: 10.15252/msb.20145136

Park, J. H., Choi, M. R., Park, K. S., Kim, S. H., Jung, K. H., and Chai, Y. G. (2012). The characterization of gene expression during mouse neural stem cell differentiation in vitro. *Neurosci. Lett.* 506, 50–54. doi: 10.1016/j.neulet.2011.10.046

Patra, K. C., and Hay, N. (2014). The pentose phosphate pathway and cancer. *Trends Biochem. Sci.* 39, 347–354. doi: 10.1016/j.tibs.2014.06.005

Petersen, T. N., Brunak, S., von Heijne, G., and Nielsen, H. (2011). SignalP 4.0: discriminating signal peptides from transmembrane regions. *Nat. Methods* 8, 785–786. doi: 10.1038/nmeth.1701

Proud, C. G. (2001). Regulation of eukaryotic initiation factor eIF2B. *Prog. Mol. Subcell. Biol.* 26, 95–114. doi: 10.1007/978-3-642-56688-2_4

Robertson, I. B., Horiguchi, M., Zilberberg, L., Dabovic, B., Hadjiolova, K., and Rifkin, D. B. (2015). Latent TGF-beta-binding proteins. *Matrix Biol.* 47, 44–53. doi: 10.1016/j.matbio.2015.05.005

Ruijter, J. M., Thygesen, H. H., Schoneveld, O. J., Das, A. T., Berkhout, B., and Lamers, W. H. (2006). Factor correction as a tool to eliminate between-session variation in replicate experiments: application to molecular biology and retrovirology. *Retrovirology* 3:2. doi: 10.1186/1742-4690-3-2

Savas, J. N., Ribeiro, L. F., Wierda, K. D., Wright, R., DeNardo-Wilke, L. A., Rice, H. C., et al. (2015). The sorting receptor SorCS1 regulates trafficking of neurexin and AMPA receptors. *Neuron* 87, 764–780. doi: 10.1016/j.neuron.2015.08.007

Sen, N. D., Zhou, F., Harris, M. S., Ingolia, N. T., and Hinnebusch, A. G. (2016). eIF4B stimulates translation of long mRNAs with structured 5' UTRs and low closed-loop potential but weak dependence on eIF4G. *Proc. Natl. Acad. Sci. U.S.A.* 113, 10464–10472. doi: 10.1073/pnas.1612398113

Sonenberg, N., and Hinnebusch, A. G. (2009). Regulation of translation initiation in eukaryotes: mechanisms and biological targets. *Cell* 136, 731–745. doi: 10.1016/j.cell.2009.01.042

Stincone, A., Prigione, A., Cramer, T., Wamelink, M. M., Campbell, K., Cheung, E., et al. (2015). The return of metabolism: biochemistry and physiology of the pentose phosphate pathway. *Biol. Rev. Camb. Philos. Soc.* 90, 927–963. doi: 10.1111/brv.12140

Stockwell, S. R., Platt, G., Barrie, S. E., Zoumpoulidou, G., Te Poele, R. H., Aherne, G. W., et al. (2012). Mechanism-based screen for G1/S checkpoint activators identifies a selective activator of EIF2AK3/PERK signalling. *PLOS ONE* 7:e28568. doi: 10.1371/journal.pone.0028568

Tsiperson, V., Li, X., Schwartz, G. J., Raine, C. S., and Shafit-Zagardo, B. (2010). GAS6 enhances repair following cuprizone-induced demyelination. *PLOS ONE* 5:e15748. doi: 10.1371/journal.pone.0015748

van der Knaap, M. S., Breiter, S. N., Naidu, S., Hart, A. A., and Valk, J. (1999). Defining and categorizing leukoencephalopathies of unknown origin: MR imaging approach. *Radiology* 213, 121–133. doi: 10.1148/radiology.213.1.r99se01121

van der Knaap, M. S., Leegwater, P. A., Konst, A. A., Visser, A., Naidu, S., Oudejans, C. B., et al. (2002). Mutations in each of the five subunits of translation initiation factor eIF2B can cause leukoencephalopathy with vanishing white matter. *Ann. Neurol.* 51, 264–270. doi: 10.1002/ana.10112

van der Knaap, M. S., Pronk, J. C., and Scheper, G. C. (2006). Vanishing white matter disease. *Lancet Neurol.* 5, 413–423. doi: 10.1016/S1474-4422(06)70440-9

van Kollenburg, B., Thomas, A. A., Vermeulen, G., Bertrand, G. A., van Berkel, C. G., Pronk, J. C., et al. (2006). Regulation of protein synthesis in lymphoblasts from vanishing white matter patients. *Neurobiol. Dis.* 21, 496–504. doi: 10.1016/j.nbd.2005.08.009

von Heijne, G. (1990). The signal peptide. *J. Membr. Biol.* 115, 195–201.

Voorma, H. O., Thomas, A. A., and Van Heugten, H. A. (1994). Initiation of protein synthesis in eukaryotes. *Mol. Biol. Rep.* 19, 139–145. doi: 10.1007/BF00986956

Wong, K., Armstrong, R. C., Gyure, K. A., Morrison, A. L., Rodriguez, D., Matalon, R., et al. (2000). Foamy cells with oligodendroglial phenotype in childhood ataxia with diffuse central nervous system hypomyelination syndrome. *Acta Neuropathol.* 100, 635–646. doi: 10.1007/s004010000234

Zelentsova, K., Talmi, Z., Abboud-Jarrous, G., Sapir, T., Capucha, T., Nassar, M., et al. (2016). Protein S regulates neural stem cell quiescence and neurogenesis. *Stem Cells* 35, 679–693. doi: 10.1002/stem.2522

Zeng, L., Zhang, P., Shi, L., Yamamoto, V., Lu, W., and Wang, K. (2013). Functional impacts of NRXN1 knockdown on neurodevelopment in stem cell models. *PLOS ONE* 8:e59685. doi: 10.1371/journal.pone.0059685

Zhang, Y., Chen, K., Sloan, S. A., Bennett, M. L., Scholze, A. R., O'Keeffe, S., et al. (2014). An RNA-sequencing transcriptome and splicing database of glia, neurons, and vascular cells of the cerebral cortex. *J. Neurosci.* 34, 11929–11947. doi: 10.1523/JNEUROSCI.1860-14.2014

3

P2X7 Receptor Activation Modulates Autophagy in SOD1-G93A Mouse Microglia

Paola Fabbrizio [1,2], Susanna Amadio [1], Savina Apolloni [1†] and Cinzia Volonté [1,3*†]*

[1]IRCCS Santa Lucia Foundation, Experimental Neuroscience, Rome, Italy, [2]Department of Systems Medicine, Tor Vergata University, Rome, Italy, [3]CNR, Institute of Cell Biology and Neurobiology, Rome, Italy

***Correspondence:**
Savina Apolloni
s.apolloni@hsantalucia.it
Cinzia Volonté
cinzia.volonte@cnr.it

†These authors have contributed equally to this work.

Autophagy and inflammation play determinant roles in the pathogenesis of Amyotrophic Lateral Sclerosis (ALS), an adult-onset neurodegenerative disease characterized by deterioration and final loss of upper and lower motor neurons (MN) priming microglia to sustain neuroinflammation and a vicious cycle of neurodegeneration. Given that extracellular ATP through P2X7 receptor constitutes a neuron-to-microglia alarm signal implicated in ALS, and that P2X7 affects autophagy in immune cells, we have investigated if autophagy can be directly triggered by P2X7 activation in primary microglia from superoxide dismutase 1 (SOD1)-G93A mice. We report that P2X7 enhances the expression of the autophagic marker microtubule-associated protein 1 light chain 3 (LC3)-II, via mTOR pathway and concomitantly with modulation of anti-inflammatory M2 microglia markers. We also demonstrate that the autophagic target SQSTM1/p62 is decreased in SOD1-G93A microglia after a short stimulation of P2X7, but increased after a sustained challenge. These effects are prevented by the P2X7 antagonist A-804598, and the autophagy/phosphoinositide-3-kinase inhibitor wortmannin (WM). Finally, a chronic *in vivo* treatment with A-804598 in SOD1-G93A mice decreases the expression of SQSTM1/p62 in lumbar spinal cord at end stage of disease. These data identify the modulation of the autophagic flux as a novel mechanism by which P2X7 activates ALS-microglia, to be considered for further investigations in ALS.

Keywords: ALS, ATP, BzATP, LC3B-II, microglia, P2X7

INTRODUCTION

Neuroinflammation plays a determinant role in the pathogenesis of Amyotrophic Lateral Sclerosis (ALS; Cozzolino et al., 2012), where degenerating motor neurons (MN) contribute to produce danger signals that activate microglia to initiate a self-propagating cytotoxic cascade resulting in muscle weakness, atrophy, spasticity, compromised breathing and finally death in patients (Henkel et al., 2009; Ilieva et al., 2009). Extracellular ATP functions as one of such endogenous alarm signal, by operating through purinergic P2 receptors, particularly the P2X7 ionotropic subtype (Beamer et al., 2016; Illes and Verkhratsky, 2016). Activation of P2X7 in microglia purified

Abbreviations: ALS, Amyotrophic Lateral Sclerosis; ARG-1, arginase-1; BzATP, 2′-3′-O-(benzoyl-benzoyl) ATP; COX-2, cyclooxygenase-2; LC3-II, microtubule-associated protein 1 light chain 3-II; MAPK, mitogen-activated protein kinase; NF-κB, nuclear factor kappa-B; NOX2, NADPH oxidase 2; SOD1, superoxide dismutase 1; TNF-α, tumor necrosis factor alpha; WM, wortmannin; WT, wild-type.

from superoxide dismutase 1 (SOD1)-G93A mouse model of ALS exacerbates pro-inflammatory responses exemplified by the levels of cyclooxygenase-2 (COX-2), mitogen-activated protein kinase (MAPK), NADPH oxidase 2 (NOX2), nuclear factor kappa-B (NF-κB), tumor necrosis factor alpha (TNF-α), with a consequent cytotoxicity towards neurons (D'Ambrosi et al., 2009; Apolloni et al., 2013b; Parisi et al., 2016). In the SOD1-G93A mouse model of ALS, we have previously reported that disease onset is significantly anticipated and disease progression is worsened in mice genetically lacking P2X7, thus suggesting that the receptor has beneficial effects at least during a precise stage of the disease that is however still undefined (Apolloni et al., 2013a). Only when administered to SOD1-G93A mice from late pre-onset, the P2X7 antagonist Brilliant Blue G enhances MN survival and reduces microgliosis in lumbar spinal cord (Apolloni et al., 2014), with a slight beneficial effect on disease progression and mice survival (Cervetto et al., 2013; Apolloni et al., 2014; Bartlett et al., 2017). By confirming the dual role previously recognized for P2X7 (Monif et al., 2010, 2016), these findings have highlighted how crucial is the lag of P2X7 modulation for eliciting beneficial effects in ALS.

Autophagy is an ubiquitous and highly conserved homeostatic mechanism by which eukaryotic cells degrade damaged organelles, intracellular components and protein aggregates. It consists in the sequestration of target cellular materials (cargo) into autophagosomes and their subsequent transfer to and catabolism by the autophagosome-lysosome (autolysosome vacuole) system (Klionsky and Emr, 2000; Glick et al., 2010). Abnormalities in autophagy have been implicated in chronic neurodegenerative conditions comprising ALS, where both patients and animal models are characterized by overexpression of the autophagy marker microtubule-associated protein 1 light chain 3 (LC3)-II in spinal cord (Song et al., 2012). Moreover, sequestosome-1, a protein also known as ubiquitin-binding protein SQSTM1/p62 encoded by the SQSTM1 gene, which is an autophagosome receptor recruiting proteins for selective autophagy, is found accumulated in spinal cord of SOD1-G93A mice, concomitantly with increased extent of autophagic vacuoles (Zhang et al., 2011).

In mouse microglia and in human epithelial cells, P2X7 has been reported to negatively regulate autophagy by impairing lysosomal function (Takenouchi et al., 2009a; Haanes et al., 2012). Conversely, P2X7 apparently increases autophagy in monocytes and macrophages during mycobacterial infections (Biswas et al., 2008), and acts as a positive regulator of autophagy in dystrophic muscle cells (Young et al., 2015). This not only confirms the twofold role often played by P2X7, but also highlights a context-specific effect of P2X7 on autophagy pathways.

The purpose of this work is thus to learn if P2X7 might participate to ALS pathogenesis by directly modulating autophagy, and particularly the expression of the autophagosome component LC3-II and of the autophagy receptor p62 in SOD1-G93A microglia. We demonstrate that activation of P2X7 deregulates LC3-II and p62 protein content. Moreover, *in vivo* treatment with the specific P2X7 antagonist A-804598 reduces p62 levels in lumbar spinal cord of SOD1-G93A mice at end stage of disease. These data identify the modulation of autophagy as a novel mechanism by which P2X7 activates ALS-microglia.

MATERIALS AND METHODS

Reagents

ATP, 2′-3′-O-(benzoyl-benzoyl) ATP (BzATP) and all other reagents, unless otherwise stated, were obtained from Sigma Aldrich (Milan, Italy). PD98059 was purchased from Calbiochem (San Diego, CA, USA). A-839977 and A-804598 were from Tocris Bioscience (Bristol, UK).

Antibodies

Arginase-1 (ARG-1) rabbit antibody (1:700) was from Abcam (Cambridge, UK); CD163 rabbit antibody (1:100) was from Santa Cruz Biotechnology, Inc. (Dallas, TX, USA); LC3B antibody (1:500); mTOR (1:500) and p-mTOR (1:500) rabbit antibodies were from Cell Signaling Technology Inc. (Beverly, MA, USA). SQSTM1/p62 mouse monoclonal antibody (1:500) was obtained from Abcam (Cambridge, UK). GAPDH mouse antibody (1:2500) from Calbiochem (San Diego, CA, USA) was used for protein normalization. HRP-linked anti-rabbit and anti-mouse antibodies were from Jackson Immunoresearch (West Grove, PA, USA).

Mice

All animal procedures have been performed according to the European Guidelines for the use of animals in research (86/609/CEE) and the requirements of Italian laws (D.L. 26/2014). The ethical procedure has been approved by the Animal Welfare Office, Department of Public Health and Veterinary, Nutrition and Food Safety, General Management of Animal Care and Veterinary Drugs of the Italian Ministry of Health (protocol number 319/2015PR). All efforts were made to minimize animal suffering and use the minimum number of animals necessary to obtain reliable results. Adult B6.Cg-Tg(SOD1-G93A)1Gur/J mice were obtained from Jackson Laboratories (Bar Harbor, ME, USA), bred in our indoor animal facility and genotyped as described (Apolloni et al., 2013a). To determine behavioral scores we employed a scale from 5 to 1, where 5 defines healthy mice without symptoms of paralysis, 4 indicates mice with slight signs of destabilized gait and paralysis of the hind limbs, 3 describes mice with clear paralysis and destabilized gait, 2 depicts mice with fully developed paralysis of the hind limbs that only crawl on the forelimbs, finally 1 defines mice with fully developed paralysis of the hind limbs that predominantly lie on the side and/or are not able to straighten up within 30 s when they are turned on their back. After reaching a score of 1 (end stage, ~23 weeks of age), the animals were euthanized, according to the guidelines for preclinical testing and colony management (Ludolph et al., 2010). SOD1-G93A mice were considered at onset when exhibited a statistically significant 10% decline of rotarod performance, with respect to wild-type (WT) mice. Rotarod performance was evaluated by a rotarod apparatus (Ugo Basile 7650 model) at a constant speed of

15 r.p.m. over a maximum period of 180 s (Apolloni et al., 2013a).

SOD1-G93A mice at 100 days of age (pre-onset) were randomly grouped into vehicle-treated or CNS penetrant P2X7 specific antagonist A-804598-treated mice (Donnelly-Roberts et al., 2009; Catanzaro et al., 2014; Iwata et al., 2016) given by intraperitoneal injection at 30 mg/kg five times a week until end stage of disease. Because there is sex diversity in response to pharmacological treatments (Pizzasegola et al., 2009) and the P2X7 antagonist Brilliant Blue G has prolonged survival only in female SOD1-G93A mice (Bartlett et al., 2017; Sluyter et al., 2017), we have chosen to study female mice.

Primary Microglia Cultures

Primary microglia cultures from brain cortex were prepared as previously described (Apolloni et al., 2013b). Briefly, neonatal SOD1-G93A mice and their non-transgenic littermates mice were sacrificed and, after removing the meninges, cortices were minced and digested with 0.25% trypsin and 0.01% DNaseI. After dissociation and passage through 70 μm filters, cells were resuspended in DMEM/F-12 media with GlutaMAX^{TM} (Gibco, Invitrogen, UK), plus 10% fetal bovine serum (FBS), 100 Units/ml gentamicine and 100 μg/ml streptomycin/penicillin. After approximately 15 days, a mild trypsinization was done for 30 min at 37°C. The resultant adherent microglial cells (pure >99%) were washed twice with DMEM/F-12 and kept at 37°C in a 5% CO_2 and 95% air atmosphere and used for experiments at least 48 h after.

Western Blotting

Microglia in serum-free medium were harvested in SDS Laemmli sample buffer. Protein lysates were obtained by homogenization of mice lumbar spinal cords segments in homogenization buffer as described (Apolloni et al., 2014). Analysis of protein components (10 μg for tissue extracts) was performed by Mini-PROTEAN® TGX^{TM} Gels (BioRad, USA) and transfer onto nitrocellulose membranes (Amersham Biosciences, USA). After saturation with blocking agent, blots were incubated overnight at 4°C with the specified antibody, then for 1 h with HRP-conjugated secondary antibodies and visualized using ECL Advance Western blot detection kit (Amersham Biosciences, USA). Signal intensity quantification was performed by Kodak Image Station analysis software.

Immunofluorescence and Confocal Microscopy

Primary microglia were fixed for 20 min in 4% paraformaldehyde, permeabilized in PBS containing 0.1% Triton X-100 and then incubated for 1.5 h at 37°C with anti-LC3B (1:200), anti-SQSTM1/p62 (1:100) or anti-CD11b (1:200, AbD Serotec, USA) in 1% BSA in PBS. Cells were stained for 1 h with Cy2-conjugated donkey anti-rabbit IgG or Cy3-conjugated donkey anti-mouse IgG or Cy5-conjugated donkey anti-rat IgG (1:200, Jackson Immunoresearch). Höechst 33342 was used for nuclei staining (1:1000) in PBS. Immunofluorescence was analyzed by means of a confocal laser scanning microscope (Zeiss, LSM700, Germany) equipped with four laser lines: 405 nm, 488 nm, 561 nm and 639 nm. The brightness and contrast of the digital images were adjusted using the Zen software.

Data Analysis

Data are presented as mean ± standard error of the mean (SEM). Statistical differences were verified by student's t-test if the normality test was passed, or by the Mann–Whitney rank sum test, if the normality test failed, considering acceptable a value for asymmetry between −2 and +2 in order to prove normal univariate distribution. One-way analysis of variance (ANOVA) followed by *Post Hoc* Tukey's test was used for multiple comparisons. For onset and survival data, differences in the values of A804598- and vehicle-treated mice were assessed using Kaplan-Meier analysis paired with log-rank tests. The software package GraphPad Prism 7.03 (GraphPad Software, San Diego, CA, USA) was used for all statistical analysis with differences considered significant for $p < 0.05$.

RESULTS

P2X7 Activation Induces LC3B-II Formation in SOD1-G93A Primary Microglia

LC3-II localizing to the autophagosome membrane is widely used to monitor autophagy. In particular, the generation of LC3-II from LC3-I is generally identified by immunoblot analysis, because the amount of LC3-II is strictly correlated with the number of autophagosomes and is considered a more reliable marker than LC3-I to be detected (Mizushima and Yoshimori, 2007; Barth et al., 2010; Klionsky et al., 2016). To directly investigate the modulation of the autophagic pathway by P2X7 activation in SOD1-G93A microglia, we thus compared the expression of LC3B-II, with respect to GAPDH.

As shown by immunoblotting in **Figure 1A**, SOD1-G93A microglia constitutively express augmented levels of LC3B-II with respect to WT microglia. Activation of P2X7 by the specific agonist BzATP (10–100–300 μM) increases LC3B-II protein content in a dose-dependent manner in SOD1-G93A microglia with maximal effect obtained at 300 μM (**Figure 1B**). The time-course experiment shows that the levels of LC3B-II are maximally induced after 15 min, with a consequent decline observed at 1–6 h (**Figure 1B**). Notably, the physiologic purinergic agonist ATP used at the mM concentrations required to activate P2X7 exerts similar effects (**Figure 1C**). LC3B-II induction by BzATP and ATP is decreased by the specific P2X7 antagonist A-804598 (10 μM; **Figure 1D**), thus indicating P2X7 involvement. Autophagy in SOD1-G93A microglia is moreover qualitatively assessed by immunofluorescence confocal analysis, by observing cells with accumulation of LC3 dots, because cells presenting a diffuse distribution of LC3 are generally described as

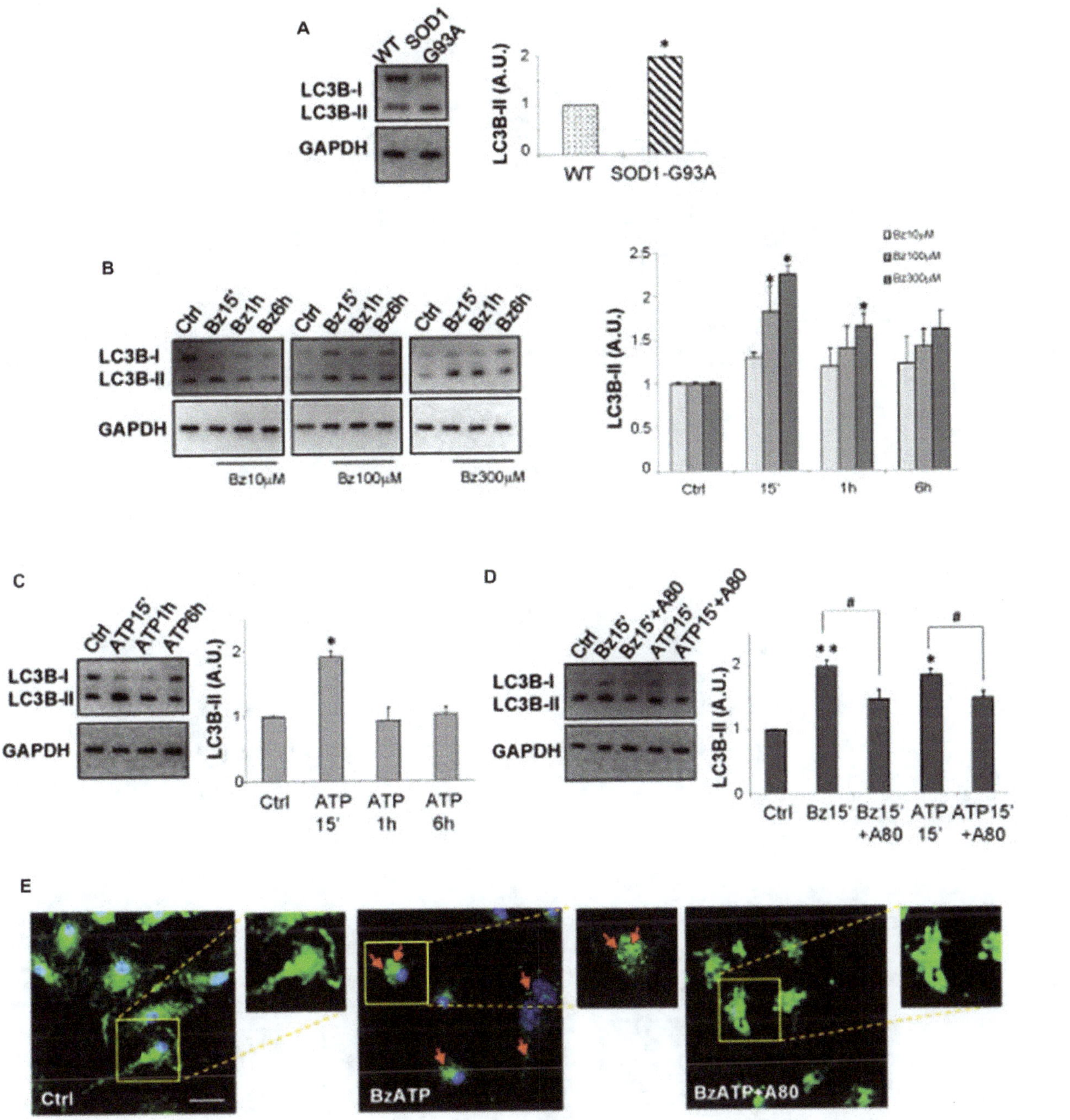

FIGURE 1 | P2X7 activation modulates microtubule-associated protein 1 light chain 3 (LC3)-II in superoxide dismutase 1 (SOD1)-G93A microglia. **(A)** Equal amounts of total protein from wild-type (WT) and SOD1-G93A primary microglia were subjected to western blotting and immunoreactions with anti-LC3B-II and anti-GAPDH to normalize proteins. **(B–D)** Equal amounts of total protein from SOD1-G93A primary microglia were subjected to western blotting and immunoreactions with anti-LC3B-II and anti-GAPDH. In **(B)**, cells were exposed to 10–100–300 μM 2′-3′-O-(benzoyl-benzoyl) ATP (BzATP) for 15 min, 1 h or 6 h. In **(C)** cells were exposed to 3 mM ATP for 15 min, 1 h or 6 h. Cells were treated for 15 min with 300 μM BzATP or 3 mM ATP, in the absence or presence of 10 μM A-804598 **(D)**. **(E)** SOD1-G93A microglia treated with 300 μM BzATP in the absence or presence of 10 μM A-804598 were analyzed by means of fluorescence microscopy after staining with anti-LC3B (arrows indicate LC3 vacuoles; scale bar 50 μm; nuclei in blue). Data represent mean ± SEM of $n = 3$ independent experiments. Statistical significance was calculated by student's *t*-test or Mann-Whitney rank sum test or analysis of variance (ANOVA) followed by *Post Hoc* Tukey's test, and referred to WT **(A)** or Ctrl **(B–D)** ($^*p < 0.05$, $^{**}p < 0.01$); to BzATP15′-treated cells or ATP15′-treated cells ($^{\#}p < 0.05$).

non-autophagic, whereas cells with increased LC3 vacuoles are classified as autophagic (Tasdemir et al., 2008; Klionsky et al., 2016). Confirming the western blot data, we observed that 300 μM BzATP increases cytoplasmic LC3-positive dots in microglia with respect to control, and the effect is reduced by the P2X7 antagonist A-804598 (**Figure 1E**).

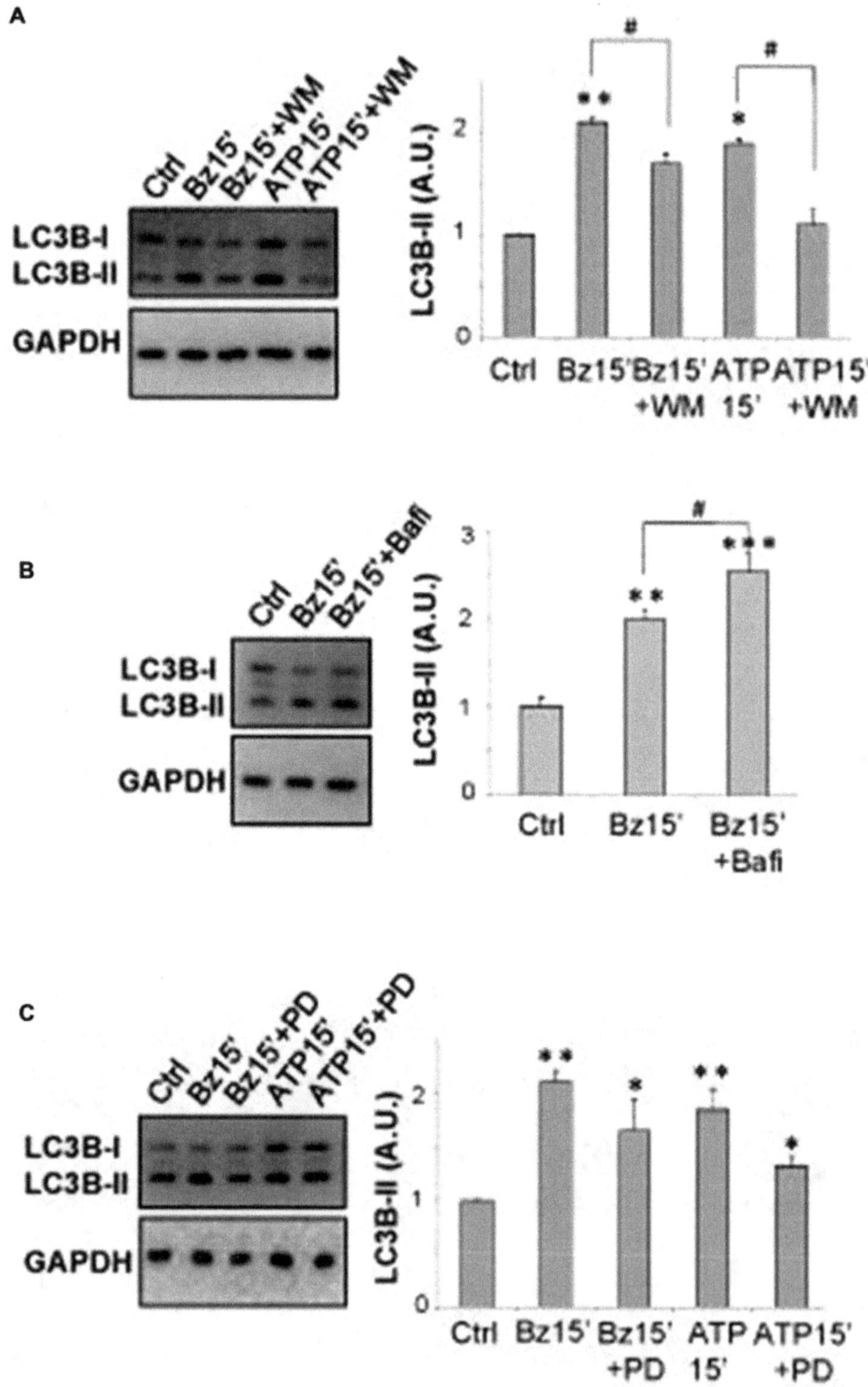

FIGURE 2 | P2X7 induces autophagic flux in SOD1-G93A microglia. Equal amounts of protein from SOD1-G93A primary microglia were subjected to western blotting and immunoreactions with anti-LC3B-II and anti-GAPDH. In **(A)** SOD1-G93A microglia were treated for 15 min with 300 μM BzATP or 3 mM ATP in the absence or presence of 100 nM wortmannin (WM) **(A)**. In **(B)** SOD1-G93A microglia were treated for 15 min with 300 μM BzATP in the absence or presence of 25 nM Bafilomycin-A1. **(C)** SOD1-G93A microglia were treated for 15 min with 300 μM BzATP or 3 mM ATP, in the absence or presence of 100 μM PD98059. Data represent mean ± SEM of $n = 3$ independent experiments. Statistical significance was calculated by ANOVA followed by *Post Hoc* Tukey's test and referred to Ctrl (*$p < 0.05$, **$p < 0.01$, ***$p < 0.001$); to BzATP15'-treated cells or to ATP15'-treated cells ($^{\#}p < 0.05$).

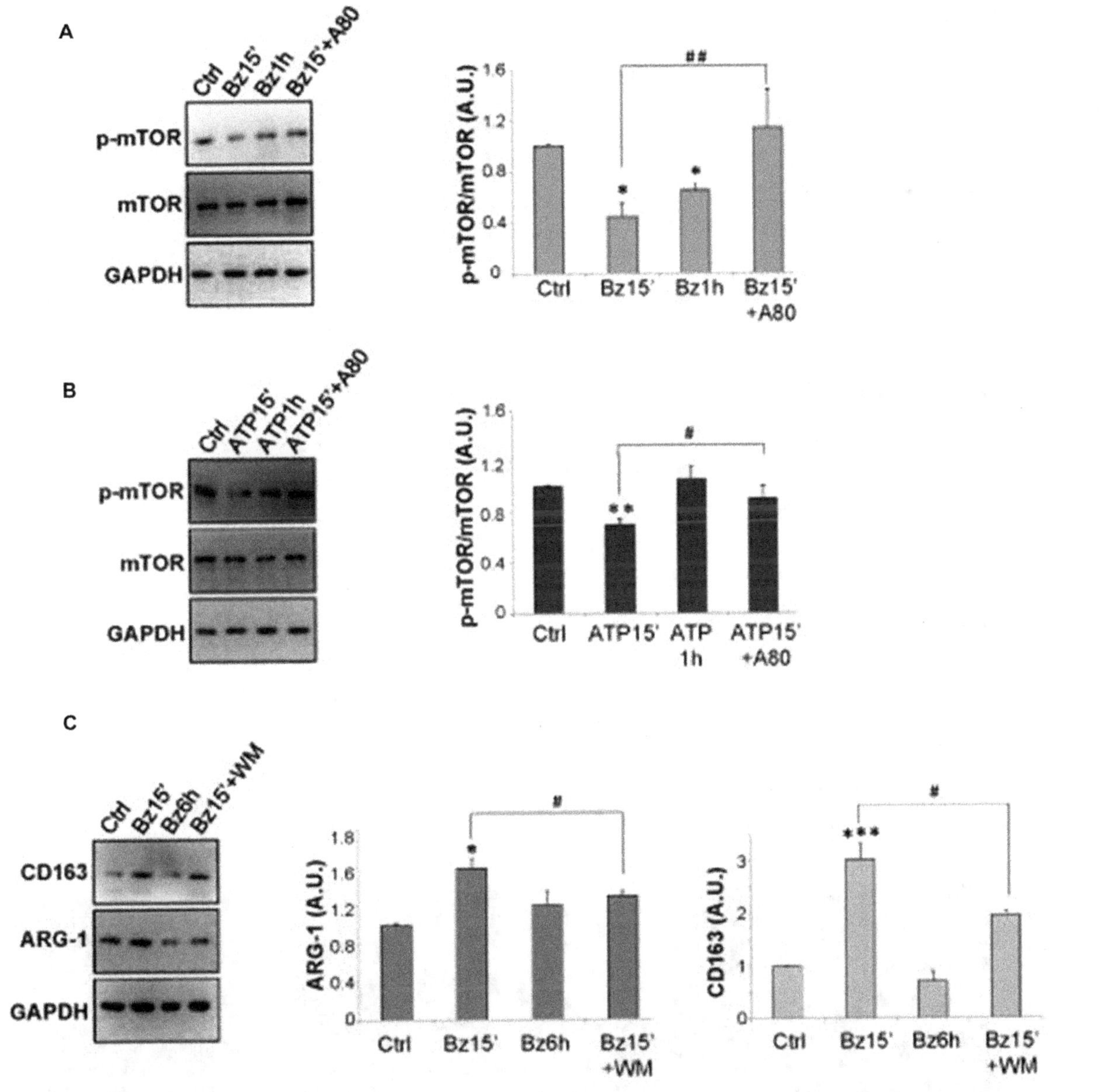

FIGURE 3 | mTOR phosphorylation is regulated by P2X7 activation in SOD1-G93A microglia. SOD1-G93A microglia were treated with 300 μM BzATP **(A)** or 3 mM ATP **(B)** for 15 min or 1 h with or without 10 μM A-804598 and subjected to western blotting and immunoreactions with anti-p-mTOR and anti-mTOR or anti-GAPDH. SOD1-G93A cells exposed to 300 μM BzATP for 15 min with or without WM or 6 h were subjected to western blotting and immunoreactions with anti-Arginase-1 (ARG-1) **(C)** or anti-CD163 and anti-GAPDH. Data represent mean ± SEM of $n = 3$ independent experiments. Statistical significance was calculated by student's *t*-test or Mann-Whitney rank sum test or ANOVA followed by *Post Hoc* Tukey's test, and referred to Ctrl ($^{*}p < 0.05$, $^{**}p < 0.01$, $^{***}p < 0.001$); to BzATP15'- or ATP15'-treated cells ($^{\#}p < 0.05$, $^{\#\#}p < 0.01$).

To demonstrate the autophagic mechanism involved in LC3-II induction, next we show that LC3B-II up-regulation by BzATP and ATP is attenuated by wortmannin (WM; 100 nM), an inhibitor of the initial stage of conventional autophagy/phosphoinositide-3-kinase (Klionsky et al., 2016; **Figure 2A**). A correct way to define if a treatment augments a true autophagic flux is to determine the levels of LC3B-II in the presence of the lysosomal proteases inhibitor bafilomycin-A1 (Klionsky et al., 2016). While the increase in LC3-II by a treatment simply indicates the accumulation of autophagosomes without proving autophagic degradation, an enhancement of autophagy indeed occurs if LC3-II further accumulates in the presence of bafilomycinA1 (Mizushima and Yoshimori, 2007). Here we show that the addition of bafilomycin-A1 (25 nM) indeed further increases the up-regulation of LC3B-II induced by 300 μM BzATP with respect to control cells, consistently with an accelerated true autophagic flux (**Figure 2B**).

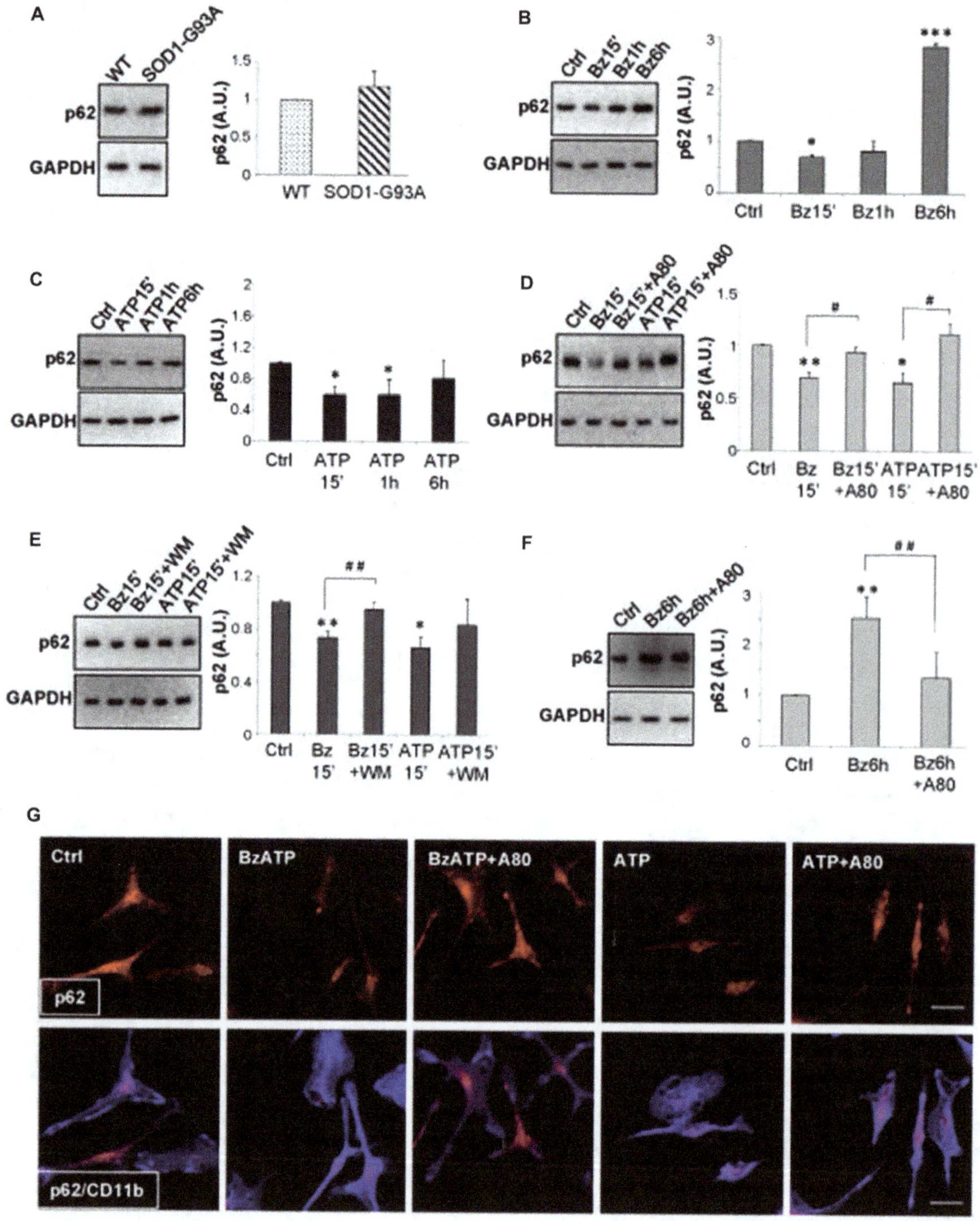

FIGURE 4 | Activation of P2X7 affects SQSTM1/p62 levels in SOD1-G93A microglia. **(A)** Equal amounts of protein from SOD1-G93A and WT microglia were subjected to western blotting and immunoreactions with anti-SQSTM1/p62 and anti-GAPDH. SOD1-G93A microglia were exposed to 300 μM BzATP **(B)** or 3 mM ATP **(C)** for 15 min, 1 h or 6 h and equal amounts of total protein were subjected to western blotting and immunoreactions with anti-SQSTM1/p62 and anti-GAPDH. SOD1-G93A microglia were exposed to 300 μM BzATP or 3 mM ATP for 15 min with or without A-804598 10 μM **(D)** or WM 100 nM **(E)** and equal amounts of protein were subjected to western blotting and immunoreactions with anti-SQSTM1/p62 and anti-GAPDH. **(F)** SOD1-G93A microglia exposed to 300 μM BzATP for 6 h with or without 10 μM A-804598 were subjected to western blotting and immunoreactions with anti-SQSTM1/p62 and anti-GAPDH. **(G)** SOD1-G93A microglia treated for 15 min with 300 μM BzATP or 3 mM ATP in the absence or presence of 10 μM A-804598 were analyzed by means of fluorescence microscopy after staining with anti-SQSTM1/p62 (red) and CD11b (blue) (scale bar 50 μm). Data represent mean ± SEM of $n = 3$ independent experiments. Statistical significance was calculated by student's *t*-test or ANOVA followed by *Post Hoc* Tukey's test referred to Ctrl ($^{*}p < 0.05$, $^{**}p < 0.01$, $^{***}p < 0.001$); to BzATP15'-treated cells or ATP15'-treated cells (**D,E**; $^{\#}p < 0.05$, $^{\#\#}p < 0.01$), or BzATP6h-treated cells (**F**; $^{\#\#}p < 0.01$).

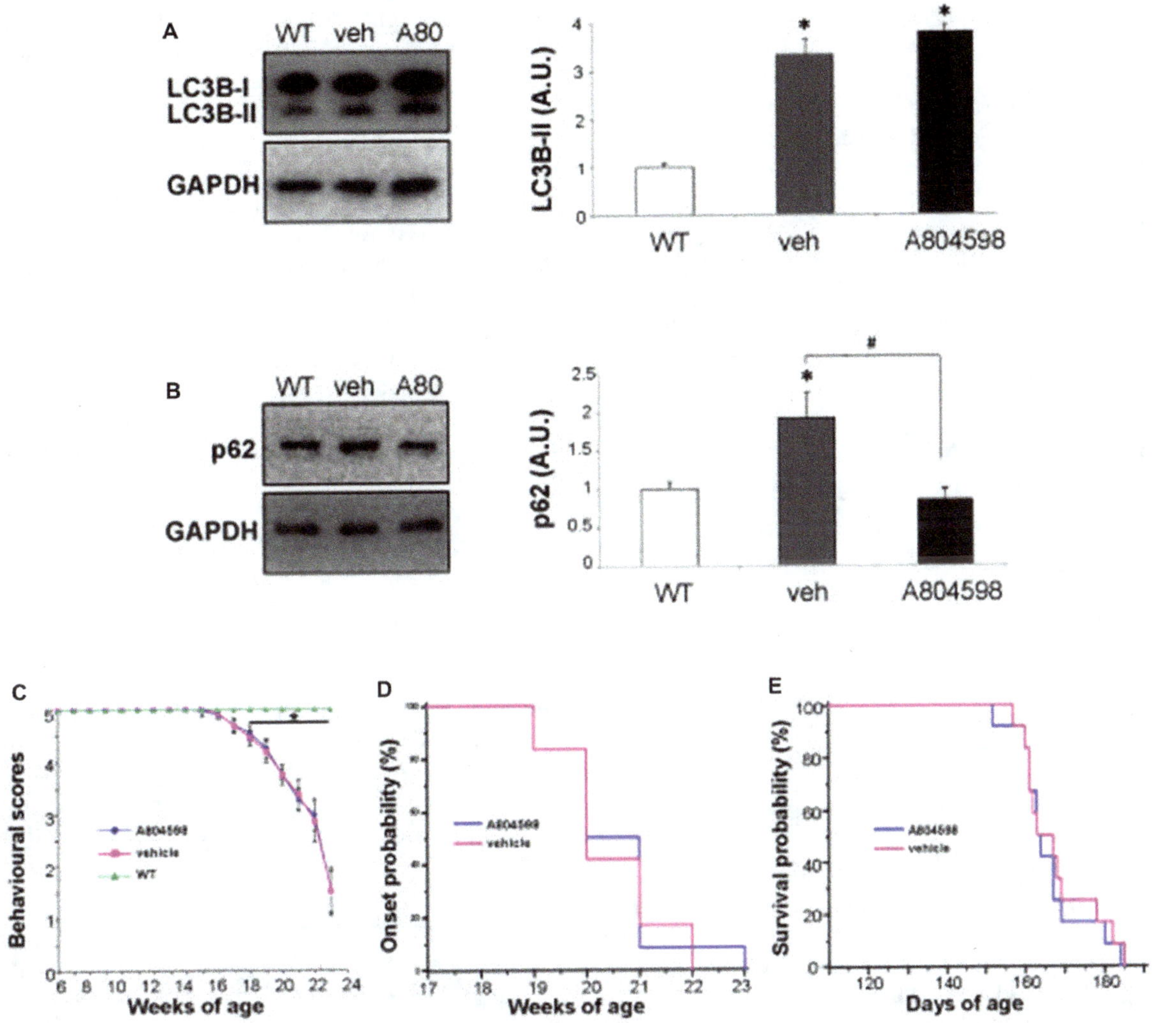

FIGURE 5 | *In vivo* chronic treatment with P2X7 antagonist A-804598 decreases SQSTM1/p62 expression in lumbar spinal cord of SOD1-G93A mice. Equal amounts of protein from lumbar spinal cord lysates of WT mice (~5 months) and vehicle- or A-804598-treated SOD1-G93A mice at end stage (n = 4/group) were subjected to western blotting with anti-LC3B-II **(A)**, anti-SQSTM1/p62 **(B)** and anti-GAPDH. **(C)** Behavioral scores are not affected by A-804598 treatment (n = 12) with respect to vehicle mice (n = 12). WT mice (n = 3). Data represent mean ± SEM. Statistical significance was calculated by student's t-test or ANOVA followed by *Post Hoc* Tukey's test referred to WT (**A–C**; $^*p < 0.05$) or to vehicle mice (**B**; $^\#p < 0.05$). **(D)** The disease onset of SOD1-G93A mice, as shown by Kaplan–Meier survival curves was not affected by A-804598. **(E)** A-804598-treated mice show no significant differences in median survival with respect to vehicle-treated SOD1-G93A, as shown by Kaplan–Meier survival curves (163.5 days for A-804598 vs. 163 days for vehicle).

We have previously demonstrated that BzATP modulates ERK1/2 phosphorylation in SOD1-G93A primary microglia (Apolloni et al., 2013b). Because the MAPK activation pathway is a positive regulator of autophagy induction (Young et al., 2015) and previous data have reported that p-ERK1/2 is specifically involved in induction of LC3B-II expression by activated P2X7 in non-transgenic microglia (Takenouchi et al., 2009b), we have stimulated SOD1-G93A microglia with 300 μM BzATP or 3 mM ATP in the presence of the MEK pathway inhibitor PD98059 (100 μM) that *per se* does not affect basal LC3B-II content (data not shown). Although a partial and not statistically significant inhibition is observed on LC3B-II up-regulation by ATP, LC3B-II induction mediated by P2X7 receptor is not apparently dependent on ERK1/2 phosphorylation, not being significantly inhibited by PD98059 (**Figure 2C**).

P2X7 Inhibits mTOR Phosphorylation and Modulates M2 Markers in SOD1-G93A Microglia

Inhibition of mTOR pathway is generally required for the induction of conventional autophagy (Klionsky et al., 2016). Thus, we have analyzed the effects of P2X7 activation on mTOR phosphorylation, showing by western blot analysis that 300 μM BzATP, or 3 mM ATP, is able to transiently decrease the protein content of p-mTOR/mTOR in SOD1-G93A microglia, with maximal inhibition observed after 15 min, as compared to control cells (**Figures 3A,B**). These effects are partially prevented by the P2X7 antagonist A-804598.

Microglia activation is known to occur by the alternating acquisition of different phenotypes concisely defined as M2 with anti-inflammatory properties, and M1 with pro-inflammatory features (Franco and Fernández-Suárez, 2015). Since the autophagic flux impacts on the M2/M1 microglia phenotypes (Su et al., 2016; Xia et al., 2016), we investigated the effects of BzATP stimulation on the levels of the M2 markers CD163 and ARG-1, in the absence or presence of WM. As shown in **Figure 3C**, 300 μM BzATP in 15 min increases the content of ARG-1 and CD163 proteins that returns approximately to basal levels in 6 h and is partially reverted by WM.

Activation of P2X7 Affects SQSTM1/p62 Protein Content in SOD1-G93A Microglia and in Lumbar Spinal Cord of SOD1-G93A Mice

A well-known substrate of autophagy is SQSTM1/p62 protein that can be degraded within the autolysosomes as an additional evidence of autophagic activity (Mizushima et al., 2010). To ascertain whether the increased LC3B-II levels induced by P2X7 activation in SOD1-G93A microglia represent a true increase in the autophagic flux, we have examined the content of SQSTM1/p62 protein. We have observed that while SOD1-G93A microglia contain levels of p62 comparable to WT microglia (**Figure 4A**), 300 μM BzATP (**Figure 4B**), or 3 mM ATP (**Figure 4C**), transiently decreases SQSTM1/p62 protein content in SOD1-G93A microglia with about 50% maximal inhibition obtained in 15 min (**Figures 4B,C**). A pre-incubation with the P2X7 antagonist A-804598 (**Figure 4D**) or with WM (**Figure 4E**), partially abolishes BzATP- and ATP-mediated SQSTM1/p62 decrease in SOD1-G93A microglia. As WM is also a class III phosphatidylinositol 3-kinase inhibitor, this result moreover indicates that P2X7-mediated increase in autophagic flux occurs via the phosphatidylinositol 3-kinase-mediated pathway. The reduction of SQSTM1/p62 by a short treatment with BzATP or ATP in SOD1-G93A microglia is confirmed also by double immunofluorescence with microglia specific marker anti-CD11b, as the intensity of the SQSTM1/p62 immunostaining is reduced after a short-term stimulation of P2X7 with respect to unstimulated cells, and the effect is reversed by the simultaneous presence of A-804598 (**Figure 4G**). Interestingly, a sustained activation of P2X7 by BzATP for 6 h markedly increases the levels of SQSTM1/p62 protein with respect to unstimulated cells, and this effect is abolished by the presence of A-804598 (**Figure 4F**), thus indicating that a prolonged P2X7 activation might cause an engulfment of the autophagic flux.

It is known that the SQSTM1/p62 autophagy substrate accumulates, together with LC3B-II, in lumbar spinal cord of ALS mice during disease progression, when the autophagic flux is impaired (Zhang et al., 2011). We have therefore measured the level of LC3B-II and SQSTM1/p62 proteins in lumbar spinal cord of SOD1-G93A mice after having pharmacologically inhibited *in vivo* the P2X7 receptor by a chronic treatment in female SOD1-G93A mice with the blood-brain permeant A-804598, demonstrated to reach brain concentrations in rodents after oral or i.p. doses (Able et al., 2011; Iwata et al., 2016), administered at 30 mg/Kg from pre-onset to end stage of disease. We find that while the protein levels of LC3B-II (**Figure 5A**) and SQSTM1/p62 (**Figure 5B**) are confirmed to be both increased at end stage in vehicle-treated SOD1-G93A mice with respect to WT, in A-804598-treated ALS mice with respect to vehicle, the LC3B-II protein content appears unmodified, while SQSTM1/p62 is inhibited to basal levels (**Figures 5A,B**). As shown, neither behavioral scores (**Figure 5C**), disease onset (**Figure 5D**), nor survival (**Figure 5E**) are however affected by A-804598 when administered in female SOD1-G93A mice.

DISCUSSION

Recent work has provided strong indications not only that autophagy plays a key role in ALS, but also that P2X7 receptor contributes to the neuroinflammatory pathway of ALS on one hand, and to the autophagic flux of microglia, on the other. The aim of this work was thus to investigate if P2X7 activation participates to ALS pathogenesis by directly modulating autophagic mechanisms in microglia, in other words to establish the occurrence of a P2X7-neuroinflammation-

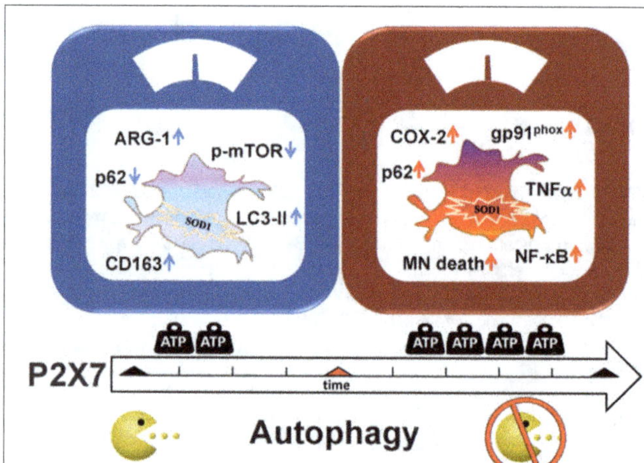

FIGURE 6 | Schematic representation of P2X7 actions in SOD1-G93A microglia. Treatment of SOD1-G93A primary microglia with P2X7 agonists BzATP or ATP modulates inflammatory and autophagic markers. After a short stimulation (left panel), P2X7 induces an increase in M2 markers (ARG-1 and CD163) concomitantly with the stimulation of the autophagic process by augmenting LC3-II and decreasing p62 and p-MTOR levels. Conversely, a sustained activation with P2X7 agonists (right panel), causes inhibition of the autophagic flux with up-regulation of p62 that is parallel to previous results showing an increase of pro-inflammatory M1 markers gp91^{phox} (Apolloni et al., 2016b), cyclooxygenase-2 (COX-2; D'Ambrosi et al., 2009), and moreover nuclear factor kappa-B (NF-κB; Parisi et al., 2016), tumor necrosis factor alpha (TNF-α) and motor neurons (MN) death (D'Ambrosi et al., 2009).

autophagy axis in ALS-microglia. We demonstrate that the modulation of the autophagic flux appears as a novel mechanism by which P2X7 contributes to activate SOD1-G93A microglia and to interfere with ALS features (**Figure 6**). In detail, we have illustrated here that a short time window of P2X7 stimulation by purinergic agonists augments LC3-II via the mTOR pathway and decreases SQSTM1/p62 levels, concurrently with an up-regulation of the M2 markers CD163 and ARG-1 in SOD1-G93A primary microglia. On the contrary, a sustained activation of P2X7 leads to an impairment of the autophagic flux with increased SQSTM1/p62 protein content. This might occur concomitantly with the previously characterized polarization of SOD1-G93A microglia toward the M1 phenotype exemplified by gp91^{phox}, COX-2, TNF-α and NF-κB activation, and toxicity towards MN (D'Ambrosi et al., 2009; Apolloni et al., 2016b; Parisi et al., 2016). Although as yet unaware of the molecular mechanisms linking the decrease in autophagy to the insurgence of pro-inflammatory actions sustained by P2X7, our results are in line with the involvement of autophagy in the inflammatory response (Levine et al., 2011; Apolloni et al., 2016a). Indeed, when autophagy is stimulated by LPS in microglia, the expression of inducible nitric oxide synthase is suppressed (Han et al., 2013). Conversely, when the autophagic flux is inhibited, the M1 phenotype is induced with higher expression of TNF-α, inducible nitric oxide synthase and COX-2 mediated by the NF-κB pathway, and M2 genes are instead inhibited with down-regulation of ARG-1, BDNF and IL-10 (Xia et al., 2016). In both ALS patients and transgenic mice, it is well known that microglia switches from a beneficial M2 phenotype occurring during the early phases of disease, to a detrimental M1 phenotype that is more preponderant when the disease accelerates and motor neuron injury escalates (Henkel et al., 2009; Beers et al., 2011; Zhao et al., 2013; Hooten et al., 2015). These and our findings would thus suggest a parallelism between M2 inflammatory state and activation of autophagy, as well as M1 inflammatory condition and inhibition of autophagy, both consistently with the dual role played by P2X7 in ALS pathogenesis.

Moreover, we suggest that the short P2X7-dependent activation of autophagy demonstrated *in vitro* in SOD1-G93A microglia could be related to the beneficial role of P2X7 activation occurring during the early phase of the disease, when genetic ablation of P2X7 elicits detrimental consequences on ALS phenotype (Apolloni et al., 2013a). The persistent stimulation of P2X7 causing autophagic flux inhibition in SOD1-G93A microglia *in vitro* is instead perhaps resembling what occurs during the symptomatic phase of the disease, when the pharmacological inhibition of P2X7 exerts beneficial effects on ALS progression (Apolloni et al., 2014; Sluyter et al., 2017).

The initial stimulation of autophagic flux found in SOD1-G93A microglia by P2X7 activation with inhibition of mTOR phosphorylation differs from what reported in non-transgenic microglia where P2X7 signaling has been reported to negatively regulate autophagic flux and stimulate the phosphorylation of mTOR (Takenouchi et al., 2009b). The induction of microglial autophagy is instead in line with data obtained in dystrophic muscle cells, where P2X7 indeed increases autophagy (Young et al., 2015), thus indicating a phenotype-specific purinergic-mediated autophagy in ALS. In addition, we have demonstrated that LC3-II induction in SOD1-G93A microglia is apparently independent from ERK1/2 activation, again in line with what reported in muscle cells, where MAP kinases play only a marginal role in ATP-induced autophagy (Young et al., 2015). This is again different from what reported in non-transgenic microglia (Takenouchi et al., 2009b), where inhibition of autophagy is instead dependent on the MAPK pathway. By emphasizing the divergence between non-transgenic and SOD1-G93A microglia, these results however reinforce the synergism between autophagy and inflammation in P2X7-activated microglia.

Finally, in this article we have shown that inhibition of P2X7 through the antagonist A-804598 in SOD1-G93A mice suppresses SQSTM1/p62 up-regulation in lumbar spinal cord, thus confirming P2X7 as an *in vivo* modulator of the ALS pathological mechanism of autophagy, although further experiments might contribute to identify the cell phenotypes that are more responsible for altered autophagy. Survival, behavioral scores and disease onset are however not affected by A-804598 when administered in female SOD1-G93A mice at the time and dose adopted in our work. This suggests that inhibition of SQSTM1/p62 overexpression is perhaps necessary, but not sufficient, to ameliorate ALS disease. Given the various controversial male vs. female differences encountered on the effects of P2X7 antagonists or genetic ablation (Sluyter et al., 2017), we cannot exclude that further dose- and sex-dependent preclinical studies might provide different results.

Ectonucleotidase activities are responsible for the degradation of extracellular ATP, and SOD1-G93A microglia have a reduced extracellular ATP hydrolyzing activity with respect to non-transgenic microglia (D'Ambrosi et al., 2009; Butovsky et al., 2012; Volonté et al., 2016). This would imply that higher extracellular ATP concentrations might be present *in vivo* in the proximity of SOD1-G93A damaged MN to activate P2X7 on neighboring cells. While an initial activation of P2X7 might ensure a positive impact on microglia by stimulating autophagy, a persistent engagement of the receptor might in turn inhibit autophagy. These results thus extend to autophagy the dual role previously suggested for P2X7 in the SOD1-G93A model (Volonté et al., 2016). In order to further validate the *in vitro* and *in vivo* results about P2X7-dependent modulation of autophagy, because previous studies suggested that SOD1-G93A microglia has a regional heterogeneity during disease progression (Nikodemova et al., 2014), future experiments will investigate the actions of P2X7 also in primary microglia from SOD1-G93A spinal cord. Moreover, because WT-SOD1 overexpression in mice *per se* causes mitochondrial vacuolization, axonal degeneration and premature motor neuron loss in mice (Jaarsma et al., 2000), we will also study how this condition might for itself influence the effects of P2X7 in microglia.

In conclusion, our results extend the role of P2X7 as a means for interfering with autophagy in SOD1-G93A microglia, and a new mechanism of intervention against autophagic dysfunction in ALS pathogenesis.

AUTHOR CONTRIBUTIONS

PF, SAm and SAp performed experiments, collected and analyzed data. SAp and CV designed the experiments and wrote the manuscript. All authors approved the manuscript for publication.

ACKNOWLEDGMENTS

This work was supported by AriSLA-Agenzia Italiana per la Ricerca sulla Sclerosi Laterale Amiotrofica (grant PATH-for-ALS). We thank Dr. Veronica Verdile for technical help, Dr. Vania Gelmetti for suggestions and critical reading of the manuscript and Dr. Daniele Peluso for statistical advice.

REFERENCES

Able, S. L., Fish, R. L., Bye, H., Booth, L., Logan, Y. R., Nathaniel, C., et al. (2011). Receptor localization, native tissue binding and *ex vivo* occupancy for centrally penetrant P2X7 antagonists in the rat. *Br. J. Pharmacol.* 162, 405–414. doi: 10.1111/j.1476-5381.2010.01025.x

Apolloni, S., Amadio, S., Montilli, C., Volonté, C., and D'Ambrosi, N. (2013a). Ablation of P2X7 receptor exacerbates gliosis and motoneuron death in the SOD1-G93A mouse model of amyotrophic lateral sclerosis. *Hum. Mol. Genet.* 22, 4102–4116. doi: 10.1093/hmg/ddt259

Apolloni, S., Parisi, C., Pesaresi, M. G., Rossi, S., Carrì, M. T., Cozzolino, M., et al. (2013b). The NADPH oxidase pathway is dysregulated by the P2X7 receptor in the SOD1-G93A microglia model of amyotrophic lateral sclerosis. *J. Immunol.* 190, 5187–5195. doi: 10.4049/jimmunol.1203262

Apolloni, S., Fabbrizio, P., Amadio, S., and Volonté, C. (2016a). Actions of the antihistaminergic clemastine on presymptomatic SOD1-G93A mice ameliorate ALS disease progression. *J. Neuroinflammation* 13:191. doi: 10.1186/s12974-016-0658-8

Apolloni, S., Fabbrizio, P., Parisi, C., Amadio, S., and Volonté, C. (2016b). Clemastine confers neuroprotection and induces an anti-inflammatory phenotype in SOD1(G93A) mouse model of amyotrophic lateral sclerosis. *Mol. Neurobiol.* 53, 518–531. doi: 10.1007/s12035-014-9019-8

Apolloni, S., Amadio, S., Parisi, C., Matteucci, A., Potenza, R. L., Armida, M., et al. (2014). Spinal cord pathology is ameliorated by P2X7 antagonism in a SOD1-mutant mouse model of amyotrophic lateral sclerosis. *Dis. Model. Mech.* 7, 1101–1109. doi: 10.1242/dmm.017038

Barth, S., Glick, D., and Macleod, K. F. (2010). Autophagy: assays and artifacts. *J. Pathol.* 221, 117–124. doi: 10.1002/path.2694

Bartlett, R., Sluyter, V., Watson, D., Sluyter, R., and Yerbury, J. J. (2017). P2X7 antagonism using Brilliant Blue G reduces body weight loss and prolongs survival in female SOD1G93A amyotrophic lateral sclerosis mice. *PeerJ* 5:e3064. doi: 10.7717/peerj.3064

Beamer, E., Gölöncsér, F., Horváth, G., Bekő, K., Otrokocsi, L., Koványi, B., et al. (2016). Purinergic mechanisms in neuroinflammation: an update from molecules to behavior. *Neuropharmacology* 104, 94–104. doi: 10.1016/j.neuropharm.2015.09.019

Beers, D. R., Zhao, W., Liao, B., Kano, O., Wang, J., Huang, A., et al. (2011). Neuroinflammation modulates distinct regional and temporal clinical responses in ALS mice. *Brain Behav. Immun.* 25, 1025–1035. doi: 10.1016/j.bbi.2010.12.008

Biswas, D., Qureshi, O. S., Lee, W. Y., Croudace, J. E., Mura, M., and Lammas, D. A. (2008). ATP-induced autophagy is associated with rapid killing of intracellular mycobacteria within human monocytes/macrophages. *BMC Immunol.* 9:35. doi: 10.1186/1471-2172-9-35

Butovsky, O., Siddiqui, S., Gabriely, G., Lanser, A. J., Dake, B., Murugaiyan, G., et al. (2012). Modulating inflammatory monocytes with a unique microRNA gene signature ameliorates murine ALS. *J. Clin. Invest.* 122, 3063–3087. doi: 10.1172/jci62636

Catanzaro, J. M., Hueston, C. M., Deak, M. M., and Deak, T. (2014). The impact of the P2X7 receptor antagonist A-804598 on neuroimmune and behavioral consequences of stress. *Behav. Pharmacol.* 25, 582–598. doi: 10.1097/FBP.0000000000000072

Cervetto, C., Frattaroli, D., Maura, G., and Marcoli, M. (2013). Motor neuron dysfunction in a mouse model of ALS: gender-dependent effect of P2X7 antagonism. *Toxicology* 311, 69–77. doi: 10.1016/j.tox.2013.04.004

Cozzolino, M., Pesaresi, M. G., Gerbino, V., Grosskreutz, J., and Carrì, M. T. (2012). Amyotrophic lateral sclerosis: new insights into underlying molecular mechanisms and opportunities for therapeutic intervention. *Antioxid. Redox Signal.* 17, 1277–1330. doi: 10.1089/ars.2011.4328

D'Ambrosi, N., Finocchi, P., Apolloni, S., Cozzolino, M., Ferri, A., Padovano, V., et al. (2009). The proinflammatory action of microglial P2 receptors is enhanced in SOD1 models for amyotrophic lateral sclerosis. *J. Immunol.* 183, 4648–4656. doi: 10.4049/jimmunol.0901212

Donnelly-Roberts, D. L., Namovic, M. T., Surber, B., Vaidyanathan, S. X., Perez-Medrano, A., Wang, Y., et al. (2009). [3H]A-804598 ([3H]2-cyano-1-[(1S)-1-phenylethyl]-3-quinolin-5-ylguanidine) is a novel, potent, and selective antagonist radioligand for P2X7 receptors. *Neuropharmacology* 56, 223–229. doi: 10.1016/j.neuropharm.2008.06.012

Franco, R., and Fernández-Suárez, D. (2015). Alternatively activated microglia and macrophages in the central nervous system. *Prog. Neurobiol.* 131, 65–86. doi: 10.1016/j.pneurobio.2015.05.003

Glick, D., Barth, S., and Macleod, K. F. (2010). Autophagy: cellular and molecular mechanisms. *J. Pathol.* 221, 3–12. doi: 10.1002/path.2697

Haanes, K. A., Schwab, A., and Novak, I. (2012). The P2X7 receptor supports both life and death in fibrogenic pancreatic stellate cells. *PLoS One* 7:e51164. doi: 10.1371/journal.pone.0051164

Han, H.-E., Kim, T.-K., Son, H.-J., Park, W. J., and Han, P. L. (2013). Activation of autophagy pathway suppresses the expression of iNOS, IL6 and cell death of LPS-stimulated microglia cells. *Biomol. Ther. (Seoul)* 21, 21–28. doi: 10.4062/biomolther.2012.089

Henkel, J. S., Beers, D. R., Zhao, W., and Appel, S. H. (2009). Microglia in ALS: the good, the bad, and the resting. *J. Neuroimmune Pharmacol.* 4, 389–398. doi: 10.1007/s11481-009-9171-5

Hooten, K. G., Beers, D. R., Zhao, W., and Appel, S. H. (2015). Protective and toxic neuroinflammation in amyotrophic lateral sclerosis. *Neurotherapeutics* 12, 364–375. doi: 10.1007/s13311-014-0329-3

Ilieva, H., Polymenidou, M., and Cleveland, D. W. (2009). Non-cell autonomous toxicity in neurodegenerative disorders: ALS and beyond. *J. Cell Biol.* 187, 761–772. doi: 10.1083/jcb.200908164

Illes, P., and Verkhratsky, A. (2016). Purinergic neurone-glia signalling in cognitive-related pathologies. *Neuropharmacology* 104, 62–75. doi: 10.1016/j.neuropharm.2015.08.005

Iwata, M., Ota, K. T., Li, X. Y., Sakaue, F., Li, N., Dutheil, S., et al. (2016). Psychological stress activates the inflammasome via release of adenosine triphosphate and stimulation of the purinergic type 2×7 receptor. *Biol. Psychiatry* 80, 12–22. doi: 10.1016/j.biopsych.2015.11.026

Jaarsma, D., Haasdijk, E. D., Grashorn, J. A., Hawkins, R., van Duijn, W., Verspaget, H. W., et al. (2000). Human Cu/Zn superoxide dismutase (SOD1) overexpression in mice causes mitochondrial vacuolization, axonal degeneration, and premature motoneuron death and accelerates motoneuron disease in mice expressing a familial amyotrophic lateral sclerosis mutant SOD1. *Neurobiol. Dis.* 7, 623–643. doi: 10.1006/nbdi.2000.0299

Klionsky, D. J., Abdelmohsen, K., Abe, A., Abedin, M. J., Abeliovich, H., Acevedo Arozena, A., et al. (2016). Guidelines for the use and interpretation of assays for monitoring autophagy (3rd edition). *Autophagy* 12, 1–222. doi: 10.1080/15548627.2015.1100356

Klionsky, D. J., and Emr, S. D. (2000). Autophagy as a regulated pathway of cellular degradation. *Science* 290, 1717–1721. doi: 10.1126/science.290.5497.1717

Levine, B., Mizushima, N., and Virgin, H. W. (2011). Autophagy in immunity and inflammation. *Nature* 469, 323–335. doi: 10.1038/nature09782

Ludolph, A. C., Bendotti, C., Blaugrund, E., Chio, A., Greensmith, L., Loeffler, J. P., et al. (2010). Guidelines for preclinical animal research in ALS/MND: a consensus meeting. *Amyotroph. Lateral Scler.* 11, 38–45. doi: 10.3109/17482960903545334

Mizushima, N., and Yoshimori, T. (2007). How to interpret LC3 immunoblotting. *Autophagy* 3, 542–545. doi: 10.4161/auto.4600

Mizushima, N., Yoshimori, T., and Levine, B. (2010). Methods in mammalian autophagy research. *Cell* 140, 313–326. doi: 10.1016/j.cell.2010.01.028

Monif, M., Burnstock, G., and Williams, D. A. (2010). Microglia: proliferation and activation driven by the P2X7 receptor. *Int. J. Biochem. Cell Biol.* 42, 1753–1756. doi: 10.1016/j.biocel.2010.06.021

Monif, M., Reid, C. A., Powell, K. L., Drummond, K. J., O'Brien, T. J., and Williams, D. A. (2016). Interleukin-1β has trophic effects in microglia and its release is mediated by P2X7R pore. *J. Neuroinflammation* 13:173. doi: 10.1186/s12974-016-0621-8

Nikodemova, M., Small, A. L., Smith, S. M., Mitchell, G. S., and Watters, J. J. (2014). Spinal but not cortical microglia acquire an atypical phenotype with high VEGF, galectin-3 and osteopontin and blunted inflammatory responses in ALS rats. *Neurobiol. Dis.* 69, 43–53. doi: 10.1016/j.nbd.2013.11.009

Parisi, C., Napoli, G., Amadio, S., Spalloni, A., Apolloni, S., Longone, P., et al. (2016). MicroRNA-125b regulates microglia activation and motor neuron death in ALS. *Cell Death Differ.* 23, 531–541. doi: 10.1038/cdd.2015.153

Pizzasegola, C., Caron, I., Daleno, C., Ronchi, A., Minoia, C., Carri, M. T., et al. (2009). Treatment with lithium carbonate does not improve disease progression in two different strains of SOD1 mutant mice. *Amyotroph. Lateral Scler.* 10, 221–228. doi: 10.1080/17482960902803440

Sluyter, R., Bartlett, R., Ly, D., and Yerbury, J. J. (2017). P2X7 receptor antagonism in amyotrophic lateral sclerosis. *Neural Regen. Res.* 12, 749–750. doi: 10.4103/1673-5374.206643

Song, C. Y., Guo, J. F., Liu, Y., and Tang, B. S. (2012). Autophagy and its comprehensive impact on ALS. *Int. J. Neurosci.* 122, 695–703. doi: 10.3109/00207454.2012.714430

Su, P., Zhang, J., Wang, D., Zhao, F., Cao, Z., Aschner, M., et al. (2016). The role of autophagy in modulation of neuroinflammation in microglia. *Neuroscience* 319, 155–167. doi: 10.1016/j.neuroscience.2016.01.035

Takenouchi, T., Fujita, M., Sugama, S., Kitani, H., and Hashimoto, M. (2009a). The role of the P2X7 receptor signaling pathway for the release of autolysosomes in microglial cells. *Autophagy* 5, 723–724. doi: 10.4161/auto.5.5.8478

Takenouchi, T., Nakai, M., Iwamaru, Y., Sugama, S., Tsukimoto, M., Fujita, M., et al. (2009b). The activation of P2X7 receptor impairs lysosomal functions and stimulates the release of autophagolysosomes in microglial cells. *J. Immunol.* 182, 2051–2062. doi: 10.4049/jimmunol.0802577

Tasdemir, E., Maiuri, M. C., Galluzzi, L., Vitale, I., Djavaheri-Mergny, M., D'Amelio, M., et al. (2008). Regulation of autophagy by cytoplasmic p53. *Nat. Cell Biol.* 10, 676–687. doi: 10.1038/ncb1730

Volonté, C., Apolloni, S., Parisi, C., and Amadio, S. (2016). Purinergic contribution to amyotrophic lateral sclerosis. *Neuropharmacology* 104, 180–193. doi: 10.1016/j.neuropharm.2015.10.026

Xia, C. Y., Zhang, S., Chu, S. F., Wang, Z. Z., Song, X. Y., Zuo, W., et al. (2016). Autophagic flux regulates microglial phenotype according to the time of oxygen-glucose deprivation/reperfusion. *Int. Immunopharmacol.* 39, 140–148. doi: 10.1016/j.intimp.2016.06.030

Young, C. N., Sinadinos, A., Lefebvre, A., Chan, P., Arkle, S., Vaudry, D., et al. (2015). A novel mechanism of autophagic cell death in dystrophic muscle regulated by P2RX7 receptor large-pore formation and HSP90. *Autophagy* 11, 113–130. doi: 10.4161/15548627.2014.994402

Zhang, X., Li, L., Chen, S., Yang, D., Wang, Y., Zhang, X., et al. (2011). Rapamycin treatment augments motor neuron degeneration in SOD1(G93A) mouse model of amyotrophic lateral sclerosis. *Autophagy* 7, 412–425. doi: 10.4161/auto.7.4.14541

Zhao, W., Beers, D. R., and Appel, S. H. (2013). Immune-mediated mechanisms in the pathoprogression of amyotrophic lateral sclerosis. *J. Neuroimmune Pharmacol.* 8, 888–899. doi: 10.1007/s11481-013-9489-x

4

In Search for Reliable Markers of Glioma-Induced Polarization of Microglia

*Kacper A. Walentynowicz[1], Natalia Ochocka[1], Maria Pasierbinska[1,2], Kamil Wojnicki[1], Karolina Stepniak[1], Jakub Mieczkowski[1], Iwona A. Ciechomska[1] and Bozena Kaminska[1]**

[1]Laboratory of Molecular Neurobiology, Neurobiology Center, Nencki Institute of Experimental Biology, Polish Academy of Sciences, Warsaw, Poland, [2]Glia Sp. z o.o, Warsaw, Poland

***Correspondence:**
Bozena Kaminska
bozenakk@nencki.gov.pl

Immune cells accumulating in the microenvironment of malignant tumors are tumor educated and contribute to its growth, progression, and evasion of antitumor immune responses. Glioblastoma (GBM), the common and most malignant primary brain tumor in adults, shows considerable accumulation of resident microglia and peripheral macrophages, and their polarization into tumor-supporting cells. There are controversies regarding a functional phenotype of glioma-associated microglia/macrophages (GAMs) due to a lack of consistent markers. Previous categorization of GAM polarization toward the M2 phenotype has been found inaccurate because of oversimplification of highly complex and heterogeneous responses. In this study, we characterized functional responses and gene expression in mouse and human microglial cultures exposed to fresh conditioned media [glioma-conditioned medium (GCM)] from human U87 and LN18 glioma cells. Functional analyses revealed mutual communication reflected by strong stimulation of glioma invasion by microglial cells and increased microglial phagocytosis after GCM treatment. To define transcriptomic markers of GCM-activated microglia, we performed selected and global gene expression analyses of stimulated microglial cells. We found activated pathways associated with immune evasion and TGF signaling. We performed computational comparison of the expression patterns of GAMs from human GBMs and rodent experimental gliomas to select genes consistently changed in different datasets. The analyses of marker genes in GAMs from different experimental models and clinical samples revealed only a small set of common genes, which reflects variegated responses in clinical and experimental settings. *Tgm2* and *Gpnmb* were the only two genes common in the analyzed data sets. We discuss potential sources of the observed differences and stress a great need for definitive elucidation of a functional state of GAMs.

Keywords: microglia, glioma, functional phenotype, transcriptomics, glioma-associated microglia/macrophages

INTRODUCTION

Tumor microenvironment consists of various non-neoplastic cells that play an important role in tumor growth, progression, and immune response evasion. In glioblastoma (GBM), one of the most aggressive and prevalent primary brain malignancy in adults, tumor-infiltrating microglia and peripheral macrophages are major immune cell population within the tumor. Glioma-associated microglia/macrophages (GAMs) account for up to 30% of tumor mass in human GBMs (1–3) and in experimental glioma models (4–7). Glioma cells secrete various cytokines and chemokines acting as chemoattractants and polarizing factors on the resident microglia [for reviews, see Ref. (8, 9)].

Their role in glioma progression and immunosuppression has been shown in different experimental glioma models (10, 11). Pharmacological or genetic ablation of GAMs reduced tumor growth in mice experimental models of glioma (10, 12, 13). While controversy remains on the magnitude of peripheral macrophage recruitment to gliomas at different steps of tumor progression, there is growing evidence that tumor-specific education of GAMs is expected to determine their effector functions during tumor progression.

In tumor microenvironment, infiltrating macrophages adopt different activation states between antitumor M1 and protumor M2 phenotypes, and these functional phenotypes are defined by differential expression of surface markers, secreted cytokines, and roles in immunoregulation (14). M2 macrophages express frequently genes involved in tissue repair and resolution and have immunosuppressive and immunoregulatory properties. Clinical and mouse model data correlate the accumulation of macrophages with protumor activities. To define a functional phenotype of GAMs *in vivo*, these cells (usually defined as $CD11b^+$ cells) have been isolated using immunomagnetic beads or immunosorting and subjected to global gene expression profiling (2, 3, 5, 6). Most of those studies used microarray gene expression profiling, whereas RNA-seq was used to profile $CD11b^+$ in human samples. The resulting definitions of transcriptomic-based functional phenotypes of GAMs from human and experimental rodent gliomas are conflicting and indicate either a mixture of M1/M2 phenotypes (3, 5), the M0 phenotype (2), or the M2 phenotype (6). Categorization of GAM polarization using M1/M2 markers reported for peripheral macrophages (14) has been found inaccurate to describe discrete states due to oversimplification of highly complex and heterogeneous responses. There is no consensus and reliable gene expression-based markers.

We have previously characterized rat primary microglia cultures responses to rat C6 glioma-conditioned medium (GCM) and found that factors secreted by glioma cells induce the proinvasive, immunosuppressive polarization of microglia (15, 16). Selected genes were upregulated in cultured microglia, and their expression was also increased in GAMs isolated as $CD11b^+$ from C6 gliomas in rats (6). While animal models are a better tool to recreate complex cell–cell interactions between a tumor and its microenvironment, mechanistic analyses of events underlying activation are feasible in cell co-cultures. In this study, we characterized interactions between human glioma and microglial cells and defined gene expression profiles in search for markers of microglia polarization. In addition, we re-evaluated published datasets consisting of differentially expressed genes in GAMs from rat and mouse gliomas, and human GBM samples to identify overlapping or common genes. As a little commonality has been found, we discuss potential sources of variability and prospects of finding reliable markers.

MATERIALS AND METHODS

Microglial Cultures

Primary microglial cultures were prepared from cerebral cortices of P0–P2 old C57BL/6J mice. Following stripping off the meninges and enzymatic brain dissociation the cells were collected and seeded onto the culture flasks. After 48 h, cell cultures were washed three times with phosphate-buffered saline (PBS) to remove debris. Primary cultures were kept in Dulbecco's modified Eagle's medium (DMEM) supplemented with 10% fetal bovine serum (FBS), 2 mM L-glutamine, 100 U/mL penicillin, and 100 μg/mL streptomycin (Gibco, MD, USA). Following establishment of glial culture monolayer, microglial cells were isolated by gentle shaking for 1 h at 100 RPM at 37°C. Microglial cells were then collected by centrifugation, counted, checked for viability, and seeded at density 60×10^4 onto 60-mm plastic dishes (for non-adherent cultures). Microglial cultures were used for experiments 48 h after seeding.

Human immortalized SV40 microglia (Applied Biological Materials Inc.) were cultured in Prigrow III Medium (Applied Biological Materials Inc.) supplemented with 10% fetal bovine serum (Gibco, MD, USA) and antibiotics (100 U/mL penicillin, 100 μg/mL streptomycin) in a humidified atmosphere of CO_2/air at 37°C. Mouse microglial BV2 cells were cultured in DMEM with GlutaMAX™ (DMEM GlutaMAX™) supplemented with 2% fetal bovine serum (Gibco, MD, USA) and antibiotics (100 U/mL penicillin, 100 μg/mL streptomycin) in a humidified atmosphere of CO_2/air (5%/95%) at 37°C (Heraeus, Hanau, Germany).

Glioma Cell Cultures

Human GBM cell lines LN18, U87-MG (ATCC, Manassas, VA, USA) were cultured in DMEM supplemented with 10% fetal bovine serum (Gibco, MD, USA) and antibiotics (100 U/mL penicillin, 100 μg/mL streptomycin) in a humidified atmosphere of CO_2/air (5%/95%) at 37°C (Heraeus, Hanau, Germany).

Patient-derived glioma cell cultures WG4 (GBM WHO grade IV) and IPIN20160420 (GBM WHO grade IV) were developed as previously described (17). Cells were cultured in DMEM/Nutrient Mixture F-12, GlutaMAX™ medium (DMEM/F-12, GlutaMAX™) supplemented with 10% fetal bovine serum (FBS, Gibco, MD, USA) and antibiotics (100 U/mL penicillin, 100 μg/mL streptomycin) in a humidified atmosphere of CO_2/air (5%/95%) at 37°C (Heraeus, Hanau, Germany).

GCM and Microglia Treatment

Human glioma cells were seeded onto 60-mm dish at density 75×10^4 in DMEM or DMEM/F12, GlutaMAX™ supplemented with 10% fetal calf serum, and 100 U/mL penicillin and 100 μg/mL streptomycin. Following 24 h, the medium was changed to a microglia culture medium (DMEM or Prigrow III Medium 10% FBS with antibiotics), and glioma cells were cultured for another 24 h. Fresh GCM was collected and used to treat microglial cells. Cells exposed to GCM were collected after 3 and 6 h.

Functional Tests on Microglia

Morphological changes were assessed in microglial cultures using a co-culture system. Microglial cells were seeded onto round glass cover slips at 8×10^4/well in a 24-well plate and incubated for 48 h. Glioma cells were seeded into the Falcon cell culture inserts with 0.4-μm pores (Falcon #353095) at 6×10^4/insert. After 24 h, the inserts were transferred into the plate with microglial cells and co-cultured for 48 h. Microglial cells were fixed with 2% paraformaldehyde in PBS, permeabilized with 0.1% TRITON X-100, and stained for 30 min at room temperature using Rhodamine–Phalloidin (1:1,000 in PBS). Cells were co-stained with DAPI (4′,6-diamidino-2-phenylindole; 0.001 mg/mL, Sigma) to visualize nuclei. Pictures were acquired using fluorescence microscope (Leica DM4000B, 40× objective). Images were analyzed using ImageJ software, and a ratio of Phalloidin stained area versus DAPI was used for a statistical analysis of cell size.

Phagocytic activity of microglia was assessed using flow cytometry. Microglial cells seeded onto 60-mm dishes and following 48 h, cells were treated for 24 h with fresh human GCM or LPS 100 ng/mL. Fluorescent latex beads were added (50 beads/cell; Sigma L3030) 3 h before read out. Cells were gently scrapped in cold PBS, washed twice and fluorescence was determined by FACSCalibur Flow Cytometer (BD Biosciences). In one run, 10,000 events were collected per sample and a mean fluorescence intensity (MFI) was calculated.

Gene Expression Analysis

RNA was isolated using RNeasy Mini Kit (QIAGEN, USA), followed by quality and quantity assessment using NanoDrop 2000 Spectrophotometer (Thermo Fisher, USA), and samples were stored at −80°C. RT-PCR was performed using SuperScript III Reverse Transcriptase (Invitrogen, USA) on 500 ng of total RNA and stored at −20°C. Gene expression profiling was performed by quantitative real-time PCR using 10 ng of cDNA in duplicates, using FAST SYBR Green PCR MasterMix (Life Technologies, USA) in 10 μL reaction with a listed set of primers (Table S1 in Supplementary Material). Amplified product was normalized to the endogenous expression of *18S*, and represented as delta Ct values.

Gene expression profiling of human SV40 microglia stimulated with GCM from human glioma LN18 and U87-MG cells was performed with GeneQuery™ Human Microglial Sensome qPCR Array (# GQH-MGS, ScienCell). GeneQuery™ qPCR array is a 96-well plate pre-made format; it contains in each well one validated primer set that recognizes and efficiently amplifies a specific target gene's cDNA. The primer set recognizes all known transcript variants of the target gene and amplifies only one gene. The array contains eight control genes. The annealing temperature in qPCR analysis was 65°C (with 2 mM Mg^{2+} and no DMSO). Gene expression profiling was performed by qPCR using 10 ng of cDNA from assorted samples from three duplicates for each group (untreated, GCM U87-MG, GCM LN18). The expression of selected genes were tested by qPCR with commercial primers for the following genes *P2RY12, SIGLEC1, SELPLG, IFITM5, PTAFR, TNFRSF1B*, and *GPR84*, purchased from ScienCell Research Laboratories (sequences undisclosed). Each primer set provided lyophilized and ready to use.

Global gene expression analysis using Affymetrix MG-430 PM Strip system (Affymetrix, USA) using GCM treated primary microglia for 6 h ($n = 3$). Samples were run on Bioanalyzer RNA 6000 Pico kit (Agilent, USA) to assess RNA integrity and concentration. The protocol was followed according to the manufacture's manual using 100 ng of total RNA and samples with RIN > 9.5. Hybridization was performed at 45°C for 16 h and fluidics, and imaging was done as described in the GeneAtlas System User's Guide (P/N 08-0246). All samples were run at the same time. Data were analyzed using R Studio software and Bioconductor package, and data were deposited into GEO Repository (GSE113370).

Invasion Assay

Microglial cells were plated onto 24-well plates at the density of 4×10^4. After 24 h, the invasion assay was performed using tissue culture inserts (6.5 mm Transwell® with 8.0 μm Pore Polycarbonate Membrane Insert, Corning, NY, USA) coated with the Growth Factor Reduced Matrigel™ Matrix (BD Biosciences, San Diego. CA, USA). 50 μL of the Matrigel™ Matrix (1 mg/mL) diluted in fresh DMEM medium was dried under sterile conditions (37°C) for 5–6 h. The medium was replaced to fresh one 1 h before seeding the glioma cells into inserts (to a medium for microglial cultures). The LN18, U87-MG, WG4, and IPIN20160420 cells were seeded at a density of 2×10^4/insert on matrigel-covered membrane in serum-reduced medium (2% FBS). The cultures were placed in a 37°C incubator with humidified air containing 5% CO_2. After 18 h, cells were fixed in ice-cold methanol and cell nuclei stained with DAPI (0.001 mg/mL, Sigma). The membranes from Transwell® inserts were cut out, and images were acquired using fluorescence microscope (Leica DM4000B, 10× objective) of the five independent fields (bottom, top, left, right side, and a center). Numbers of migrating cell nuclei were counted using ImageJ software. All experiments were performed at least three times, in duplicates.

Statistical Analysis

Each experiment was performed at least three times, on independent passages/cultures, at least in duplicates. Statistical analyses were performed using GraphPad Prism v6.01 (GraphPad Software, Inc., San Diego, CA, USA). The data were plotted as mean ± SD. Differences between the means of the treatments were evaluated using one-way analysis of variance (one-way ANOVA) followed by *post hoc* Dunnett's multiple comparison test or one-sided paired sample *t*-test (*p*-value) for two groups analysis. *p*-Values <0.05 were considered to be statistically significant and significance is marked as $^*p < 0.05$, $^{**}p < 0.01$, and $^{***}p < 0.001$.

Comparison of Different Gene Expression Data Sets

Mouse (E-MTAB-2660) and rat (E-MTAB-5050) microarray gene expression data were obtained from the Array Express database. Human RNA-seq expression profiles were obtained from GEO database (GSE80338). The raw microarray data were pre-processed using RMA as described (6). Annotation of probe sets was performed with information provided in the Ensembl database. For RNA-seq data, the already computed FPKM values were used. All data analyses were performed in the R statistical environment and relevant Bioconductor software.

RESULTS

Functional Analyses of Microglia Responses to Human Glioma Cells

Microglia can respond to various insults or environmental stimuli by changing its morphology, activating phagocytosis, and modulating gene expression (16). Changes from thin/ramified to more amoeboid/round shapes are associated with functional activation *in vitro* and *in vivo* and are useful indicator of activated microglia. We used a co-culture system, in which cells are separated by inserts which allows only the exchange of soluble factors. Murine microglial cells were co-cultured with glioma cells to assess morphological changes induced by conditioned media from U87-MG and LN18 glioma cell cultures (**Figures 1A–D**). In the presence of glioma cells, we observed profound changes in morphology of microglia and transformation of bipolar cells to the amoeboid shape cells (**Figures 1B,C**). Insets show representative images of cells with changed morphology in higher magnification. We quantified percentages of amoeboid microglia under various conditions (**Figure 1D**). U87-MG glioma cells showed the stronger ability than LN18 to induce morphological changes of murine microglia that display more amoeboid/round shapes. Next, we employed flow cytometry to assess phagocytosis of fluorescent beads in primary microglial cultures upon the exposure to GCM from three glioma cultures or LPS, a potent immunomodulator. Phagocytosis was determined as mean fluorescence intensity (MFI) of cells. Graph shows FACS measurements from a representative experiment. There was a significant increase in phagocytosis following the treatment with GCM from primary GBM patient-derived cell cultures (IPIN20160420), and a consistent trend in the increase of phagocytosed beads in microglia treated with U87-MG GCM.

Interactions between microglia and human glioma cells were further investigated by a matrigel invasion assay in a co-culture system. We have previously demonstrated that the presence of rat microglia doubles glioma invasion (18, 19). First, we examined the ability of different microglial cells and other non-microglial cells to increase invasion of U87-MG glioma cells. As shown in the **Figure 1G**, all tested microglial cells: murine BV2 cells, primary murine microglial cultures, and human SV40 microglial cells significantly increased invasion of U87-MG glioma cells. Non-microglial cells, such as normal human astrocytes, human HEK-293 cells, and murine 3T3 fibroblasts, did not increase invasion of U87-MG glioma cells. As the primary microglial cultures produced most variable results, invasion assay experiments were performed on microglial BV2 cells. As the primary microglial cultures produced most variable results, invasion assay experiments were performed on microglial BV2 cells, and also for comparison on human SV40 cells (Figure S1 in Supplementary Material). We investigated the ability of BV2 microglial cells to support invasion of human U87-MG and LN18 glioma cells, as well as two patient-derived glioma cultures (IPIN20160420 and WG4; **Figure 1H**). Each glioma cell line had a specific invasion potential under basal conditions, so the increase of invasion in the presence of microglial cells was expressed as a fold change in relation to a cell line control performed for each treatment group. The results show a variable extent of dependence of glioma invasion on microglial cells. Three of four tested glioma cells showed increased in invasion in presence of murine microglial cells. The most prominent increase of invasion was observed in a case of U87-MG and WG4 cells, whereas IPIN20160420 showed moderate increase (**Figure 1H**). Altogether, our results show that both established and primary cultures of glioma interact in co-culture system, and U87-MG glioma cells exert strong and reproducible effect on murine microglia. Both murine and human microglial cells support invasion of U87 MG glioma cells, while astrocytes and other cells do not have such effect.

The Expression of Pre-Defined Marker Genes in Murine and Human Microglial Cells Treated With GCM From Human Glioma Cells

We have demonstrated the induction of certain genes in rat microglia cultures exposed for 6 h to GCM from C6 rat glioma cells (15, 16). Therefore, we sought to investigate whether those markers of microglia activation are modulated in murine microglia by GCM from human glioma cells. Primary murine microglia cultures were treated with fresh GCM from LN18 and U87-MG human glioma cells for 3 and 6 h. The expression of *Mmp14*, *cMyc*, *Arg1*, *Ifnb1*, *Cxcl14*, *Irf7*, and *Smad7* was assessed with qRT-PCR, and the results are plotted as delta Ct values relative to the endogenous *18S* expression (**Figure 2**). Increases in the *Mmp14* and *Arg1* expression were detected in microglia exposed to GCM from U87-MG cells for 3 h. Small increases of *Arg1*, *Cxcl14*, and *Smad7* mRNA levels at 6 h were observed; however, due to large variations between the biological repeats, these changes did not reach significance. None of the tested genes was significantly upregulated by GCM from LN18 glioma cells. The basal expression of selected genes varied between 3 and 6 h. The expression of endogenous *18S* was used as a reference for the amount of cDNA as its expression did not change following treatment.

We used GeneQuery™ Human Microglial Sensome qPCR Array examining 89 genes to perform gene expression profiling of human SV40 microglial cells polarized by GCM from human LN18 and U87-MG glioma cells. The results, presented as a heatmap, demonstrate moderate similarity in the profiles of upregulated genes in microglia stimulated with GCM U87-MG or LN18. The expression of genes strongly upregulated by either treatment is indicated by arrows and gene names (**Figure 3A**). The expression of selected, upregulated genes (incorporating *c-MYC*, *SMAD7*, and *MMP14*) was validated by qPCR in independent cultures of human SV40 microglial cells treated for 6 h with GCM and corroborated the qPCR Array results. A number of genes were significantly upregulated, and we observed the stronger effect after GCM LN18 treatment (**Figure 3B**).

Global Gene Expression Analysis in GCM-Treated Mice Microglia

As the preselected gene approach did not appear useful in identification of genes modulated during polarization of microglia by human glioma cells, to identify differentially expressed genes, we performed gene expression using Affymetrix arrays. Microarray analysis was performed on murine microglial cells exposed to fresh GCM from LN18 and U87-MG for 6 h. Volcano plots represent differentially expressed genes relative to untreated microglia (**Figure 4A**). Following the analysis, 36 unique genes were found

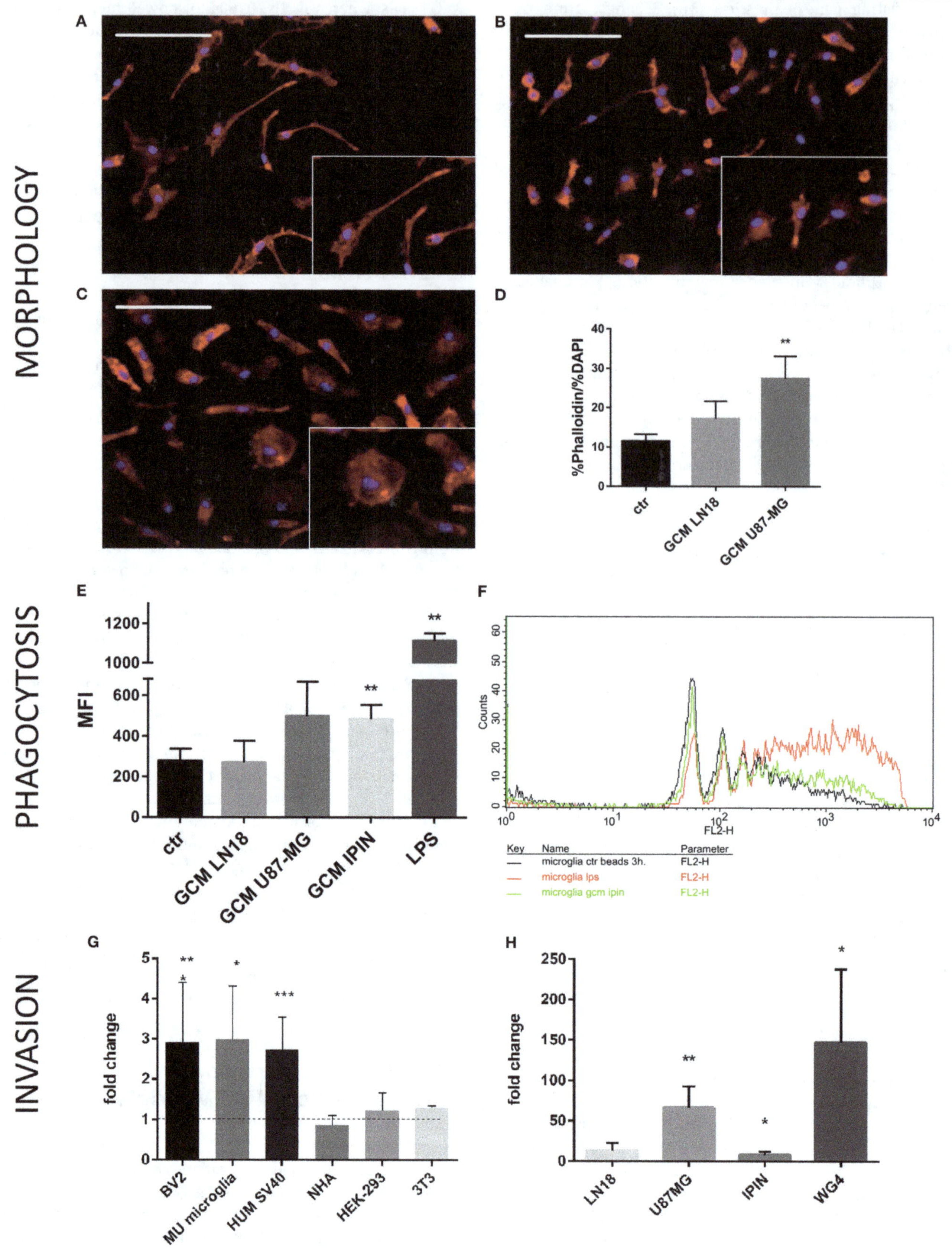

FIGURE 1 | Continued

FIGURE 1 | Functional analyses of glioma-induced polarization of murine microglia. Primary murine microglia cultures were co-cultured with human U87-MG or LN18 glioma cells. **(A–C)** Representative images show morphological changes induced in primary murine microglia cultures following co-culture with human U87-MG or LN18 glioma cells. Morphological alterations were visualized by F-actin staining; cell nuclei were co-stained with DAPI. Insets show in higher magnification numerous microglia with amoeboid shape in co-cultures with U87-MG cells. **(D)** Changes were quantified by ratio of percentage of Phalloidin staining to percentage of DAPI staining that is proportional to cell size. **(E,F)** Murine microglia were treated for 24 h with conditioned media [glioma-conditioned medium (GCM)] from human U87-MG, LN18, and IPIN glioma cells or LPS (100 ng/mL), incubated for 3 h with fluorescent beads and subsequently analyzed by flow cytometry. Phagocytosis of fluorescent beads in microglia is represented as mean fluorescence intensity (MFI); graph shows statistically significant groups **(F)** using one-way ANOVA with Dunnett's multiple comparisons test; $n \geq 3$. **(G)** Matrigel assay was performed to determine invasion of human U87-MG glioma cells in the presence of different microglial (BV2—immortalized murine microglia, Mu microglia—primary microglial cultures, SV40 immortalized human microglia) or non-microglial cells (NHA—normal human astrocytes, HEP293 cells, and NIH3T3 fibroblasts). Data are calculated as fold change in relation to basal invasion in the absence of microglia. Matrigel invasion data are calculated as means ± SD, $n \geq 3$ and were analyzed by one-sided paired sample *t*-test; $n \geq 3$. Differences at $p < 0.05$ were considered significant (***$p < 0.001$, **$p < 0.01$, and *$p < 0.05$). **(H)** Invasiveness of various human glioma cells (IPIN and WG4 are primary patient-derived glioma cultures) is increased to different extent in the presence of BV2 murine microglial cells.

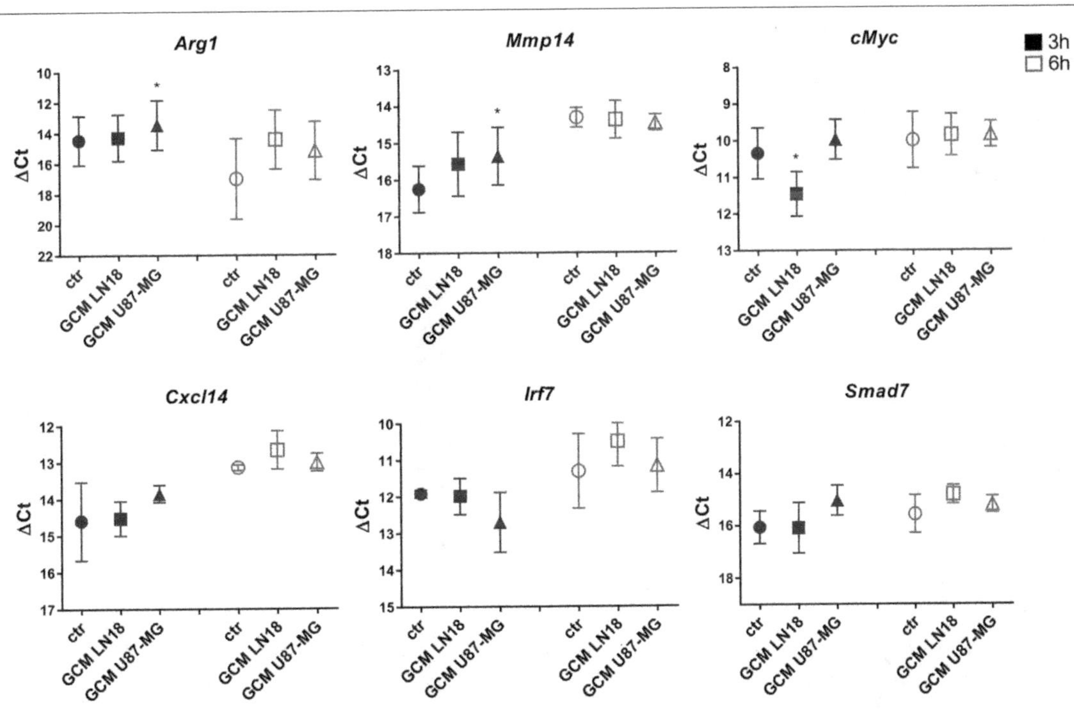

FIGURE 2 | The expression of selected genes in glioma-conditioned medium (GCM) stimulated primary murine microglia cultures. Gene expression was determined by qRT-PCR in microglial cultures left untreated (circles), treated with GCM from LN18 (squares), or U87-MG (triangles) for 3 h (black) and 6 h (white). Data are shown as delta Ct values relative to *18S* expression. Assessment of statistical significance was performed using one-way ANOVA test, followed by Dunnett's multiple comparison test. The results are calculated as means ± SD, $n \geq 3$. Differences at $p < 0.05$ were considered as significant (***$p < 0.001$, **$p < 0.01$, and *$p < 0.05$).

to be significantly differentially expressed, with 20 genes common in both groups. There were 15 uniquely expressed genes in GCM LN18 and only 1 in GCM U87-MG (*Acss1*) stimulated microglia. Overall, the expression of genes was similar in LN18 and U87 GCM-treated microglia (**Figure 4B**). We performed validation of selected genes by qPCR and found five of six upregulated by GCM from two cell lines (Figure S2 in Supplementary Material).

Genes upregulated in both groups are associated with immune response, chemotaxis, and signal transduction regulation as defined by Gene Ontology (GO) functions (**Figure 4C**). Interestingly, we found two upregulated genes that were related to transforming growth factor beta (TGFβ) signaling. Western blot analysis for a phosphorylated Smad2, an active component of TGFβ signaling pathway, was performed on total extracts from microglia polarized with GCM from glioma cells. We found the increased levels of active, phosphorylated Smad2 proteins accumulating at 6 h after GCM treatment (Figure S3 in Supplementary Material).

Tgm2 and *Gpnmb* Are Universal Markers of GAMs

Our results show surprisingly low commonality in transcriptomic responses in different models of glioma-microglia interactions. Therefore, we performed reassessment of publicly available datasets for genome-wide analysis of gene expression in GAMs isolated from mouse (5) and rat gliomas (6), and compared the results to human GAMs data sets (2, 3).

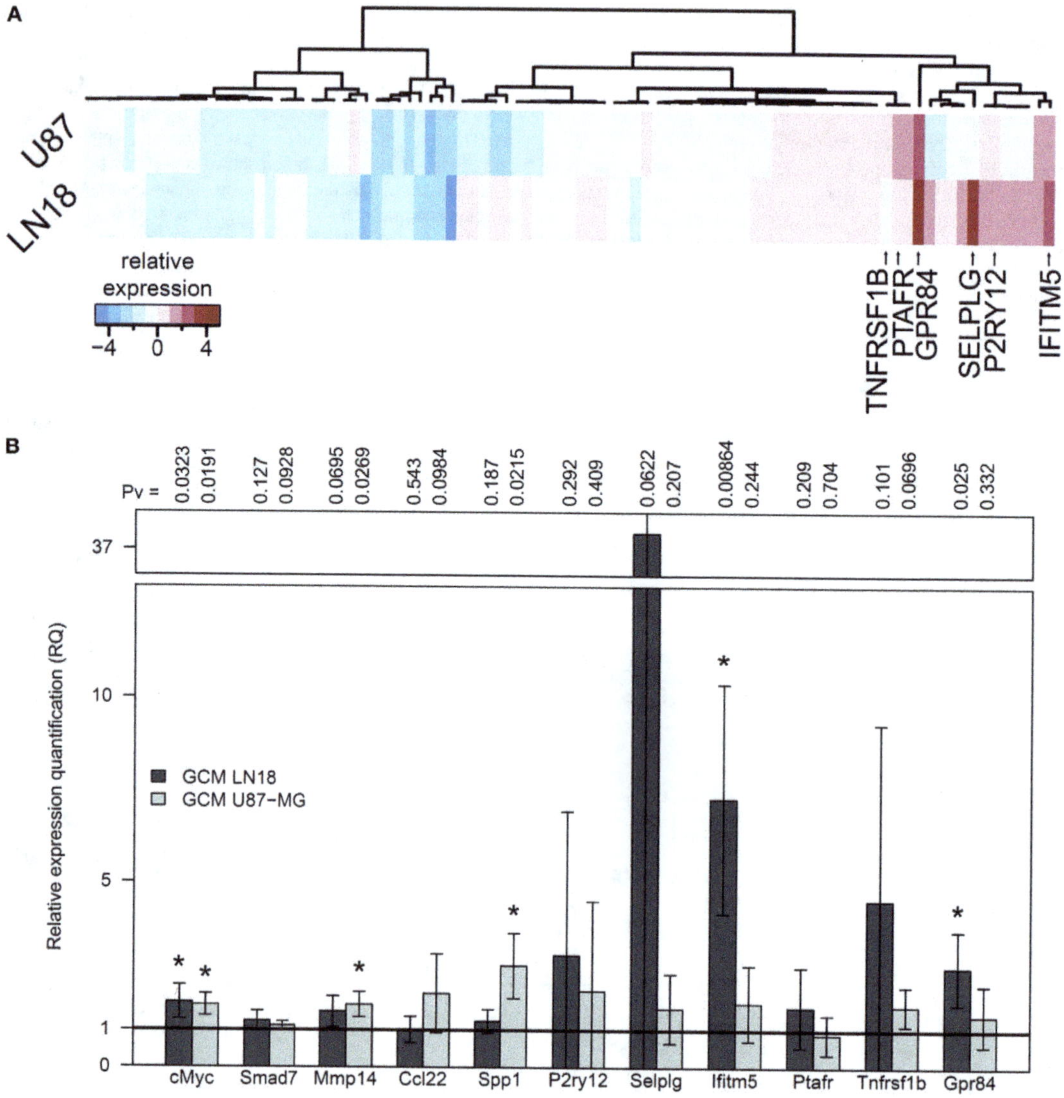

FIGURE 3 | Gene expression in human SV40 microglia polarized by factors secreted by human glioma cells. **(A)** Human Microglial Sensome Gene expression profiling in human SV40 microglia treated by glioma-conditioned medium (GCM) from human LN18 and U87-MG glioma cells was determined with GeneQuery™ Human Microglial Sensome qPCR Array 6 h after GCM addition. The analysis was performed on assorted samples from three replicates for each group (untreated, stimulated with GCM U87-MG or LN18). Heatmap was generated to represent differentially expressed genes and samples. **(B)** The expression of selected genes (*cMYC, SMAD7, MMP14, CCL22,* and *SPP1*) was validated by qPCR in independent cultures of human SV40 microglia treated with GCM from U87-MG or LN18 cells for 6 h. The results are shown as relative quantification (RQ) of GCM stimulated samples compared with untreated controls. Assessment of statistical significance was performed using one-sample t-test. The results are calculated as means ± SD, $n \geq 3$. Differences at $p < 0.05$ were considered as significant (***$p < 0.001$, **$p < 0.01$, and *$p < 0.05$).

Although we found hundreds of upregulated genes in either mouse or rat GAMs (**Figure 5**), the correlation analysis computed using expression changes of orthologs in GAMs versus control microglia did not point to any model which more closely mimics gene expression levels in human GAMs (Figure S4 in Supplementary Material). Nevertheless, we observed that expression changes are more prominent in a mouse model (**Figures 5A,B**). We identified 890 genes with higher expression in mouse GAMs, when compared with control microglia samples, and 621 of such genes in rat GAMs; 277 of the identified genes were common for both models (**Figure 5C**). GO analysis revealed that out four of five top enriched GO terms are common for mouse and rat models (**Figures 5D,E**). Interestingly, GO terms identified only for mouse model were connected mostly with cellular responses, while GO terms identified in a rat model were mainly connected with chromatin and cell cycle. In summary, these results show that despite some similarities, the mouse and rat models represent different pathways of GAM activation.

To determine whether any universal GAM marker might be identified, we set up a comparative analysis of four publically available data sets: (a) MACS sorted $CD14^{+}$ from human GBM

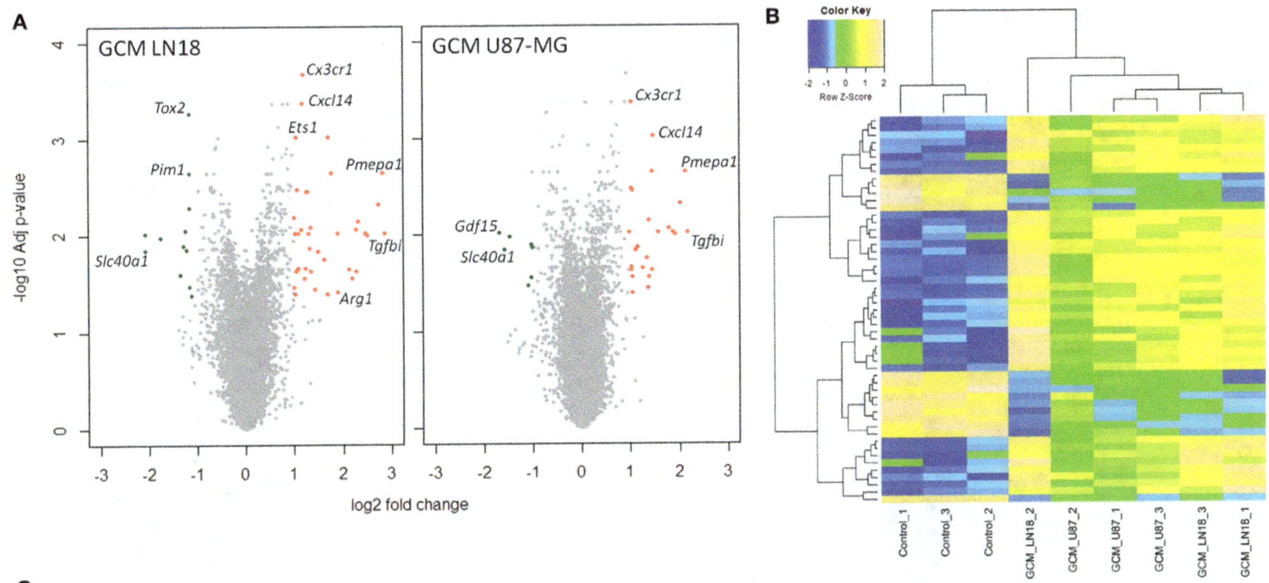

C

Gene	GCM LN18 (FC)	GCM U87-MG (FC)	Adj p.val	GO biological function
Cxcl14	2.3	2.0	4.34E-04	immune response, cell chemotaxis
Olfml3	3.2	2.8	9.77E-04	multicellular organismal development
Pmepa1	7.0	4.3	2.28E-03	negative regulation of signal transduction, negative regulation of transforming growth factor beta receptor signaling pathway
Ctla2a/b	3.5	2.8	2.28E-03	regulation of regulatory T cell differentiation, negative regulation of inflammatory response
Lrrc16a	4.9	3.5	7.34E-03	positive regulation of cell migration, positive regulation of actin filament polymerization
Tgfbi	4.9	3.5	8.73E-03	cell adhesion, extracellular matrix organization, chondrocyte differentiation, angiogenesis
Ch25h	2.1	2.0	2.23E-02	lipid metabolic process, oxidation-reduction process
Tgm2	4.3	2.5	2.26E-02	positive regulation of I-κB kinase/NF-κB signaling , positive regulation of inflammatory response , apoptotic cell clearance
Gatm	2.5	2.0	2.38E-02	creatine biosynthetic process, tissue regeneration, response to oxidative stress

FIGURE 4 | Global gene analysis in primary microglia cultures stimulated with glioma-conditioned medium (GCM) from human glioma cells. Primary microglia cultures were treated for 6 h with GCM. Total RNAs were isolated and subjected to Affymetrix microarray analysis. Data were pre-processed as described in Section "Materials and Methods" and significantly regulated genes are depicted. **(A)** Volcano plots represent log2 fold change versus −log10 adjusted *p*-value in microglia treated with GCM from LN18 and U87-MG glioma cells; upregulated genes are marked in red; downregulated genes in green. **(B)** Heatmap was generated to show significantly differentially expressed genes and sample clustering. **(C)** Selected top significantly upregulated genes in microglia treated with GCM LN18 or U87-MG were listed in a table with fold change expression (FC), adjusted *p*-value, and associated Gene Ontology (GO) biological functions. Significance was assigned to adj *p*-val < 0.05.

versus postmortem tissue and low grade gliomas profiled with microarrays (2); (b) MACS sorted CD11b$^+$ from human GBM versus postmortem and epilepsy tissue analyzed by RNA-seq (3); (c) MACS sorted CD11b$^+$ from murine GL261 GBM profiled with microarrays (5); (d) FACS sorted CD11b$^+$/CD45^{low} from rat C6 gliomas profiled with microarrays (6). Those data were compared with the *in vitro* model presented here—primary murine microglia cultures stimulated with LN18 or U87-MG GCM and profiled with microarrays.

We have found that across all data sets, 129 genes are significantly upregulated (inclusion criteria $p < 0.05$, FC > 2) in at least 2 models, whereas only 4 genes were found to be upregulated in 3 models, and none appeared in 4 or more data sets (**Table 1**). Downregulated genes (inclusion criteria $p < 0.05$, FC < −2) showed nearly no overlap—only one gene was shared between two data sets. The lack of commonly downregulated genes could be caused by very low number of downregulated genes compared with upregulated across investigated data sets (27 down and 17 up in human CD14$^+$; 43 down and 292 up in human CD11b$^+$, 78 down and 539 up in murine CD11b$^+$; 36 down and 287 up in rat CD11b$^+$CD45^{low}). Therefore, in further analysis, we focused on the upregulated genes.

Interestingly, all of the four genes overexpressed in three data sets—*Tgm2, Gpnmb, Il2ra,* and *Ccr7* (**Table 1**), encode proteins that are expressed at the plasma membrane, which makes them good marker candidates. Transglutaminase 2 (Tgm2) was upregulated in human CD11b$^+$, murine CD11b$^+$, and murine microglia stimulated with LN18 GCM. Rat microglia (CD11b$^+$CD45^{low}) also expressed *Tgm2* at high level

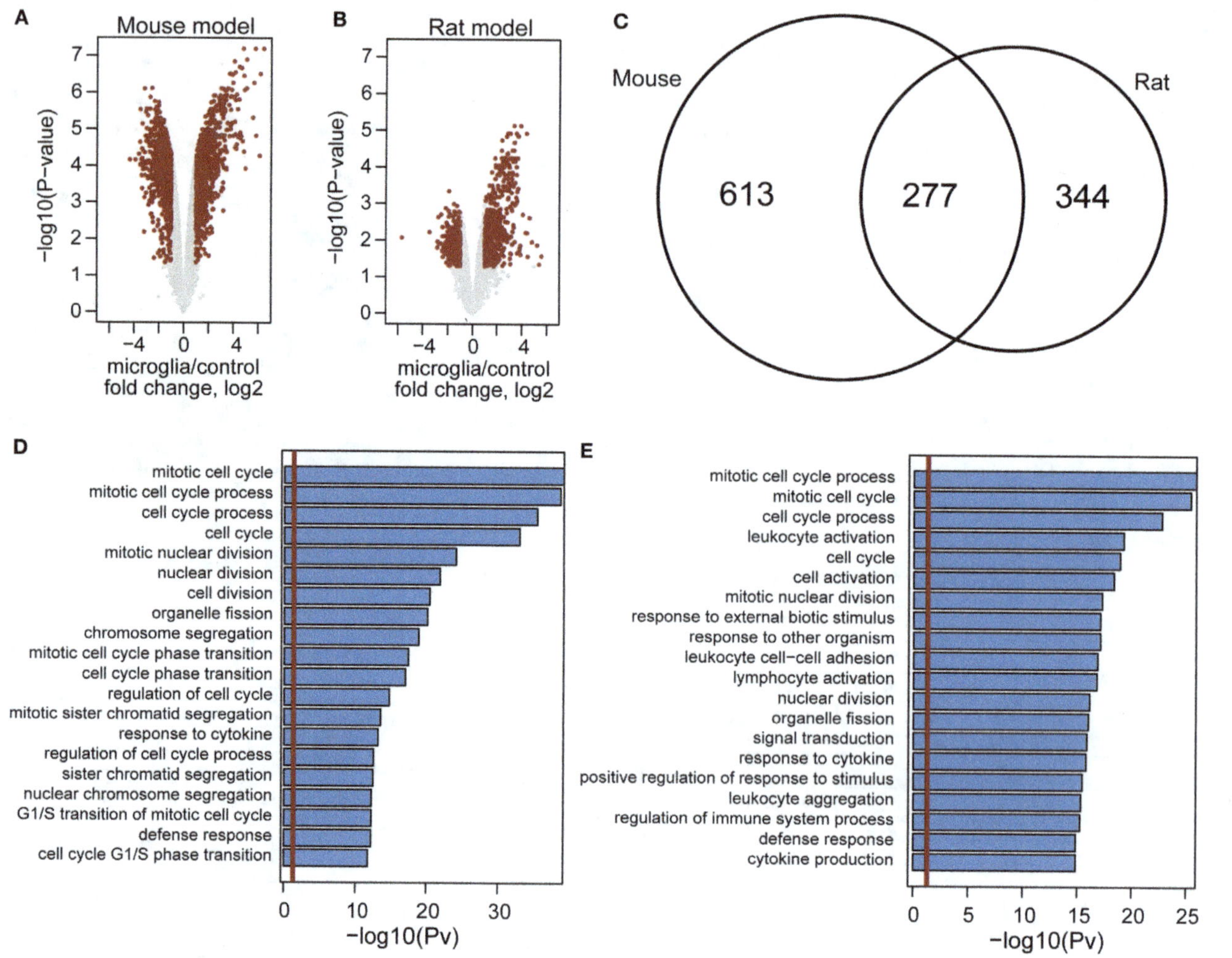

FIGURE 5 | Genome-wide comparison of gene expression in microglia in mouse and rat glioma models. Volcano plot of gene expression changes in glioma-associated microglia/macrophages from mouse **(A)** and rat **(B)** models. The data for all genes are plotted as log2 fold change between microglia and control samples versus the −log10 of the corresponding adjusted *p*-values. Genes selected as significantly different are highlighted as red dots. Venn diagram for overlap of genes upregulated in mouse (left) and rat (right) model **(C)**. Enrichment analysis of Gene Ontology terms within mouse **(D)** and rat **(E)** upregulated genes. Bar plots represent the statistical significance of the enrichment [−log10 (*p*-val)] **(D,E)**.

(FC = 8.12), although the upregulation was not significant ($p = 0.098$). Tgm2 is a cross-linking enzyme involved in matrix stabilization, which is necessary for effective phagocytosis of apoptotic cells by macrophages (20). The role of Tgm2 was also shown in tumorigenesis, as it is highly expressed by $CD44^{high}$ (stem cells marker) glioma tissue and Tgm2 inhibition lead to decrease of $CD44^{high}$ cells (21). In turn, upregulation of *Gpnmb* was found for all three investigated species (murine and human $CD11b^+$, and rat $CD11b^+CD45^{low}$). The importance of transmembrane glycoprotein NMB (Gpnmb) was already implied by Szulzewsky et al. (5), who showed its overexpression in mouse and human $CD11b^+$ GAMs, and demonstrated an association between increased *Gpnmb* expression and poor patient prognosis.

Interleukins and chemokines are cytokine families frequently overexpressed by GAMs in various glioma models. Nevertheless, we have found that only one member of each family—interleukin 2 receptor alpha (Il2ra) and C-C chemokine receptor type 7 (Ccr7) is overexpressed in three of five analyzed models. Importantly, in contrast to *Tgm2* and *Gpnmb*, overexpression of *Il2ra* and *Ccr7* was found only in human and murine GAMs and did not occur in either a rat experimental model or an *in vitro* model.

Enriched Functional Groups in GAMs

In addition to screening for a specific GAM marker, we performed a search for enriched GO biological processes. Enriched GO terms that contained genes represented in all data sets are presented in **Table 2** (for complete functional analysis results see Table S2 in Supplementary Material). Notably, out of 129 genes upregulated in at least 2 data sets, 68 genes were shared between human $CD11b^+$ and rat $CD11b^+CD45^{low}$ data sets, and only 27 and 22 genes between human $CD11b^+$ and murine $CD11b^+$, and rat $CD11b^+CD45^{low}$ and murine $CD11b^+$ data sets,

TABLE 1 | Top markers of glioma infiltrating microglia/macrophages.

	Human CD14$^+$		Human CD11b$^+$		Murine CD11b$^+$		Rat CD11b$^+$CD45^{low}		Primary murine microglia stimulated with			
									LN18 glioma-conditioned medium (GCM)		U87 GCM	
Gene ID	Gabrusiewicz et al. (2)		Szulzewsky et al. (3)		Szulzewsky et al. (5)		Gieryng et al. (6)		Walentynowicz et al.			
	FC	adj. *p*-value	FC	adj. *p*-value	FC	adj. *p*-value	FC	adj. *p*-value	FC	adj. *p*-value	FC	adj. *p*-value
Tgm2	–	–	**6.50**	4.60E−02	**6.94**	8.33E−04	**8.12**	9.80E−02*	**4.29**	2.60E−02	**2.46**	2.60E−02
Gpnmb	–	–	**8.71**	1.22E−06	**36.02**	2.61E−05	**11.54**	4.16E−02	–	–	–	–
Il2ra/m	Upregulated		**4.44**	1.79E−07	**5.24**	2.77E−05						
Ccr7	Upregulated		**2.83**	3.05E−02	**3.75**	2.29E−03	–	–	–	–	–	–

Only genes that were upregulated in the minimum of three models were included.
**Raw data for Gabrusiewicz et al. (2) were not available, thus the gene expression is indicated based on a heatmap from Figure 6 (2).*

respectively. In addition, the most enriched GO term—mitotic cell cycle—contains mainly genes from the human CD11b$^+$ and rat CD11b$^+$CD45^{low} data sets.

This result is not biased by the size of data sets, since comparable numbers of genes were included for the analysis of human CD11b$^+$ and rat CD11b$^+$CD45^{low}, whereas the murine CD11b$^+$ data set was even bigger (**Table 2**). Therefore, we provide further evidence that the immune infiltrates of human GBMs share more similarities with rat C6 gliomas than with murine GL261 gliomas (6).

Gene Ontology term "Regulation of cell adhesion" consisted of genes which roles in the glioma progression were widely studied: *Il10, Mmp14,* and *Spp1*. Interestingly, upregulation of all of the three genes was found only for the murine CD11b$^+$ GAMs, *IL10* and *SPP1* were also found in CD14$^+$ human GAMs and *MMP14* in CD11b$^+$ human GAMs. In rat, CD11b$^+$CD45^{low} *Spp1* was also upregulated; however, the increase did not cross the significance threshold (logFC = 5.39, *p*-adj = 0.074), *Il10* and *Mmp14* were also represented in the rat data set, but no change was noted for those genes (Il10 logFC = 0.98, *p*-adj = 0.536; Mmp14 logFC = 1.27, *p*-adj = 0.03).

Other less known genes (*Lgals1/3, Ctla2a, Il2rn, Il1ra,* and *Prdm1*) play a role in T-cell activity and immune response regulation. Cytotoxic T-lymphocyte-associated protein 2 alpha (Ctl2a) is expressed by activated T-cells; however, its function has not been widely studied (22). Galectins are family of proteins regulating T-cell death, where Galectin 1 (Lgals1) is a strong inducer of T-cell apoptosis, and Galectin 3 (Lgals3) either blocks or induces T-cell death depending whether it occurs intra- or extracellularly (23). *Il1rn* is a gene encoding interleukin 1 receptor antagonist, which functions as an inhibitor of the inflammatory response mediators IL1α and IL1β, whereas *Il2ra* encodes an alpha chain of IL2 receptor and its homodimer form a low-affinity receptor. Positive regulatory domain I-binding factor 1 (Prdm1) in turn, is a transcription factor critical for production of IL-10 cytokine, which expression is restricted to antigen-exposed CD4$^+$ and CD8$^+$ T-cells (24, 25).

Six genes of chemokine family were found in "Cell chemotaxis" GO category; however, this group appeared to be relatively model specific. Human and murine GAMs (CD11b$^+$) shared upregulation of *Ccr7* and *Cxcl2,* murine and rat GAMs shared only *Cxcl10* overexpression. Chemokines unique for each model were human CD14b$^+$—*CCL13*; human CD11b$^+$—*CCL20*; murine CD11b$^+$—*Ccl8, Ccl11, Ccl17, Cxcl16, Cxcl4*; rat CD11b$^+$CD45^{low}—*Cxcl13*. Among genes of GO terms "Regulation of response to external stimulus" and "Blood vessel morphogenesis," we found *Ptgs2* encoding prostaglandin endoperoxide synthase 2 (COX-2), an enzyme producing inflammatory prostaglandins. *PTGS2* is overexpressed in many cancers and was shown to play role in enhancing tumor progression and angiogenesis [reviewed in Ref. (34)].

DISCUSSION

The concept of GAMs playing a key role in GBM pathobiology has been verified in recent studies that demonstrated reduction of glioma growth after microglia ablation or pharmacological inhibition (10, 12, 13). We could use this knowledge to steer GAMs from a tumor supportive and anti-inflammatory phenotype toward an antitumor phenotype. However, translation of such results to a clinic has been found difficult as clinical trials with molecules validated in murine models failed to be successful (26). We still need to separate roles of microglia and macrophages and identify signaling pathways underlying their polarization. Use of CD11bCD45 allows for separation of microglia and bone marrow-derived macrophages that infiltrate the tumor mass in rodent models but this cannot be achieved in human glioma samples, as CD11bCD45 discrimination is challenging in flow cytometry analysis.

We intended to understand mechanisms underlying polarization of microglia driven by factors secreted by human glioma cells in a controlled system. Murine microglia exposed to human glioma undergo morphological alterations, increased phagocytosis, and supported glioma invasion (**Figures 1A–H**). Several of those responses were mimicked by immortalized BV2 and human SV40 microglial cells. We found that microglia functional responses and gene expression profiles differ when exposed to different human glioma cells. Many previously described markers did not manifest in murine microglia polarized by human glioma cells. It is worth

TABLE 2 | Co-occurrence of genes from the same functional groups across all data sets.

Data set		Human CD14+	Human CD11b+	Murine CD11b+	Rat CD11b+CD45low
Genes upregulated in minimum 2 sets/total number of upregulated genes in the set		11/17	96/292	60/539	88/287
Enriched Gene Ontology (GO) biological process	All	Gabrusiewicz et al. (2)	Szulzewsky et al. (3)	Szulzewsky et al. (5)	Gieryng et al. (6)
Mitotic cell cycle GO:0000278 adj. *p*-value 9.88×10^{-9}	*Bub1 Aurkb Bub1 Bub1b Ccnb1 Cdc45 Cdc6 Cdca8 Cdk1 Cdkn3 Cdt1 Cenpf Cep55 Chek1 Cit Clspn Dtl E2f7 E2f8 Espl1 Fanci Foxm1 Gpnmb Gpr132 Kif11 Kif18b Kif23 Kif4 Mad2l1 Mki67 Mybl2 Ncapg Ndc80 Nek2 Nuf2 Nusap1 Pim1 Rad51 Ska1 Ska3 Top2a Tpx2 Ube2c*		*Aurkb Bub1 Bub1b Ccnb1 Cdc45 Cdc6 Cdca8 Cdk1 Cdkn3 Cdt1 Cenpf Cep55 Chek1 Cit Clspn Dtl E2f7 E2f8 Espl1 Fanci Foxm1 Gpnmb Kif11 Kif18b Kif23 Mad2l1 Mki67 Mybl2 Ncapg Ndc80 Nek2 Nuf2 Nusap1 Rad51 Ska1 Ska3 Top2a Tpx2 Ube2c*	*Gpnmb Gpr132 Pim1*	*Aurkb Bub1 Bub1b Ccnb1 Cdc45 Cdc6 Cdca8 Cdk1 Cdkn3 Cdt1 Cenpf Cep55 Chek1 Cit Clspn Dtl E2f7 E2f8 Espl1 Fanci Foxm1 Gpnmb Gpr132 Kif11 Kif18b Kif23 Mad2l1 Mki67 Mybl2 Ncapg Ndc80 Nek2 Nuf2 Nusap1 Pim1 Rad51 Ska1 Ska3 Top2a Tpx2 Ube2c*
Regulation of cell adhesion GO:0030155 adj. *p*-value 2.25×10^{-2}	*Prdm1 Ccr7 Ctla2a Has2 Il10 Il1rn Il2ra Lgals1 Lgals3 Mmp14 Nrp1 Plau Ccl5 Spp1 Tgm2 Tnc Vcam1 Gpnmb*	*Ccl5 Ccr7 Il10 Il1rn Il2ra Spp1*	*Ccr7 Gpnmb Has2 Il2ra Lgals1 Lgals3 Mmp14 Nrp1 Plau Tgm2 Tnc*	*Ccl5 Ccr7 Ctla2a Gpnmb Has2 Il10 Il1rn Il2ra Lgals1 Mmp14 Prdm1 Spp1 Tgm2 Tnc Vcam1*	*Gpnmb Lgals3 Nrp1 Plau Prdm1 Vcam1*
Cell chemotaxis GO:0060326 adj. *p*-value 2.50×10^{-2}	*Ccl5 Ccl18 Ccr7 Cxcl10 Cxcl2 Cxcl9 Jaml Lgals3 Nrp1 Pde4b Spp1 Vcam1*	*Ccl5 Ccl18 Ccr7 Cxcl9 Spp1*	*Ccl18 Ccr7 Cxcl2 Lgals3 Nrp1*	*Ccl5 Ccr7 Cxcl10 Cxcl2 Cxcl9 Pde4b Spp1 Vcam1*	*Cxcl10 Lgals3 Nrp1 Pde4b Vcam1*
Regulation of response to external stimulus GO:0032101 adj. *p*-value 2.63×10^{-2}	*Acp5 Aoah Ccl5 Ccr7 Cd109 Ctla2a Cxcl10 Cxcl2 Cxcl9 Gbp4 Il10 Il2ra Nrp1 Plau Prdm1 Ptgs2 Tgm2 Usp18*	*Ccl5 Ccr7 Cxcl9 Il10 Il2ra Ptgs2*	*Acp5 Ccr7 Cd109 Cxcl2 Il2ra Nrp1 Plau Tgm2*	*Acp5 Ccl5 Ccr7 Cd109 Ctla2a Cxcl10 Cxcl2 Cxcl9 Gbp4 Il10 Il2ra Prdm1 Ptgs2 Tgm2 Usp18*	*Aoah Cxcl10 Gbp4 Nrp1 Plau Prdm1 Usp18*
Blood vessel morphogenesis GO:0048514 adj. *p*-value 5.44×10^{-2}	*Ahr Ccl5 Clic4 Cxcl10 E2f7 E2f8 Foxm1 Has2 Hif1a Lgals3 Nrp1 Plau Prdm1 Prrx1 Ptgs*	*Ccl5 Hif1a Ptgs2*	*Ahr E2f7 E2f8 Foxm1 Has2 Lgals3 Nrp1 Plau Prrx1*	*Ahr Ccl5 Clic4 Cxcl10 Has2 Hif1a Prdm1 Prrx1 Ptgs2*	*Clic4 Cxcl10 E2f7 E2f8 Foxm1 Lgals3 Nrp1 Plau Prdm1*

Only genes that were upregulated in at least two data sets are included.

noting that in this study microglial cells were treated for several hours to define early responses and this could be partially responsible for small changes in the pre-defined markers. Nonetheless, activated pathways and biological processes show similarities in microglial responses. One of most established markers of polarized microglia in glioma is Arginase 1 (*Arg1*). *Arg1* reduces levels of L-arginine that is essential for normal T-cell function. The increased Arg1 expression was associated with the immunosuppressive phenotype in microglia/macrophages in rat and mice glioma models (10, 15, 27). A recent study reported that inhibition of *Arg1* with the specific inhibitor in murine gliomas leads to the improved antitumor immune response and reduced tumor size (28). We report the increased *Arg1* expression in murine microglia exposed to human glioma GCM, suggesting the activation of immune-suppressive phenotype in these cells. Interestingly, *Arg1* expression is not significantly upregulated in published data sets of GAMs.

Nonetheless, global gene expression analysis revealed genes associated with immune response and immunosuppression (*Pmepa1*, *Ctla2a/b*, *Larc15a*, *Tgfbi*, and *Tgm2*). Activation of TGFβ signaling pathway has been corroborated by Western blot results and detection of the increased levels of phosphorylated Smad2 in microglia after treatment with GCM from LN18 and U87-MG cultures for 6 h. Signaling associated with TGFβ pathway and increased *Tgfbi* expression have been reported in microglia from GL261 gliomas (5). *Tgm2* expression is also induced by TGFβ pathway, participates in immune responses, and contributes to functions of monocytes/macrophages from migration to differentiation (29).

Interestingly, while GCM from all tested glioma cell lines induced the proinvasive activation of microglia, and in case of LN18 and U87MG increases in gene expression, GCM from U87-MG was particularly potent in induction of phagocytosis, morphological transformation, and tumor invasion support. All those cellular events are regulated in microglia by integrin signaling and mediated by activation of focal adhesion kinase/FAK-PI3K-Akt signaling. We have previously reported that tumor derived osteopontin, a ligand of integrins αvβ3, induces FAK-PI3K-Akt signaling in rat microglia (16). In fact, we found that U87-MG cells express and secrete larger quantities of osteopontin than other glioma cells (29), so it may explain the strongest effects of GCM from U87-MG cells on integrin signaling related events in microglia.

A Critical Assessment of Gene Expression Markers of Glioma-Induced Microglia Polarization

Classification of macrophage functions according to M1 and M2 phenotypes have been based on *in vitro* studies (30), but those phenotypes failed to be reproduced *in vivo*. It has been suggested that GAMs do not undergo an M2 polarization, but rather present a mixture of M1, and M2a,b,c-specific genes (5).

Our comparative analysis of GAM transcriptomics across different *in vivo* models—human, mouse, and rat, failed to reproduce consistent microglia phenotypes that could be classified according to previously reported gene signatures and showed remarkably low similarity between models. A mouse GL261 model is most commonly used in studies on glioma pathobiology, and thus, it is best characterized. However, our results show a stronger similarity of human GAMs gene expression profiles with transcriptomes of microglia infiltrating rat C6 gliomas, than murine GL261 gliomas (Figure S4 in Supplementary Material; **Table 2**). Both rat C6 and murine GL261 gliomas are cell line allografts that allow recapitulating the immune system response (31, 32), which might be limited in human xenograft models utilizing immunodeficient animals. C6 cell line produces tumors in outbred Wistars rat and has been reported to mimic best the gene expression changes of human glioma, among other rat cell lines (33). On the other hand, mouse GL261 cells form aggressive tumors in syngeneic C57BL/6 strains and present several molecular alterations that are commonly found in human gliomas: *p53* and *K-ras* mutations, MHCI, and MHCII downregulation (32).

The discrepancies and the low commonality in presented data sets might stem from different sources. The analyzed data sets were collected by different groups, using different methodologies (RNA-seq, microarrays) and different platforms for high-throughput transcriptome analysis, which could lead to an increased rate of false negatives and did not allow us to detect less pronounced effects. Even more important constrain which may hamper our analyses stems from methodological differences in GAM isolation, namely, FACS-based immunosorting versus immunomagnetic beads which may result in different degree of sample purity. Magnetic-activated cell sorting is thought to exert less shear stress on sorted cells than FACS. However, FACS offers a possibility to separate cells based on several markers simultaneously, allows for adjustments in gating strategy, and yields better purity.

Analyzed data sets were collected utilizing distinct markers: CD14 or CD11b or CD11bCD45$^{low/high}$ cells, which may result in different subpopulations with discrete functions. Unification of protocols, better separation of different subpopulations, or single-cell sequencing may bring more conclusive results.

Summarizing, despite of all difficulties we have identified two gene candidates for an universal GAM marker in experimental gliomas—*Tgm2* and *Gpnmb*, which should be further investigated. In addition, our functional analysis shows that pathways regulating the similar cell functions are upregulated across all models. It suggests that studies on universal mechanisms orchestrating an adaptation of the tumor-supporting phenotype by microglia and macrophages should be focused on functional groups rather than on individual genes.

AUTHOR CONTRIBUTIONS

KWalentynowicz and BK design the study, interpreted data, and wrote the manuscript. KWalentynowicz performed most of functional and microarray experiments. NO performed transcriptomic data analyses on public data sets. MP and KWojnicki performed invasion and SqPCR assays. KS performed microarray experiments. JM performed computational analyses. IC developed primary GBM cell cultures. All the authors wrote the manuscript.

SUPPLEMENTARY MATERIAL

FIGURE S1 | Matrigel assay was performed to determine invasion of different human glioma cells in the presence of SV40 immortalized human microglia. Data are calculated as fold change in relation to basal invasion in the absence of microglia. Matrigel invasion data are calculated as means ± SD, $n = 3$ and were analyzed by one-sided paired sample *t*-test; $n \geq 3$. Differences at $p < 0.05$ were considered as significant ($^{***}p < 0.001$, $^{**}p < 0.01$, and $^{*}p < 0.05$).

FIGURE S2 | Microarray gene expression validation. Five genes were selected for validation of the microarray gene expression analysis (*Tgm2, Cx3cr1, Tgfbi, Tmem37*, and *Socs2*). Expression is presented as delta Ct values in control, glioma-conditioned medium (GCM) LN18, and GCM U87-MG treated for 6 h. Validation was performed on separate experiments $n = 4$; $^{**}p < 0.001$, $^{*}p < 0.05$.

FIGURE S3 | Western blot analysis of total extracts from microglia polarized with glioma cells for active components of transforming growth factor beta signaling pathway revealed the increased levels of active, phosphorylated SMAD2 proteins accumulating at 3 and 6 h after glioma-conditioned medium (GCM) treatment. Immunoblots were stripped and re-probed first with an antibody recognizing total SMAD2 followed by detection of actin (used as a loading control).

FIGURE S4 | Correlation of gene expression patterns in various datasets. A heatmap shows Spearman's correlation coefficients between microglia activation (log2 fold changes) for samples collected from human, mouse, and rat samples.

REFERENCES

Komohara Y, Ohnishi K, Kuratsu J, Takeya M. Possible involvement of the M2 anti-inflammatory macrophage phenotype in growth of human gliomas. *J Pathol* (2008) 216:15–24. doi:10.1002/path.2370

Gabrusiewicz K, Rodriguez B, Wei J, Hashimoto Y, Healy LM, Maiti SN, et al. Glioblastoma-infiltrated innate immune cells resemble M0 macrophage phenotype. *JCI Insight* (2016) 1:1–32. doi:10.1172/jci.insight.85841

Szulzewsky F, Arora S, de Witte L, Ulas T, Markovic D, Schultze JL, et al. Human glioblastoma-associated microglia/monocytes express a distinct RNA profile compared to human control and murine samples. *Glia* (2016) 64:1416–36. doi:10.1002/glia.23014

Badie B, Schartner JM. Flow cytometric characterization of tumor-associated macrophages in experimental gliomas. *Neurosurgery* (2000) 46:957–62. doi:10.1227/00006123-200004000-00035

Szulzewsky F, Pelz A, Feng X, Synowitz M, Markovic D, Langmann T, et al. Glioma-associated microglia/macrophages display an expression profile different from

M1 and M2 polarization and highly express Gpnmb and Spp1. *PLoS One* (2015) 10(2):e0116644. doi:10.1371/journal.pone.0116644

Gieryng A, Pszczolkowska D, Bocian K, Dabrowski M, Rajan WD, Kloss M, etal. Immune microenvironment of experimental rat C6 gliomas resembles human glioblastomas. *Sci Rep* (2017) 7:17556. doi:10.1038/s41598-017-17752-w

SasakiA,YokooH,TanakaY,HommaT,NakazatoY,OhgakiH.Characterization of microglia/macrophages in gliomas developed in S-100β-v-erbB transgenic rats. *Neuropathology* (2013) 33:505–14. doi:10.1111/neup.12015

Hambardzumyan D, Gutmann DH, Kettenmann H. The role of microglia and macrophages in glioma maintenance and progression. *Nat Neurosci* (2016) 19:20–7. doi:10.1038/nn.4185

Gieryng A, Pszczolkowska D, Walentynowicz KA, Rajan WD, Kaminska B. Immune microenvironment of gliomas. *Lab Investig* (2017) 97:498–518. doi:10.1038/labinvest.2017.19

Gabrusiewicz K, Ellert-Miklaszewska A, Lipko M, Sielska M, Frankowska M, Kaminska B. Characteristics of the alternative phenotype of microglia/macrophages and its modulation in experimental gliomas. *PLoS One* (2011) 6:e23902. doi:10.1371/journal.pone.0023902

Sielska M, Przanowski P, Wylot B, Gabrusiewicz K, Maleszewska M, Kijewska M, et al. Distinct roles of CSF family cytokines in macrophage infiltration and activation in glioma progression and injury response. *J Pathol* (2013) 230:310–21. doi:10.1002/path.4192

Markovic DS, Vinnakota K, van Rooijen N, Kiwit J, Synowitz M, Glass R, et al. Minocycline reduces glioma expansion and invasion by attenuating microglial MT1-MMP expression. *Brain Behav Immun* (2011) 25:624–8. doi:10.1016/j.bbi.2011.01.015

Xu S, Wei J, Wang F, Kong LY, Ling XY, Nduom E, et al. Effect of miR-142-3p on the M2 macrophage and therapeutic efficacy against murine glioblastoma. *J Natl Cancer Inst* (2014) 106. doi:10.1093/jnci/dju162

Murray PJ, Allen JE, Biswas SK, Fisher EA, Gilroy DW, Goerdt S, et al. Macrophage activation and polarization: nomenclature and experimental guidelines. *Immunity* (2014) 41:14–20. doi:10.1016/j.immuni.2014.06.008

Ellert-Miklaszewska A, Dabrowski M, Lipko M, Sliwa M, Maleszewska M, Kaminska B. Molecular definition of the pro-tumorigenic phenotype of glioma-activated microglia. *Glia* (2013) 61:1178–90. doi:10.1002/glia.22510

Ellert-Miklaszewska A, Wisniewski P, Kijewska M, Gajdanowicz P, Pszczolkowska D, Przanowski P, et al. Tumour-processed osteopontin and lactadherin drive the protumorigenic reprogramming of microglia and glioma progression. *Oncogene* (2016) 35(50):6366–77. doi:10.1038/ onc.2016.55

Ciechomska IA, Przanowski P, Jackl J, Wojtas B, Kaminska B. BIX01294, an inhibitor of histone methyltransferase, induces autophagy-dependent differentiation of glioma stem-like cells. *Sci Rep* (2016) 6:38723. doi:10.1038/srep38723

Sliwa M, Markovic D, Gabrusiewicz K, Synowitz M, Glass R, Zawadzka M, et al. The invasion promoting effect of microglia on glioblastoma cells is inhibited by cyclosporin A. *Brain* (2007) 130:476–89. doi:10.1093/brain/ awl263

Wesolowska A, Kwiatkowska A, Slomnicki L, Dembinski M, Master A, Sliwa M, et al. Microglia-derived TGF-b as an important regulator of glioblastoma invasion—an inhibition of TGF-b-dependent effects by shRNA against human TGF-b type II receptor. *Oncogene* (2008) 27:918–30. doi:10.1038/ sj.onc.1210683

Szondy Z, Sarang Z, Molnar P, Nemeth T, Piacentini M, Mastroberardino PG, et al. Transglutaminase 2-/- mice reveal a phagocytosis-associated crosstalk between macrophages and apoptotic cells. *Proc Natl Acad Sci U S A* (2003) 100:7812–7. doi:10.1073/pnas.0832466100

Fu J, Yang QY, Sai K, Chen FR, Pang JCS, Ng HK, et al. TGM2 inhibition attenuates ID1 expression in CD44-high glioma-initiating cells. *Neuro Oncol* (2013) 15:1353–65. doi:10.1093/neuonc/not079

Denizot F, Brunet JF, Roustan P, Harper K, Suzan M, Luciani MF, et al. Novel structures CTLA-2α and CTLA-2β expressed in mouse activated T cells and mast cells and homologous to cysteine proteinase proregions. *Eur J Immunol* (1989) 19:631–5. doi:10.1002/eji.1830190409

Stillman BN, Hsu DK, Pang M, Brewer CF, Johnson P, Liu F-T, et al. Galectin-3 and galectin-1 bind distinct cell surface glycoprotein recep- tors to induce T cell death. *J Immunol* (2006) 176:778–89. doi:10.4049/ jimmunol.176.2.778

Neumann C, Heinrich F, Neumann K, Junghans V, Mashreghi M-F, Ahlers J, et al. Role of Blimp-1 in programing Th effector cells into IL-10 producers. *J Exp Med* (2014) 211:1807–19. doi:10.1084/jem.20131548

Kallies A, Hawkins ED, Belz GT, Metcalf D, Hommel M, Corcoran LM, et al. Transcriptional repressor Blimp-1 is essential for T cell homeostasis and self-tolerance. *Nat Immunol* (2006) 7:466–74. doi:10.1038/ni1321

Butowski N, Colman H, De Groot JF, Omuro AM, Nayak L, Wen PY, et al. Orally administered colony stimulating factor 1 receptor inhibitor PLX3397 in recurrent glioblastoma: an Ivy Foundation Early Phase Clinical Trials Consortium phase II study. *Neuro Oncol* (2016) 18(4):557–64. doi:10.1093/neuonc/nov245

Zhang I, Alizadeh D, Liang J, Zhang L, Gao H, Song Y, et al. Characterization of arginase expression in glioma-associated microglia and macrophages. *PLoS One* (2016) 11(12):e0165118. doi:10.1371/journal.pone.0165118

Steggerda SM, Bennett MK, Chen J, Emberley E, Huang T, Janes JR, et al. Inhibition of arginase by CB-1158 blocks myeloid cell-mediated immune suppression in the tumor microenvironment. *J Immunother cancer* (2017) 5:101. doi:10.1186/s40425-017-0308-4

Chrobok NL, Sestito C, Wilhelmus MMM, Drukarch B, van Dam AM. Is monocyte- and macrophage-derived tissue transglutaminase involved in inflammatory processes? *Amino Acids* (2017) 49:441–52. doi:10.1007/ s00726-016-2334-9

Mills CD, Kincaid K, Alt JM, Heilman MJ, Hill AM. M-1/M-2 macrophages and the Th1/Th2 paradigm. *J Immunol* (2000) 164:6166–73. doi:10.4049/jimmunol.164.12.6166

Barth RF, Kaur B. Rat brain tumor models in experimental neuro-oncology: the C6, 9L, T9, RG2, F98, BT4C, RT-2 and CNS-1 gliomas. *J Neurooncol* (2009) 94:299–312. doi:10.1007/s11060-009-9875-7

Szatmári T, Lumniczky K, Désaknai S, Trajcevski S, Hídvégi EJ, Hamada H, et al. Detailed characterization of the mouse glioma 261 tumor model for experimental glioblastoma therapy. *Cancer Sci* (2006) 97:546–53. doi:10.1111/j.1349-7006.2006.00208.x

Sibenaller ZA, Etame AB, Ali MM, Barua M, Braun TA, Casavant TL, et al. Genetic characterization of commonly used glioma cell lines in the rat animal model system. *Neurosurg Focus* (2005) 19:E1. doi:10.3171/ foc.2005.19.4.2

Ricciotti E, Fitzgerald GA. Prostaglandins and inflammation. *Arterioscler Thromb Vasc Biol* (2011) 31:986–1000. doi:10.1161/ATVBAHA.110.207449

5

It's all about Timing: The Involvement of Kir4.1 Channel Regulation in Acute Ischemic Stroke Pathology

Meagan Milton and Patrice D. Smith*

Department of Neuroscience, Carleton University, Ottawa, ON, Canada

***Correspondence:**
Patrice D. Smith
patrice.smith@carleton.ca

An acute ischemic stroke is characterized by the presence of a blood clot that limits blood flow to the brain resulting in subsequent neuronal loss. Acute stroke threatens neuronal survival, which relies heavily upon proper function of astrocytes. Neurons are more susceptible to cell death when an astrocyte is unable to carry out its normal functions in supporting the neuron in the area affected by the stroke (Rossi et al., 2007; Takano et al., 2009). For example, under normal conditions, astrocytes initially swell in response to changes in extracellular osmotic pressure and then reduce their regulatory volume in response to volume-activated potassium (K^+) and chloride channels (Vella et al., 2015). This astroglial swelling may be overwhelmed, under ischemic conditions, due to the increased levels of glutamate and extracellular K^+ (Lai et al., 2014; Vella et al., 2015). The increase in extracellular K^+ contributes to neuronal damage and loss through the initiation of harmful secondary cascades (Nwaobi et al., 2016). Reducing the amount of extracellular K^+ could, in theory, limit or prevent neuronal damage and loss resulting in an improved prognosis for individuals following ischemic stroke. Kir4.1, an inwardly rectifying K^+ channel, has demonstrated an ability to regulate the rapid reuptake of this ion to return the cell to basal levels allowing it to fire again in rapid transmission (Sibille et al., 2015). Despite growing interest in this area, the underlying mechanism suggesting that neuroprotection could occur through modification of the Kir4.1 channel's activity has yet to be described. The purpose of this review is to examine the current literature and propose potential underlying mechanisms involving Kir4.1, specially the mammalian target of rapamycin (mTOR) and/or autophagic pathways, in the pathogenesis of ischemic stroke. The hope is that this review will instigate further investigation of Kir4.1 as a modulator of stroke pathology.

Keywords: Kir4.1, ischemia, astrocytes, autophagy, mammalian target of rapamycin

GENERAL PROPERTIES OF KIR4.1

Kir4.1, initially named BIR10, was first identified by the Adelman's group (Bond et al., 1994). It is predominately expressed on glial cells and is responsible for developmental regulation of extracellular K^+ dynamics (described below) (D'Adamo et al., 2011, 2013). Research has demonstrated that the channel forms both homomeric Kir4.1-Kir4.1 tetramers and heteromeric Kir4.1-Kir5.1 tetramers (Hibino et al., 2010; D'Adamo et al., 2011, 2013). The two types of tetramers have different sensitivities to pH. Between 6.5 and 8.0, the homomeric Kir4.1 channel is inhibited, whereas the heteromeric Kir4.1-Kir5.1 is suppressed significantly (Pessia et al., 2001; D'Adamo et al., 2011, 2013). In addition, Kir4.1 and Kir5.1 are coexpressed on locus coeruleus neurons where they appear to be involved in neuronal carbon dioxide (CO_2) chemosensitivity (D'Adamo et al., 2011). Taken together, these results suggest that pH, CO_2, and Kir5.1 subunits modulate Kir4.1 activity.

Previous research has demonstrated that Kir4.1 channels are implicated in the pathophysiology of several disease/disorders. Abnormalities, specifically missense variations, within this channel have been linked to epilepsy (Hibino et al., 2010; D'Adamo et al., 2013). For example, a missense variation in Kir4.1 (T262S) was found to be the reason that DBA/2 mice are more susceptible to induced seizures than C57BL/6 mice (Ferraro et al., 2004; D'Adamo et al., 2013). In patients with either focal or generalized epilepsy, the R271C mutation in the Kir4.1 channel has been associated with a resistance to seizures (Ferraro et al., 2004; D'Adamo et al., 2013). Another condition, autism spectrum disorders (ASD), is associated with Kir4.1 channel mutations. In children with ASD, two specific mutations (R18Q and V84M) within the Kir4.1 channel have been identified. R18Q has been associated with several ASD symptoms, such as absence of speech and severe social interaction deficits, whereas the V84M appears to confer an increased expression of poor social gaze and withdrawal behaviors that are also characteristic of ASD (Sicca et al., 2011; Guglielmi et al., 2015). Kir4.1 abnormalities are also believed to underlie the comorbidity between epilepsy and ASD (D'Adamo et al., 2013; Guglielmi et al., 2015). Finally, Kir4.1 activity appears to be altered in response to ischemia (Nwaobi et al., 2016). Research has shown that Kir4.1 expression and Kir-mediated currents are reduced from day 1 to day 14 post-injury following global and focal ischemia (Pivonkova et al., 2010; Steiner et al., 2012). Furthermore, Kir4.1 channels appear more at the soma of the astrocyte rather than on the astrocytic processes when these reductions occur (Stewart et al., 2010; Nwaobi et al., 2016). It has been suggested that this shift changes Kir4.1 focus to proliferation instead of K^+ spatial buffering (Nwaobi et al., 2016).

THE ROLE OF KIR4.1 IN ASTROCYTIC FUNCTIONING

Within the central nervous system, astrocytes are involved in controlling ion and water homeostasis, moving metabolite and waste products, and participating in the formation of the blood-brain barrier (Takano et al., 2009; Nwaobi et al., 2016). These biophysical properties appear to be affected by Kir4.1; specifically, the homeostasis of extracellular K^+, the regulation of extracellular glutamate, and the mediation of water and volume levels are three of the processes that Kir4.1 is involved (Nwaobi et al., 2016). First, K^+ spatial buffering is a process carried out by astrocytes to ensure that extracellular K^+ concentrations are regulated following an action potential (Nwaobi et al., 2016). An action potential may increase local K^+ concentrations by 1 mM under normal neuronal activity or by >10–12 mM under ischemic conditions (Moody et al., 1974; Ransom et al., 2000). The increased K^+ concentrations prevent the neuron from firing in response to further stimulation and in the case of ischemic conditions, contribute to neuronal loss (Nwaobi et al., 2016). Thus, it is important for K^+ to be removed from the extracellular space through a mechanism such as the Kir4.1 channel. This is supported by the fact that at the sites of local accumulation, Kir4.1 has been shown to allow K^+ influx to return the extracellular K^+ concentrations to baseline. Recent work has demonstrated that local extracellular K^+ concentrations were greater following the blockage of Kir4.1 using barium (Ransom et al., 2000; Larsen et al., 2014; Nwaobi et al., 2016). Furthermore, previous work has shown a slower recovery rate of extracellular K^+ concentrations and enhanced undershoot recovery following brainstem stimulation in glial-conditional Kir4.1 knock out animals (Neusch et al., 2006). Taken together, these results implicate the Kir4.1 channel's involvement in extracellular K^+ homeostasis and neuronal survival following ischemia.

Second, astrocytes are involved in glutamate uptake through two transporters, GLAST (EAAT1) and GLT-1 (EAAT2), in an energetically unfavorable process (Rothstein et al., 1996). Sodium, hydrogen and K^+ electrochemical gradients work in concert with the two transporters to bring glutamate into the astrocyte in an efficient manner (Barbour et al., 1988; Nwaobi et al., 2016). With respect to extracellular K^+, high concentrations decrease K^+ unbinding and depolarize the glial membrane reducing the amount of glutamate reuptake resulting in an increase in neuronal excitability (Barbour et al., 1988). The contribution of Kir4.1 to astrocytic glutamate uptake has been previously studied. Researchers found a 33.1 and 57.0 percent decrease in glutamate uptake following pharmacological inhibition and siRNA-mediated Kir4.1 knockdown in cortical astrocytes (Kucheryavykh et al., 2007). The authors attributed these decreases to be the result of a loss in the hyperpolarized resting membrane potential of the astrocyte. In addition, a TBOA (*threo*-beta-benzyloxyaspartate)-sensitive glutamate uptake reduction of >50 percent was seen in Kir4.1 knockout animals when compared to wild-type animals (Nwaobi et al., 2016). The results support Kir4.1's role in glutamate uptake by allowing the astrocyte to maintain a K^+ electrochemical gradient that promotes K^+ unbinding and glial depolarization.

Lastly, a third property of astrocytes that appears to involve Kir4.1 functioning is water and volume regulation. Aquaporins, specifically aquaporin 4 (AQP4), are found at the astroglial endfeet where they are responsible for bringing water into specific cells and removing excess water to alleviate swelling to prevent cell lysis (Vella et al., 2015). Under conditions of water and/or food deprivation, AQP4 has demonstrated an ability to alter its expression levels in order to maintain the brain's normal water content and prevent cell loss (Ye et al., 2016). Studies examining the relationship between Kir4.1 and AQP4 have found that they co-localize with one another. For

---

Abbreviations: AKT, protein kinase B; AMPA, α-amino-3-hydroxy-5-methyl-4-isoxazolepropionic acid; ASD, autism spectrum disorders; ATG, autophagy-related gene; ATP, adenosine triphosphate; AQP4, aquaporin 4; CO_2, carbon dioxide; GLAST, L-glutamate/L-aspartate transporter; GLT-1, glutamate transporter 1; K^+, potassium; KA, kainite receptor; mTOR, mammalian target of rapamycin; mTORC1, mammalian target of rapamycin complex 1; NMDA, *N*-methyl-D-aspartate; PI3K, phosphatidylinositol 3 kinase.

example, one study demonstrated that co-immunoprecipitates between the two were found in Müller cells and that loss or mislocalization of Kir4.1 in the post-ischemic retina played a role in glial cell swelling (Pannicke et al., 2004). However, a follow-up study was able to induce the same cell swelling through the inhibition, using physostigmine, of Kir4.1 to prevent K^+ efflux without changing AQP4 expression or function (Nwaobi et al., 2016). This suggests that changing Kir4.1 channel's activity alone, through the application of different pharmacological agents, could reduce the amount of cell lysis or autophagic cell death seen in ischemic conditions.

In addition, due to the interaction between Kir4.1 and APQ4 channels, an important factor that may modulate Kir4.1 activity is water movement through different compartments. With a specific focus on glutamate, a review published by Rothman and Olney (1986) described how neurotoxicity results in the simultaneous build-up of intracellular sodium, K^+, and water that may lead to lysis/autophagic injury. Glutamate, a neurotransmitter impacted by K^+ gradients and Kir4.1 as described above, acts on many different receptors to exert its effects. The two most important ones for the purpose of this review are α-amino-3-hydroxy-5-methyl-4-isoxazolepropionic acid (AMPA) and kainite (KA) receptors. These two receptors are highly permeable to sodium and K^+ (Chen et al., 1998). Over-activation of these receptors, as well as the *N*-methyl-D-aspartate (NMDA) receptor, results in response to brain ischemia (Rothman and Olney, 1987; Coyle and Puttfarcken, 1993). The over-activation leads to excessive ion influx, osmotic swelling, free radical generation, and cell death (Rothman and Olney, 1987; Coyle and Puttfarcken, 1993). Specifically, Rothman and Olney (1986) found that the recovery of water regulation might also reverse the lysis/autophagic injury. Based on these findings, water accumulation may also be considered a "stressor" that damages organelles. Due to the Kir4.1 channel's interaction with APQ4, and its involvement in water/volume regulation, further studies are required to examine how levels of intracellular K^+ contribute to such conditions and how the proposed mTOR pathway involving Kir4.1 (described below; **Figure 1**) is linked to cell survival.

POTENTIAL MECHANISMS INVOLVED IN MEDIATING KIR4.1 FUNCTIONAL BENEFITS

Kir4.1 channel activity following ischemia and its contribution to the biophysical properties of astrocytes has previously been described. How Kir4.1 is involved in preventing neuronal loss following ischemia through an intrinsic pathway has yet to be outlined. Specifically, it is possible that Kir4.1 may play a role in the activation of the mammalian target of rapamycin complex 1 (mTORC1). mTORC1 is one of two mTOR complexes that contains a kinase mTOR component (Heras-Sandoval et al., 2014). It is mediated upstream by the phosphatidylinositol 3 kinase/protein kinase B (PI3K/AKT) pathway that is involved in cell survival and the inhibition of apoptosis (Heras-Sandoval et al., 2014). Furthermore, researchers have shown that mTORC1 is involved in the negative regulation of autophagy (Noda and Ohsumi, 1998; Schmelzle and Hall, 2000; Cuyàs et al., 2014).

Autophagy involves the breakdown of damaged organelles and misfolded proteins *via* a stress-induced catabolic pathway that maintains proper cellular homeostasis (Heras-Sandoval et al., 2014; Sahni et al., 2017). Autophagy does not seem to always result in cell survival. Autophagic programmed cell death occurs in response to stressors, such as water accumulation or nutrient deprivation, due to induced autophagy (Heras-Sandoval et al., 2014). Induced autophagy has been shown to occur in response to altered expression of autophage-related gene (Atg) 5 and 6 within the cell leading to cellular lysis (Amelio et al., 2011; Majid, 2014). Research has also demonstrated that autophagy is impacted significantly more in nutrient-deprived situations, such as K^+-deprivation, as the process is associated with energy re-usage in cells (Ye et al., 2016; Sahni et al., 2017). For example, within cerebellar granule cells, K^+-deprivation has not only induced autophagy but has been linked to programmed cell death as conditions move into K^+-starvation (K^+ reduced to 5 mM) (Canu et al., 2005; Kaasik et al., 2005; Sahni et al., 2017). Kir4.1 is dependent on adenosine triphosphate (ATP) (Nwaobi et al., 2016). Under K^+-starvation, Kir4.1 may be inactive as a result of ATP depletion in response to brain ischemia and low pH due to the acidosis that occurs in response to ischemia (Pessia et al., 2001; Hu and Song, 2017). As a result, Kir4.1 is no longer activated in a PI3K-dependent manner (as suggested below) and mTORC1 no longer prevents autophagic cell death.

The point at which PI3K attempts to activate Kir4.1 appears to be dependent on timing. This may be because recent evidence has pointed not only to the dual role of autophagy following ischemia (Chen et al., 2014; Majid, 2014) but implicates the potential role of K^+ in preventing autophagy (Canu et al., 2005; Kaasik et al., 2005; Sahni et al., 2017). Initially, Koike et al. (2008) demonstrated that the induction of autophagy, following hypoxia-ischemia injury, results in neuronal death. On the other hand, Carloni et al. (2010) described a pro-survival signaling complex involving autophagy to prevent neuronal death. More recently, it was suggested that the role autophagy plays following ischemia is determined by the time at which it is induced (Chen et al., 2014). Ravikumar et al. (2010) stated that a protective role for autophagy might be seen during ischemic preconditioning, whereas following ischemia/reperfusion the process might aggravate cerebral ischemic injury. Based on these findings, He et al. (2012) hypothesized that inducing autophagy at different time points during early and late stage ischemia may account for the different outcomes. For example, infarct size was reduced significantly and eliminated water content increases in the brain after treatment with 3-MA (a known autophagy inhibitor) prior to reperfusion (Chen et al., 2014). On the other hand, Carloni et al. (2010) found that treatment with rapamycin decreased brain injury and increased autophagy when administered prior to hypoxia-ischemia. Furthermore, the neuroprotective effects of ischemic postconditioning, previously described as being mimicked (Yan et al., 2011), are weakened when rapamycin is applied at the onset of reperfusion rather than at the onset of hypoxia-ischemia (Gao et al., 2012).

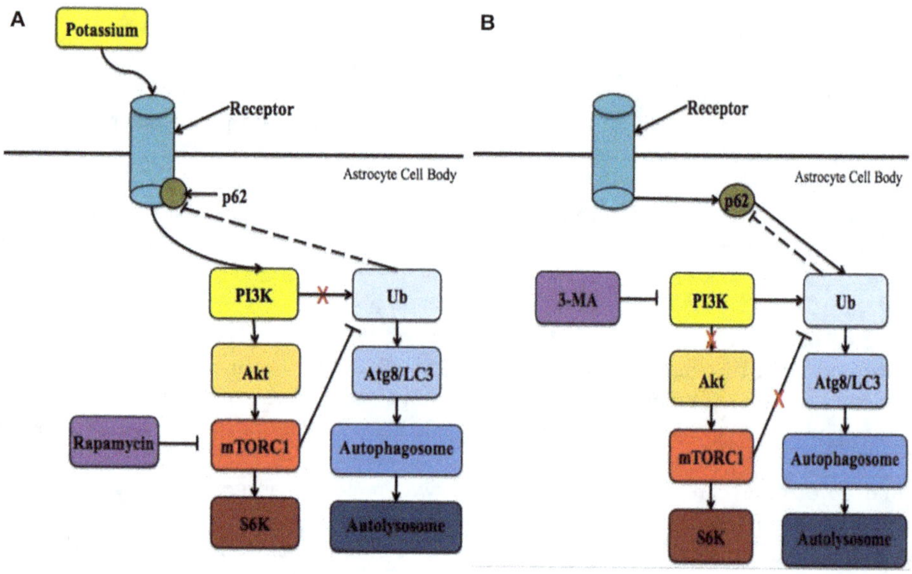

FIGURE 1 | The proposed intrinsic mechanism involving Kir4.1 in neuronal survival following ischemia. **(A)** Following ischemia, the levels of extracellular potassium increase significantly. PI3K is recruited to activate Kir4.1 leading to the initiation of mTORC1 and the overall mTOR pathway. In addition, it is possible that p62 is recruited to regulate Kir4.1. As a result, more cells in the ischemic penumbra survive and brain injury is reduced. Treatment with rapamycin has been demonstrated to reduce these effects. **(B)** During potassium starvation Kir4.1 is not activated due to ATP-depletion and mTORC1 no longer negatively regulates autophagy. Instead, autophagic cell death via protein breakdown in the autolysosome is initiated in a PI3K-dependent manner. Under these conditions, p62 interacts with ubiquitin (Ub) and binds with Atg8/LC3 within the phagophore. Lysosomes then interact with the autophagosome to form the autolysosome. 3-MA is an inhibitor of the PI3K pathway that prevents the initiation of autophagic cell death. Ubiquitin is the only molecule that is depicted as potentially feeding back in the mechanism, however, many other molecules are known to feedback in both the mTOR and autophagic pathways, respectively.

The mammalian target of rapamycin (mTOR) pathways is one of several cellular pathways that are involved in the maintenance of neuronal survival. It is also inhibited by rapamycin. As mentioned above, the timing at which PI3K attempts to activate Kir4.1, resulting in mTORC1 activation, may determine which pathway is activated leading to either cell survival or death. It is possible that targeting Kir4.1 activity prior to reperfusion may increase PI3K activity, and subsequent AKT phosphorylation, resulting in the activation of the mTORC1. It is important to note that the channel's involvement in the mTOR pathway has only been previously implicated by a single study. Zaika et al. (2016) showed that using insulin and insulin-like growth factor-1, within the cortical collecting duct, activates the heteromeric Kir4.1/Kir5.1 channel in a phosphoinositide 3-kinase (PI3K)-dependent manner. It is possible that Kir4.1 works through the mTOR pathway to prevent the induction of autophagy, through its PI3K-dependent activation (Zaika et al., 2016), following the onset reperfusion leading to the reduction of infarct size and a decrease in water content. This hypothesis is supported by the fact that during the onset of reperfusion ATP levels are high (Murphy and Steenbergen, 2008; Kalogeris et al., 2012), which corresponds to the time when Kir4.1 is optimally activated (Takumi et al., 1995; Kucheryavykh et al., 2009) and APQ4 levels remain unaltered (Shin et al., 2011). On the other hand, when energy levels are low, corresponding to Kir4.1 inactivation, the PI3K/AKT pathway is less active leading to the adoption of a quiescent state by the cells (Cheung and Rando, 2013). Thus, it is hypothesized that Kir4.1 is activated in PI3K-dependent manner, leading to cell survival through the mTOR pathway. Future research could focus on determining the upstream effectors within this intrinsic pathway and suggests that Kir4.1 represents a potential therapeutic target for the treatment of ischemic insult. The two potential mechanisms involved in mediating Kir4.1 functional benefits are summarized in **Figure 1**.

It is also important to note that p62, an adapter molecule in autophagy, may also play a role in this pathway. Voltage-gated K^+ channels (K_v) are regulated by p62 (Sahni et al., 2017). It has been previously demonstrated that PKCζ interacts with p62 to increase the phosphorylation of the β subunit of K_v ($K_v\beta$) (Puls et al., 1997; Ishii et al., 2013). As a result, K_v channels, specifically $K_v1.5$, are inhibited under acute hypoxia (Ishii et al., 2013). While Kir4.1 is not a voltage-gated channel, as it lacks a voltage-sensing domain, PKC isoforms have been shown to modulate both inwardly rectifying and voltage-gated channels (Dini et al., 2006). For example, the heteromeric Kir4.1-Kir5.1 channel is inhibited in a PKC-dependent manner (Rojas et al., 2007). Assuming p62 and Kir4.1 do interact, the timing at which p62 attempts to interact with Kir4.1 may determine which pathway is activated leading to either cell survival or death. Immediately following ischemic onset, p62 may be recruited to regulate Kir4.1, which is activated in a PI3K-dependent manner, leading to cell survival through the mTOR pathway (**Figure 1A**). Thus, it is possible that targeting Kir4.1 prior to reperfusion, with different pharmacological agents, may increase the likelihood of p62 associating with Kir4.1 and subsequent cell survival. On the other hand, the point at which p62 attempts to interact with Kir4.1 would be dependent on timing. This is because as more ATP is used, Kir4.1 is no longer active and p62 would be able to associate with ubiquitin to initiate autophagic cell death under nutrient-deprived situations (**Figure 1B**). Future research should examine whether or not the homomeric Kir4.1 channel is inhibited by PKC isoforms, specifically PKCζ, and if there is an interaction between Kir4.1 and p62.

CONCLUSION

Previous evidence points to an intrinsic pathway, involving the regulation of Kir4.1, within the central nervous system that regulates neuronal survival. Further understanding of this pathway within astrocytes, and how it impacts neuronal viability (example: the impact on autophagy), could improve treatment options following ischemia. In this regard, further research is critical to identifying the potential role of Kir4.1 channel, and its subsequent effectors, in ischemia in order to guide the development of novel treatment options for stroke. These treatment strategies could be focused on altering Kir4.1 channel activity, which may be useful in improving the clinical outcomes after ischemic stroke.

AUTHOR CONTRIBUTIONS

All authors listed have made a substantial, direct, and intellectual contribution to the work, and approved it for publication.

REFERENCES

Amelio, I., Melino, G., and Knight, R. A. (2011). Cell death pathology: cross-talk with autophagy and its clinical implications. *Biochem. Biophys. Res. Commun.* 414, 277–281. doi: 10.1016/j.bbrc.2011.09.080

Barbour, B., Brew, H., and Attwell, D. (1988). Electrogenic glutamate uptake in glial cells is activated by intracellular potassium. *Nature* 335, 433–435. doi: 10.1038/335433a0

Bond, C. T., Pessia, M., Xia, X. M., Lagrutta, A., Kavanaugh, M. P., and Adelman, J. P. (1994). Cloning and expression of a family of inward rectifier potassium channels. *Recept. Channels* 2, 183–191.

Canu, N., Tufi, R., Serafino, A. L., Amadoro, G., Ciotti, M. T., and Calissano, P. (2005). Role of the autophagic-lysosomal system on low potassium-induced apoptosis in cultured cerebellar granule cells. *J. Neurochem.* 92, 1228–1242. doi: 10.1111/j.1471-4159.2004.02956.x

Carloni, S., Girelli, S., Scopa, C., Buonocore, G., Longini, M., and Balduini, W. (2010). Activation of autophagy and Akt/CREB signaling play an equivalent role in the neuroprotective effect of rapamycin in neonatal hypoxia-ischemia. *Autophagy* 6, 366–377. doi: 10.4161/auto.6.3.11261

Chen, Q., Olney, J. W., Lukasiewicz, P. D., Almli, T., and Romano, C. (1998). Fenamates protect neurons against ischemic and excitotoxic injury in chick embryo retina. *Neurosci. Lett.* 242, 163–166. doi: 10.1016/S0304-3940(98)00081-0

Chen, W., Sun, Y., Liu, K., and Sun, X. (2014). Autophagy: a double-edged sword for neuronal survival after cerebral ischemia. *Neural Regen. Res.* 9, 1210–1216. doi: 10.4103/1673-5374.135329

Cheung, T. H., and Rando, T. A. (2013). Molecular regulation of stem cell quiescence. *Nat. Rev. Mol. Cell Biol.* 14, 329–340. doi: 10.1038/nrm3591

Coyle, J. T., and Puttfarcken, P. (1993). Oxidative stress, glutamate, and neurodegenerative disorders. *Science* 262, 689–695. doi: 10.1126/science.7901908

Cuyàs, E., Corominas-Faja, B., Joven, J., and Menendez, J. A. (2014). "Cell cycle regulation by the nutrient-sensing mammalian target of rapamycin (mTOR) pathway," in *Cell Cycle Control: Mechanisms and Protocols*, eds E. Noguchi and M. C. Gadaleta. (New York, NY: Springer), 113–144.

D'Adamo, M. C., Catacuzzeno, L., Di Giovanni, G., Franciolini, F., and Pessia, M. (2013). K^+ channelepsy: progress in the neurobiology of potassium channels and epilepsy. *Front. Cell. Neurosci.* 7:134. doi: 10.3389/fncel.2013.00134

D'Adamo, M. C., Shang, L., Imbrici, P., Brown, S. D., Pessia, M., and Tucker, S. J. (2011). Genetic inactivation of Kcnj16 identifies Kir5.1 as an important determinant of neuronal PCO2/pH sensitivity. *J. Biol. Chem.* 286, 192–8.

Dini, G., Ilkanich, E. V., and Janigro, D. (2006). "Potassium channels, cell cycle, and tumorigenesis in the central nervous system," in *The Cell Cycle in the Central Nervous System*, ed D. Janigro, (Totowa, NJ: Humana Press), 177–191.

Ferraro, T. N., Golden, G. T., Smith, G. G., Martin, J. F., Lohoff, F. W., Gieringer, T. A., et al. (2004). Fine mapping of a seizure susceptibility locus on mouse Chromosome 1: nomination of Kcnj10 as a causative gene. *Mamm. Genome* 15, 239–251. doi: 10.1007/s00335-003-2270-3

Gao, L., Jiang, T., Guo, J., Liu, Y., Cui, G., Gu, L., et al. (2012). Inhibition of autophagy contributes to ischemic postconditioning-induced neuroprotection against focal cerebral ischemia in rats. *PLoS ONE* 7:e46092. doi: 10.1371/journal.pone.0046092

Guglielmi, L., Servettini, I., Caramia, M., Catacuzzeno, L., Franciolini, F., D'Adamo, M. C., et al. (2015). Update on the implication of potassium channels in autism: K(+) channelautism spectrum disorder. *Front. Cell. Neurosci.* 9:34. doi: 10.3389/fncel.2015.00034

He, S., Wang, C., Dong, H., Xia, F., Zhou, H., Jiang, X., et al. (2012). Immune-related GTPase M (IRGM1) regulates neuronal autophagy in a mouse model of stroke. *Autophagy* 8, 1621–1627. doi: 10.4161/auto.21561

Heras-Sandoval, D., Pérez-Rojas, J. M., Hernández-Damián, J., and Pedraza-Chaverri, J. (2014). The role of PI3K/AKT/mTOR pathway in the modulation of autophagy and the clearance of protein aggregates in neurodegeneration. *Cell. Signal.* 26, 2694–2701. doi: 10.1016/j.cellsig.2014.08.019

Hibino, H., Inanobe, A., Furutani, K., Murakami, S., Findlay, I., and Kurachi, Y. (2010). Inwardly rectifying potassium channels: their structure, function, and physiological roles. *Physiol. Rev.* 90, 291–366. doi: 10.1152/physrev.00021.2009

Hu, H. J., and Song, M. (2017). Disrupted ionic homeostasis in ischemic stroke and new therapeutic targets. *J. Stroke Cerebrovasc. Dis.* 26, 2706–2719. doi: 10.1016/j.jstrokecerebrovasdis.2017.09.011

Ishii, T., Warabi, E., Siow, R. C., and Mann, G. E. (2013). Sequestosome1/p62: a regulator of redox-sensitive voltage-activated potassium channels, arterial remodeling, inflammation, and neurite outgrowth. *Free Radic. Biol. Med.* 65, 102–116. doi: 10.1016/j.freeradbiomed.2013.06.019

Kaasik, A., Rikk, T., Piirsoo, A., Zharkovsky, T., and Zharkovsky, A. (2005). Up-regulation of lysosomal cathepsin L and autophagy during neuronal death induced by reduced serum and potassium. *Eur. J. Neurosci.* 22, 1023–1031. doi: 10.1111/j.1460-9568.2005.04279.x

Kalogeris, T., Baines, C. P., Krenz, M., and Korthuis, R. J. (2012). Cell biology of ischemia/reperfusion injury. *Int Rev Cell Mol Biol.* 298:229–317. doi: 10.1016/B978-0-12-394309-5.00006-7

Koike, M., Shibata, M., Tadakoshi, M., Gotoh, K., Komatsu, M., Waguri, S., et al. (2008). Inhibition of autophagy prevents hippocampal pyramidal neuron death after hypoxic-ischemic injury. *Am. J. Pathol.* 172, 454–469. doi: 10.2353/ajpath.2008.070876

Kucheryavykh, L. Y., Kucheryavykh, Y. V, Inyushin, M., Shuba, Y. M., Sanabria, P., Cubano, L. A., et al. (2009). Ischemia increases TREK-2 channel expression in astrocytes: relevance to glutamate clearance. *Open Neurosci. J.* 3, 40–47. doi: 10.2174/1874082000903010040

Kucheryavykh, Y. V., Kucheryavykh, L. Y., Nichols, C. G., Maldonado, H. M., Baksi, K., Reichenbach, A., et al. (2007). Downregulation of Kir4.1 inward rectifying potassium channel subunits by RNAi impairs potassium transfer and glutamate uptake by cultured cortical astrocytes. *Glia* 55, 274–281. doi: 10.1002/glia.20455

Lai, T. W., Zhang, S., and Wang, Y. T. (2014). Excitotoxicity and stroke: identifying novel targets for neuroprotection. *Prog. Neurobiol.* 115, 157–188. doi: 10.1016/j.pneurobio.2013.11.006

Larsen, B. R., Assentoft, M., Cotrina, M. L., Hua, S. Z., Nedergaard, M., Kaila, K., et al. (2014). Contributions of the Na^+/K^+-ATPase, NKCC1, and Kir4.1 to hippocampal K^+ clearance and volume responses. *Glia* 62, 608–622. doi: 10.1002/glia.22629

Majid, A. (2014). Neuroprotection in stroke : past, present, and future. *ISRN Neurol.* 2014:515716. doi: 10.1155/2014/515716

Moody, W. J., Futamachi, K. J., and Prince, D. A. (1974). Extracellular potassium activity during epileptogenesis. *Exp. Neurol.* 42, 248–263. doi: 10.1016/0014-4886(74)90023-5

Murphy, E., and Steenbergen, C. (2008). Mechanisms underlying acute protection from cardiac ischemia-reperfusion injury. *Physiol. Rev.* 88, 581–609. doi: 10.1152/physrev.00024.2007

Neusch, C., Papadopoulos, N., Müller, M., Maletzki, I., Winter, S. M., Hirrlinger, J., et al. (2006). Lack of the Kir4.1 channel subunit abolishes K^+ buffering properties of astrocytes in the ventral respiratory group: impact on extracellular K^+ regulation. *J. Neurophysiol.* 95, 1843–1852. doi: 10.1152/jn.00996.2005

Noda, T., and Ohsumi, Y. (1998). Tor, a phosphatidylinositol kinase homologue, controls autophagy in yeast. *J. Biol. Chem.* 273, 3963–3966. doi: 10.1074/jbc.273.7.3963

Nwaobi, S. E., Cuddapah, V. A., Patterson, K. C., Randolph, A. C., and Olsen, M. L. (2016). The role of glial-specific Kir4.1 in normal and pathological states of the CNS. *Acta Neuropathol.* 132, 1–21. doi: 10.1007/s00401-016-1553-1

Pannicke, T., Iandiev, I., Uckermann, O., Biedermann, B., Kutzera, F., Wiedemann, P., et al. (2004). A potassium channel-linked mechanism of glial cell swelling in the postischemic retina. *Mol. Cell. Neurosci.* 26, 493–502. doi: 10.1016/j.mcn.2004.04.005

Pessia, M., Imbrici, P., D'Adamo, M. C., Salvatore, L., and Tucker, S. J. (2001). Differential pH sensitivity of Kir4.1 and Kir4.2 potassium channels and their modulation by heteropolymerisation with Kir5.1. *J. Physiol.* 532, 359–367. doi: 10.1111/j.1469-7793.2001.0359f.x

Pivonkova, H., Benesova, J., Butenko, O., Chvatal, A., and Anderova, M. (2010). Impact of global cerebral ischemia on K+ channel expression and membrane properties of glial cells in the rat hippocampus. *Neurochem. Int.* 57, 783–794. doi: 10.1016/j.neuint.2010.08.016

Puls, A., Schmidt, S., Grawe, F., and Stabel, S. (1997). Interaction of protein kinase C ζ with ZIP, a novel protein kinase C-binding protein. *Proc. Natl. Acad. Sci. U.S.A.* 94, 6191–6196. doi: 10.1073/pnas.94.12.6191

Ransom, C. B., Ransom, B. R., and Sontheimer, H. (2000). Activity-dependent extracellular K^+ accumulation in rat optic nerve: the role of glial and axonal Na^+ pumps. *J. Physiol.* 522(Pt 3), 427–442. doi: 10.1111/j.1469-7793.2000.00427.x

Ravikumar, B., Sarkar, S., Davies, J. E., Futter, M., Garcia-Arencibia, M., Green-Thompson, Z. W., et al. (2010). Regulation of mammalian autophagy in physiology and pathophysiology. *Physiol. Rev.* 90, 1383–1435. doi: 10.1152/physrev.00030.2009

Rojas, A., Cui, N., Su, J., Yang, L., Muhumuza, J. P., and Jiang, C. (2007). Protein kinase C dependent inhibition of the heteromeric Kir4.1–Kir5.1 channel. *Biochim. Biophys. Acta* 1768, 2030–2042. doi: 10.1016/j.bbamem.2007.04.008

Rossi, D. J., Brady, J. D., and Mohr, C. (2007). Astrocyte metabolism and signaling during brain ischemia. *Nat. Neurosci.* 10, 1377–1386. doi: 10.1038/nn2004

Rothman, S. M., and Olney, J. W. (1986). Glutamate and the pathophysiology of hypoxic–ischemic brain damage. *Ann. Neurol.* 19, 105–111. doi: 10.1002/ana.410190202

Rothman, S. M., and Olney, J. W. (1987). Excitotoxity and the NMDA receptor. *Trends Neurosci.* 10, 299–302. doi: 10.1016/0166-2236(87)90177-9

Rothstein, J. D., Dykes-Hoberg, M., Pardo, C. A., Bristol, L. A., Jin, L., Kuncl, R. W., et al. (1996). Knockout of glutamate transporters reveals a major role for astroglial transport in excitotoxicity and clearance of glutamate. *Neuron* 16, 675–686. doi: 10.1016/S0896-6273(00)80086-0

Sahni, S., Bae, D. H., Jansson, P. J., and Richardson, D. R. (2017). The mechanistic role of chemically diverse metal ions in the induction of autophagy. *Pharmacol. Res.* 119, 118–127. doi: 10.1016/j.phrs.2017.01.009

Schmelzle, T., and Hall, M. N. (2000). TOR, a central controller of cell growth. *Cell* 103, 253–262. doi: 10.1016/S0092-8674(00)00117-3

Shin, J. A., Choi, J. H., Choi, Y. H., and Park, E. M. (2011). Conserved aquaporin 4 levels associated with reduction of brain edema are mediated by estrogen in the ischemic brain after experimental stroke. *Biochim. Biophys. Acta - Mol. Basis Dis.* 1812, 1154–1163. doi: 10.1016/j.bbadis.2011.05.004

Sibille, J., Dao Duc, K., Holcman, D., and Rouach, N. (2015). The neuroglial potassium cycle during neurotransmission: role of Kir4.1 channels. *PLOS Comput. Biol.* 11:e1004137. doi: 10.1371/journal.pcbi.1004137

Sicca, F., Imbrici, P., D'Adamo, M. C., Moro, F., Bonatti, F., Brovedani, P., et al. (2011). Autism with seizures and intellectual disability: possible causative role of gain-of-function of the inwardly-rectifying K^+ channel Kir4.1. *Neurobiol. Dis.* 43, 239–247. doi: 10.1016/j.nbd.2011.03.016

Steiner, E., Enzmann, G. U., Lin, S., Ghavampour, S., Hannocks, M. J., Zuber, B., et al. (2012). Loss of astrocyte polarization upon transient focal brain ischemia as a possible mechanism to counteract early edema formation. *Glia* 60, 1646–1659. doi: 10.1002/glia.22383

Stewart, T. H., Eastman, C. L., Groblewski, P. A., Fender, J. S., Verley, D. R., Cook, D. G., et al. (2010). Chronic dysfunction of astrocytic inwardly rectifying K^+ channels specific to the neocortical epileptic focus after fluid percussion injury in the rat. *J. Neurophysiol.* 104, 3345–3360. doi: 10.1152/jn.00398.2010

Takano, T., Oberheim, N., Cotrina, M. L., and Nedergaard, M. (2009). Astrocytes and ischemic injury. *Stroke* 40, S8–S12. doi: 10.1161/STROKEAHA.108.533166

Takumi, T., Ishii, T., Horio, Y., Morishige, K. I., Takahashi, N., Yamada, M., et al. (1995). A novel ATP-dependent inward rectifier potassium channel expressed predominantly in glial cells. *J. Biol. Chem.* 270, 16339–16346. doi: 10.1074/jbc.270.27.16339

Vella, J., Zammit, C., Di Giovanni, G., Muscat, R., and Valentino, M. (2015). The central role of aquaporins in the pathophysiology of ischemic stroke. *Front. Cell. Neurosci.* 9:108. doi: 10.3389/fncel.2015.00108

Yan, W., Zhang, H., Bai, X., Lu, Y., Dong, H., and Xiong, L. (2011). Autophagy activation is involved in neuroprotection induced by hyperbaric oxygen preconditioning against focal cerebral ischemia in rats. *Brain Res.* 1402, 109–121. doi: 10.1016/j.brainres.2011.05.049

Ye, Q., Wu, Y., Gao, Y., Li, Z., Li, W., and Zhang, C. (2016). The "selfish brain" is regulated by aquaporins and autophagy under nutrient deprivation. *Mol. Med. Rep.* 13, 3842–3848. doi: 10.3892/mmr.2016.4988

Zaika, O., Palygin, O., Tomilin, V., Mamenko, M., Staruschenko, A., and Pochynyuk, O. (2016). Insulin and IGF-1 activate Kir4.1/5.1 channels in cortical collecting duct principal cells to control basolateral membrane voltage. *Am. J. Physiol. Ren. Physiol.* 310, F311–F321. doi: 10.1152/ajprenal.00436.2015

6

Plant Polyphenols and Exendin-4 Prevent Hyperactivity and TNF-α Release in LPS-Treated *In vitro* Neuron/Astrocyte/Microglial Networks

Francesca Gullo [1]***, Michela Ceriani*** [1]***, Alessia D'Aloia*** [2]***, Enzo Wanke*** [2]***, Andrew Constanti*** [3]***, Barbara Costa*** [1†] ***and Marzia Lecchi*** [1*†]

[1] Department of Biotechnology and Biosciences and Milan Center for Neuroscience, University of Milano-Bicocca, Milan, Italy, [2] Department of Biotechnology and Biosciences, University of Milano-Bicocca, Milan, Italy, [3] Department of Pharmacology, School of Pharmacy, University College London, London, United Kingdom

***Correspondence:**
Marzia Lecchi
marzia.lecchi1@unimib.it

† These authors jointly supervised this work.

Increasing evidence supports a decisive role for neuroinflammation in the neurodegenerative process of several central nervous system (CNS) disorders. Microglia are essential mediators of neuroinflammation and can regulate a broad spectrum of cellular responses by releasing reactive oxygen intermediates, nitric oxide, proteases, excitatory amino acids, and cytokines. We have recently shown that also in *ex-vivo* cortical networks of neurons, astrocytes and microglia, an increased level of tumor necrosis factor-alpha (TNF-α) was detected a few hours after exposure to the bacterial endotoxin lipopolysaccharide (LPS). Simultaneously, an atypical "seizure-like" neuronal network activity was recorded by multi-electrode array (MEA) electrophysiology. These effects were prevented by minocycline, an established anti-inflammatory antibiotic. We show here that the same inhibitory effect against LPS-induced neuroinflammation is exerted also by natural plant compounds, polyphenols, such as curcumin (CU, curcuma longa), crocin (CR, saffron), and resveratrol (RE, grape), as well as by the glucagon like peptide-1 receptor (GLP-1R) agonist exendin-4 (EX-4). The drugs tested also caused *per-se* early transient (variable) changes of network activity. Since it has been reported that LPS-induced neuroinflammation causes rearrangements of glutamate transporters in astrocytes and microglia, we suggest that neural activity could be putatively increased by an imbalance of glial glutamate transporter activity, leading to prolonged synaptic glutamatergic dysregulation.

Keywords: sterile inflammation, LPS, TNF-α, GLP-1, plant polyphenols, exendin-4 (EX-4), neocortical cultures, multi-electrode array (MEA)

INTRODUCTION

Microglia are an integral part of central nervous system (CNS) networks, forming the innate defensive system, and their pathological potential has been extensively investigated (Kettenmann et al., 2011). In different pathologies, microglia acquire distinct functional states and, during the disease progression, modify and change their activated phenotype. Activated microglia specifically interact with neurons and influence their survival either in a positive or in a negative direction. They can physically contact injured neurons and remove synapses, a process termed synaptic stripping (Kettenmann et al., 2013). Conceptually, microglial cells not only can affect neural networks through removal of cellular and subcellular elements and secreting cytokines, trophic factors, and neurotransmitters (Kettenmann et al., 2011), but also by receiving messages; in fact, they express a variety of receptors for neurotransmitters, neuropeptides, and neuromodulators. Thus they have also the capacity to sense neuronal activity (Pocock and Kettenmann, 2007).

Current pharmacological interventions against neuroinflammation have symptomatic benefits to certain degrees, but do not prevent progressive neurodegeneration. More recently, it has been shown that some natural substances protect the substantia nigra dopaminergic neurons (involved in the onset of Parkinson's disease) through their anti-inflammatory action. Among these are some polyphenols such as curcumin, CU (Ojha et al., 2012), resveratrol, RE (Zhang et al., 2010; Gao et al., 2014; Lofrumento et al., 2014), and peptide molecules such as exendin-4, EX-4 (Kim et al., 2009; Athauda and Foltynie, 2016). Moreover, also crocin, CR, extracted from saffron, resulted in an effective protection from macular degeneration, a retinal disease principally related to oxidative stress and chronic inflammation (Nam et al., 2010; Marangoni et al., 2013). In many of these reports, sterile inflammation was induced, microglia-released cytokine concentration increased, but the properties of neuronal network activity were never explored.

By using low concentrations of the purified bacterial endotoxin lipopolysaccharide (LPS) over 6–8 h, we reproduced *in vitro* a "sterile" CNS neuroinflammation in a 10,000-cell network of neurons, astrocytes and microglia, grown on a multielectrode array (MEA) dish, where neurons were regularly bursting for weeks (Gullo et al., 2014). We found that an "atypical" neuronal excitability took place causing long-lasting bursts resembling epileptiform seizures. These slow changes of neuronal excitability were accompanied by a simultaneous increase in microglia-released tumor necrosis factor-alpha (TNF-α) concentration, suggesting a crucial role of microglia, but not of astrocytes, in this process. Both these effects were blocked by pre-treatment with minocycline, an anti-inflammatory antibiotic drug, which was inactive when applied alone.

Here, we wished to examine whether the action of minocycline on our neuron/astrocyte/microglial co-culture system could be reproduced by the above mentioned natural molecules as anti-inflammatory tools. By simultaneously using an electrophysiological recording technique, such as microelectrode arrays (MEA), and TNF-α ELISA assays, we found that all of these molecules, depending on the concentration used, were transiently able to modify the balanced network activity and to imitate the blocking effects of minocycline against LPS neuroinflammation. Furthermore, all of the agents tested resulted in an anti-inflammatory action at concentrations lower than those generally reported in the literature.

MATERIALS AND METHODS

Cell Cultures

Primary cultures of cortical neurons were prepared from post-natal mice (P1–P3) as previously described (Gullo et al., 2009). All the procedures concerning animal handling and sacrifice followed the Principles of Laboratory Animal Care (2010/63/UE Directive), and were approved by the University of Milan-Bicocca Ethics Committee and the Italian Ministry of Health (D.Lgs 26/2014). Briefly, the cerebral cortex (excluding the hippocampus) was removed from decapitated mice, cut into 1 mm^3 pieces, and digested by trypsin (0.15%) and DNAse (10 μg/ml) at 37°C for 20 min. After enzyme digestion, cells were mechanically dissociated by means of trituration, and plated at densities of 600–900 $\times$ 10^3 cells/ml on glass coverslips (for immunocytochemistry) or MEA Petri dishes pre-coated with polyethyleneimine 0.1% (wt/vol) and laminin 20 μg/ml (30 μm diameter ITO electrodes spaced 200 μm apart, Multichannels System, Germany). After 3 h incubation, the plating medium was replaced by neurobasal medium (NB) with B27 (Invitrogen, Italy), glutamine 1 mM and basic fibroblast growth factor (bFGF) 10 ng/ml, and the culture was maintained at 37°C in 5% CO_2. One-half of the medium volume was replaced every 3 days. The cultures in MEA dishes were covered with gas-permeable covers (MEA-MEM, Ala Scientific Instruments, Inc., USA) throughout the culture period (12–22 days-*in-vitro*, DIV).

MEA Electrophysiology: Drug Application-General Aspects

As previously described (Gullo et al., 2009), we report results obtained within a few hours after positioning the MEA dish into the incubator, which can be considered at the steady-state. The recording area in our MEA dishes was ~2 mm^2, and in this area, the average number of neurons was in the order of ~5,000, plus about the same number of astrocytes (see Figure 1 of Supplementary Material in Gullo et al., 2010, at http://www/frontiersin.org/neuralcircuits/paper/10.3389/fncir.2010.00011/). The average space between cells was therefore relatively large. Neuroinflammation was induced by incubating cultures with LPS, isolated and purified from *E. coli* R515 by Enzo Life Sciences (Alexis Biochemicals, code 581-007). This product does not contain detectable protein or DNA contaminants with agonistic Toll-like receptor (TLR) activity. Since it is a strong activator of TLR4 but it does not activate TLR2 or the other TLRs, it was considered suitable for our experiments. The drugs to be tested were kept as frozen stock solutions in distilled water (or DMSO <0.1%) at −20°C, until diluted as appropriate with MEA culture medium before each experiment. The drugs used were: crocin (CR), the polyphenols resveratrol (RE) and curcumin (CU), purchased from Sigma, Italy; exendin-4 (EX-4) and the GLP-1 receptor antagonist, exendin (9–39), purchased from Tocris, UK. All experiments were performed by adding the drugs in volumes that were always <1% of the total conditioned media volume bathing the neurons. When indicated, a washout was carried out with a solution pre-conditioned by the same network under control conditions.

MEA Electrophysiology: Recordings, Waveform Acquisition, and Sorting

We used the same procedures previously described in Gullo et al. (2009, 2010). Briefly, analog signals sampled at 40 kHz were recorded at 36°C in CO_2-controlled incubators using MEA-1060BC or 1060INV pre-amplifiers (bandwidth 1–8,000 Hz, Multichannel Systems, Germany) connected to a MEA Workstation (bandwidth 100–8,000 Hz, Plexon Inc., USA). Data were sorted into timestamp files by the MEAWorkstation Sorter software (MEAWS, see details below) and cleaned of artifacts using the OFFLine Sorter program (Plexon Inc., USA). Next, during the PCA-based waveform sorting and for multi-unit electrodes, we applied one of the following procedures: (i) spike removal with a Mahalanobis threshold in the range 1.8 to 1.4; we concurrently checked that the *P*-value of multivariate ANOVA sorting quality statistics was <0.01 among the identified units; (ii) when the previous procedure led to excessive spike invalidation, we manually removed the spikes invading the adjacent unit ellipsoids (the latter method was very effective in decreasing the *P*-values, with only a limited number of erased spikes).

Neuronal Cluster Classification

The method of classification into excitatory or inhibitory neurons is described in detail in Gullo et al. (2009, 2010) and Becchetti et al. (2012). Briefly, for each identified unit and each burst, the following characteristics were computed in defined time segments: the autocorrelation function (ACF), the burst duration (BD), the spike number (SN), the spike rate (SR), the intra-burst spike rate (IBSR), the inter-burst intervals (IBIs), and the Fano factor (FF; Teich, 1989; Baddeley et al., 1997). We classified the neurons on the basis of an unsupervised learning approach consisting of data reducing principal component analysis (PCA) based on FF as a feature (Becchetti et al., 2012), followed by the K-means clustering procedure. The large differences in these burst metrics was the basis for adopting FF as the best feature to clusterize neurons. As previously described (Becchetti et al., 2012), these procedures normally separated two statistically different clusters composed of numbers of excitatory (~60–90) and inhibitory (~15–25) neurons, whose ratio always fitted the ratio present in the neocortex i.e., from 4 to 5 (Gullo et al., 2010; Sahara et al., 2012). To give an approximate idea of the computing time involved in the analysis of a 60-electrode MEA dish with ~100 identified neurons, the following are the typical parameters: 20 h-long experiment, 0.5 GB memory size, 10 min to obtain results.

Advanced Tools for Characterizing Firing and BD Histograms

The global network burst structure was analyzed with standard techniques as well as procedures recently developed by us (Gullo et al., 2012). Briefly, we applied a running window of variable duration (from 10 ms to 1 s) in order to search for the start and end of each burst and simultaneously collecting all of the spikes which were precisely tagged to the engaged excitatory and inhibitory neurons, already designated as previously explained above. In conclusion, for each spike in a generic burst, we knew exactly which neuron fired it and how many other spikes were fired by all of the other neurons. This had the consequence that we could also compute for the two identified neuronal clusters (defined above) the average propensity for firing, namely, the average number of spikes per burst, here called "excitability" (EXC) in each time segment of interest (shown in **Figure 1A**). Normally, it is much more interesting to compare EXC data normalized to control for both clusters, namely, normEXC. To investigate the heterogeneity of the burst length, BD, we studied its distribution in the form of cumulative probabilities in each time segment of interest, cumBD, as previously done in Gullo et al. (2014). To compare experiments, two types of plots are shown as follows: (i) time plots where *normEXC* describes how clusters change firing in the presence of drugs with respect to control (open, □, and closed, ■, symbols for excitatory and inhibitory clusters, respectively). Aligned characters, attached to the time axis, are control (c), washout (w), or drug names (short names were: CU, CR, RE, EX-4) indicating the presence or absence of drugs (note axis breaks necessary to zoom to particular time segments); (ii) *cumBD* plots (bin 0.2 s) vs. burst duration describing how drugs modify the distribution of the burst lifetimes (thick continuous lines for control; open stars for drugs; open rightward triangles, ▷, for LPS and open upward triangles, Δ, for washout. In our previous paper (Gullo et al., 2014), we found that, at 6–7 h after LPS application, its action was either present or blocked if the values of two variables, namely normEXC and $cumBD_{95}$ (the value of BD at cumBD of 95%), resulted in an increase with respect to control beside fixed thresholds. The normEXC threshold was +45 ± 10% (with respect to control value, in this case 1) and cumBD threshold was +40 ± 5% (with respect to the BD value in control computed at $cumBD_{95}$). Therefore, to highlight if drugs were able or unable to block LPS action in the normEXC and $cumBD_{95}$ plots, in the figures these thresholds were represented by a thin straight line and an arrow respectively.

TNF-α Concentration Measurements

In each MEA dish, small (150 μl) aliquots of the incubation medium were collected in control and during different times (6 and 12 h, representing the peak-time of LPS-induced TNF-α release, Gullo et al., 2014) after the addition of LPS (3 μg/ml) or polyphenols/crocin/exendin-4 + LPS. Samples (50 μl) were analyzed in triplicate for murine TNF-α with enzyme-linked immunosorbent assay (ELISA) kits (KCM3012, Invitrogen, Italy) according to manufacturer's instructions. The data were expressed as pg/ml following interpolation on the basis of a standard curve. As suggested by the manufacturer, the inter-assay and intra-assay variability were 3.5–4.3 and 2.6–8.2%, with a lower limit of detection <3 pg/ml. During the analysis we did not correct the data for the small change in the total incubation medium volume resulting from the aliquots removal. Data were analyzed using GraphPad 7.0 software employing ANOVA followed by Dunnet's test for group comparison. $P < 0.05$ was considered statistically significant.

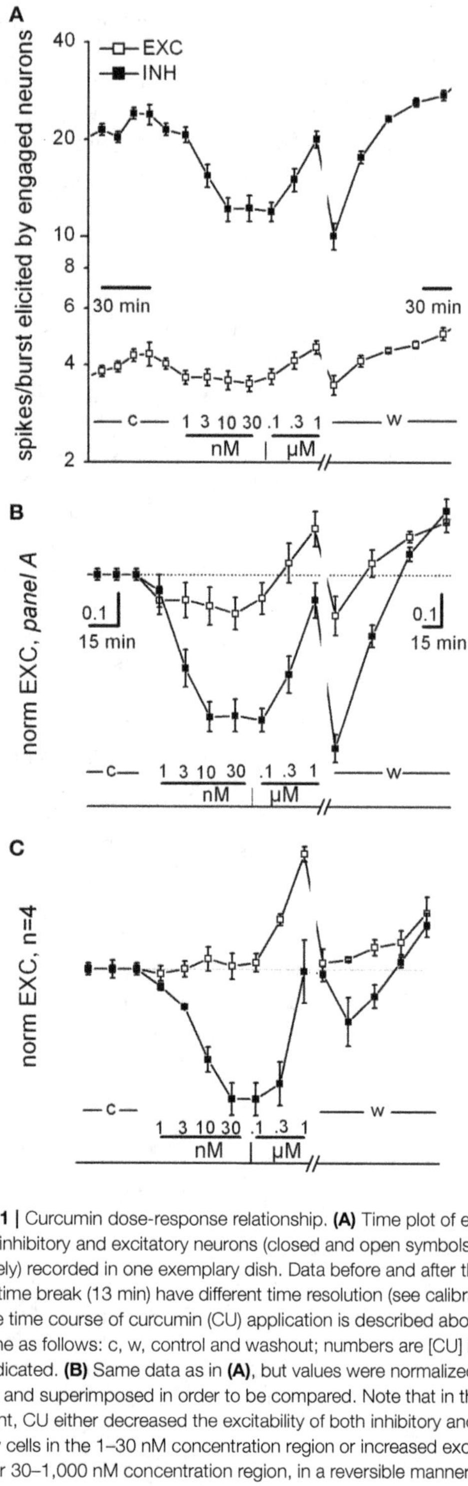

FIGURE 1 | Curcumin dose-response relationship. **(A)** Time plot of excitability (EXC) for inhibitory and excitatory neurons (closed and open symbols, respectively) recorded in one exemplary dish. Data before and after the washout time break (13 min) have different time resolution (see calibration bars). The time course of curcumin (CU) application is described above the bottom line as follows: c, w, control and washout; numbers are [CU] in nM and μM as indicated. **(B)** Same data as in **(A)**, but values were normalized to those of control and superimposed in order to be compared. Note that in this experiment, CU either decreased the excitability of both inhibitory and excitatory cells in the 1–30 nM concentration region or increased excitability in the further 30–1,000 nM concentration region, in a reversible manner. **(C)** Plot as in **(B)**, but average of data from four independent experiments from different dishes but similar DIVs.

Data Analysis and Statistical Significance

All of the data are expressed as mean values ± SEM, with n indicating the number of experiments. We used OriginPro 9.1 to test normality of data such as cumBD and FSH. Moreover, for statistical significance among cumBD and FSH distributions from different experiments, we used the non-parametric tests Wilcoxon signed rank test or Kruskal–Wallis test both available from OriginPro 9.1. The data were analyzed and the figures prepared using OriginPro 7.0 or OriginPro 9.1 software.

RESULTS

To study the activity of reverberating cortical networks, we followed the procedures described in Gullo et al. (2014). In our cultures, neurons, astrocytes and microglial cells survived together for weeks. In these conditions, neurons were able to generate spikes, which were stably recorded by MEA system (60 recording electrodes, 200 μm apart). The spikes were acquired during each burst (every ~10 s) consisting of an ~1 s-brief period of synaptic-mediated global network activity among all excitable cells. Indeed, each elicited spike was the result of the temporal and spatial summation of a minimum of ~100 postsynaptic potentials reaching each of the ~3,500 neurons present in the dish. On the whole, each burst was a reverberating event of the spontaneous network activity consisting of a total of ~600 spikes elicited by neurons (~75 and ~25 excitatory and inhibitory cells, at ~4 and 12/burst, respectively). This means that data recorded from a single burst were the outcome of ~60,000 synaptic events, but it should be kept in mind that this was a significant (n = ~100 electrodes) sample compared to the ~3,500 interconnected neurons present in the whole network.

When the culture dishes reached maturation (from 10 up to 22 DIV) we performed experiments, lasting more than ~15 h, during which no significant changes in network activity were observed under control conditions. These cultures were therefore useful for acquiring data in the long-lasting test on the LPS effects, in which we recorded the atypical "seizure-like" activity (Gullo et al., 2014). To study the neuronal activity, we analyzed firing data by using two variables fully describing the temporal changes: (a) plots of normEXC (excitability normalized to control) of the clusters of excitatory and inhibitory neurons and (b) plots of cumBD (cumulative distribution of the burst durations). Since we found that the drugs were able to induce relatively fast, but transient effects (compared to control data and long-lasting LPS effects, see Gullo et al., 2014), we analyzed results in graphs whose time axis contained appropriate breaks and drug labels. Moreover, to decide if the LPS action should be considered successful or blocked by drugs, we fixed thresholds in normEXC and $cumBD_{95}$ (the value of BD at cumBD of 95%) plots as a thin straight line and a vertical arrow, respectively (see Section Advanced Tools for Characterizing Firing and BD Histograms). Drugs were always tested, before applying LPS, by a brief preconditioning (30–60 min) interval in which a drug was added to the incubation medium.

Are Drugs Intrinsically Affecting Excitability?

For each tested drug, we performed a dose-response curve using concentrations based on published data, and then we verified if the substances *per-se* introduced relevant excitability changes with respect to control and washout.

In **Figure 1**, we show some specific exemplary experiments for CU; similar experiments were done also for the other drugs. As shown in **Figure 1A** (an exemplary experiment out of 4 similar results), the CU dose-response relationships consistently confirmed that in the low nanomolar region, both excitatory and inhibitory neurons decreased their activity immediately (over a few minute time course). On the contrary, in the region from 30 nM up to 1 μM the CU effect was opposite. Finally, the effects were almost completely reversible on washout. This plot shows the time course of the average spikes/burst elicited by neurons engaged in the activity ("excitability," EXC, see details in Section Methods) by the two inhibitory (in control ~20) and excitatory (~4.5) neuronal clusters. EXC is characterized by values that are very different between the clusters and can be changed by pharmacological manipulations (see Becchetti et al., 2012). The neuronal clusters responded immediately and differently after each drug application (10 min) starting from 1 nM up to 1 μM (see legend).

To compare different experiments, it was necessary to normalize data to the control data, and in **Figures 1B,C** these plot-types (now of "normEXC") are shown superimposed. It can be noticed that from the response in B, the IC_{50} for inhibitory neurons (closed square) was ~3 nM with a saturation after 10 nM and a recovery during washout. In **Figure 1C** we show the average of 4 similar experiments and the normEXC dose-response curves were found to have two different shapes in the two concentration regions from 1 to 30 nM and from 0.1 to 1 μM, respectively. In the first region only inhibitory cells were inhibited, and the IC_{50} was 4.5 ± 0.6 nM; on the contrary, in the second region, inhibitory cells recovered their activity and excitatory cells increased their activity. On average, at 1 μM, the global effect was an increase of activity that recovered on washout. It has previously been shown that an early and fast effect of CU, characterized by an IC_{50} in the micromolar range, consists of inhibiting glutamate (glu) release from synaptosomes from rat cortex nerve endings (Lin et al., 2011, 2012).

On the whole, these results are only partially in line with those obtained by Lin et al. (2011) by using a different technique. Indeed, since in the cortex microcircuits, inhibitory neurons receive glutamatergic synaptic inputs (in both feedback and feedforward loops, see Isaacson and Scanziani, 2011), we expected to observe a consistent decrease in their activity only in the lower CU concentration region. Compared to those of Lin et al. (2011) (glutamate release from synaptosomes), our methods were characterized by the functional synapse integrity and the astrocyte-mediated $[glu]_o$ uptake (Wanke et al., 2016), suggesting a much higher sensitivity. Moreover, our dose-response curves cannot be considered at steady-state because each drug concentration was changed every 10 min.

Transient and Long Term Effects of CU: LPS-Induced Inflammation Is Blocked at 1 μM CU

Since our experiments with LPS-induced inflammation imply very long time-segments of the order of about 18 h, we investigated if the inhibitory CU effects seen in **Figure 1** were indeed transient and therefore eventually negligible over this time-frame. Moreover, we wanted to use CU concentrations lower than those that have been recently used in animals (~100 μg/Kg/day, see Ojha et al., 2012; Hoppe et al., 2013; Zhu et al., 2014) approximatively equivalent to ~100 μM. Indeed, since we did not have any blockade induced by the blood brain barrier (see Wang et al., 2016) and the drug action should be directly on neurons, we decided to fix 1 μM as a reasonable concentration, which is in line with the cited experiments performed by Lin et al. (2011).

Experiments were done by sequentially recording activity in control (3 h), during a preconditioning in CU (about 1 h at 1 μM) and successively in LPS (at least 7 h at 3 μg/ml); a final washout of 5 h was also recorded. The fast effects of a single drug application were studied in detail by averaging data every 15 min. After 3 days, in the same dish we performed a standard experiment by applying only LPS to be sure that the effects of the induced inflammation were still working. The results of such a dual experiment are shown in **Figures 2A–D**, respectively. In the experiment of **Figure 2A** that plots normEXC, 1 μM CU produced a fast and transient activity increase which actually decayed in ~45 min. In the inset are shown the original data before normalization: we specifically analyzed the average number of spikes elicited in each burst and in each cluster of neurons; this analysis was performed counting for each neuron the spikes engaged in each burst. In the first 15 min, the drug caused an immediate firing increase of the inhibitory and excitatory clusters from 25.4 ± 1.5 to 38.9 ± 2.2 and from 5.5 ± 0.2 to 8.73 ± 0.4 spikes/burst (the bursts were 82, number of inhibitory and excitatory cells, 21 and 74, respectively) (see other details in legend). The subsequent LPS application, in fact, caused a further *decline* of normEXC for both clusters of neurons and the trend of normEXC for excitatory and inhibitory clusters was +14 ± 3 and +9 ± 1% respectively, suggesting that LPS was unable to exert its expected excitatory inflammatory function as previously observed (Gullo et al., 2014). Indeed, as shown in **Figure 2B**, the effects of a standard LPS application (3 μg/ml), done in the same dish 3 DIV later, illustrate that, after 4 and up to 8 h, a strong increase of the activity of excitatory neurons (+86 ± 6%) took place, although we did not observe any effect on inhibitory cells.

The cumBD data associated with the **Figures 2A,B** experiments are shown in **Figures 2C,D**, respectively. In C four curves are superimposed as follows: (i) control (thick line), (ii) CU at 15 min (line + asterisk), (iii) CU at 60 min (thin line), (iv) LPS at 7 h (line + right-triangles). The corresponding 90% BD values were: 2.6, 3.8, 2.6, 3.7 s, respectively. These results suggest that, as compared to control, CU only transiently increased $cumBD_{90\%}$, and LPS over 7 h produced only a 37% increase, confirming that the expected LPS action did not take place. Statistical significance by Kruskal–Wallis test applied to control data and CU at 60 min resulted in a $P < 0.12$, and at 7 h in LPS data resulted in $P < 0.05$. In contrast, as shown in **Figure 2D**, a typical effect of LPS on cumBD (in the same dish) was to further right-shift, with respect to control (line, 95% at 7.2 s), the curves at 5 h (right-triangles) and at 9 h (upward-triangles) to 12 s (+66%) and 17 s (+136%) values, respectively, with "seizure-like" burst durations reaching 22 s.

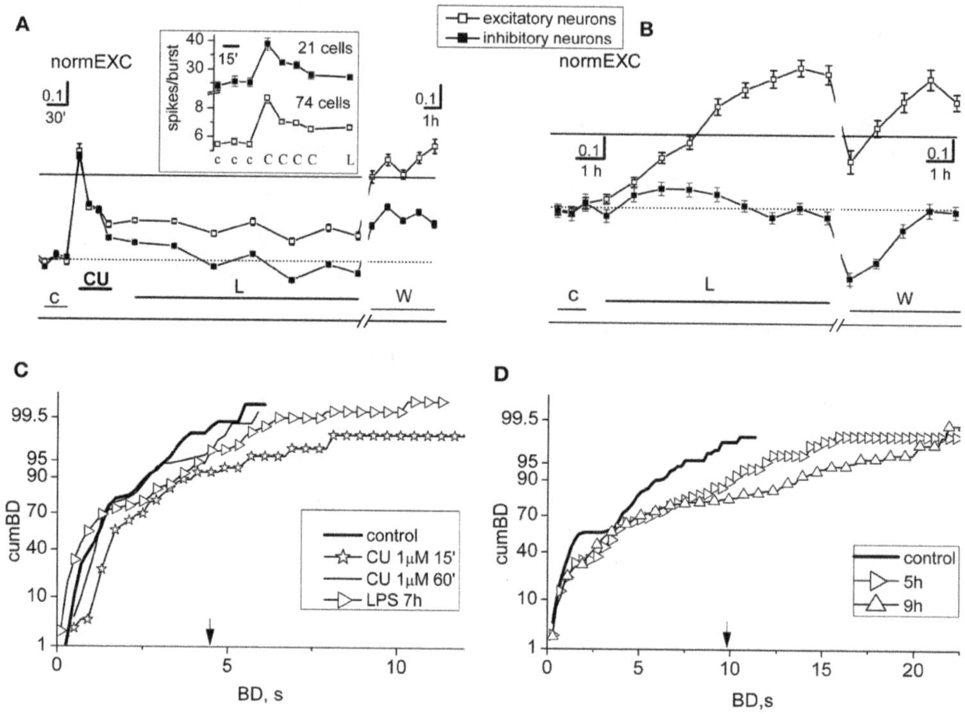

FIGURE 2 | Short- and long-term effects of curcumin in networks treated with LPS. Results from two different experiments, performed in the same dish, with preconditioning or not-preconditioning with 1 μM curcumin (CU) (done 3 DIV later) before the application of LPS (3 μg/ml) are shown in **(A,C)** and **(B,D)**, respectively. **(A)** Plots of normEXC in control (c), 1 μM CU (CU), LPS (L), and during washout (w). Inset: for the brief (~3 h) initial time segment in control, CU and L, plot of average number of spikes/burst observed in neurons of both clusters. The continuous line is the fixed threshold. Other details: the total number of spikes from excitatory (inhibitory, in parenthesis) neurons in control and CU (in bold): 304.2 ± 18, **568.5 ± 32** (481.9 ± 33, **764.7 ± 45.4**); the total number of engaged excitatory (inhibitory, in parenthesis) neurons was in control: 52 ± 1.8, **63.8 ± 1.1** (17.8 ± 0.3, **19.3 ± 0.2**). **(B)** In the same dish as in **(A)**, but 3 DIV later, plots of normEXC in control (c), LPS (L) and during washout (w). The continuous line is the fixed threshold. **(C)** Plots of cumBD in different time regions of the experiment shown in **(A)**. Data of cumBD are superimposed as explained in text. **(D)** Plots of superimposed cumBD curves in different time regions of the experiment shown in **(B)**. Note that pre-conditioning with CU prevented the late, prolonged excitatory response to LPS. The vertical arrow indicates the fixed threshold.

The Kruskal–Wallis test of significance applied to control data and LPS data resulted in *P*-values which were always smaller than 10^{-4}.

On the whole, these results demonstrate that 1 μM CU by itself suddenly produced changes of activity in bursting, by promoting long BD values and significant increases in the excitability of both neuronal clusters. These transient modifications rapidly returned to quasi-control values as if the effects were partially compensated, only producing scarce or null effects in the early hours after adding LPS. On all of the occasions ($n = 10$) in which we performed these types of experiments, the results were similar. In other experiments ($n = 5$) in which we used smaller CU concentrations (i.e., 0.1 μM), the results were not clear and these data were discarded. Few ($n = 4$) experiments performed in dishes whose LPS responses were not significant were also discarded.

Transient and Long Term Effects of CR: LPS-Induced Inflammation Is Blocked at 20 μM CR

Crocin and crocetin have been found to block the release of NO and various chemokines under the action of LPS (Nam et al., 2010). As for curcumin, we did fast dose-response experiments initially to test putative effects of crocin (CR) on the activity of our networks (**Supplementary Figure 1**). In the range 1–100 μM, some inhibitory effects were seen only beyond 20 μM (n = 3). We here show that during the first hour of incubation, experiments using CR at 3 ($n = 3$) and 20 μM ($n = 3$) had either negligible or transient inhibitory effects, respectively. These CR effects were accompanied by other different actions caused by the further LPS presence.

In **Figure 3** are shown the results from two exemplary experiments at 3 (**Figures 3A,C**) and 20 (**Figures 3B,D**) μM CR, respectively. At 3 μM it can be observed in **Figure 3A** that the early CR-induced activity was barely modified and LPS induced a very slow, but sizeable increase of activity up to ~52% at eighth hours. An analysis to test the cumBD plots, shown in **Figure 3C**, suggests that a significant increase by 46% of BD was present at 7 h LPS as compared to control.

On the contrary, at 20 μM CR, the effects were found to be largely depressive. After the CR application (see **Figure 3B**), we detected an early negative peak of 45 and 16%, respectively in inhibitory and excitatory cells, but this effect decayed in the next hour. During the next 7 h in LPS, the activity remained

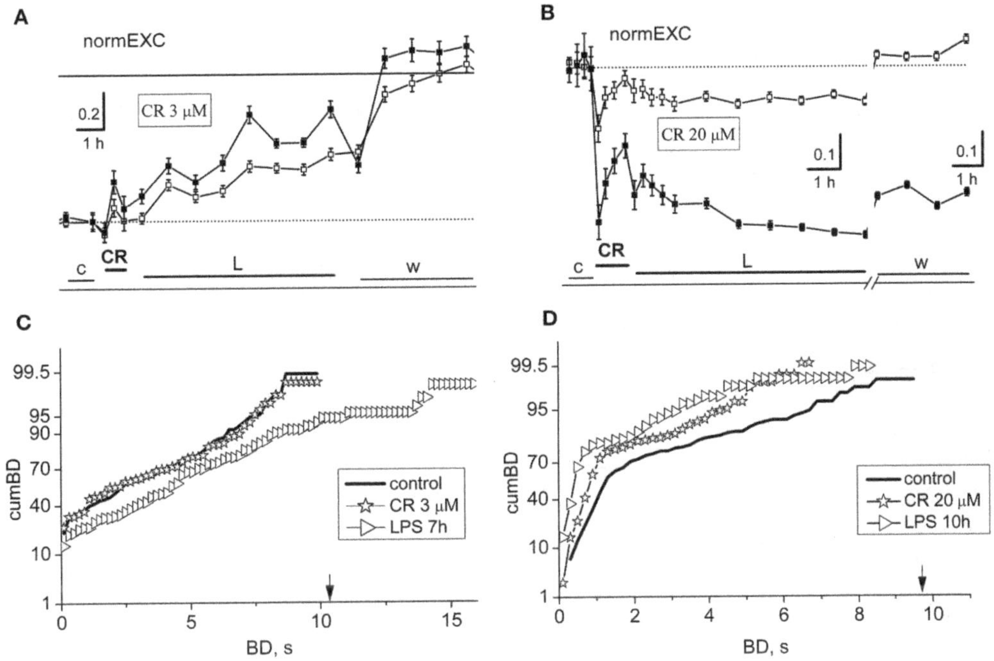

FIGURE 3 | Short- and long-term effects of crocin (CR) in networks treated with LPS. Data in **(A,C)** and those in **(B,D)** belong to the same exemplary experiment. **(A,B)** Plots of normEXC data for two exemplary experiments in which the dishes were treated with crocin (CR) at 3 and 20 μM, respectively. Open and closed symbols indicate excitatory and inhibitory neurons. The continuous line is the fixed threshold. **(C)** Plot of cumBD in control (thick line), in the presence of 3 μM CR (asterisks), and during the seventh hour after the addition of LPS (3 μg/ml; right triangles). These two last curves were significantly different from the control curve with $P < 0.05$. **(D)** Plot of cumBD in control (thick line), in the presence of 20 μM CR (asterisks), and during the 10th h after the addition of LPS (right triangles). The vertical arrow indicates the fixed threshold. These last curves were significantly different from the control curve with $P < 0.05$. The highest dose of CR prevented the late excitatory LPS responses.

stable but depressed. Furthermore, by analyzing in **Figure 3D** the cumBD plots, it can be concluded that the curves shifted to the left, suggesting that the burst durations decreased, in line with a weaker activity of the network.

On the whole, it can be concluded that CR is able to block the LPS-induced inflammation only at concentrations higher than 15–20 μM. Interestingly, at these concentrations, although the fast earlier transient inhibitory effects were present, a strong general inhibition of activity remained and was washed out only during recovery.

Transient and Long Term Effects of RE: LPS-Induced Inflammation Is Blocked at 1 μM RE

We studied the action of resveratrol (RE) at concentrations in the lower micromolar range, much lower than those reported in many papers present in the literature (Gambini et al., 2015). As reported above for CU, also RE has been described to inhibit the release of glutamate from nerve terminals (Chang and Wang, 2009). These authors showed that the threshold to detect the start of the inhibition was between 10 and 100 nM, but at 1 μM the effect reached the IC_{50}. When tested in our dishes, 200 nM RE (**Figure 4B**) exerted immediately a decrease in the neuronal firing, but the network slowly recovered after ~30 min.

Due to this reason, we decided to search for a concentration range in which effects were either unable or able to cause changes in the plots of normEXC and the cumBD in opposite directions. In **Figures 4A,B** are shown two exemplary experiments in which RE applied at 50 and 200 nM induced a negligible or a fast inhibitory effect, respectively, in the normEXC plots. The average duration of these intrinsic responses of RE were never longer than ~45 min. The subsequent application of LPS was tested every hour over 8 h.

In **Figures 4A,B** it can be observed that the development of the response was divergent, i.e.: normEXC data either increased in **Figure 4A** (at 50 nM) or decreased in **Figure 4B** (at 200 nM), respectively, suggesting that the drug preconditioning produced opposite results during the LPS action. Similar results were also seen in other 4 dishes at 50 nM and 3 dishes at 200 nM. The fluctuations of the activity shown in **Figure 4A**, during control and RE, were not significant up to the first hour of LPS application, whereas from 2 to 8 h in LPS, the activity increased as expected by 71% for excitatory cells and 127% for inhibitory cells (see in Gullo et al., 2014). On the other hand, in **Figure 4B**, the normEXC slowly decayed during the following 8 h.

To complete the analysis, we computed the distribution of the BD durations of these experiments and the results are shown in **Figures 4C,D**, respectively. Although the profile properties of the control curves (thick line) were different in the two experiments, the data acquired during the presence of RE alone show, in both cases, a cumBD curve (see the asterisk symbols) very similar to control. On the contrary, comparing the last hours of LPS

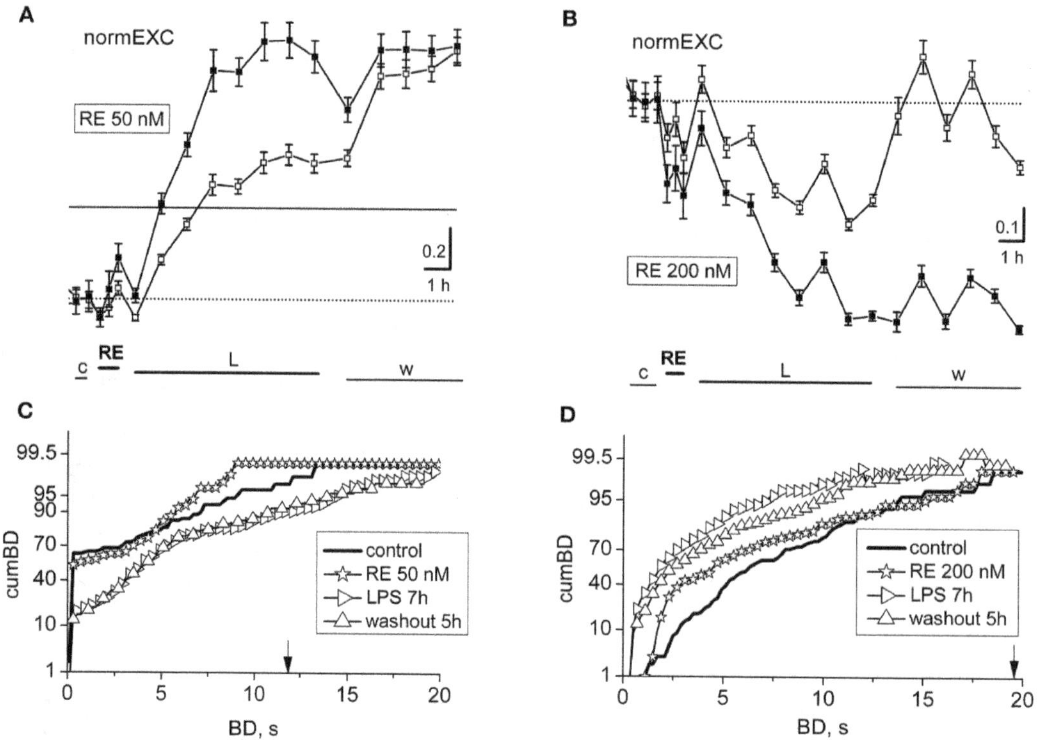

FIGURE 4 | Effects of resveratrol (50 or 200 nM) on LPS responses. Data in **(A,B)** and **(C,D)** derive from two experiments performed on a single dish at 14 or 20 DIV. **(A,B)** Plots of normalized excitability following application of resveratrol (RE) 50 nM **(A)** or 200 nM **(B)** and LPS (3 μg/ml). The continuous line is the fixed threshold. Note that in **(A)**, the late excitatory responses to LPS were not prevented, whereas in **(B)** the responses were inhibited. **(C,D)** Plots of cumBD in the same conditions as in **(A,B)**. The vertical arrow indicates the fixed threshold.

(right-pointing triangles) data in **Figure 4C** suggest that the 90% cumBD showed an 87% increase, and in 4 D a dramatic decrease, thus confirming that the RE doses used in these experiments either did not block or completely blocked the LPS action, respectively.

In Agreement with Data Reported in Hippocampal Pyramidal Neurons, EX-4 Triggered Fast Transient Pro-Excitatory Responses

In hippocampal pyramidal neurons *in vivo*, it has been shown that a brief (1 s) application of GLP-1 (7–36) amide, a naturally produced active fragment of GLP-1, first rapidly increased (i.e., 1 s) and then slowly decreased (>10 s) single unit firing activity (Oka et al., 1999). Interestingly, these effects were inhibited either by the specific GLP-1 receptor antagonist, exendin (9–39) or by the specific non-NMDA glutamate receptor antagonist 6-cyano-7-nitroquinoxaline-2,3-dione (CNQX), suggesting that the initial GLP-1 action involved a fast and transient release of glutamate.

Our *in vitro* networks of cortical neurons from neonatal mice contain fractions of excitatory and inhibitory cortical neurons, which always fit the standard ratio of 4:1 known to be present in adult cortex (Sahara et al., 2012). As for curcumin and crocin, we did fast dose-response experiments to test putative effects of EX-4 on the activity of our networks (**Supplementary Figure 1**). Surprisingly, we found that in about ~80% (n = 24) of the experiments done using EX-4, the fast effects consisted, as expected, of a transient, brief pro-excitability response, but there were also some networks (~20%, n = 6) whose responses consisted of a transient *decreased* excitability. This was probably caused by unknown heterogeneity factors intrinsic to our cultured preparation of dissociated cells, and therefore we excluded these experiments from the results (although they were characterized by late LPS responses where atypical activity was present). EX-4, from 10 up to 500 nM, elicited an early fast excitatory response and prevented the late LPS-induced neuroinflammation. Amongst the EX-4 experiments in which the early effects of the drug were pro-excitable, in **Figure 5** are shown the results of two experiments, performed in the same dish (at two different DIVs) with 10 and 500 nM EX-4 and one experiment done at 50 nM. In **Figures 5A–C** are compared the normEXC data and it is evident that concentrations differing by 5 and 50 times indeed produced fast transient excitatory effects, suggesting a dose-dependent mechanism of action. On the contrary, the long-term normEXC curves at the end of the LPS application did not show any significant change, thus suggesting that activity of the network was not different with respect to control. We analyzed in **Figures 5D–F** the related

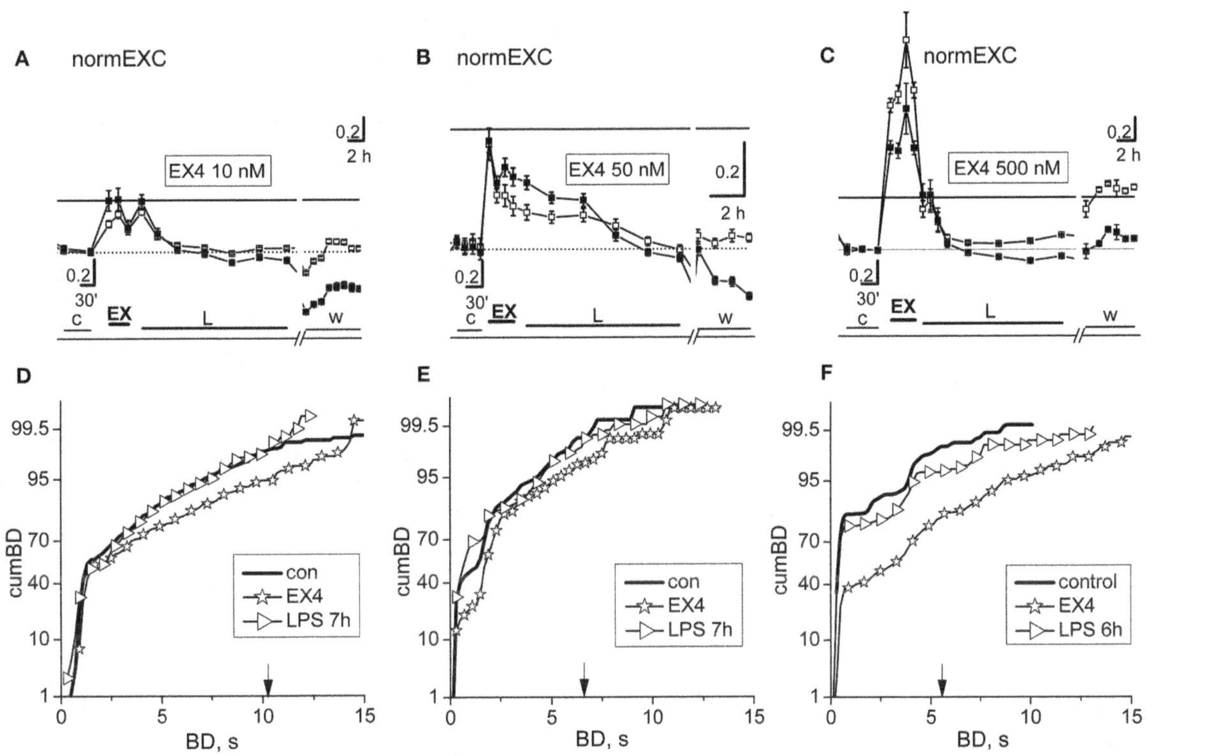

FIGURE 5 | The fast and slow network responses to 10, 50, and 500 nM exendin-4 or exendin-4 + LPS. Data in **(A,D)** and **(C,F)** were obtained from the same dish in two different experiments performed 3 DIV apart. **(A–C)** NormEXC plot obtained with 10, 50, and 500 nM exendin-4 (EX-4) preconditioning, and 3 μg/ml LPS. The continuous line is the fixed threshold. Note that the transient and fast increase of normEXC became progressively higher with increasing EX-4 concentration. **(D–F)** CumBD plots of the experiments shown in **(A–C)**. Control, thick line; EX-4, asterisk; rightward triangles are for the late LPS effects. The vertical arrow indicates the fixed threshold. Note that preconditioning with EX-4 at each concentration tested, effectively prevented the development of the late excitatory responses to LPS.

cumBD histograms in control, during the early EX-4 action and at the end of the long-term LPS action. In the three cases, the early EX-4 histograms showed a conspicuous rightward shift of the curves. On the contrary, the final effect of LPS was either negligible or without atypical bursts showing a BD increase at 90% in the cumBD curves.

The GLP-1 Receptor Antagonist Exendin (9-39) Blocked the Fast Transient EX-4 Effect and the Protective Effect against LPS Neuroinflammation

In order to confirm that the EX-4 action shown in **Figure 5** is really due to the activation of GLP-1Rs, we performed experiments ($n = 6$) in which we preconditioned the EX-4 application (50 nM) with the specific GLP-1R antagonist exendin (9–39) at 100 nM. The results of such an exemplary experiment are shown in **Figure 6**. In **Figure 6A**, the normEXC plot shows that the early application of exendin (9–39) (e) induced a very short (10 min) and small negative deflection, followed by the application of EX-4 (E), and of LPS, which increased the slow normEXC up to 57 ± 3%. After 7 h, the activity stopped increasing. In **Figure 6B** the cumBD plots show that the 90% value was 8.1 and 12.2 s, in control and LPS, respectively, thus suggesting that a normal excitatory action of LPS took place as if EX-4 was not present. On the whole, these experiments confirm that the results shown in **Figure 5** robustly indicate EX-4 as a potent agent to prevent LPS-induced inflammation.

The Levels of TNF-α, Measured at 6 and 12 h after Preconditioning with the Polyphenols or EX-4 in LPS-Treated Dishes, Is Negligible Compared to the Levels Found in Non-preconditioned Dishes

As reported in our previous paper (Gullo et al., 2014) where we showed that minocycline blocked the microglial-release of TNF-α induced by LPS treatment, we performed, for all the polyphenols and EX-4, appropriate experiments to test if the drug concentrations used in the experiments (**Figures 2–5**) were able to block also TNF-α production. As shown in **Figure 7**, the amount of this cytokine was, both at 6 and 12 h after LPS treatment, significantly smaller than that found when the preconditioning drugs were absent. On the whole, these results confirm that the drugs used here were not only able to block the atypical seizures induced by LPS, but also the release of TNF-α in the network dishes.

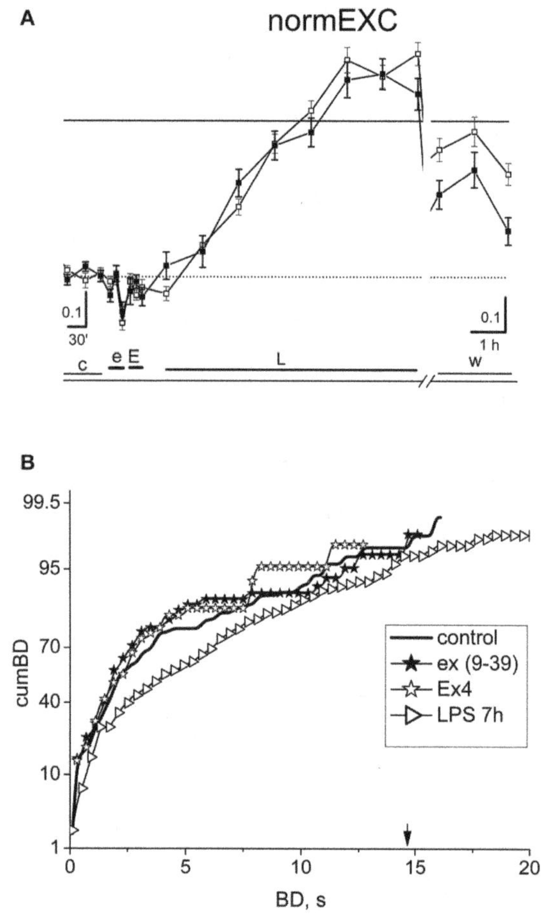

FIGURE 6 | The effect of the specific antagonist of EX-4, exendin (9-39). **(A)** Plot of the normEXC in an experiment in which 100 nM exendin (9–39) (e) was applied for 30 min before 50 nM EX-4 (30 min, E) and the final long-term application of LPS (3 μg/ml). The continuous line is the fixed threshold. **(B)** Plot of cumBD in control (thick line), in the presence of exendin (9–39) (closed asterisk), or EX-4 added on top (open asterisk) and in LPS after 7 h (rightward triangles). The vertical arrow indicates the fixed threshold. The four curves were tested with the non-parametric Wilcoxon signed rank test analysis (exendin (9-39) vs. control, $P < 10^{-5}$; LPS 7 h vs. control, $P < 10^{-10}$; EX4 vs. exendin (9–39), $P = 0.32$). The GLP1-R antagonist blocked the protective effect of EX-4 observed in **Figure 5**, allowing the late excitatory responses to LPS to be expressed.

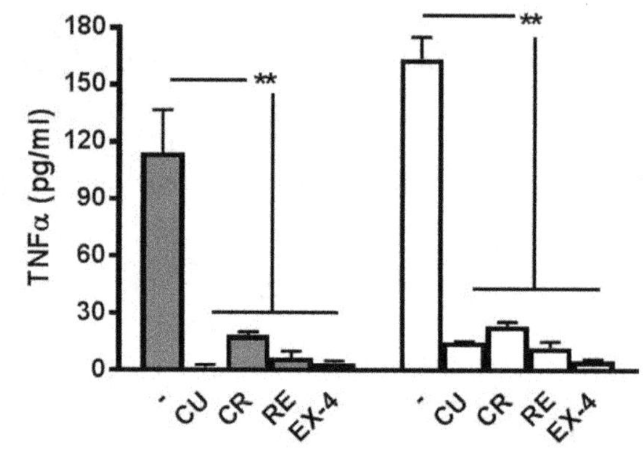

FIGURE 7 | Release of TNF-α at 6 and 12 h after LPS application. The cytokine TNF-α is released from LPS-treated networks but not from networks pre-treated with polyphenols or exendin-4. Graph shows TNF-α concentration 6 h (filled bars) and 12 h (open bars) post-LPS (3 μg/ml) treatment with or without curcumin (CU, 1 μM) crocin (CR, 20 μM), resveratrol (RE, 200 nM) or exendin-4 (EX-4, 30 nM) pre-conditioning. Data are expressed as pg/ml and represent mean ± SEM of three independent experiments. $^{**}P < 0.01$ (ANOVA, Dunnet's test).

DISCUSSION

The present study is the first to demonstrate that polyphenols and the incretin GLP-1 hormone receptor agonist exendin-4 can mimic the anti-inflammatory action of minocycline (Yrjänheikki et al., 1998). In particular, they were able to block the seizure-like atypical activity caused by LPS and the concurrent TNF-α release from microglia, triggered by the Toll-like receptor 4 (TRL4) pathway, in a cortical co-culture of neurons, astrocytes, and microglia. The similar hyperactivity caused by LPS incubation was also observed with the application of a very low (60 pM) TNF-α concentration for ~12 h (Gullo et al., 2014), probably because of a marked change in the expression of glutamate transporters in astrocytes and microglia, as previously described (Persson et al., 2005; O'Shea et al., 2006; Takaki et al., 2012). The putative role of altered glutamate uptake in mediating components of the sterile inflammation response is currently under investigation in our cortical co-colture system (EW, FG, and ML), which includes functional astocytes (Wanke et al., 2016).

Roles of Microglia

Microglia are highly ramified cells which rapidly scan the local environment and react to its modification (Davalos et al., 2005). Under physiological conditions, microglia react rapidly to neuronal activity by modulating the contacts that their processes continuously establish with synaptic elements (Wake et al., 2009). Amongst the microglial-released molecules with a putative role in neurotransmission, TNF-α was shown to control basal synaptic functions (Santello et al., 2011) as well as plasticity (Stellwagen and Malenka, 2006; Costello et al., 2011), even if this role was attributed to TNF-α produced by astrocytes. However, the astrocytes have often been thought to release TNF-α because cultures of astrocytes are consistently contaminated by microglia (Saura, 2007). In fact, other data reveal no TNFα-encoding transcript in astrocytes (Foo et al., 2011; Zamanian et al., 2012). Thus, the TNF-α that controls several aspects of synaptic transmission is produced by microglial cells. Similarly, TLR4, the specific LPS receptor, was detected in mixed astrocyte/neuronal cultures, but when microglia were removed from astrocyte cultures, TLR4 expression was no longer present (Lehnardt et al., 2002).

Polyphenols and Their Features of Action

Apart from the obvious concentration-dependence, the effects of polyphenols can be subdivided into various types of action according to: (i) time-dependence (fast or long-term), (ii) cell-dependence (nervous or non-nervous system), and (iii) disease-dependence (i.e., Parkinson's, Alzheimer's diseases, or cancer). CU was found to be neuroprotective in the MPTP model of Parkinson's disease (Ojha et al., 2012) by inhibiting the generation of pro-inflammatory cytokines through the prevention of NF-κB traslocation into the nucleus (Karunaweera et al., 2015). In addition, CU attenuated inflammation in experimental traumatic brain injury (Zhu et al., 2014) and decreased human TNF-α levels (Sahebkara et al., 2016). The nanoencapsulated CU version was also found to reduce β-amyloid-induced cognitive impairment in rats (Hoppe et al., 2013). In the same way as CU, RE was also able to protect dopamine neurons (Zhang et al., 2010) from LPS-induced inflammation. Moreover, RE was found to protect cortical neurons from oxygen-glucose deprivation-induced apoptosis (Gao et al., 2014). In general, RE studies in *in vitro* and *in vivo* showed interesting data of metabolism and bioavailability in various animal models and in humans (Smoliga and Blanchard, 2014; Gambini et al., 2015). Interestingly, RE was shown to ameliorate the clinical severity observed in an animal model of multiple sclerosis, by maintaining the integrity of the blood-brain barrier (Wang et al., 2016). In summary, we believe that the principal mechanisms by which polyphenols were able to protect against LPS neuroinflammation in our system are those related to the TLR4 pathway and microglia activation.

The Action of GLP-1 in the CNS

There is evidence that the gut hormone GLP-1 is also produced in the brain (Korol et al., 2015; Thiebaud et al., 2016). In particular, it was shown that the GLP-1 receptor agonist exendin-4 (EX-4) protected dopaminergic neurons by inhibiting TNF-α release from microglia (Kim et al., 2009). A reduction of brain TNF-α levels was also described in an animal model of Alzheimer disease (AD) exposed to EX-4 treatment (Solmaz et al., 2015). In LPS-treated rats, GLP-1 protected hippocampal neurons from synaptic impairments (Iwai et al., 2014) and inhibited IL-1β production from cultured astrocytes (Iwai et al., 2006). Furthermore, *in vitro* administration of EX-4 in rats increased the expression of the astrocytic glutamate transporter GLT-1 in the hippocampus, lowering extracellular glutamate concentration (Kobayashi et al., 2013). This is in line with the lower long-term excitability data shown in **Figures 5B,C** during LPS inflammation. The results shown in **Figures 5, 6** split the presumed fast AMPAR-induced transient firing increase following GLP-1R activation (by EX-4), as characterized by Oka et al. (1999), from the long-term effects related to the switch-off of the LPS-induced inflammation.

Our data show, quite importantly, and for the first time, that these effects are not peculiar to the hippocampus *in vivo*, but can also be reproduced *in vitro*, in networks from neocortical dissociated cells containing neurons, astrocytes, and microglia. Moreover, the data shown in **Figure 6** demonstrate that the GLP-1R antagonist exendin (9–39) functionally blocked the ability of EX-4 to counteract the LPS sterile inflammation. Thus in summary, we think that the intriguing connection between the fast electrophysiological responses and the slow microglia pathway merits a specialized investigation, difficult to be carried out by means of our procedures.

Ion Channels in Microglia

Only recently, ion channels have also been described to be functionally expressed in microglia (for a minireview, see Madry and Attwell, 2015). Among the various types of identified channels, the voltage-gated proton channel Hv1 (not present in neurons) was thought to have a dramatic importance during damage induced by ischemic stroke when reactive oxygen species (ROS) are generated (Wu et al., 2012). Accordingly, knock-out mice lacking the Hv1 channel were protected from ROS-mediated neuronal death and brain damage. Furthermore, since acute neuroinflammation provokes intracellular acidification (Tyrtyshnaia et al., 2016), it cannot be excluded that voltage-gated proton channels or acid-sensing channels (Yu et al., 2015) could also be involved in this type of process. Indeed, a polyphenol such as epigallocatechin-3-gallate (EGCG) is known to inhibit the Hv1 currents in microglial BV2 cells (Jin et al., 2012). Interestingly, the reduction in TNF-α release produced by some antidepressants was supposed to be related to the same effect (Song et al., 2012). The putative role of this ionic mechanism in the LPS-induced sterile inflammation is also currently under investigation in our laboratory (EW, FG, and ML).

AUTHOR CONTRIBUTIONS

AC, ML, BC, and EW conceived and designed the research; FG, MC, and AD performed the experiments; EW and BC analyzed the data; EW and AC drafted the manuscript.

REFERENCES

Athauda, D., and Foltynie, T. (2016). The glucagon-like peptide 1(GLP) receptor as a therapeutic target in Parkinson's disease: mechanisms of action. *Drug Discov. Today* 21, 802–818. doi: 10.1016/j.drudis.2016.01.013

Baddeley, R., Abbot, L. F., Booth, M. C., Sengpiel, F., Freeman, T., Wakeman, E. A., et al. (1997). Responses of neurons in primary and inferior temporal visual cortices to natural scenes. *Proc. R. Soc Lond. B Biol. Sci.* 264, 1775–1783. doi: 10.1098/rspb.1997.0246

Becchetti, A., Gullo, F., Bruno, G., Dossi, E., Lecchi, M., and Wanke, E. (2012). Exact distinction of excitatory and inhibitory neurons in neural networks: a study with GFP-GAD67 neurons optically and electrophysiologically recognized on multielectrode arrays. *Front. Neural Circuits* 6:63. doi: 10.3389/fncir.2012.00063

Chang, Y., and Wang, S. J. (2009). Inhibitory effect of glutamate release from rat cerebrocortical nerve terminals by resveratrol. *Neurochem. Int.* 54, 135–141. doi: 10.1016/j.neuint.2008.11.001

Costello, D. A., Lyons, A., Denieffe, S., Browne, T. C., Cox, F. F., and Lynch, M. A. (2011). Long term potentia-tion is impaired in membrane gly-coprotein CD200-deficient mice: a role for Toll-like receptor activation. *J. Biol. Chem.* 286, 34722–34732. doi: 10.1074/jbc.M111.280826

Davalos, D., Grutzendler, J., Yang, G., Kim, J. V., Zuo, Y., Jung, S., et al. (2005). ATP mediates rapid microglial response to local brain injury *in vivo*. *Nat. Neurosci.* 8, 752–758. doi: 10.1038/nn1472

Foo, L. C., Allen, N. J., Bushong, E. A., Ventura, P. B., Chung, W.-S., Zhou, L., et al. (2011). Development of a method for the purification and culture of rodent astrocytes. *Neuron* 71, 799–811. doi: 10.1016/j.neuron.2011.07.022

Gambini, J., Inglés, M., Olaso, G., Lopez-Grueso, R., Bonet-Costa, V., Gimeno-Mallench, L., et al. (2015). Properties of Resveratrol: *In vitro* and *In vivo* studies about metabolism, bioavailability, and biological effects in animal models and humans. *Oxid. Med. Cell. Longev.* 2015:837042. doi: 10.1155/2015/837042

Gao, D., Huang, T., Jiang, X., Hu, S., Zhang, L., and Fei, Z. (2014). Resveratrol protects primary cortical neuron cultures from transient oxygen glucose deprivation by inhibiting MMP 9. *Mol. Med. Rep.* 9, 2197–2204. doi: 10.3892/mmr.2014.2086

Gullo, F., Amadeo, A., Donvito, G., Lecchi, M., Costa, B., Constanti, A., et al. (2014). Atypical "seizure-like" activity in cortical reverberating networks *in vitro* can be caused by LPS-induced inflammation: a multi-electrode array study from a hundred neurons. *Front. Cell. Neurosci.* 8:361. doi: 10.3389/fncel.2014.00361

Gullo, F., Maffezzoli, A., Dossi, E., and Wanke, E. (2009). Short latency cross-and autocorrelation identify clusters of interacting neurons recorded from multi-electrode arrays. *J. Neurosci. Methods* 181, 186–198. doi: 10.1016/j.jneumeth.2009.05.003

Gullo, F., Maffezzoli, A., Dossi, E., Lecchi, M., and Wanke, E. (2012). Classifying heterogeneity of spontaneous up-states: a method for revealing variations in firing probability, engaged neurons and Fano factor. *J. Neurosci. Methods* 203, 407–417. doi: 10.1016/j.jneumeth.2011.10.014

Gullo, F., Mazzetti, S., Maffezzoli, A., Dossi, E., Lecchi, M., Amadeo, A., et al. (2010). Orchestration of "presto" and "largo" synchrony in up-down activity of cortical networks. *Front. Neural Circuits* 4:11. doi: 10.3389/fncir.2010.00011

Hoppe, J. B., Coradini, K., Frozza, R. L., Oliveira, C. M., Meneghetti, A. B., Bernardi, A., et al. (2013). Free and nanoencapsulated curcumin suppress β-amyloid-induced cognitive impairments in rats: involvement of BDNF and Akt/GSK-3β signaling pathway. *Neurobiol. Learn. Mem.* 106, 134–144. doi: 10.1016/j.nlm.2013.08.001

Isaacson, J. S., and Scanziani, M. (2011). How inhibition shapes cortical activity. *Neuron* 72, 231–243. doi: 10.1016/j.neuron.2011.09.027

Iwai, T., Ito, S., Tanimitsu, K., Udagawa, S., and Oka, J. (2006). Glucagon-like peptide-1 inhibits LPS-induced IL-1beta production in cultured rat astrocytes. *Neurosci. Res.* 55, 352–360. doi: 10.1016/j.neures.2006.04.008

Iwai, T., Sawabe, T., Tanimitsu, K., Suzuki, M., Sasaki-Hamada, S., and Oka, J. (2014). Glucagon-like peptide-1 protects synaptic and learning functions from neuroinflammation in rodents. *J. Neurosci. Res.* 92, 446–454. doi: 10.1002/jnr.23335

Jin, S., Park, M., and Song, J. H. (2012). (-)-Epigallocatechin-3-gallate inhibits voltage-gated proton currents in BV2 microglial cells. *Eur. J. Pharmacol.* 698, 154–160. doi: 10.1016/j.ejphar.2012.11.036

Karunaweera, N., Raju, R., Gyengesi, E., and Muench, G. (2015). Plant polyphenols as inhibitors of NF-κB induced cytokine production—a potential anti-inflammatory treatment for Alzheimer's disease? *Front. Mol. Neurosci.* 8:24. doi: 10.3389/fnmol.2015.00024

Kettenmann, H., Hanisch, U. K., Noda, M., and Verkhratsky, A. (2011). Physiology of microglia. *Physiol. Rev.* 91, 461–553. doi: 10.1152/physrev.00011.2010

Kettenmann, H., Kirchhoff, F., and Verkhratsky, A. (2013). Microglia: new roles for the synaptic stripper. *Neuron* 77, 10–18. doi: 10.1016/j.neuron.2012.12.023

Kim, S., Moon, M., and Park, S. (2009). Exendin-4 protects dopaminergic neurons by inhibition of microglial activation and matrix metalloproneinase-3 expression in an animal model of Perkinson's disease. *J. Endocrinol.* 202, 431–439. doi: 10.1677/JOE-09-0132

Kobayashi, K., Iwai, T., Sasaki-Hamada, S., Kamanaka, G., and Oka, J. (2013). Exendin (5-39), an antagonist of GLP-1 receptor, modulates synaptic transmission via glutamate uptake in the dentate gyrus. *Brain Res.* 1505, 1–10. doi: 10.1016/j.brainres.2013.01.012

Korol, S. V., Jin, Z., Babateen, O., and Birnir, B. (2015). GLP-1 and Exendin-4 transiently enhance GABAA receptor-mediated synaptic and tonic currents in rat hippocampal CA3 pyramidal neurons. *Diabetes* 64, 79–89. doi: 10.2337/db14-0668

Lehnardt, S., Lachance, C., Patrizi, S., Lefebvre, S., Follett, P. L., Jensen, F. E., et al. (2002). The toll-like receptor TLR4 is necessary for lipopolysaccharide-induced oligo-dendrocyte injury in the CNS. *J. Neurosci.* 22, 2478–2486.

Lin, T. Y., Lu, C. W., Huang, S. K., and Wang, S. J. (2012). Curcumin inhibits glutamate release from rat prefrontal nerve endings by affecting vesicle mobilization. *Int. J. Mol. Sci.* 13, 9097–9109. doi: 10.3390/ijms13079097

Lin, T. Y., Lu, C. W., Wang, C. C., Wang, Y. C., and Wang, S. J. (2011). Curcumin inhibits glutamate release in nerve terminals from rat prefrontal cortex: possible relevance to its antidepressant mechanism. *Prog. Neuro Psychopharm. Biol. Psychol.* 35, 1785–1793. doi: 10.1016/j.pnpbp.2011.06.012

Lofrumento, D. D., Nicolardi, G., Cianciulli, A., De Nuccio, F., La Pesa, V., Carofiglio, V., et al. (2014). Neuroprotective effects of resveratrol in an MPTP mouse model of Parkinson's-like disease: possible role of SOCS-1 in reducing pro-inflammatory responses. *Innate Immun.* 20, 249–260. doi: 10.1177/1753425913488429

Madry, C., and Attwell, D. (2015). Receptors, ion channels, and signaling mechanisms underlying microglial dynamics. *J. Biol. Chem.* 290, 12443–12450. doi: 10.1074/jbc.R115.637157

Marangoni, D., Falsini, B., Piccardi, M., Ambrosio, L., Minnella, A. L., Savastano, M. C., et al. (2013). Functional effect of Saffron supplementation and risk genotypes in early age-related macular degeneration: a preliminary report. *J. Trans. Med.* 11:228. doi: 10.1186/1479-5876-11-228

Nam, K. N., Park, Y. M., Jung, H. J., Lee, J. Y., Min, B. D., Park, S. U., et al. (2010). Anti-inflammatory effects of crocin and crocetin in rat brain microglial cells. *Eur. J. Pharmacol.* 648, 110–116. doi: 10.1016/j.ejphar.2010.09.003

Ojha, R. P., Rastogi, M., Devi, B. P., Agrawal, A., and Dubey, G. P. (2012). Neuroprotective effect of curcuminoids against inflammation-mediated dopaminergic neurodegeneration in the MPTP model of Parkinson's disease. *J. Neuroimmune Pharmacol.* 7, 609–618. doi: 10.1007/s11481-012-9363-2

Oka, J. I., Goto, N., and Kameyama, T. (1999). Glucagon-like peptide-1 modulates neuronal activity in the rat's hippocampus. *Neuroreport* 10, 1643–1646. doi: 10.1097/00001756-199906030-00004

O'Shea, R. D., Lau, C. L., Farso, M. C., Diwakarla, S., Zagami, C. J., Svendsen, B. B., et al. (2006). Effects of lipopolysaccharide on glial phenotype and activity of glutamate transporters: evidence for delayed up-regulation and redistribution of GLT-1. *Neurochem. Int.* 48, 604–610. doi: 10.1016/j.neuint.2005.12.028

Persson, M., Brantefjord, M., Hansson, E., and Rönnbäck, L. (2005). Lipopolysaccharide increases microglial GLT-1 expression and glutamate uptake capacity *in vitro* by a mechanism dependent on TNF-alpha. *Glia* 51, 111–120. doi: 10.1002/glia.20191

Pocock, J. M., and Kettenmann, H. (2007). Neurotransmitter receptors on microglia. *Trends Neurosci.* 30, 527–535. doi: 10.1016/j.tins.2007.07.007

Sahara, S., Yanagawa, Y., O'Learly, D. M., and Stevens, C. F. (2012). The fraction of cortical GABAergic neurons is constant from near the start of cortical neurogenesis to adulthood. *J. Neurosci.* 32, 4755–4761. doi: 10.1523/JNEUROSCI.6412-11.2012

Sahebkara, A., Cicero, A. F. G., Simental-Mendíad, L. E., Aggarwale, B. B., and Gupta, S. C. (2016). Curcumin downregulates human tumor necrosis factor-α: a systematic review and meta-analysis ofrandomized controlled trials. *Pharmacol. Res.* 107, 234–242. doi: 10.1016/j.phrs.2016.03.026

Santello, M., Bezzi, P., and Volterra, A. (2011). TNFα controls glutamater-gic gliotransmission in the hip-pocampal dentate gyrus. *Neuron* 69, 988–1001. doi: 10.1016/j.neuron.2011.02.003

Saura, J. (2007). Microglial cells in astroglial cultures: a cautionary note. *J. Neuroinflammation* 4:26. doi: 10.1186/1742-2094-4-26

Smoliga, J. M., and Blanchard, O. (2014). Enhancing the delivery of resveratrol in humans: If low availability is the problem, what is the solution? *Molecules* 19, 17154–17172. doi: 10.3390/molecules191117154

Solmaz, V., Çınar, B. P., Yiğittürk, G., Çavuşoğlu, T., Taşkıran, D., and Erbaş, O. (2015). Exenatide reduces TNF-α expression and improves hippocampal neuron numbers and memory in streptozotocin treated rats. *Eur. J. Pharmacol.* 765, 482–487. doi: 10.1016/j.ejphar.2015.09.024

Song, J. H., Marszalec, W., Kai, L., Yeh, J. Z., and Narahashi, T. (2012). Antidepressants inhibit proton currents and tumor necrosis factor-α production in BV2 microglial cells. *Brain Res.* 1435, 15–23. doi: 10.1016/j.brainres.2011.11.041

Stellwagen, D., and Malenka, R. C. (2006). Synaptic scaling mediated by glial TNF-*α*. *Nature* 440, 1054–1059. doi: 10.1038/nature04671

Takaki, J., Fujimori, K., Miura, M., Suzuki, T., Sekino, Y., and Sato, K. (2012). L-glutamate released from activated microglia downregulates astrocytic L-glutamate transporter expression in neuroinflammation: the 'collusion' hypothesis for increased extracellular L-glutamate concentration in neuroinflammation. *J. Neuroinflammation* 9:275. doi: 10.1186/1742-2094-9-275

Teich, M. C. (1989). Fractal character of the auditory neural spike train. *IEEE Trans. Biomed. Eng.* 36, 150–160. doi: 10.1109/10.16460

Thiebaud, N., Llewellyn-Smith, I. J., Gribble, F., Reimann, F., Trapp, S., and Fadool, D. A. (2016). The incretin hormone glucagon-like peptide 1 increases mitral cell excitability by decreasing conductance of a voltage-dependent potassium channel. *J. Physiol.* 594, 2607–2628. doi: 10.1113/JP272322

Tyrtyshnaia, A. A., Lysenko, L. V., Madamba, F., Manzhulo, I. V., Khotimchenko, M. Y., and Kleschevnikov, A. M. (2016). Acute neuroinflammation provokes intracellular acidification in mouse hippocampus. *J. Neuroinflammation* 13:283. doi: 10.1186/s12974-016-0747-8

Wake, H., Moorhouse, A. J., Jinno, S., Kohsaka, S., and Nabekura, J. (2009). Resting microglia directly monitor the functional state of synapses *in vivo* and determine the fate of ischemic terminals. *J. Neurosci.* 29, 3974–3980. doi: 10.1523/JNEUROSCI.4363-08.2009

Wang, D., Li, S. P., Fu, J. S., Zhang, S., Bai, L., and Guo, L. (2016). Resveratrol defends blood-brain barrier integrity in experimental autoimmune encephalomyelitis mice. *J. Neurophysiol.* 116, 2173–2179. doi: 10.1152/jn.00510.2016

Wanke, E., Gullo, F., Dossi, E., Valenza, G., and Becchetti, A. (2016). Neuron-glia crosstalk revealed in reverberating networks by simultaneous extracellular recording of spikes and astrocytes' glutamate transporter and K^+ currents. *J. Neurophysiol.* 28, 2706–2719. doi: 10.1152/jn.00509.2016

Wu, L. J., Wu, G., Akhavan Sharif, M. R., Baker, A., Jia, Y., Fahey, F. H., et al. (2012). The voltage-gated proton channel Hv1 enhances brain damage from ischemic stroke. *Nat. Neurosci.* 15, 565–573. doi: 10.1038/nn.3059

Yrjänheikki, J., Keinänen, R., Pellikka, M., Hökfelt, T., and Koistinaho, J. (1998). Tetracyclines inhibit microglial activation and are neuroprotective in global brain ischemia. *Proc. Natl. Acad. Sci. U.S.A.* 95, 15769–15774. doi: 10.1073/pnas.95.26.15769

Yu, X. W., Hu, Z. L., Ni, M., Fang, P., Zhang, P. W., Shu, Q., et al. (2015). Acid-sensing ion channels promote the inflammation and migration of cultured rat microglia. *Glia* 63, 483–496. doi: 10.1002/glia.22766

Zamanian, J. L., Xu, L., Nouri, N., Zhou, L., Giffard, R. G., and Barres, B. A. (2012). Genomic analysis of reactive astrogliosis. *J. Neurosci.* 32, 6391–6419. doi: 10.1523/JNEUROSCI.6221-11.2012

Zhang, F., Shi, J. S., Zhou, H., Wilson, B., Hong, J. S., and Gao, H. M. (2010). Resveratrol protects dopamine neurons against lipopolysaccharide-induced neurotoxicity through its anti-inflammatory actions. *Mol. Pharmacol.* 78, 466–477. doi: 10.1124/mol.110.064535

Zhu, H. T., Bian, C., Yuan, J. C., Chu, W. H., Xiang, X., Chen, F., et al. (2014). Curcumin attenuates acute inflammatory injury by inhibiting the TLR4/MyD88/NF-κB signaling pathway in experimental traumatic brain injury. *J. Neuroinflammation* 11:59. doi: 10.1186/1742-2094-11-59

7

Astrocyte-Mediated Neuromodulatory Regulation in Preclinical ALS

Kathleen Jordan[1]*, Joseph Murphy*[1]*, Anjanya Singh*[1,2] *and Cassie S. Mitchell*[1*]

[1] *Laboratory for Pathology Dynamics, Department of Biomedical Engineering, Georgia Institute of Technology, Emory University School of Medicine, Atlanta, GA, United States,* [2] *School of Medicine, University of Texas Southwestern Medical Center, Dallas, TX, United States*

Correspondence:
Cassie S. Mitchell
cassie.mitchell@bme.gatech.edu

Amyotrophic Lateral Sclerosis (ALS) is a neurodegenerative disease characterized by progressive degradation of motoneurons in the central nervous system (CNS). Astrocytes are key regulators for inflammation and neuromodulatory signaling, both of which contribute to ALS. The study goal was to ascertain potential temporal changes in astrocyte-mediated neuromodulatory regulation with transgenic ALS model progression: glutamate, GTL-1, GluR1, GluR2, GABA, ChAT activity, VGF, TNFα, aspartate, and IGF-1. We examine neuromodulatory changes in data aggregates from 42 peer-reviewed studies derived from transgenic ALS mixed cell cultures (neurons + astrocytes). For each corresponding experimental time point, the ratio of transgenic to wild type (WT) was found for each compound. ANOVA and a student's *t*-test were performed to compare disease stages (early, post-onset, and end stage). Glutamate in transgenic SOD1-G93A mixed cell cultures does not change over time ($p > 0.05$). GLT-1 levels were found to be decreased 23% over WT but only at end-stage ($p < 0.05$). Glutamate receptors (GluR1, GluR2) in SOD1-G93A were not substantially different from WT, although SOD1-G93A GluR1 decreased by 21% from post-onset to end-stage ($p < 0.05$). ChAT activity was insignificantly decreased. VGF is decreased throughout ALS ($p < 0.05$). Aspartate is elevated by 25% in SOD1-G93A but only during end-stage ($p < 0.05$). TNFα is increased by a dramatic 362% ($p < 0.05$). Furthermore, principal component analysis identified TNFα as contributing to 55% of the data variance in the first component. Thus, TNFα, which modulates astrocyte regulation via multiple pathways, could be a strategic treatment target. Overall results suggest changes in neuromodulator levels are subtle in SOD1-G93A ALS mixed cell cultures. If excitotoxicity is present as is often presumed, it could be due to ALS cells being more sensitive to small changes in neuromodulation. Hence, seemingly unsubstantial or oscillatory changes in neuromodulators could wreak havoc in ALS cells, resulting in failed microenvironment homeostasis whereby both hyperexcitability and hypoexcitability can coexist. Future work is needed to examine local, spatiotemporal neuromodulatory homeostasis and assess its functional impact in ALS.

Keywords: glutamate, GABA, GLT-1, ChAT, VGF, TNFα, aspartate

INTRODUCTION

Amyotrophic Lateral Sclerosis (ALS) is a neurodegenerative disease characterized by the progressive degradation of motoneurons, which results in muscle paralysis, respiratory failure, and ultimately death. Its multifactorial nature, population heterogeneity, and inherent complexity has made both clinical survival prediction (Pfohl et al., 2018) and intervention optimization (Khamankar et al., 2018) extremely difficult.

Because the multi-factorial nature of ALS is difficult to study in humans, experimental mouse models are critical for etiological and treatment elucidation. The superoxide dismutase 1 glycine 93 to alanine (SOD1-G93A) transgenic mouse is currently the predominant means of investigating the preclinical pathophysiology of ALS (Pfohl et al., 2015) because it of its rich publication history and reproducibility. ALS pathophysiology has multi-scalar disturbances that impact numerous processes, including inflammation, apoptosis, energetics, excitability, genetic transcription, cellular chemistry, oxidative stress, proteomics, and systemic function (Irvin et al., 2015; Kim et al., 2016). While ALS is a motoneuron disease, non-neuronal tissues are also affected, including astrocytes, glia, and muscle. In fact, inflammation, a process that is predominantly driven by non-neural tissue, is the most published etiology examined in SOD1-G93A ALS mice (Kim et al., 2016).

Evidence suggests that non-neuromuscular cells, such as astrocytes, play one of the earliest roles in ALS and are being considered as therapeutic targets (Vargas and Johnson, 2010; Pehar et al., 2017). Astrocytes are the most abundant subtype of glial cells found in the central nervous system (CNS), comprising 20–50% of the brain volume (Rossi and Volterra, 2009). One function of astrocytes is to stabilize the neural microenvironment after an injury through the release of cytokines (Rossi and Volterra, 2009). Ideally, the inflammatory regulators of astrocytes stimulate the healing process and lay down a protective glial scar (Sofroniew and Vinters, 2010). However, inflammation is also negatively implicated across neuropathology, where its consequences can expand the lesion volume (Mitchell and Lee, 2008) or perpetuate local dysregulation. For example, in SOD1-G93A ALS, there is a dynamic imbalance of cytokines that inappropriately amplifies inflammation (Jeyachandran et al., 2015). Glial fibrillary acid protein (GFAP) levels in SOD1-G93A mice are almost double those of wild type (WT) mice prior to onset and continue to significantly increase with disease progression. Furthermore, ITPR2 gene expression is significantly upregulated before and after the induction of inflammation (Staats et al., 2016).

Another key function of astrocytes is to assist in reuptake of neuromodulatory agents. It has been proposed that the hyperexcitability of motoneurons is caused by overstimulation by the main excitatory neurotransmitter, glutamate, which leads to a large influx of calcium (*Ca2+*) and sodium (Na+) into the cell through glutamate transporter 1 (GLT-1) (Van Den Bosch et al., 2006; Do-Ha et al., 2017). Intracellular *Ca2+* levels are further increased as Na+ is passed out through the Na+-*Ca2+* exchanger (Van Den Bosch et al., 2006). Motoneurons possess little ability to counteract the effects of *Ca2+* influx, so overstimulation can easily lead to excitotoxicity and cell death (Do-Ha et al., 2017). Glutamate excitotoxicity and impaired intracellular calcium signaling in astrocytes is theorized to significantly impact disease progression in ALS and other types of neuropathology (Staat and Van Den Bosch, 2009; Kawamata et al., 2014). Yet, such neuromodulatory regulation mediated by astrocytes is much lesser-studied than inflammation.

Astrocyte neuromodulatory dysregulation, resulting in failed neuromodulatory homeostasis, is thought to contribute significantly to neuronal depolarization, hyperexcitability, excitotoxicity, and subsequent neuronal death in ALS (Lin et al., 2013). Increased levels of glutamate and reduced levels of excitatory amino acid transporter 2 (EAAT2) have been found in the CNS of ALS patients, suggesting EAAT2 dysfunction and glutamate excitotoxicity is involved in the disease progression (Rothstein et al., 1990, 1995). It is still unknown as to what causes the glutamate to initially collect within the CNS. One possible explanation is that, prior to functional onset of the disease, the astrocytes fail to reuptake the extracellular glutamate at the proper homeostatic rate, resulting in a slow accumulation. Another explanation is that the motoneurons, or another nearby cell, releases glutamate at a significantly higher rate than normal. As part of the latter explanation, astrocytes do not compensate for the increased glutamate release and, rather, simply continue to reabsorb at a normal rate, resulting in glutamate accumulation over time. Many studies (Li et al., 2015) have sought to increase GLT-1 levels in astrocytes to increase astrocyte reuptake, albeit unsuccessfully.

The inhibitory neurotransmitter gamma-Aminobutyric acid (GABA) also has therapeutic potential. Mildly reduced expression of GABAergic markers and interneurons have been found in some SOD1 mice and ALS patients (Nihei et al., 1993; Hossaini et al., 2011). GABA transmission can decrease levels of glutamate and protect against excessive neuronal damage (Brockington et al., 2013). However, clinical trials of gabapentin, a pharmaceutical anti-epileptic drug that modulates GABA, did not extend life span or slow the rate of muscle decline or respiratory function (Diana et al., 2017). Elucidation of temporal relationships among glutamate levels, transporter proteins, such as GLT-1 and glutamate receptors (GluR), and GABA is needed to determine therapeutic timing and efficacy. For example, if glutamate-related treatment must occur well before ALS symptom onset to have a functional impact, its clinical treatment value is greatly diminished.

The adjunctive regulation of intracellular *Ca2+* by astrocytes is vital for cell signaling. Astrocytes are able to signal neurons by *Ca2+* dependent release of glutamate (Appel et al., 2001; Rossi and Volterra, 2009). Intracellular concentrations of *Ca2+* are characteristically elevated in response to pathological signaling (Rossi and Volterra, 2009; Guerra-Gomes et al., 2018). The glutamate-mediated excitotoxicity of astrocytes relies on intracellular concentrations of *Ca2+* but has also been found to be accompanied by the cytokine TNFα (Rossi and Volterra, 2009). The blockage of the formation of TNFα and endoplasmic reticulum (ER) *Ca2+* overload have a significant negative effect on astrocyte glutamate release (Rossi and Volterra, 2009;

Kawamata and Manfredi, 2010). Other cytokines, such as VGF nerve growth factor (VGF), insulin-like growth factor 1 (IGF-1), aspartate, and choline acetyltransferase (ChAT) can also be linked to intracellular calcium levels (Palmieri et al., 2001; Fernández et al., 2004; Kandinov et al., 2013).

An examination of dynamic, temporal interactions among key players in astrocyte-mediated neuromodulatory regulation is necessary to better evaluate ALS etiology and therapy. Of course, pathological extracellular increases of glutamate, calcium, and other co-factors does lead, or minimally contribute, to co-existing pathology, like oxidative stress, inflammation, and excitotoxicity. Such pathological overlap makes detangling the multi-factorial ALS etiology all the more difficult (Kim et al., 2016). Due to the large number of variables, a single all-encompassing *in vivo* experiment is not feasible. The goal of this study is to determine temporal trends of intrinsic astrocyte-mediated compounds that contribute to neuromodulatory regulation over the disease progression of preclinical ALS, with the primary focus on regulation of glutamate, GABA, and related compounds. Specifically, this metadata analysis is comprised of temporal neuromodulatory data compiled from 42 peer-reviewed studies that utilized mixed cultures of astrocytes and neurons predominantly derived from *in vivo* SOD1-G93A ALS mice and normal WT mice.

METHODS

A metadata analysis was performed to construct a macroscopic view of astrocyte-mediated neuromodulatory regulation over the course of transgenic ALS mouse model disease progression. The general method involved (1) mining, selecting and recapturing published data from preclinical ALS experiments examining astrocytes; (2) normalizing recaptured data to enable aggregation across studies; (3) analyzing aggregate data using appropriate statistical methods.

Data Source Identification and Inclusion Criteria

Keywords were used to identify potential data sources in PubMed/Medline. All potential data sources were initially searched using key words "Amyotrophic Lateral Sclerosis" OR "ALS" AND "transgenic mouse." Searches were limited to articles published in English and with publication dates through June 2018. Primary search articles were downloaded into a Filemaker Pro relational database (Mitchell et al., 2015a; Kim et al., 2016) and had relevant data recaptured using our lab's highly accurate biocuration process (Mitchell et al., 2015a). Secondary searches on relevant sub-topics were performed within the Filemaker Pro database including all synonyms (see **Table 1**). We specifically utilized the "high-copy" transgenic SOD1-G93A mouse model, which is not only more common but also has less implicit outcome heterogeneity (Pfohl et al., 2015). Recaptured data included measures of glutamate, GTL-1, GluR1, GluR2, GABA, ChAT activity, VGF, TNFα, aspartate, and IGF-1, which were identified using searches of figure axis labels and figure/table captions. As astrocytic compensation can only be studied in the

TABLE 1 | Keywords for "*Astrocytes*" and terms associated with each subtopic.

Category	Keywords
Glutamate	Nitric oxide, NO* conc*, synth*, glutamate, glutamate* conc*, GLT1, GLT1* transporter*, GluRA, GluR1, GluR2, GluR3, GluR4, GluR*, excitotoxicity, excito*
Calcium	Membrane potential, cyto* cal* conc*, cal* buffer* capacity, mito* cal* conc*, cal* uptake, cal* conc*, cal* transient, permeability transition pore, acetylcholine, voltage-gated, calcium channel*, calcium pump*, GABA, ER([Ca]), endoplasmic reticulum, AMPA receptor*, VDAC, CypD, FCCP, IGF-1, VGF, TNF-alpha, TNF*, aspartate, ChAT*

*The subtopics were broad topics that were used to categorize articles discussing ALS studies. The keywords were used in searches in the database to determine article numbers and potential data points for each subtopic. The searches were performed on figure axis labels and captions. *indicates that pieces of the word/phrase or synonyms were included in the search.*

presence of neurons, most data was taken from mixed cultures. Only studies that presented quantified data for both transgenic and age-matched WT control mice were included. A total of 42 articles with quantifiable experimental data had data extracted for analysis. A diagram of the literature review structure is shown in the **Supplementary Figure 1**.

Data Normalization and Aggregation

Recaptured quantified data was used to construct ratios of ALS-representative mouse models to WT (such as, SOD1-G93A/WT) for each included metric at each temporal disease stage. The genetic background, onset, survival, tissue origin/measurement procedure of each included study is detailed in **Supplementary Table 1**. Temporal data points were aggregated into three disease stages; pre-onset (0–96 days), post-onset (97–116 days), and end stage (117+ days). Disease stages were determined by finding the average age of onset (97 days) and survival duration (117+ days) for the included mice in the present study. Non-aggregated raw data can be found in **Supplementary Figure 2**. Six of the post-onset glutamate data points were empirically extrapolated using standard statistical regression to obtain the necessary sample size determined by a standard power analysis. **Table 2** illustrates the breakdown of the 264 data points (e.g., ratios of transgenic to WT for each included factor for each temporal disease stage). Aggregation is an inherent limitation to meta-analysis, and this is further discussed in the context of this study in the Limitations.

Not every included journal article expressed the mouse sample size corresponding to each of their data points. Thus, each data point (e.g., ratio of transgenic to WT for the included factor measured at a specific temporal disease stage) in the present metadata analysis is weighted equally on a per-article basis rather than the corresponding mouse sample size comprising the data point.

To determine the ALS disease stages, the average onset (97 days) and time of death (117 days) was calculated using all included onset and time of death data from the original experimental studies (see **Supplementary Data Sheet**). Note that these averages are within what is expected for high copy SOD1-G93A ALS mice (Pfohl et al., 2015).

TABLE 2 | Number of data points used in each sub-category.

Sub-Category	Time bin	Sample size	References
Glutamate	Pre-onset Post-onset End-Stage	22 12* 15	Alexander et al., 2000; Guo et al., 2000, 2003; Bendotti et al., 2001; Raiteri et al., 2004; Niessen et al., 2007; Choi et al., 2009; Gu et al., 2010; Milanese et al., 2011; Albano et al., 2013; Valbuena et al., 2016; Tefera and Borges, 2018
GLT-1	Pre-onset Post-onset End-Stage	22 10 16	Alexander et al., 2000; Bendotti et al., 2001; Deitch et al., 2002; Chen et al., 2004; Rothstein et al., 2005; Boston-Howes et al., 2006; Pardo et al., 2006; Yang et al., 2009; Gu et al., 2010; Benkler et al., 2013; Morel et al., 2013
GluR1	Pre-onset End-Stage	24 7	Petri et al., 2005; Spalloni et al., 2006; Martinez et al., 2008; Zhao P. et al., 2008 Caioli et al., 2011
GluR2	Pre-onset End-Stage	17 5	Petri et al., 2005; Spalloni et al., 2006; Tortarolo et al., 2006; Zhao P. et al., 2008
GABA	Pre-onset End-Stage	4 11	Raiteri et al., 2004; Tsai et al., 2010; Caioli et al., 2013; Tefera and Borges, 2018
ChAT Activity	Pre-onset Post-onset End-Stage	24 12 23	Crochemore et al., 2005, 2009; Kalmar et al., 2012
VGF	Pre-onset End-Stage	12 6	Zhao Z. et al., 2008
TNFα	Pre-onset End-Stage	8 3	Fang et al., 2010; Yang and Cheng, 2010; Song et al., 2013; Cai et al., 2015; Jeyachandran et al., 2015; Lee et al., 2015
Aspartate	Pre-onset	7	Alexander et al., 2000; Niessen et al., 2007; Choi et al., 2009; Tefera and Borges, 2018
IGF-1	Pre-onset	4	Kaspar et al., 2005; Wu et al., 2006; Messi et al., 2007; Fergani et al., 2011

* *To reach the desired sample size, 6 of the 12 data points were linearly extrapolated for this time bin.*

Statistical Analysis

The distributions of all data sets were tested with Shapiro-Wilks tests to assure sufficient distribution normality for the corresponding statistical tests. To determine statistical significance of differences between disease stages, an ANOVA was performed using the Bonferroni correction. To compare protein levels, a Student's t-test with a Bonferroni correction was performed for each disease stage. Error bars on figures correspond to the standard deviation. All statistical tests were performed with built-in functions in MATLAB (version R2016a). Correlation matrices were constructed to visualize the relationships between the analyzed measures. A principal component analysis (PCA) was used to evaluate contributions to data variance and visualized reduced data structure. The correlation matrices and PCA were generated in MATLAB (version R2016a).

RESULTS

A total of nine astrocyte-mediated neuromodulatory regulators were examined from 42 peer-reviewed articles. As stated in the Introduction, nearly all of these factors still have secondary ties to inflammation. However, the focus of the present study was not on the inflammatory roles of astrocytes but rather neuromodulatory regulation of GABA, glutamate and calcium related co-factors. **Table 2** lists the measures along with the included sample size (number of data points). The glutamate excitotoxicity sub-categories included glutamate, GLT-1, GluR1, GluR2, and GABA concentrations. The calcium homeostasis sub-categories included ChAT activity, VGF, TNFα, aspartate, and IGF-1 levels.

Glutamate and Glutamate Transporter Proteins

Glutamate and glutamate transporter levels from mixed cell cultures from SOD1-G93A ALS mice are compared to WT. There was no significant difference in glutamate concentrations over disease progression (**Figure 1A**). Glutamate concentrations were also not significantly increased from WT concentrations at any disease stage. GLT-1 and GluR1 levels decrease over disease progression. At end-stage, GLT-1 levels were 23% lower than end-stage WT values ($p = 0.008$) (**Figure 1B**). GluR1 concentrations at end-stage were 21% decreased from pre-onset levels ($p = 0.035$) (**Figure 1C**). There was no significant difference in GluR2 concentrations over disease progression (**Figure 1C**).

GABA

Next, GABA levels were compared in ALS mice and WT mice (**Figure 2**). GABA concentrations showed no significant difference between WT and ALS transgenic values. Moreover, there was no significant change in GABA over the course of ALS disease stage or progression.

Other Neuromodulatory Factors

Other calcium-related cytokines and neuromodulatory factors were examined over temporal SOD1-G93A disease progression (**Figure 3**). Choline acetyltransferase (ChAT), a transferase enzyme, is responsible for producing the neurotransmitter, acetylcholine (ACH). ChAT activity did not significantly vary over disease progression (**Figure 3A**). The lack of a significant trend in ChAT, which otherwise qualitatively appears to be depressed in SOD1-G93A compared to WT, is confounded by the large variance in experimental sample population. Nerve growth factor (VGF) is a secreted protein and neuropeptide precursor that may play a role in regulating energy homeostasis, metabolism and synaptic plasticity. VGF levels are depressed in SOD1-G93A ALS mixed cultures. VGF levels are significantly decreased from WT levels at both pre-onset ($p = 0.0036$) and end-stage ($p < 0.0001$), with end-stage being further significantly decreased from pre-onset ($p = 0.0002$; **Figure 3B**). Tumor necrosis factor alpha (TNFα) is a cytokine involved in inflammation, apoptosis, and synaptic function. TNFα expression in ALS mixed cultures significantly increased 362%

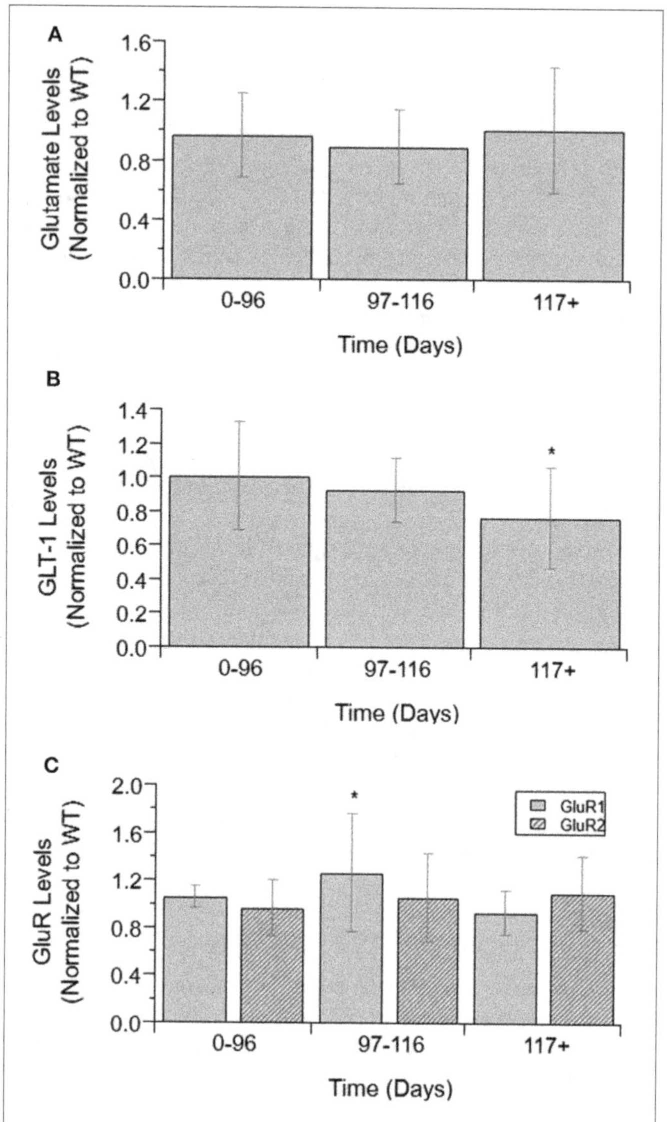

FIGURE 1 | Glutamate, GLT-1, and GluR complex levels in mixed cultures over ALS disease progression. **(A)** Glutamate level average in SOD1-G93A mice normalized to wild-type (ratio presented as SOD1-G93A/WT) in three temporal stages: 0–96 days, 97–116 days, and 117+ days. There was no significant change in glutamate levels between any of the temporal stages. **(B)** GLT-1 average in SOD1-G93A normalized to wild-type (ratio presented as SOD1-G93A/WT) in three temporal stages: 0–96 days, 97–116 days, and 117+ days. GLT-1 is significantly decreased from WT (*$p = 0.0075$) at 117 + days. **(C)** GluR1 (solid bars) and GluR2 (stripped bars) average in SOD1-G93A normalized to wild-type (ratio presented as SOD1-G93A/WT) in three temporal stages: 0–96 days, 97–116 days, and 117+ days. There was no significant change in GluR2 levels between temporal stages. GluR1 at 117+ days was significantly decreased from 0–96 (*$p = 0.0346$). Error bars represent SD.

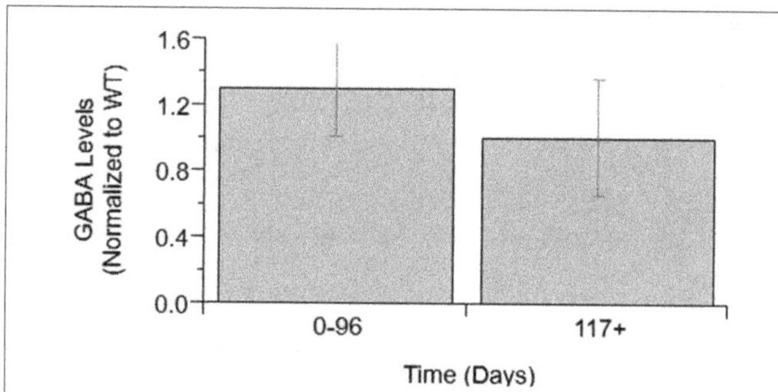

FIGURE 2 | GABA levels over ALS disease progression. GABA level average normalized to wild-type (ratio presented as transgenic/WT) in two temporal stages: 0–96 days and 117+ days. Error bars represent SD.

from WT at end-stage (**Figure 3C**). Aspartate concentrations also significantly increased 25% from WT (**Figure 3D**).

Glutamate, GABA, and Cytokine Relationships

A series of cross-correlation matrices were used to examine the correlations between various neuromodulatory regulators examined in this study (**Figure 4**). A cross-correlation matrix with correlations was constructed for each disease stage: 0–96 days (pre-onset); 97–116 days (post-onset); and 117+ days (end-stage). Highly negatively correlated factors are closer to −1, uncorrelated factors are closer to 0, and highly positively correlated factors are closer to 1; the illustrated gray scale represents the degree and sign of corresponding correlation. Only factors with a sample size calculated as having sufficient statistical power were included in each disease stage matrix. Thus, not every matrix included all factors. The pre-onset matrix included the most factors. The notable result from this analysis is that ChAT and VGF levels are high positively correlated at pre-onset ($p = 0.0003$; **Figure 5**).

A principal component analysis (PCA) was used to determine the individual components, which contribute the most to variance in the data (**Figure 5**). PCA is a type of exploratory and dimensionality reduction analysis. It is mathematically defined as an orthogonal linear transformation that transforms the data to a new coordinate system such that the greatest variance by some projection of the data comes to lie on the first coordinate (called the first principal component), the second greatest variance on the second coordinate, etc. The biplot displays how much the variables contribute to the variance of the first two principal components while also identifying reduced variable clusters (**Figure 5A**). The length of each factor's vector corresponds to the variance explained. Component 1 accounted for 55% of the variance and was almost completely comprised of TNFα (**Figure 5B**). Component 2 accounted for 17% of the variance and was mostly comprised of GABA and aspartate (**Figure 5C**). Interestingly, TNFα was nearly orthogonal to GluR1, indicating a lack of relationship.

Finally, an analysis was performed to determine which normalized cytokine co-factor levels were significantly different than either glutamate or GABA in mixed cultures across each temporal disease stage (**Figure 6**). VGF levels were significantly decreased from glutamate at pre-onset and end-stage ($p = 0.0331$ and $p = 0.0028$; **Figure 6A**). TNFα was significantly increased from glutamate levels at pre-onset ($p = 0.0018$; **Figure 6B**). GABA levels were not significantly increased from glutamate at any ALS disease stage (**Figure 6C**). ChAT activity

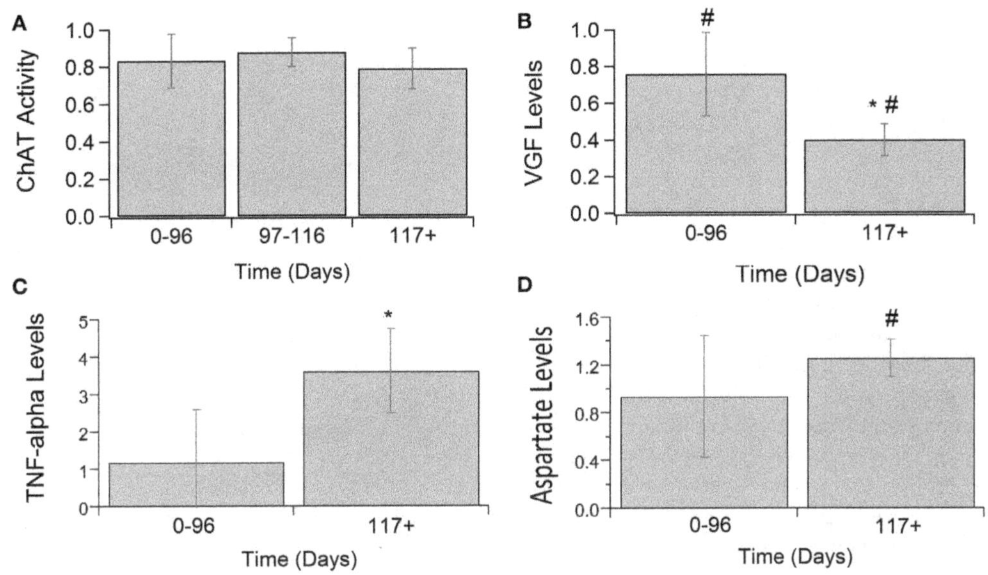

FIGURE 3 | Calcium-related cytokine levels over ALS disease progression. ChAT activity **(A)**, VGF **(B)**, TNFα **(C)**, and aspartate **(D)** averages in SOD1-G93A normalized to wild type (ratio presented as SOD1-G93A/WT) in three stages: 0–96 days, 97–116 days, and 117 + days. SOD1-G93A VGF levels were significantly decreased compared to WT at 0–96 days (#$p = 0.0036$) and 117+ days (#$p < 0.0001$). SOD1-G93A VGF levels at end-stage are also significantly lower than early-stage (*$p = 0.0002$). TNFα (*$p = 0.0488$) and aspartate (#$p = 0.0350$) levels were significantly increased at end-stage compared to WT. Error bars represent SD.

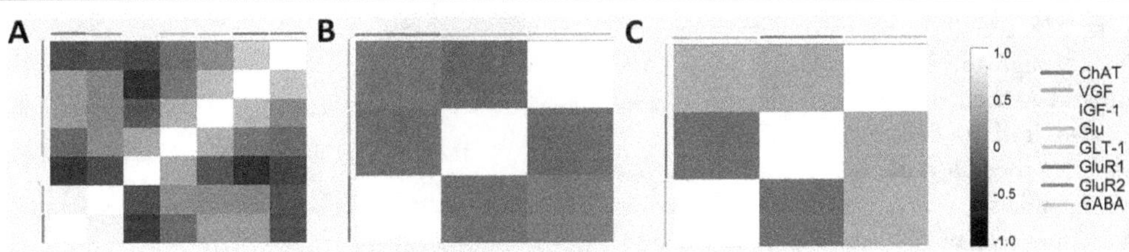

FIGURE 4 | Temporal relationships between non-inflammatory regulators of astrocytes visualized as a cross-correlation matrix. Average concentrations of glutamate, GABA, and related cytokines for ALS mice were normalized to wild-type mice with data aggregated across three temporal stages: **(A)** 0–96 days, **(B)** 97–116 days, and **(C)** 117+ days. Highly positive correlations were found between VGF and ChAT at 0–96 days. Note that only factors with sufficient sample sizes (calculated using standard statistical power analysis) were included.

was significantly lower than GABA at pre-onset ($p = 0.0033$) and end-stage ($p = 0.009$; **Figure 6D**). Aspartate concentrations were not significantly lower than GABA at any disease stage (**Figure 6E**).

DISCUSSION

The metadata analysis results do not paint a picture of widespread, overt glutamate-mediated excitotoxicity in preclinical ALS as measured in mixed cell cultures. The key neuromodulators, namely glutamate, glutamate receptors, and transports, GABA, and ChAT, showed no so significant changes in ALS compared to WT or significant changes over the course of ALS disease progression. GLT-1 and GluR1 were significantly decreased at end stage, whereas aspartate was significantly increased at end stage. TNFα was significantly and drastically increased throughout ALS disease duration, whereas VGF was significantly decreased throughout the ALS disease course. Other than TNFα, directly measured changes in neuromodulators are modest at best. Below we discuss what these results suggest about the role of astrocyte-mediated neuromodulatory regulation in ALS etiology and potential treatments.

Evaluating the Case for and Against ALS Excitotoxicity

Astrocytes play a critical role in reabsorbing excess extracellular glutamate and maintaining homeostasis for the motoneurons, which normally prevents excitotoxicity. There has been ongoing debate as to the presence or source of possible excitotoxicity (e.g., Leroy and Zytnicki, 2015; Rosenblum and Trotti, 2017; Martínez-Silva et al., 2018), especially given some animal model experiments suggest ALS motoneurons are hyperexcitable (Jiang et al., 2017) whereas other work suggests they are hypoexcitable (Martínez-Silva et al., 2018).

Previous clinical work illustrated evidence of increased glutamate levels in the CNS of ALS patients (Rothstein et al., 1990). Lack of resorption results in glutamate being left in the extracellular fluid resulting in continuous over-stimulation of the postsynaptic neurons (Rosenblum and Trotti, 2017). However, there was no identified increase in the excitatory transmitter, glutamate, in the preclinical ALS model mixed cell cultures examined in this study. As improved technology enables further examination, future experimental studies should focus on the specific localization of neurotransmitters to determine if key neurotransmitter changes are a function of spatial resolution, specific cell type, or a function of the type of ALS model.

Astrocytes use the enzyme glutamine synthetase (GS) to convert glutamate to glutamine as part of the glutamate/GABA-glutamine cycle (Norenberg and Martinez-hernandez, 1979; Bak et al., 2006). It is possible the constant levels of glutamate indicate SOD1-G93A astrocytes are reuptaking glutamate at higher rates and immediately converting it to maintain homeostasis. However, if glutamate was being taken in at higher rates it would require significant increases in GLT-1 and GluR to transport it into the cells. No such increases were found in this analysis. Rather, GLT-1 and GluR1 levels actually decreased compared to WT near end stage.

There are a few possible explanations for the decrease in GLT-1 and/or GluR1 levels. First, GLT-1 decreases could possibly be attributed to the difference in metabolic needs of astrocytes and neurons. Since GLT-1 activity in astrocytes leads to an influx of *Ca2+* through the energy powered Na+-*Ca2+* exchanger (Bazargani and Attwell, 2016), metabolic impairment in astrocytes would also lead to decreased GLT-1 activity. A second explanation for the decrease of GLT-1 and GluR1 over ALS disease progression could simply be due to the lower number of living or fully functioning cells in later stages of ALS. Finally, *if* there are potential cell-specific or spatially localized increases in excitation [spatial differences were not able to be examined in the present work], decreases in GLT-1 could suggest failed regulation or an inadequate compensatory response.

Beyond glutamate-related findings, there were several other neuromodulatory findings in the present study that provide further insight. For example, GABA, an inhibitory neurotransmitter, is not upregulated over the course of ALS in mixed cell cultures. If glutamate and other excitatory transmitters were in excess (either due to over-production or do to impaired re-uptake), it would be expected that GABA follow a similar trend due to innate compensatory mechanisms to balance excitation and inhibition. In the present study, neither GABA nor glutamate were significantly different in preclinical ALS compared to WT. Notably, aspartate, another excitatory neurotransmitter similar to glutamate, did show a slight increase in end stage preclinical ALS. However, ChAT, which is used to make the neurotransmitter acetylcholine, showed no significant difference.

In summary, there were no major changes in key neuromodulators that would point to very obvious glutamate-related excitotoxicity. Rather, the evidence presented here for mixed ALS cell cultures would suggest that direct neuromodulatory changes are subtle to modest, at least in

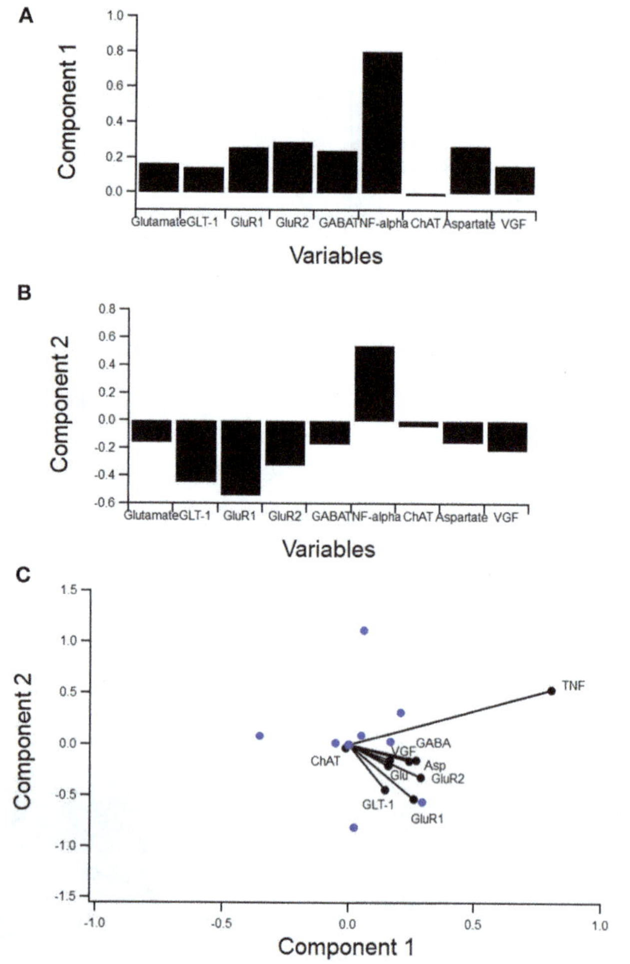

FIGURE 5 | PCA to determine variance contribution of non-inflammatory regulators of astrocytes. Biplot showing variable contributions to the first two PCA-determined components **(A)** broken down to the first two components and the comprising variables **(B,C)**. Component 1 accounted for 55% of the variance and was almost completely comprised of TNFα **(B)**. Component 2 accounted for 17% of the variance and was mostly comprised of GABA and aspartate **(C)**.

SOD1 G93A ALS mixed cell cultures. This would suggest that, if excitotoxicity is present, it could be due to the fact that ALS neurons and astrocytes have properties that make them simply more susceptible or sensitive to even small changes in neuromodulation. Thus, even small or seemingly insignificant oscillatory increases or decreases in neuromodulators, which could easily be rectified with compensatory regulation in normal cells, could potentially wreak havoc in ALS cells. In fact, as discussed in more detail in the Section Astrocyte GluR as a Potential Pre-onset ALS Treatment Target, motoneurons are mathematically and functionally more susceptible to such instability (Mitchell and Lee, 2012; Irvin et al., 2015). Other recent work has also illustrated that motoneuron microcircuits are more prone to homeostatic dysregulation (Brownstone and Lancelin, 2018). Given the large variance in experimental data, small changes that were statistically insignificant may actually

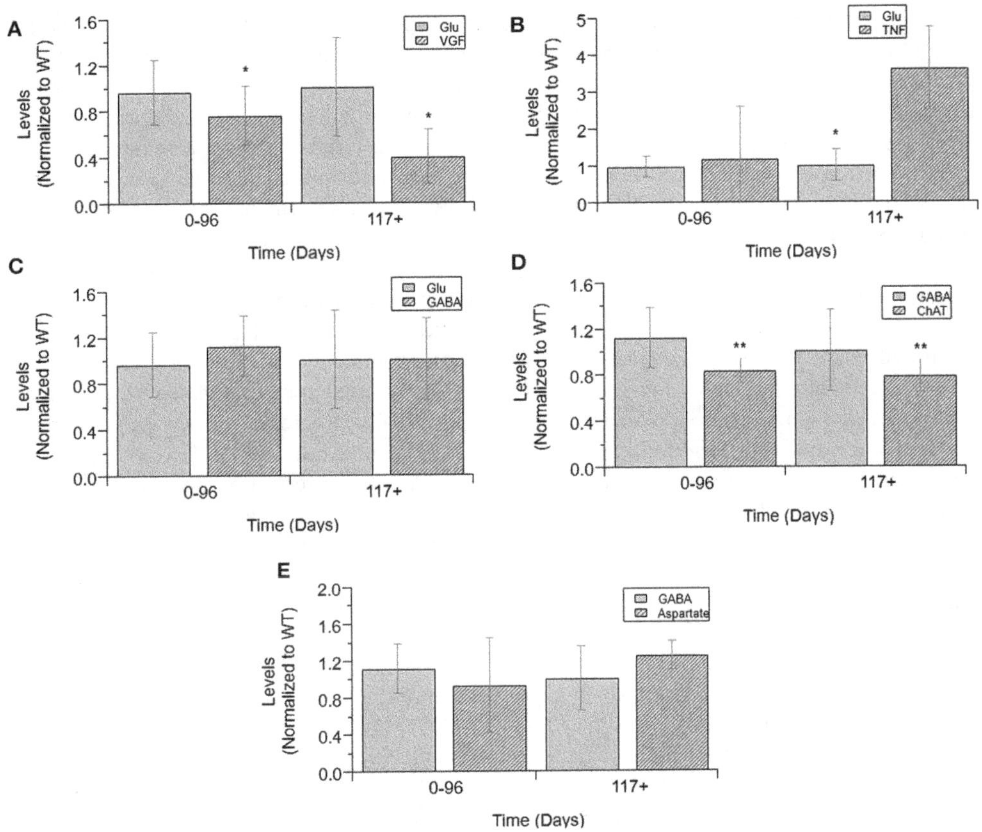

FIGURE 6 | Cytokine levels compared to glutamate and GABA levels in mixed cultures over ALS disease progression. VGF **(A)**, TNFα **(B)**, GABA **(C)**, and glutamate **(D)** averages normalized to wild-type (ratio presented as transgenic/WT) in three temporal stages: 0-96 days, 97-116 days, and 117+ days. ChAT activity **(D)**, aspartate **(E)**, and GABA average normalized to wild-type (ratio presented as transgenic/WT) in two temporal stages: 0–96 days and 117+ days, as insufficient data was available for 97–116 days. VGF levels were significantly decreased from glutamate levels at both stages ($^*p < 0.05$). TNFα levels were significantly increased from glutamate levels at 117+ days ($^*p = 0.0018$). GABA levels were significantly decreased from ChAT levels at 0–96 days ($^{**}p = 0.0033$) and at 117+ days ($^{**}p = 0.009$). Error bars represent SD.

be physiologically significant to ALS cells. If ALS cells are not able to sufficiently control their microenvironment, there could be spatial "pockets" or temporal phases of both hyperexcitability and hypoexcitability. This could mean that both hyperexcitability and hypoexcitability could co-exist in ALS, which could explain the discrepancies seen in previous literature.

Astrocyte GLT-1 as a Potential Post-onset ALS Treatment Target

Due to the modest success of Riluzole, a drug used to treat ALS that is thought to increase glutamate reuptake, there has been significant focus on finding ways to increase the reuptake of glutamate by astrocytes (Do-Ha et al., 2017). Additionally, post-mortem studies of ALS tissue and mice have found reduced levels of EAAT2 and GLT-1, respectively (Rothstein et al., 1995; Pardo et al., 2006). It is believed that decreased GLT-1 levels are contributing to the accumulation of extracellular glutamate and increasing the levels could help prevent the hyperexcitability of the motoneurons. Few experiments have had any notable success, however, due to the complex interaction and timing of the pathophysiology (Li et al., 2015). The present study found that GLT-1 levels only decrease significantly from post-onset to end-stage (**Figure 1B**). This finding suggests that increasing GLT-1 levels before ALS end-stage may increase glutamate reuptake, thereby prolonging survival or minimally prolonging muscle function after ALS onset.

If cell-specific or spatially localized excitation is present, extrinsically increasing GLT-1 levels before ALS end-stage may increase glutamate reuptake, thereby prolonging survival or minimally prolonging muscle function after ALS onset. Recent analysis has illustrated the potential impact of treatments that can increase patient quality of life by preserving muscle function for as long as possible even if corresponding survival increases are not dramatic. For example, similar metadata analysis of oxidative stress treatments in high copy SOD1-G93A ALS mice prolong limb muscle function by 59.6% but only prolong survival by 11.2% (Bond et al., 2018).

Astrocyte GluR as a Potential Pre-onset ALS Treatment Target

The influx of *Ca2+* caused by the overstimulation of motoneurons leads to cell death as *Ca2+* is involved in many apoptotic pathways. The permeability of glutamate receptors is determined by GluR subunits. Glutamate receptors with GluR1 subunits are highly permeable to *Ca2+* and causes the mobilization of intracellular *Ca2+* when activated (D'antoni et al., 2011). GluR2 lacking glutamate receptors are highly permeable to *Ca2+* (Grosskreutz et al., 2010). Studies have shown that a lack of GluR2 increased motoneuron degeneration (Van Damme et al., 2005) in mice and increasing GluR2 levels in motoneurons prolonged survival of mice (Tateno et al., 2004). While decreased GluR2 levels were not observed in ALS mixed cultures in this study, there was a decrease in GluR1 levels at end-stage when compared to post-onset (**Figure 1C**), possibly from the high levels of cell death that have occurred by that point. Additionally, GluR1 is elevated from WT at post-onset, though this difference was not found to be significant, likely owing to small sample size. These findings suggest that reducing GluR1 levels in astrocytes at pre-onset, prior to their elevation post-onset, could reduce intracellular *Ca2+* levels and *Ca2+* dysregulation effects. Due to the limited number of data points available for intracellular *Ca2+* measured, the relationships of *Ca2+* and the GluR subunits were unable to be examined.

Of course, treatment at "pre-onset" in clinical terms would mean treatment in familial or genetic cases of ALS, referred to as FALS, where disease predisposition or the presence or pre-symptomatic ALS could be identified well before symptoms. Such a treatment is not realistic at this point because precise, reliable tests for explicit ALS identification [prior to symptom onset] are not yet available, even for FALS. The ability to identify more reliable biomarkers, genetic or otherwise, that predict future onset of ALS would open up the possibility of prophylactic treatment pre-onset. Such an approach would be similar to the familial amyloid-β treatments being tested prophylactically in a known Colombian family where genetic tests can predict with certainty who will get a very specific form of familial Alzheimer's Disease well before cognitive symptoms appear (Reardon, 2018). Early preclinical evidence illustrates that infamous amyloid-β treatments, when used alone, may require prophylactic treatment to be successful (Huber et al., 2018). While many FALS markers have been identified, albeit SOD1, C9orf72, or the even more rare mutations like FUS/TLS, VAPB, etc. (Ajroud-Driss and Siddique, 2015), no such marker has yet been identified that definitively, with great sensitivity and specificity, can pre-determine a future FALS onset.

Elucidating Etiology of *Ca2+* Increase in ALS

Measuring intracellular calcium can be very complicated. *Ca2+* transients are spatially localized within the astrocyte (Bazargani and Attwell, 2016). Due to methodological constraints, the full intracellular *Ca2+* store may not be found by simply looking at the soma, which complicates experimental *Ca2+* examination in astrocytes at the present time. Given the complications of local, spatial, and temporal *Ca2+* experimental examination, it is difficult to determine what mechanism is most directly related to the increases in *Ca2+* that have been previously identified in ALS (Irvin et al., 2015).

In healthy astrocytes, glutamate transporter activity leads to an increased entry of *Na+* as GLT-1 couples glutamate and *Na+* (Bazargani and Attwell, 2016). In addition to the *Na+-K+* ATPase pump, the *Na+-Ca2+* exchange provides a way for this *Na+* to exit while *Ca2+* enters. Therefore, an increase in glutamate uptake through GLT-1 activity can lead to increase in intracellular *Ca2+*. ALS astrocytes do not appear to increase glutamate uptake over disease progression and, consequently, do not increase intracellular *Ca2+* concentrations. Neurons, however, can become increasingly more susceptible to glutamate excitotoxicity over time and, thus, see a rise in intracellular *Ca2+*.

Interestingly, early pre-onset elevation of intracellular *Ca2+* in SOD1-G93A mixed cell cultures occurs without a concurrent increase in intracellular glutamate. Thus, the mechanisms causing failed astrocytic *Ca2+* regulation is not explicitly tied to glutamate. Exocytosis of GABA is, in part, dependent on the influx of *Ca2+* through voltage-gated calcium channels (VGCCs) (Sitges and Chiu, 1995). However, we found no increase in GABA in ALS preclinical model mixed cell cultures. Thus, unless there are localized changes in excitatory neurotransmitters that have not yet been experimentally measured, the increase in *Ca2+* appears to not be directly tied to changes in excitatory neurotransmitter balance—at least not in preclinical ALS mixed cell cultures analyzed in the present study. Interestingly, a recent study showed that clinical patients with the C9ORF72 mutation had motoneurons that were more vulnerable to *Ca2+* permeable AMPA receptors (Selvaraj et al., 2018).

Ca2+ increases could come from a variety of mechanisms beyond direct or significant changes in neuromodulatory transmission. As proposed in prior work (Mitchell and Lee, 2012; Mitchell et al., 2015b; Hollinger et al., 2016; Kim et al., 2016), failed homeostasis in nearly ten different major pathways could explain how a variety of different perturbations that result in ALS and identified homeostatic regulatory instability. *Ca2+* regulation is also greatly impacted by changes in energetic pathways (namely mitochondria and endoplasmic reticulum *Ca2+* stores) and oxidative stress (Bond et al., 2018), and to a lesser degree, by apoptotic signaling (Irvin et al., 2015).

Motoneurons are more prone to instability due to their long length (Mitchell and Lee, 2012), which requires transport and signaling over very long distances in the axon. Any delays or offset in signaling within cells or even between cells, such as between neurons and astrocytes, could contribute to oscillatory instability. The oscillations seen in CNS experimental data, and especially *Ca2+*, support the homeostatic instability theory (Irvin et al., 2015) of ALS. Given the very small margin of error for homeostatic regulation in motoneurons (Mitchell and Lee, 2012), localized examination of *Ca2+* with sufficient spatio-temporal resolution in the future could provide key evidence for experimentally confirming the contributions, or lack thereof, of neuromodulatory regulation mechanisms in neurons and astrocytes.

Puzzling Non-correlation Between Calcium-Permeable GluR1 and TNFα

In general, TNFα production and secretion is believed to be highly dependent upon calcium levels (Watanabe et al., 1996). One mechanism proposed for TNFα-induced neuron death is through the rapid TNFα-induced surface expression changes of AMPA-type glutamate receptors, such as GluR1 (Ferguson et al., 2008). Dysregulation of glutamate receptor trafficking can alter neuronal calcium permeability and contribute to excitotoxic vulnerability (Olmos and Llado, 2014; Kia et al., 2018).

Therefore, the lack of a statistically significant relationship between TNFα and GluR1 levels in SOD1-G93A mixed cultures was unexpected (**Figure 6**). Moreover, the PCA (**Figure 5A**) showed TNFα and GluR1 levels to be practically decoupled given TNFα is almost orthogonal to GluR1 on the biplot (**Figure 5A**), which also indicates a lack of relationship.

Of course, TNFα has a complex relationship with many cellular processes. It could be that the data variance identified and imparted by TNFα in the PCA is more related to its other roles, including inflammation, synaptic function, and especially caspase-8 initiated apoptosis (Kia et al., 2018). On a more detailed level, TNFα and NF-κB have been shown to participate in oscillatory positive feedback loops wherein NF-κB can regulate TNFα transcription. This process plays a key role in regulating inflammatory responses across a variety of cell types, and has been found to be central in the FUS mutation of ALS (Pekalski et al., 2013) as well as other neurodegenerative diseases. Another possible role of TNFα could be its relationship with glial derived neurotrophic factor, GDNF. Recent work has shown that TNFα contributes to rises in astrocytic GDNF that have a protective effect on neuron damage (Brambilla et al., 2016); in fact, the TNFα –GDNF mechanism is currently being sought as a potential therapeutic target. Most recently, astrocytes in the ALS FUS model were shown to induce motoneuron death directly via the release of TNFα (Kia et al., 2018).

Thus, while more research is necessary to examine the puzzling lack of relationship seen between TNFα and GluR1 levels, it appears both GluR1 and TNFα are each having an impact on ALS pathology, but the direct relationship between TNFα and GluR1 is simply not substantial in SOD1-G93A ALS mixed cell cultures.

Aspartate Increases Near End-Stage

Aspartate is another excitatory neurotransmitter, similar to glutamate, that is found in the brain. In the present metadata analysis of preclinical ALS mixed cell cultures, the aspartate increase is not seen until end-stage. This supports the finding of normal or slightly decreased aspartate levels in cerebral spinal fluid serum of ALS patients with a mild disease course (Niebroj-Dobosz and Janik, 1999). In severely progressing patients, aspartate levels were increased, and GABA levels were normal or increased. When the imbalance of these amino acids favors the excitatory over the inhibitory, it appears to contribute to a more rapid decline in survival.

Failed Regulation Could Contribute to Spatial Spread of ALS

An important characteristic of ALS is the spread of motoneuron degeneration and corresponding cell death. Recent research has found that transplanting familial ALS astrocytes into healthy mice causes ALS-like degeneration in non-specific healthy neurons (Qian et al., 2017). The spread has been partially attributed to a prion-like mechanism (Grad et al., 2015). More research is needed to better understand the mechanistic etiology of the spread of ALS. Nonetheless, one of the most prominent theories has been related to excitotoxic spillover. For example, gliotransmitters, such as increased glutamate or decreased GABA, can propagate the rise of intracellular calcium levels in astrocytes and adjacent neurons. Glutamate can be released from astrocytes through *Ca2+*-activated bestrophin-1 anion channels. Changes in intracellular calcium levels in astrocytes can also affect the activity of membrane transporters, such as GluRs (Bazargani and Attwell, 2016). Yet, in the present metadata analysis in preclinical ALS mixed cell cultures, we did not see glaring evidence of severe excitotoxic spillover in preclinical ALS mixed cell cultures, as key excitatory (glutamate) and inhibitory transmitters (GABA) were not elevated. However, the identified decrease in inhibitory GLT-1 and increase of excitatory aspartate that occurs with disease progression does suggest there is at least a mild form of neuromodulatory regulation, or lack thereof, that is contributing to ALS etiology. Additionally, the presented analysis does illustrate potential evidence of upregulation, spillover or spread of other chemokines, especially TNFα. Dysregulated or delayed neuromodulator regulatory signaling between neurons and astrocytes should be further explored in future experimental work; evidence of dysregulation could be subtler than what is typically seen with primary excitotoxicity but could nonetheless help to explain the underlying "spread" mechanism(s) of ALS.

Other Non-inflammatory Functions of Astrocytes

ALS-causing genes are also linked to lipid homeostasis, glucose homeostasis, mitochondrial formation, ATP production, and other metabolic functions (Ngo and Steyn, 2015). Extracellular concentrations of ATP in the CNS regulate activation and migration of immune and glial cells and increase in response to trauma and inflammation (Gandelman et al., 2010). In astrocytes, it can cause pro-inflammatory signaling which causes an increased production of nitric oxide and various chemokines (Gandelman et al., 2010). It has been found that G93A astrocytes degrade ATP at a faster rate than non-transgenic astrocytes and display ATP-dependent proliferation. It has also been shown that high extracellular concentrations of ATP can cause non-transgenic astrocytes to induce motoneuron death (Gandelman et al., 2010). Evidence shows that a complicated cycle of bioenergetic deficits may worsen disease progression over time (Ngo and Steyn, 2015). An analysis of extracellular ATP was not completed as part of this study due to lack of data meeting inclusion criteria. However, a similar analysis of ATP in the CNS shows deficiency throughout the SOD1-G93A ALS life span (Irvin et al., 2015). Like glutamate, the temporal trends of metabolic homeostasis need to be determined to identify the most effective time for potential therapeutic intervention.

LIMITATIONS

The biggest limitations to any metadata analysis is the ability to aggregate data sources in a way that sufficiently increases sample size but not so aggregated that group definitions are too diverse to provide meaningful insight. The adage of "apples to apples" and "apples to oranges" is a good analogy for metadata analysis. To carry this analogy into an easy to understand example, a specific metadata analysis group defined as "apples" must literally include all apples, although different individual apples in the group may look slightly different (red, green, yellow, etc.). Transitioning to the present work, "glutamate," for example, include metrics that are slightly different: acutely released glutamate (Milanese et al., 2011; Albano et al., 2013; etc.), steady state extracellular measurements of glutamate (Alexander et al., 2000), whole brain MRI levels of glutamate (Niessen et al., 2007; Choi et al., 2009, etc.), among many other slight variations of direct or indirect glutamate measurement, which are contained within the "glutamate" data aggregate as shown in **Table 2** and the **Supplementary Data Sheet.** It would be preferable to have more specifically defined groups based on different glutamate locations, functions, and/or experimental methodology. However, the number of available studies and their corresponding heterogeneity does not allow more specifically defined sub-groups at the time of this writing; there simply would not be enough statistical power. Thus, the present work utilized more generalized groupings and aggregation schemes (see Methods) that met the standardized criteria and statistical power required by metadata analysis. Nonetheless, this study's presented results still provide new high-level insights as to which of the different biomedical concepts are impacting astrocyte-mediated ALS neuromodulatory regulation over the course of ALS disease progression. Such high-level insight helps researchers to better synthesize complex interactions as well as assess and prioritize biomedical concept research aimed at improving diagnostics, prognostics, and therapeutics. It is an acknowledged limitation that many more data sources are needed to construct narrow aggregate definitions that enable very specific mechanistic hypotheses to be confirmed.

AUTHOR CONTRIBUTIONS

KJ: data collection, statistical analysis, results interpretation, drafting of initial manuscript, critical review of content; JM: data collection, results interpretation, critical review of content; AS: data collection, statistical analysis, critical review of content; CM: framing of study, project oversight, results interpretation, drafting of final manuscript, critical review of content.

REFERENCES

Ajroud-Driss, S., and Siddique, T. (2015). Sporadic and hereditary amyotrophic lateral sclerosis (ALS). *Biochim Biophys Acta* 1852, 679–684. doi: 10.1016/j.bbadis.2014.08.010

Albano, R., Liu, X., and Lobner, D. (2013). Regulation of system x(c)- in the SOD1-G93A mouse model of ALS. *Exp. Neurol.* 250, 69–73. doi: 10.1016/j.expneurol.2013.09.008

Alexander, G. M., Deitch, J. S., Seeburger, J. L., Valle, L. D., and Heiman-Patterson, T. D. (2000). Elevated cortical extracellular fluid glutamate in transgenic mice expressing human mutant (G93A) Cu/Zn superoxide dismutase. *J. Neurochem.* 74, 1666–1673. doi: 10.1046/j.1471-4159.2000.0741666.x

Appel, S. H., Beers, D., Siklos, L., Engelhardt, J. I., and Mosier, D. R. (2001). Calcium: the darth vader of ALS. *Amyotroph. Lateral Scler. Other Motor Neuron Disord.* 2 (Suppl. 1), S47–54. doi: 10.1080/14660820152415744

Bak, L. K., Schousboe, A., and Waagepetersen, H. S. (2006). The glutamate/GABA-glutamine cycle: aspects of transport, neurotransmitter homeostasis, and ammonia transfer. *J. Neurochem.* 98, 641–653. doi: 10.1111/j.1471-4159.2006.03913.x

Bazargani, N., and Attwell, D. (2016). Astrocyte calcium signaling: the third wave. *Nat. Neurosci.* 19, 182–189. doi: 10.1038/nn.4201

Bendotti, C., Tortarolo, M., Suchak, S. K., Calvaresi, N., Carvelli, L., Bastone, A., et al. (2001). Transgenic SOD1 G93A mice develop reduced GLT-1 in spinal cord without alterations in cerebrospinal fluid glutamate levels. *J. Neurochem.* 79, 737–746. doi: 10.1046/j.1471-4159.2001.00572.x

Benkler, C., Ben-Zur, T., Barhum, Y., and Offen, D. (2013). Altered astrocytic response to activation in SOD1(G93A) mice and its implications on amyotrophic lateral sclerosis pathogenesis. *Glia* 61, 312–326. doi: 10.1002/glia.22428

Bond, L., Bernhardt, K., Madria, P., Sorrentino, K., Scelsi, H., and Mitchell, C. S. (2018). A metadata analysis of oxidative stress etiology in preclinical amyotrophic lateral sclerosis: benefits of antioxidant therapy. *Front. Neurosci.* 12:10. doi: 10.3389/fnins.2018.00010

Boston-Howes, W., Gibb, S. L., Williams, E. O., Pasinelli, P., Brown, R. H. Jr., and Trotti, D. (2006). Caspase-3 cleaves and inactivates the glutamate transporter EAAT2. *J. Biol. Chem* 281, 14076–14084. doi: 10.1074/jbc.M600653200

Brambilla, L., Guidotti, G., Martorana, F., Iyer, A. M., Aronica, E., Valori, C. F., et al. (2016). Disruption of the astrocytic TNFR1-GDNF axis accelerates motor neuron degeneration and disease progression in amyotrophic lateral sclerosis. *Hum. Mol. Genet.* 25, 3080–3095. doi: 10.1093/hmg/ddw161

Brockington, A., Ning, K., Heath, P. R., Wood, E., Kirby, J., Fusi, N., et al. (2013). Unravelling the enigma of selective vulnerability in neurodegeneration: motor neurons resistant to degeneration in ALS show distinct gene expression characteristics and decreased susceptibility to excitotoxicity. *Acta Neuropathol.* 125, 95–109. doi: 10.1007/s00401-012-1058-5

Brownstone, R. M., and Lancelin, C. (2018). Escape from homeostasis: spinal microcircuits and progression of amyotrophic lateral sclerosis. *J Neurophysiol.* 119, 1782–1794. doi: 10.1152/jn.00331.2017

Cai, M., Choi, S. M., and Yang, E. J. (2015). The effects of bee venom acupuncture on the central nervous system and muscle in an animal hSOD1G93A mutant. *Toxins* 7, 846–858. doi: 10.3390/toxins7030846

Caioli, S., Curcio, L., Pieri, M., Antonini, A., Marolda, R., Severini, C., et al. (2011). Substance P receptor activation induces downregulation of the AMPA receptor functionality in cortical neurons from a genetic model of amyotrophic lateral sclerosis. *Neurobiol. Dis.* 44, 92–101. doi: 10.1016/j.nbd.2011.06.008

Caioli, S., Pieri, M., Antonini, A., Guglielmotti, A., Severini, C., and Zona, C. (2013). Monocyte Chemoattractant Protein-1 upregulates GABA-induced current: evidence of modified GABAA subunit composition in cortical neurons from the G93A mouse model of Amyotrophic Lateral Sclerosis. *Neuropharm* 73, 247–260. doi: 10.1016/j.neuropharm.2013.05.045

Chen, L. C., Smith, A., Ben, Y., Zukic, B., Ignacio, S., Moore, D., et al. (2004). Temporal gene expression patterns in G93A/SOD1 mouse. *Amyotroph. Lateral. Scler. Other Motor Neuron Disord.* 5, 164–171. doi: 10.1080/14660820410017091

Choi, J., Kustermann, E., Dedeoglu, A., and Jenkins, B. G. (2009). Magnetic resonance spectroscopy of regional brain metabolite markers in fals mice and the effects of dietary creatine supplementation. *Eur. J. Neurosci.* 30, 2143–2150. doi: 10.1111/j.1460-9568.2009.07015.x

Crochemore, C., Pena-Altamira, E., Virgili, M., Monti, B., and Contestabile, A. (2005). Disease-related regressive alterations of forebrain cholinergic system in SOD1 mutant transgenic mice. *Neurochem Int.* 46, 357–368. doi: 10.1016/j.neuint.2004.12.004

Crochemore, C., Virgili, M., Bonamassa, B., Canistro, D., Pena-Altamira, E., Paolini, M., et al. (2009). Long-term dietary administration of valproic acid does not affect, while retinoic acid decreases, the lifespan of G93A mice, a model for amyotrophic lateral sclerosis. *Muscle Nerve* 39, 548–552. doi: 10.1002/mus.21260

D'antoni, S., Berretta, A., Seminara, G., Longone, P., Giuffrida-Stella, A. M., Battaglia, G., et al. (2011). A prolonged pharmacological blockade of type-5 metabotropic glutamate receptors protects cultured spinal cord motor neurons against excitotoxic death. *Neurobiol. Dis.* 42, 252–264. doi: 10.1016/j.nbd.2011.01.013

Deitch, J. S., Alexander, G. M., Del Valle, L., and Heiman-Patterson, T. D. (2002). GLT-1 glutamate transporter levels are unchanged in mice expressing G93A human mutant SOD1. *J. Neurol. Sci.* 193, 117–126. doi: 10.1016/S0022-510X(01)00656-6

Diana, A., Pillai, R., Bongioanni, P., O'Keeffe, A. G., Miller, R. G., and Moore, D. H. (2017). Gamma aminobutyric acid (GABA) modulators for amyotrophic lateral sclerosis/motor neuron disease. *Cochrane Database Syst. Rev.* 1:CD006049. doi: 10.1002/14651858.CD006049.pub2

Do-Ha, D., Buskila, Y., and Ooi, L. (2017). Impairments in motor neurons, interneurons and astrocytes contribute to hyperexcitability in als: underlying mechanisms and paths to therapy. *Mol. Neurobiol.* 55, 1410–1418. doi: 10.1007/s12035-017-0392-y

Fang, L., Teuchert, M., Huber-Abel, F., Schattauer, D., Hendrich, C., Dorst, J., et al. (2010). MMP-2 and MMP-9 are elevated in spinal cord and skin in a mouse model of ALS. *J. Neurol. Sci.* 294, 51–56. doi: 10.1016/j.jns.2010.04.005

Fergani, A., Eschbach, J., Oudart, H., Larmet, Y., Schwalenstocker, B., Ludolph, A. C., et al. (2011). A mutation in the dynein heavy chain gene compensates for energy deficit of mutant SOD1 mice and increases potentially neuroprotective IGF-1. *Mol. Neurodegener.* 6:26. doi: 10.1186/1750-1326-6-26

Ferguson, A. R., Christensen, R. N., Gensel, J. C., Miller, B. A., Sun, F., Beattie, E. C., et al. (2008). Cell death after spinal cord injury is exacerbated by rapid TNF alpha-induced trafficking of GluR2-lacking AMPARs to the plasma membrane. *J. Neurosci.* 28, 11391–11400. doi: 10.1523/JNEUROSCI.3708-08.2008

Fernández, M., Sanchez-Franco, F., Palacios, N., Sanchez, I., Fernandez, C., and Cacicedo, L. (2004). IGF-I inhibits apoptosis through the activation of the phosphatidylinositol 3-kinase/Akt pathway in pituitary cells. *J. Mol. Endocrinol.* 33, 155–163. doi: 10.1677/jme.0.0330155

Gandelman, M., Peluffo, H., Beckman, J. S., Cassina, P., and Barbeito, L. (2010). Extracellular ATP and the P2X7 receptor in astrocyte-mediated motor neuron death: implications for amyotrophic lateral sclerosis. *J. Neuroinflam.* 7, 33–33. doi: 10.1186/1742-2094-7-33

Grad, L. I., Fernando, S. M., and Cashman, N. R. (2015). From molecule to molecule and cell to cell: prion-like mechanisms in amyotrophic lateral sclerosis. *Neurobiol. Dis.* 77, 257–265. doi: 10.1016/j.nbd.2015.02.009

Grosskreutz, J., Van Den Bosch, L., and Keller, B. U. (2010). Calcium dysregulation in amyotrophic lateral sclerosis. *Cell Calcium* 47, 165–174. doi: 10.1016/j.ceca.2009.12.002

Gu, R., Hou, X., Pang, R., Li, L., Chen, F., Geng, J., et al. (2010). Human adipose-derived stem cells enhance the glutamate uptake function of GLT1 in SOD1 G93A-bearing astrocytes. *Biochem. Biophys. Res. Commun.* 393, 481–486. doi: 10.1016/j.bbrc.2010.02.029

Guerra-Gomes, S., Sousa, N., Pinto, L., and Oliveira, J. F. (2018). Functional roles of astrocyte calcium elevations: from synapses to behavior. *Front. Cell Neurosci.* 11:427. doi: 10.3389/fncel.2017.00427

Guo, H., Lai, L., Butchbach, M. E., Stockinger, M. P., Shan, X., Bishop, G. A., et al. (2003). Increased expression of the glial glutamate transporter EAAT2 modulates excitotoxicity and delays the onset but not the outcome of ALS in mice. *Hum Mol Genet.* 12, 2519–2532. doi: 10.1093/hmg/ddg267

Guo, Z., Kindy, M. S., Kruman, I., and Mattson, M. P. (2000). ALS-Linked Cu/Zn-SOD Mutation impairs cerebral synaptic glucose and glutamate transport and exacerbates ischemic brain injury. *J. Cereb. Blood. Flow Metab.* 20, 463–468. doi: 10.1097/00004647-200003000-00004

Hollinger, S. K., Okosun, I. S., and Mitchell, C. S. (2016). Antecedent disease and amyotrophic lateral sclerosis: what is protecting whom? *Front. Neurol.* 7:47. doi: 10.3389/fneur.2016.00047

Hossaini, M., Cano, S. C., van Dis, V., Haasdijk, E. D., Hoogenraad, C. C., Holstege, J. C., et al. (2011). Spinal inhibitory interneuron pathology follows motor neuron degeneration independent of glial mutant superoxide dismutase 1 expression in SOD1-ALS mice. *J. Neuropath. Exp. Neurol.* 70, 662–677. doi: 10.1097/NEN.0b013e31822581ac

Huber, C. M., Yee, C., May, T., Dhanala, A., and Mitchell, C. S. (2018). Cognitive decline in preclinical alzheimer's disease: amyloid-beta versus tauopathy. *J. Alz. Dis.* 61, 265–281. doi: 10.3233/JAD-170490

Irvin, C. W., Kim, R. B., and Mitchell, C. S. (2015). Seeking homeostasis: temporal trends in respiration, oxidation, and calcium in SOD1 G93A amyotrophic lateral sclerosis mice. *Front. Cell Neurosci.* 9, 248. doi: 10.3389/fncel.2015.00248

Jeyachandran, A., Mertens, B., Mckissick, E. A., and Mitchell, C. S. (2015). Type I Vs. Type II cytokine levels as a function of SOD1 G93A mouse amyotrophic lateral sclerosis disease progression. *Front. Cell Neurosci.* 9:462. doi: 10.3389/fncel.2015.00462

Jiang, M. C., Adimula, A., Birch, D., and Heckman, C. J. (2017). Hyperexcitability in synaptic and firing activities of spinal motoneurons in an adult mouse model of amyotrophic lateral sclerosis. *Neuroscience* 362:33–46. doi: 10.1016/j.neuroscience.2017.08.041

Kalmar, B., Edet-Amana, E., and Greensmith, L. (2012). Treatment with a coinducer of the heat shock response delays muscle denervation in the SOD1-G93A mouse model of amyotrophic lateral sclerosis. *Amyotroph. Lateral. Sclerosis* 13, 378–392. doi: 10.3109/17482968.2012.660953

Kandinov, B., Grigoriadis, N. C., Touloumi, O., Drory, V. E., Offen, D., and Korczyn, A. D. (2013). Immunohistochemical analysis of sympathetic involvement in the SOD1-G93A transgenic mouse model of amyotrophic lateral sclerosis. *Amyotroph. Lateral. Scler. Frontotemporal. Degener.* 14, 424–433. doi: 10.3109/21678421.2013.780622

Kaspar, B. K., Frost, L. M., Christian, L., Umapathi, P., and Gage, F. H. (2005). Synergy of insulin-like growth factor-1 and exercise in amyotrophic lateral sclerosis. *Ann. Neurol.* 57, 649–655. doi: 10.1002/ana.20451

Kawamata, H., and Manfredi, G. (2010). Mitochondrial dysfunction and intracellular calcium dysregulation in ALS. *Mech. Aging Develop.* 131, 517–526. doi: 10.1016/j.mad.2010.05.003

Kawamata, H., Ng, S. K., Diaz, N., Burstein, S., Morel, L., Osgood, A., et al. (2014). Abnormal intracellular calcium signaling and SNARE-dependent exocytosis contributes to SOD1G93A astrocyte-mediated toxicity in amyotrophic lateral sclerosis. *J. Neurosci.* 34, 2331–2348. doi: 10.1523/JNEUROSCI.2689-13.2014

Khamankar, N., Coan, G., Weaver, B., and Mitchell, C. S. (2018). Associative increases in amyotrophic lateral sclerosis survival duration with non-invasive ventilation initiation and usage protocols. *Front. Neurol.* 9, 578. doi: 10.3389/fneur.2018.00578

Kia, A., Mcavoy, K., Krishnamurthy, K., Trotti, D., and Pasinelli, P. (2018). Astrocytes expressing ALS-linked mutant FUS induce motor neuron death through release of tumor necrosis factor-alpha. *Glia* 66, 1016–1033. doi: 10.1002/glia.23298

Kim, R. B., Irvin, C. W., Tilva, K. R., and Mitchell, C. S. (2016). State of the field: An informatics-based systematic review of the SOD1-G93A amyotrophic lateral sclerosis transgenic mouse model. *Amyotroph. Lateral. Scler. Frontotemporal. Degener.* 17, 1–14. doi: 10.3109/21678421.2015.1047455

Lee, J. K., Shin, J. H., Gwag, B. J., and Choi, E. J. (2015). Iron accumulation promotes TACE-mediated TNF-alpha secretion and neurodegeneration in a mouse model of ALS. *Neurobiol. Dis.* 80, 63–69. doi: 10.1016/j.nbd.2015.05.009

Leroy, F., and Zytnicki, D. (2015). Is hyperexcitability really guilty in amyotrophic lateral sclerosis?. *Neural Regen Res.* 10, 1413–1415 doi: 10.4103/1673-5374.165308

Li, K., Hala, T. J., Seetharam, S., Poulsen, D. J., Wright, M. C., and Lepore, A. C. (2015). GLT1 overexpression in SOD1G93A mouse cervical spinal cord does not preserve diaphragm function or extend disease. *Neurobiol. Dis.* 78, 12–23. doi: 10.1016/j.nbd.2015.03.010

Lin, C.-L. G., Kong, Q., Cuny, G. D., and Glicksman, M. A. (2013). Glutamate transporter EAAT2: a new target for the treatment of neurodegenerative diseases. *Future Med. Chem.* 4, 1689–1700. doi: 10.4155/fmc.12.122

Martinez, J. A., Francis, G. J., Liu, W. Q., Pradzinsky, N., Fine, J., Wilson, M., et al. (2008). Intranasal delivery of insulin and a nitric oxide synthase inhibitor in an experimental model of amyotrophic lateral sclerosis. *Neuroscience* 157, 908–925. doi: 10.1016/j.neuroscience.2008.08.073

Martínez-Silva, M. L., Imhoff-Manuel, R. D., Sharma, A., Heckman, C. J., Shneider, N. A., Roselli, F., et al. (2018). Hypoexcitability precedes denervation in the large fast-contracting motor units in two unrelated mouse models of ALS. *Elife.* 7:e30955. doi: 10.7554/elife.30955

Messi, M. L., Clark, H. M., Prevette, D. M., Oppenheim, R. W., and Delbono, O. (2007). The lack of effect of specific overexpression of IGF-1 in the central nervous system or skeletal muscle on pathophysiology in the G93A SOD-1 mouse model of ALS. *Exper. Neurol.* 207, 52–63. doi: 10.1016/j.expneurol.2007.05.016

Milanese, M., Zappettini, S., Onofri, F., Musazzi, L., Tardito, D., Bonifacino, T., et al. (2011). Abnormal exocytotic release of glutamate in a mouse model of amyotrophic lateral sclerosis. *J. Neurochem.* 116, 1028–1042. doi: 10.1111/j.1471-4159.2010.07155.x

Mitchell, C. S., Cates, A., Kim, R. B., and Hollinger, S. K. (2015a). Undergraduate biocuration: developing tomorrow's researchers while mining today's data. *J. Undergrad. Neurosci. Educ.* 14, A56–A65.

Mitchell, C. S., Hollinger, S. K., Goswami, S. D., Polak, M. A., Lee, R. H., and Glass, J. D. (2015b). Antecedent disease is less prevalent in amyotrophic lateral sclerosis. *Neurodegener. Dis.* 15, 109–113. doi: 10.1159/000369812

Mitchell, C. S., and Lee, R. H. (2008). Pathology dynamics predict spinal cord injury therapeutic success. *J. Neurotrauma.* 25, 1483–1497. doi: 10.1089/neu.2008.0658

Mitchell, C. S., and Lee, R. H. (2012). "Dynamic meta-analysis as a therapeutic prediction tool for amyotrophic lateral sclerosis," *in Amyotrophic Lateral Sclerosis*, ed. M. Maurer (Intech), 59–80. doi: 10.5772/32384

Morel, L., Regan, M., Higashimori, H., Ng, S. K., Esau, C., Vidensky, S., et al. (2013). Neuronal exosomal miRNA-dependent translational regulation of astroglial glutamate transporter GLT1. *J. Biol. Chem.* 8, 7105–7116. doi: 10.1074/jbc.M112.410944

Ngo, S. T., and Steyn, F. J. (2015). The interplay between metabolic homeostasis and neurodegeneration: insights into the neurometabolic nature of amyotrophic lateral sclerosis. *Cell Regen* . 4:5. doi: 10.1186/s13619-015-0019-6

Niebroj-Dobosz, I., and Janik, P. (1999). Amino acids acting as transmitters in amyotrophic lateral sclerosis (ALS). *Acta Neurol. Scand.* 100, 6–11. doi: 10.1111/j.1600-0404.1999.tb00717.x

Niessen, H. G., Debska-Vielhaber, G., Sander, K., Angenstein, F., Ludolph, A. C., Hilfert, L., et al. (2007). Metabolic progression markers of neurodegeneration in the transgenic G93A-SOD1 mouse model of amyotrophic lateral sclerosis. *Eur. J. Neurosci.* 25, 1669–1677. doi: 10.1111/j.1460-9568.2007.05415.x

Nihei, K., McKee, A. C., and Kowall, N. W. (1993). Patterns of neuronal degeneration in the motor cortex of amyotrophic lateral sclerosis patients. *Acta Neuropathologica* 86, 55–64. doi: 10.1007/BF00454899

Norenberg, M. D., and Martinez-hernandez, A. (1979). Fine structural localization of glutamine synthase in astrocytes of rat brains. *Brain Res.* 161, 303–310. doi: 10.1016/0006-8993(79)90071-4

Olmos, G., and Llado, J. (2014). Tumor necrosis factor alpha: a link between neuroinflammation and excitotoxicity. *Mediators Inflamm.* 2014:861231. doi: 10.1155/2014/861231

Palmieri, L., Pardo, B., Lasorsa, F. M., Del Arco, A., Kobayashi, K., Iijima, M., et al. (2001). Citrin and aralar1 are Ca(2+)-stimulated aspartate/glutamate transporters in mitochondria. *EMBO J.* 20, 5060–5069. doi: 10.1093/emboj/20.18.5060

Pardo, A. C., Wong, V., Benson, L. M., Dykes, M., Tanaka, K., Rothstein, J. D., et al. (2006). Loss of the astrocyte glutamate transporter GLT1 modifies disease in SOD1 G93A mice. *Exper. Neurol.* 201, 120–130. doi: 10.1016/j.expneurol.2006.03.028

Pehar, M., Harlan, B. A., Killoy, K. M., and Vargas, M. R. (2017). *Role and Therapeutic Potential of Astrocytes in Amyotrophic Lateral Sclerosis. Curr. Pharm. Des.* 23, 5010–5021. doi: 10.2174/1381612823666170622095802

Pekalski, J., Zuk, P. J., Kochanczyk, M., Junkin, M., Kellogg, R., Tay, S., et al. (2013). Spontaneous NF-kappaB activation by autocrine TNFalpha signaling: a computational analysis. *PLoS ONE* 8:e78887. doi: 10.1371/journal.pone.0078887

Petri, S., Schmalbach, S., Grosskreutz, J., Krampfl, K., Grothe, C., Dengler, R., et al. (2005). The cellular mRNA expression of GABA and glutamate receptors in spinal motor neurons of SOD1 mice. *J. Neurol. Sci.* 238, 25–30. doi: 10.1016/j.jns.2005.06.005

Pfohl, S. R., Halicek, M. T., and Mitchell, C. S. (2015). Characterization of the contribution of genetic background and gender to disease progression in the SOD1 G93A mouse model of amyotrophic lateral sclerosis: a meta-analysis. *J. Neuromuscul. Dis.* 2, 137–150. doi: 10.3233/JND-140068

Pfohl, S. R., Kim, R. B., Coan, G. S., and Mitchell, C. S. (2018). Unraveling the complexity of amyotrophic lateral sclerosis survival prediction. *Front. Neuroinform.* 12:36. doi: 10.3389/fninf.2018.00036

Qian, K., Huang, H., Peterson, A., Hu, B., Maragakis, N. J., Ming, G. L., et al. (2017). Sporadic ALS astrocytes induce neuronal degeneration *in vivo*. *Stem Cell Rep.* 8, 843–855. doi: 10.1016/j.stemcr.2017.03.003

Raiteri, L., Stigliani, S., Zappettini, S., Mercuri, N. B., Raiteri, M., and Bonanno, G. (2004). Excessive and precocious glutamate release in a mouse model of amyotrophic lateral sclerosis. *Neuropharm* 46, 782–792. doi: 10.1016/j.neuropharm.2003.11.025

Reardon, S. (2018). Pioneering Alzheimer's study in Colombia zeroes in on enigmatic protein. *Nature* 555, 567–568. doi: 10.1038/d41586-018-03848-4

Rosenblum, L. T., and Trotti, D. (2017). EAAT2 and the molecular signature of amyotrophic lateral sclerosis. *Adv. Neurobiol.* 16, 117–136. doi: 10.1007/978-3-319-55769-4_6

Rossi, D., and Volterra, A. (2009). Astrocytic dysfunction: insights on the role in neurodegeneration. *Brain Res. Bull.* 80, 224–232. doi: 10.1016/j.brainresbull.2009.07.012

Rothstein, J. D., Patel, S., Regan, M. R., Haenggeli, C., Huang, Y. H., Bergles, D. E., et al. (2005). Beta-lactam antibiotics offer neuroprotection by increasing glutamate transporter expression. *Nature* 433, 73–77. doi: 10.1038/nature03180

Rothstein, J. D., Tsai, G., Kunci, R. W., Clawson, L., Conblath, D. B., Pestronk, A., et al. (1990). Abnormal excitatory amino acid metabolism in amytrophic lateral sclerosis. *Ann. Neruol.* 28, 18–25. doi: 10.1002/ana.410280106

Rothstein, J. D., Van Kammen, M., Levey, A. I., Martin, L. J., and Kuncl, R. W. (1995). Selective loss of glial glutamate transporter GLT-1 in amyotrophic lateral sclerosis. *Ann. Neurol.* 38, 73–84. doi: 10.1002/ana.410380114

Selvaraj, B. T., Livesey, M. R., Zhao, C., Gregory, J. M., James, O., Cleary, E. M., et al. (2018). C9ORF72 repeat expansion causes vulnerability of motor neurons to Ca2+-permeable AMPA receptor-mediated excitotoxicity. *Nat Commun.* 9:347. doi: 10.1038/s41467-017-02729-0

Sitges, M., and Chiu, L. M. (1995). Characterization of the type of calcium channel primarily regulating GABA exocytosis from brain nerve endings. *Neurochem. Res.* 20, 1073–1088. doi: 10.1007/BF00995562

Sofroniew, M. V., and Vinters, H. V. (2010). Astrocytes: biology and pathology. *Acta Neuropathologica.* 119, 7–35. doi: 10.1007/s00401-009-0619-8

Song, L., Gao, Y., Zhang, X., and Le, W. (2013). Galactooligosaccharide improves the animal survival and alleviates motor neuron death in SOD1G93A mouse model of amyotrophic lateral sclerosis. *Neuroscience* 246, 281–290. doi: 10.1016/j.neuroscience.2013.05.002

Spalloni, A., Geracitano, R., Berretta, N., Sgobio, C., Bernardi, G., Mercuri, N. B., et al. (2006). Molecular and synaptic changes in the hippocampus underlying superior spatial abilities in pre-symptomatic G93A+/+ mice overexpressing the human Cu/Zn superoxide dismutase (Gly93 -> ALA) mutation. *Exp. Neurol.* 197, 505–514. doi: 10.1016/j.expneurol.2005.10.014

Staat, K. A., and Van Den Bosch, L. (2009). Astrocytes in amytrophic lateral sclerosis: direct effects on motor neuron survival. *J. Biol. Phys.* 35, 337–346. doi: 10.1007/s10867-009-9141-4

Staats, K. A., Humblet-Baron, S., Bento-Abreu, A., Scheveneels, W., Nikolaou, A., Deckers, K., et al. (2016). Genetic ablation of IP_3 receptor 2 increases cytokines and decreases survival of $SOD1^{G93A}$ mice. *Hum. Mol. Genet.* 25, 3491–3499. doi: 10.1093/hmg/ddw190

Tateno, M., Sadakata, H., Tanaka, M., Itohara, S., Shin, R. M., Miura, M., et al. (2004). Calcium-permeable AMPA receptors promote misfolding of mutant SOD1 protein and development of amyotrophic lateral sclerosis in a transgenic mouse model. *Hum. Mol. Genet.* 13, 2183–2196. doi: 10.1093/hmg/ddh246

Tefera, T. W., and Borges, K. (2018). Neuronal glucose metabolism is impaired while astrocytic TCA cycling is unaffected at symptomatic stages in the hSOD1(G93A) mouse model of amyotrophic lateral sclerosis. *J. Cereb. Blood. Flow Metab.* doi: 10.1177/0271678X18764775. [Epub ahead of print].

Tortarolo, M., Grignaschi, G., Calvaresi, N., Zennaro, E., Spaltro, G., Colovic, M., et al. (2006). Glutamate AMPA receptors change in motor neurons of SOD1G93A transgenic mice and their inhibition by a noncompetitive antagonist ameliorates the progression of amytrophic lateral sclerosis-like disease. *J. Neurosci. Res.* 83, 134–146. doi: 10.1002/jnr.20715

Tsai, K. J., Yang, C. H., Fang, Y. H., Cho, K. H., Chien, W. L., Wang, W. T., et al. (2010). Elevated expression of TDP-43 in the forebrain of mice is sufficient to cause neurological and pathological phenotypes mimicking FTLD-U. *J. Exp. Med.* 207, 1661–1673. doi: 10.1084/jem.20092164

Valbuena, G. N., Rizzardini, M., Cimini, S., Siskos, A. P., Bendotti, C., Cantoni, L., et al. (2016). Metabolomic analysis reveals increased aerobic glycolysis and amino acid deficit in a cellular model of amyotrophic lateral sclerosis. *Mol. Neurobiol.* 53, 2222–2240. doi: 10.1007/s12035-015-9165-7

Van Damme, P., Dewil, M., Robberecht, W., and Van Den Bosch, L. (2005). Excitotoxicity and amyotrophic lateral sclerosis. *Neurodegener. Dis.* 2, 147–159. doi: 10.1159/000089620

Van Den Bosch, L., Van Damme, P., Bogaert, E., and Robberecht, W. (2006). The role of excitotoxicity in the pathogenesis of amyotrophic lateral sclerosis. *Mol. Basis Dis.* 1762, 1068–1082. doi: 10.1016/j.bbadis.2006.05.002

Vargas, M. R., and Johnson, J. A. (2010). Astrogliosis in amytrophic lateral sclerosis: role and therapeutic potential of astrocytes. *Neurotherapeutics* 7, 471–481. doi: 10.1016/j.nurt.2010.05.012

Watanabe, N., Suzuki, J., and Kobayashi, Y. (1996). Role of calcium in tumor necrosis factor-alpha production by activated macrophages. *J. Biochem.* 120, 1190–1195. doi: 10.1093/oxfordjournals.jbchem.a021540

Wu, D. C., Re, D. B., Nagai, M., Ischiropoulos, H., and Przedborski, S. (2006). The inflammatory NADPH oxidase enzyme modulates motor neuron degeneration in amyotrophic lateral sclerosis mice. *Proc Natl Acad Sci USA.* 103, 12132–12137. doi: 10.1073/pnas.0603670103

Yang, X., and Cheng, B. (2010). Neuroprotective and anti-inflammatory activities of ketogenic diet on MPTP-induced neurotoxicity. *J. Mol. Neurosci.* 42, 145–153. doi: 10.1007/s12031-010-9336-y

Yang, Y., Gozen, O., Watkins, A., Lorenzini, I., Lepore, A., Gao, Y., et al. (2009). Presynaptic regulation of astroglial excitatory neurotransmitter transporter GLT1. *Neuron* 61, 880–894. doi: 10.1016/j.neuron.2009.02.010

Zhao, P., Ignacio, S., Beattie, E. C., and Abood, M. E. (2008). Altered presymptomatic AMPA and cannabinoid receptor trafficking in motor neurons of ALS model mice: implications for excitotoxicity. *Eur. J. Neurosci.* 27, 572–579. doi: 10.1111/j.1460-9568.2008.06041.x

Zhao, Z., Lange, D. J., Ho, L., Bonini, S., Shao, B., Salton, S. R., et al. (2008). Vgf is a novel biomarker associated with muscle weakness in amyotrophic lateral sclerosis (ALS), with a potential role in disease pathogenesis. *Int. J. Med. Sci.* 5, 92–99. doi: 10.7150/ijms.5.92

8

Microglial Dynamics during Human Brain Development

David A. Menassa* and Diego Gomez-Nicola*

Biological Sciences, Faculty of Natural and Environmental Sciences, Southampton General Hospital, University of Southampton, Southampton, United Kingdom

***Correspondence:**
David A. Menassa
d.menassa@soton.ac.uk;
Diego Gomez-Nicola
d.gomez-nicola@soton.ac.uk

Microglial cells are thought to colonize the human cerebrum between the 4th and 24th gestational weeks. Rodent studies have demonstrated that these cells originate from yolk sac progenitors though it is not clear whether this directly pertains to human development. Our understanding of microglial cell dynamics in the developing human brain comes mostly from postmortem studies demonstrating that the beginning of microglial colonization precedes the appearance of the vasculature, the blood–brain barrier, astrogliogenesis, oligodendrogenesis, neurogenesis, migration, and myelination of the various brain areas. Furthermore, migrating microglial populations cluster by morphology and express differential markers within the developing brain and according to developmental age. With the advent of novel technologies such as RNA-sequencing in fresh human tissue, we are beginning to identify the molecular features of the adult microglial signature. However, this is may not extend to the much more dynamic and rapidly changing antenatal microglial population and this is further complicated by the scarcity of tissue resources. In this brief review, we first describe the various historic schools of thought that had debated the origin of microglial cells while examining the evidence supporting the various theories. We then proceed to examine the evidence we have accumulated on microglial dynamics in the developing human brain, present evidence from rodent studies on the functional role of microglia during development and finally identify limitations for the used approaches in human studies and highlight under investigated questions.

Keywords: microglia, brain development, glial cells, proliferation, human brain, neurodevelopment

INTRODUCTION

Microglia are resident macrophages of the mammalian central nervous system (CNS) with a plethora of functions including innate immunity and neuroprotection (1, 2), synaptic pruning (3), and the phagocytosis of cellular debris (4). Microglia also have a dark side, suggested to be indirectly involved in pathologies such as schizophrenia (5) and more directly in Alzheimer's disease (6). They play a central role in driving neuroinflammation and also coordinating this response with the systemic compartment and with other glial cells including astrocytes (7). A careful characterization of human microglial morphology, transcriptome, regional heterogeneity, and turnover in the adult human brain is emerging greatly due to the increased availability of appropriately preserved postmortem tissue. In the adult human cortex, for example, microglial cells acquire a ramified morphology with processes surveying the neighboring environment and an amoeboid morphology with retracted processes when they are phagocytosing the debris of dying neurons or synapses or reacting to a foreign body (8). Adult human microglia are enriched for genes associated with the

TABLE 1 | A summary of main microglial topography studies in embryonic and fetal human tissue.

Reference	Year	Age (weeks)	Sample size	Anatomical areas	Markers used	Findings and conclusions
Rydberg (24)	1932	38.0	2	Cerebrum, cerebellum, spinal cord	Silver carbonate	First study on microglial cytogenesis in human fetuses and newborns. Proponent of the neuroectodermal theory, this study suggests that any glial cell could originate from amoeboid glia located along the ependymal matrix
Von Santha (23)	1932	24.0 and 29.0	2	Basal ganglia, telencephalon, white matter tracts, subcortex, medulla	Silver carbonate	Microglia are ramified in the basal ganglia, the telencephalon, and the medulla. Plump branched cells are observed in the subcortex and prospective white matter tracts. No globose forms are reported
Juba (25)	1933	10.0–34.0	7	Cerebrum, cerebellum	Silver carbonate	Microglia are amoeboid early in development and ramify when they become differentiated. Microglia enter the developing brain from mesodermal elements closely associated with the vasculature. Microglia are genetically related to monocytes
Kershman (76)	1939	8.0–29.0	22	Cerebrum, cerebellum, spinal cord	Silver carbonate, H&E	The youngest form of microglia is amoeboid first before they ramify into a mature form. Amoeboid cells enter into the brain from specific locations known as fountains located around the choroid plexus, the meningeal compartment, and blood vessels. First evidence of microglia is in the rhombencephalon at 8.0 weeks. Once in the developing brain, amoeboid microglia acquire pseudopods as they penetrate deep into the brain matter and eventually acquire a final stable ramified phenotype
Choi (62)	1981	6.0–22.0	53	Cerebrum, cerebellum, spinal cord	H&E; Toluidine Blue	6.0–12.0 weeks: hematogenous cells are found in the cerebellum, spinal cord, and cerebrum. These become rarer as the fetus gets older. The neuropils of embryos/fetuses contain macrophage-like cells inside and outside of blood vessels. Subpial, perivascular, and perineuronal regions of neural parenchyma contain small cells with fusiform nuclei, some with elongated nuclei and some with rod-shaped/oblong nuclei
Fujimoto et al. (66)	1989	7.5–26.0	7	Cerebrum	NDP-ase	Vascularization in the cerebral cortex is apparent by 9.0 weeks, vessels invade the pallium. Amoeboid NDP-ase cells appear around the vessels. The MZ has poorly ramified microglia. By 11.0 weeks, amoeboid cells present in the matrix and subcortical layers. Highly ramified NDP-ase cells are seen in layer I, the subcortex, and cortical mantle between 22.0 and 26.0 weeks
Hutchins et al. (77)	1990	13.0–24.0	22	Frontal cerebrum	RCA-1; EBM-11	Densest RCA-1 and EBM-11 cells are present in the germinal matrix at 13.0 weeks. Microglial morphology is variable between the gray and white matter between 13.0 and 18.0 weeks (amoeboid versus intermediate) and less variable and ramified across the mantle and white matter between 19.0 and 24.0 weeks
Esiri et al. (86)	1991	18.0-term	15	Telencephalon	EBM-11; HLA-DR	Amoeboid cells are present in the germinal matrix at high densities throughout this temporal window. Ramified microglia in gray and white matter are only seen from 35.0 weeks. Very little reactivity to HLA-DR is detected throughout
Gould and Howard (80)	1991	13.0–27.0	19	Cerebral white matter	A1ACT; MAC387	Macrophage clusters positive for A1ACT accumulate in prospective white matter hotspots including the periventricular areas, corpus callosum, the junction between the external and internal capsule, the optic tract, and the thalamus. Increased densities followed a caudo-rostral pattern with more rostral areas showing microglial fountains after 22.0 weeks
Hutchins et al. (90)	1992	18.0–24.0	6	Cervical spinal cord	RCA-1; EBM-11	RCA-1/CD68-positive amoeboid microglia are seen in the white matter and only RCA-1 but CD68 negative ramified microglia are seen in the gray matter
Wierzba-Bobrowicz et al. (43)	1995	7.0-term	47	Mesencephalon	RCA-1; Anti-Ferritin	RCA-1/Ferritin-positive amoeboid cells are observed in the mesencephalon between 16.0 and 40.0 weeks

(Continued)

TABLE 1 | Continued

Reference	Year	Age (weeks)	Sample size	Anatomical areas	Markers used	Findings and conclusions
Rezaie et al. (82)	1997	16.0–22.0	7	Frontal telencephalon	CD11b; CD68; CD64; CD45; RCA-1; BSB-4	Fetal RCA1 microglia co-occur with highly vascularized ICAM-1 positive vessels within the developing brain
Andjelkovic et al. (63)	1998	4.5–13.5	14	Telencephalon, diencephalon, rhombencephalon	RCA-1; Tomato Lectin; CD68	At 5.5 weeks, CD68 and lectin positive cells are observed in the cerebellum and medulla. Between 6.0 and 7.5 weeks, first blood vessels in telencephalic wall and CD68/lectin cells increase in density in line with vascularization. Vascularization by CD34 tagging is prominent first caudally (cerebellum and diencephalon), then rostrally eventually. At 7.5 weeks, ramified microglia are detected in the cerebellum only among amoeboid cells. Between 8.0 and 9.0 weeks, clusters of microglia are observed in diencephalon. Ramified microglia are also observed in the MZ and IZ and none in CP. By 11.0–13.0 weeks, ramified microglia are detected in all layers of the cortex. Two classes of cells are identified here lectin positive and CD68/lectin positive
Wierzba-Bobrowicz et al. (12)	1998	8.0–22.0	72	Mesencephalon, cerebellum, frontal lobe	RCA-1; Anti-Ferritin; HAM56	Amoeboid microglia are increased in all structures as gestation progresses. White matter ramified microglia are highest in density by 22 weeks in the mesencephalon and the lowest in the white matter of the cerebellum
Rezaie et al. (31)	1999	7.0–15.0	9	Spinal cord	RCA-1; CD68; HAM56; CD11b; CD45; CD64	Microglia present are present in the spinal cord from 9.0 weeks and enter from the meninges/connective tissue unlike the cerebrum where they migrate out from the germinal matrix toward the cortical surface
Rezaie et al. (75)	2005	12.0–24.0	45	Telencephalon	RCA-1; CD68; MHC-II	12.0–14.0 weeks of gestation, 2 populations are detected: CD68+/RCA-1+/MHC-II– amoeboid cells aligned within the subplate and RCA-1+/CD68–/MHC-II- cells in the MZ and lower CP that progressively ramify within the subplate. Microglia are absent from the germinal layers and IZ. From 14.0 to 15.0 weeks, MHC-II+/CD68+ cells in the germinal layers and corpus callosum appear and further populate the lower telencephalon from 18.0 to 24.0 weeks
Billiards et al. (84)	2006	22.0–37.0	9	Cerebral white matter	Tomato lectin; CD68; MHC-II	Ramified microglia are only detected in the prospective white matter at 22.0 weeks then intermediate or amoeboid forms are seen between 25.0 and 37.0 weeks in similar areas
Monier et al. (60)	2006	5.0–24.0	31	Diencephalon, telencephalon	IBA-1; CD68; CD45; MHC-II; LN3; HLA-DR	By 5.5 weeks, accumulation of IBA-1 microglia near the meninges and choroid plexus before blood vessel formation in the brain parenchyma. Early entry along the di-telencephalic fissure of amoeboid cells from within the choroid plexus and meninges is reported. These are fetal macrophages reported by Choi and Andjelkovic et al. (1998). IBA-1/CD34 associated cells are observed at 10.0 weeks in the white matter
Monier et al. (61)	2007	4.5–24.0	33	Telencephalon	IBA-1; CD68; CD45; MHC-II; LN3; HLA-DR	Microglia express IBA-1 at 4.5 weeks. Amoeboid cells are present in the lumen and leptomeninges then. Entry into the brain is through the ventricles, whereby microglia assume eventually tangential and radial migration routes. Other route is from the pia to the MZ. Microglia are ramified by 22.0 weeks in the cortical mantle, amoeboid in the IZ, and grouped in clusters between 22.0 and 23.5 weeks
Verney et al. (73)	2010	20.0–32.0	4	Telencephalon, diencephalon	IBA-1	Accumulation of IBA-1 microglia in clusters is seen in the white matter (centrum semiovale)

(Continued)

TABLE 1 | Continued

Reference	Year	Age (weeks)	Sample size	Anatomical areas	Markers used	Findings and conclusions
Cho et al. (89)	2013	15.0–25.0	9	Brainstem, olive, gray matter, white matter, hippocampus	H&E; HLA-DR2; CD68	CD68 positive cells are evident in the floor of the fourth ventricle, the pons and olive at 15.0–16.0 weeks, accumulating in and around the hippocampus at 22.0–25.0 weeks. At both stages, the accumulation of these cells was evident in the optic tract and the anterior limb of the internal capsule. GAP-43 developing axons are not associated with CD68 cells
Mildner et al. (88)	2016	11.0–35.0	4	Cerebellum, white matter, cortex	P2Y12; CD68; IBA-1; HLA-DR	Clusters of microglia are detected between the lateral ganglionic eminence and the caudate nucleus, as well as in nerve fibers spanning the caudoputamen. Two smaller clusters of P2Y12 microglia are also visible in the thalamic region and the optic tract. All P2Y12 microglia express IBA-1. Non-clustered microglia have moderate CD68 immunoreactivity but strong P2Y12 reactivity. P2Y12 amoeboid myeloid cells are seen in the choroid plexus in an 11.0-week fetus. Choroid plexus cells are positive for IBA-1, but show very low immunoreactivity against P2Y12

CP, cortical plate; IZ, intermediate zone; H&E, hematoxylin and eosin; MZ, marginal zone.

inflammasome, cell-adhesion, and immune modulation among others and these characteristic profiles have been helpful in defining the differences and similarities with the better characterized model of murine microglia (9, 10). Adult human microglia also demonstrate regional heterogeneity with lower densities in the gray matter compared with the white matter and substantially lower densities in the cerebellar cortex, for example, compared to regions with higher densities such as the substantia nigra (11, 12). Adult human microglia renew slowly at a median rate of 28% per year and at any given moment, only 2% of microglia are thought to be proliferating (13, 14). It is suggested that the spatiotemporal coupling between proliferation and apoptosis maintains microglial density throughout life (14). In sum, though our understanding of the human adult microglial population has progressed tremendously, we know very little about these cells during human brain development. In this short review, we first present a brief history of the discovery of microglia and then examine the various schools of thought that had debated the origins of these cells. We then proceed to present an overview of what is known about microglia during human brain development (see **Table 1** for a summary of the main studies on microglia in developing human tissue) and examine how rodent studies have helped elucidate some of the key functions of microglia during brain development. We conclude by discussing the challenges associated with the utilized approaches in human studies.

ON THE DISCOVERY OF MICROGLIA

In 1899, Franz Nissl, a German neuropathologist, reported on the presence of rod-shaped cells in human brain cases of syphilitic paralysis, a severe neuropsychiatric condition characterized by eventual dramatic cerebral atrophy (15). This was perhaps one of the earliest reported descriptions of rod-shaped/bipolar microglia, thought to correspond to an intermediate phenotype of cells. Decades later, Kreutzberg reported on the response of microglial cells to injury in a rodent model of facial nerve axotomy (16). Unlike Nissl's more general observation, Kreutzberg was specifically investigating how the microglial population reacted to injury and demonstrated a marked increase in proliferation in rodents. Mononuclear phagocytes of the CNS such as microglia and macrophages were known as the mesoglia, being of mesodermal origin, and this classification distinguished them from astrocytes and oligodendrocytes that are of neuroectodermal origin (17). In 1913, Cajal optimized Camillo Golgi's black reaction method by using the gold chloride sublimate to enhance the visualization of non-neuronal elements of the CNS that the first morphological description of the microglial cell body emerged (18). Cajal identified cell bodies without processes differentiating them from neurons and astrocytes. Due to limitations in his method, he could not observe the ramifications emanating from microglial cells. Subsequently, Pio Del Rio Hortega optimized Cajal's histochemical method and used silver carbonate to selectively label and visualize the cell body and the ramified morphology of these cells (19). He also incepted the term "microglia" in 1919 and provided a thorough morphological and functional characterization of these cells from development to injury, providing seminal findings that settled long lasting disputes over the developmental lineage of microglia (20, 21) [for translations of Hortega's work, see Ref. (22)]. In 1932, the first studies on microglia in human developmental tissue emerged (23–25). After Hortega and these initial studies, the investigation of microglial cells plunged into a long hiatus and received little attention from the neuroscience community, to be re-visited from the mid 1980s largely due to the technical advances of immunohistochemistry, with studies on the origin and identity of brain macrophages by Perry et al. (26) or Hickey and Kimura (27). Since then, the study of microglia has developed quickly and nowadays is one of the most active fields of neuroscience.

THE ORIGINS OF MICROGLIA

A critical phase in human embryonic development is gastrulation whereby the blastula becomes organized into a trilaminar structure that will yield out all the organs. The endoderm, forms the gut lining; the mesoderm generates the circulatory, cardiovascular and lymphatic systems; and the ectoderm generates both skin and the neuroectoderm, from which the nervous system develops (28, 29). Microglial origins have been a subject of debate and two dominant theories argued whether these cells originated from precursors in the neuroectoderm or the mesoderm. The neuroectodermal theory argues that microglia originate in the neurepithelium from pluripotent glioblasts and are, therefore, akin to astrocytes and oligodendrocytes in lineage (17, 30, 31). *In vitro* evidence of rodent neuroepithelial or astroglial cultures generating microglia supported this theory (32). Additionally, when bacteriophage lambda transgenic mice bone marrows (BMs) were transplanted into recipient murine neonates and adults, *in situ* hybridization and immunohistochemistry showed that the donors' BM-derived macrophages did not contribute to the adult microglial population arguing that microglia could be neuroectodermal in origin (33). The second theory argues that microglia originate in the yolk sac (YS) from hematopoietic stem cell progenitors akin to primitive macrophages and migrate to the brain rudiment as early as the fourth week of gestation in humans, the closure of the caudal neuropore occurring by 27 days post-conception [(17, 20), see also **Figure 1**]. Briefly, between the 15th and the 40th gestational days, the YS is the main hematopoietic organ and localized thickenings showing prospective blood islands are visible in the mesoderm by the 16th day and the peak of hematopoiesis is thought to be the middle of the fifth week of gestation (34–36). Rare histological reports demonstrate the occasional presence of macrophages prior to the onset of hematopoiesis in the yolk-sac by day 13, notably in the trophoblastic mesenchyme (37). By the sixth week of gestation, hepatic colonization by hematopoietic stem cells initiates subsequent hematopoiesis (38). Bone-marrow development does not occur until the 10th week of gestation, and it is well-established that it is not a hematopoietic organ until much later in term, and by the 50th post-conceptional day, YS hematopoiesis will have substantially decreased (34, 38, 39).

Evidence for the mesodermal theory comes from early studies showing the expression of macrophage markers in microglial cells such as FcγRI and CD45 suggesting an extra-cerebral origin and a shared ontology with macrophages within the mononuclear phagocytic system (49). In contrast, rodent studies have provided compelling evidence on the ontology of microglial cells. In mice, RUNX1$^+$ (Runt-related transcription factor 1 is used as a marker for precursor myeloid cells in the extra-embryonic YS) progenitors appear prior to embryonic day 8 in the YS and generate the entire microglial adult population independently of any bone-marrow derived contribution (50). These precursors are thought to be c-KIT$^+$ stem cells that eventually develop into mature CD45$^-$/c-KIT$^-$/CX3CR1$^+$ cells in a PU-1, IRF-8, and colony stimulating factor 1 (CSF-1R)-dependent manner before they acquire their identity as microglial cells upon invading the brain rudiment (50–52). In the human YS and mesenchyme sampled between 4 and 6 gestational weeks (gw), two populations of cells with different phenotypes have been distinguished: the majority carried both macrophage-associated (RFD7$^+$) and monocyte-associated markers (UCHM1$^+$) with no detectable HLA-DR antigen, and the minority expressed class II (HLA-DR and -DP) and no RFD markers, alluding to the possibility of distinct signatures in these populations prior to BM development (53). Overall, it is highly likely that human microglia derive from primitive YS macrophages similarly to rodents. This is an early process that precedes hepatic and BM hematopoiesis (54). Subsequently, mononuclear phagocytes issued from YS hematopoiesis colonize the various organs including the human brain, akin the model proposed for mouse microglial development (55).

MICROGLIAL POPULATION OF THE DEVELOPING BRAIN

The human developing forebrain, midbrain, and hindbrain become identifiable following the closure of the neuropore by 27 post-conceptional days (56–59) (see **Figure 1**). The first wave of entry of microglia into the developing brain predates the peak of neurogenesis and neuronal migration in the telencephalon (40, 60–66). Neural progenitors in the various regional germinal neuroepithelia along the ventricular zone (VZ) proliferate between 8 and 28 gw and migrate toward the cortical plate between 9 and 38 gw (40, 65). The pattern of migration predicts the identity of the neuron in the developing cortex. Generally, excitatory projection neurons originating in the VZ/subventricular zone (SVZ) migrate radially to the cortical plate (64, 67) while inhibitory GABA-ergic neurons originating in the VZ/SVZ and the ganglionic eminence migrate tangentially to the cortical plate before undergoing radial migration in the intermediate zone (IZ) though this does not appear to be universal as a subclass of interneurons migrates radially (68–71). Subsequent to assuming their cortical positions and acquiring their identity, synaptogenesis (26 gw onward) ensues to establish connectivity across the developing brain and this process carries on beyond birth (65, 72). Some, thought to be mesoderm-derived cells, make their route to the developing brain very early and three microglial entry routes have been described: the leptomeningeal route, the ventricular lumen route and the choroid plexus route. In the telencephalic wall, amoeboid microglial cells are detected in the ventricular lumen and VZ as early as gw 4.5. Concomitantly, amoeboid microglia also accumulate in the superficial marginal zone (MZ) presumed to enter from the pial surface and by 5.5 gw some amoeboid cells are also presumed to enter from the choroid

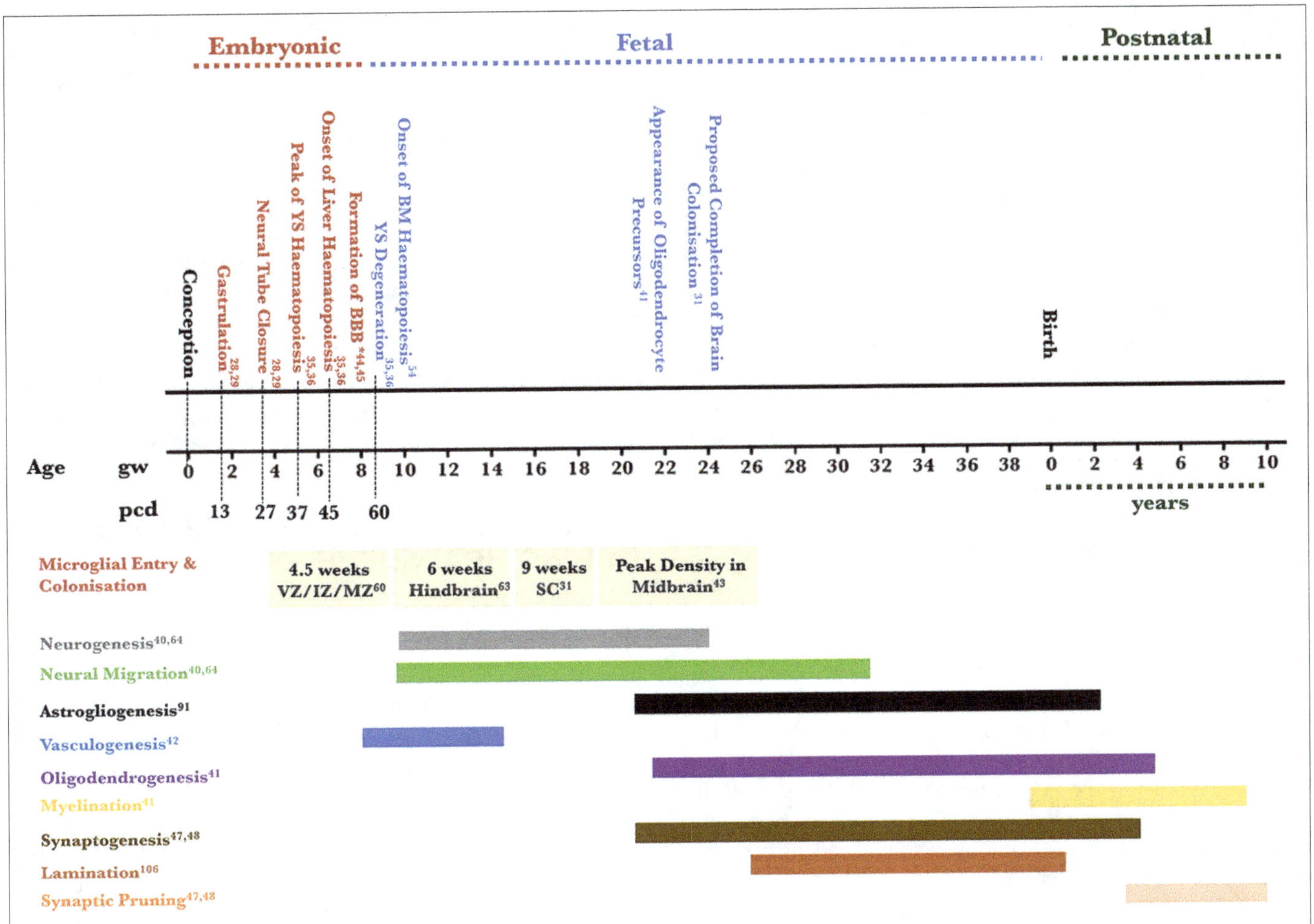

FIGURE 1 | Timeline of human microglial development. This figure places microglial entry into developing brain areas in the context of the approximate co-occurrence of cortical developmental events. Microglia are detected in the amoeboid state in the telencephalic ventricular, intermediate, and MZs and not the cortical plate at 4.5 weeks of gestation; microglia are detected in the ramified state in the hindbrain by 6 weeks; amoeboid microglia are detected in the SC by 9 weeks of gestation; the density of amoeboid microglia is highest in the mesencephalon (midbrain) by 22 weeks of gestation. The relevant references to each process are included (28, 29, 31, 35, 36, 40–45, 47, 48, 54, 60, 63, 64, 70, 91, 106). *Two barriers are included here: the BBB proper at the level of the endothelium of the cerebral blood vessels and the blood–CSF barrier at the level of the choroid plexus epithelial cells. The CSF–brain barrier is detected as early as 16 days through the appearance of strap junctions and is not shown on the timeline. Abbreviations: BBB, blood–brain barrier; BM, bone marrow; CSF, cerebrospinal fluid; IZ, intermediate zone; MZ, marginal zone; PCD, post conceptional days; GW, gestational week; SC, spinal cord; YS, yolk sac.

plexus into the parenchyma of the thalamic eminence. These cells are positive for ionized-calcium binding adaptor protein 1 (IBA-1) and are thought to be undifferentiated immature microglial cells (31, 60–63, 73). However, whether their transcriptional signature is already defined at that stage remains unresolved as some could belong to the macrophage lineage as there is some positivity for CD68, HLA-DR, and CD45. The density of microglial cells at the areas of entry during development does not correlate with their preponderance in areas of the adult brain. Therefore, it can be assumed that the cells would be migrating and proliferating or vice-versa (26). By the eighth gw and as the cortical plate thickens, intermediate rod-shaped microglial cells appear in the MZ and their density increases in the VZ, SVZ, IZ, and the subplate (74–77). Cells particularly in the IZ are suggested to be proliferating between 8 and 12 gw. By the 10th gw, and as fetal blood vessels appear in the telencephalic wall, some amoeboid microglia seem closely associated with the endothelium. Between the 9th and until the end of the 13th gw, proliferating amoeboid microglia aligned along the subplate-cortical plate boundary assume this location and do not enter the cortical plate well beyond its resolve by the third trimester (75, 78). These cells are suggested to support thalamocortical synaptic development (79). Between the 14th and 17th gw, already proliferated microglial cells cluster near or within the white matter at the junction between the thalamus and the internal capsule, the optic tract, and the junction between the internal capsule and the cerebral peduncle and later clusters are observed in the corpus callosum and the periventricular hypothalamus (60, 75, 80). By the 18th gw, intermediate microglia are suggested to migrate radially within the cortical plate to eventually develop into ramified gray matter microglia. Tangential migration occurs within the IZ yielding white matter microglia though interestingly, studies on quail retina suggest that initially, microglial cells migrate tangentially along the various brain layers then assume a radial migration pattern to reach the deep brain parenchyma (81). Microglial cells with various morphologies (intermediate or ramified) populate the cortical mantle and are distributed differentially across the various layers (60, 82, 83). Finally, between the 19th and 30th gw, tangentially oriented microglial packed clusters with an intermediate rod-shaped morphology are reported in white matter tracts such as the corpus callosum and the external capsule. Their density decreases after the 35th gw (73). It is interesting to note that the onset of myelination coincides with the temporal window during which these cells increase in size and proliferate. In humans, the appearance of oligodendrocyte precursors dates to the 22nd gw with myelination commencing when these cells differentiate and become functional from the 28th gw (31, 41). It is, therefore, tempting to speculate that increased microglial presence during this time could be phagocytic during the onset of myelination particularly as some neurons will have migrated into the white matter and would, therefore, need to be cleared. The evidence for this comes from the activated shape of these microglia in these regions within 22–30 gw (84). Furthermore, various authors have also alluded to this temporal window as being when white matter is most susceptible to injury (60, 73, 85). Therefore, increased microglial presence could be seen as protective during this phase. Evidence for this hypothesis comes from pre-term fetal tissue assessed for white matter injury (periventricular leukomalacia) where abnormally high densities of activated microglia are seen in the same regions as the aforementioned clusters (73). By the end of gestation, ramified microglia can be seen across the cortical layers and the white matter shows a higher density of cells compared to the gray matter (86).

Very few studies have focused on characterizing microglial populations in the developing midbrain, cerebellum, and spinal cord (SC). In the cerebellum, its molecular and granular layers are apparent by the 12th gw (87). Ramified microglial cells have been detected as early as 7.5 gw (63) and as the telencephalic wall thickens, very few amoeboid microglial cells remain in the cerebellum (88). It has been suggested that the caudorostral appearance of the vasculature (cerebellum to telencephalon) coincides with increases in microglial densities (12, 42, 63). Much less is known about the mesencephalon (future midbrain) though a couple of studies have highlighted that microglial ramified densities reach a record high during development in this area in comparison with other areas (12, 43, 89). Finally, in the SC, microglia are suggested to colonize the gray and white matter by the ninth gw (31). Their pattern of migration into the cord is thought to follow an outside-in pattern (from the meninges into the gray/white matters) (31). Toward 18–24 gw, SC gray matter microglia are ramified while white matter microglia are amoeboid (90).

Overall, microglial cells enter the developing brain very early preceding neurogenesis, neuronal migration, and gliogenesis in the regional neuroepithelia (40, 91). As development is a highly dynamic and metabolically demanding period, it is likely that apoptosis is coupled with proliferation as has been documented in the cortical subplate for example at the end of the first trimester (92). It is also interesting to consider the human blood–brain barrier (BBB) across cerebral endothelial cells and the CSF brain barrier across the neuroependyma, which shows mature tight junctions as early as the seventh gw as demonstrated by freeze-fracture studies on human embryos of that age (44, 45, 93, 94). Additionally, Gröntoft (and later confirmed by experiments on rabbit fetuses) clearly demonstrated that trypan blue in a 9 gw fetus did not reach the brain if injected in less than 10 min from fetal placental separation (cesarean section delivery) indicating an intact BBB (95). Thus, this may suggest that microglial populations could be expanding as a population from residing cells beyond the closure of the BBB. Furthermore, knowing that some areas of the brain develop much later than others (the cerebellar primary fissure is only visible from the 12th gw, for example) and after BBB closure has been completed, investigating microglial dynamics in this area may shed light on why the adult human cerebellum may contain much less microglial cells in comparison with gray and white matter regions.

FUNCTIONAL ROLES OF MICROGLIA DURING BRAIN DEVELOPMENT

The early colonization of the developing brain by microglia antecedes neurogenesis, neuronal migration, and many other cellular processes (see **Figure 1**) suggesting that these cells are very likely to play pivotal roles in mediating the normal occurrence of some of these events. Multiple lines of functional evidence, predominantly from rodent studies, support this. First, maternal administration of liposomal clodronate to deplete fetal microglia or a maternal immune challenge to activate microglia affect the size of the neuronal precursor pool in the SVZ during the late stages of cortical neurogenesis (96). This has been demonstrated in the rat fetus and highlights the importance of the phagocytic role of microglia in restricting the number of neuronal precursors during development. Furthermore, microglia are thought to promote the apoptosis of Purkinje cells (through superoxide anion signaling during late cerebellar development) in the mouse (97), granular cells in the rat hippocampus (98), and ganglionic cells in the avian retina (81). Second, the transient blockade of CSF-1R at the onset of yolk-sac microglial progenitor production (E6.5) in the mouse embryo alters the normal positioning of a subclass of cortical interneurons (Lhx6) and causes aberrant dopaminergic axonal outgrowth (99). This supports the role of microglial cells in the migration of some cortical interneurons to their correct laminar position and provides novel evidence of the role of these cells in axonal outgrowth. Third, the embryonic absence of microglia in PU.1 knock-out mice is associated with a decreased rate of cortical astrocyte differentiation, suggesting an association between these two cell types (46). In fact, a close association between GFAP-vimentin positive cells and microglia has been already documented in the ninth gw human SC though the functional implication of this has not been elucidated (31). Fourth, challenging the immune system of pregnant dams (by lipopolysaccharide injection at E15.5) or mutating part of the microglial TREM2 signaling pathway results in defasciculation of the corpus callosum in mouse embryos (100). Furthermore, microglia are associated during development with axonal tracts, and this has been repeatedly shown in gestational human tissue (31, 60, 73). Hence, these findings suggest a role of microglia in axonal pruning and the support of myelination. Finally, fractalkine receptor (CX3CR1) knock-out mice have delayed synaptic pruning demonstrated by a transient excess of dendritic spines concomitant with decreased microglial numbers during development. Functionally, these mice had also reduced synaptic efficiencies in hippocampal long-term potentiation (101). Thus, microglial cells have also a pivotal role during development in synaptic refinement.

Altogether, microglial cells participate actively in various processes during brain development. Functional human studies are very difficult to conduct. However, and as previously discussed, histopathological studies on human embryonic and fetal tissues concur about the distribution of microglial cells in association with some key developmental processes including neuronal migration, white matter tract formation, vasculogenesis, as well as astrocyte formation and synaptic pruning.

LIMITATIONS OF EXISTING METHODS AND FUTURE PROSPECTS

Postmortem studies on human tissue are invaluable for investigating the antigenic features and morphology of microglial cells across various developmental ages. The identified studies in this review give us a qualitative insight into microglial dynamics during development. However, there are multiple caveats. First, though the total number of cases investigated in each study varied between $n = 4$ and $n = 47$ cases, it is extremely difficult to interpret the results when a single case is used to generalize over one gestational age. This is mainly because tissue is scarce. Additionally, and in cases of spontaneous abortion, the lack of neuropathological abnormalities may not exclude the fact that pregnancy loss could be the outcome of an underlying condition and that could also bias microglial findings linked to death. To date, most neuropathological studies have focused on predominantly qualitative methods to assess microglial populations in human embryonic/fetal tissues and with immunological (CD45, MHCII, CD68, IBA-1) or histochemical markers (RCA-1) that do not offer the possibility to distinguish between microglial cells and macrophages. This is likely to be a challenge for the field as development is highly dynamic, and we do not have specific markers to distinguish between cell lineages. New methods are needed to help identify a possible developmental microglial signature in the like of RNA-sequencing and single cell sorting methods from embryonic/fetal human tissue. To the author's knowledge, only one study has emerged on this topic (102). At present, these tools are being used to identify enriched genes in human adult microglia against perivascular macrophages, for example, and to distinguish human microglial cells from rodent microglia (9, 10). Quantitative methods to assess microglial numbers and morphology in sections are needed to calculate the degree of proliferation, apoptosis, and to establish the microglial/macrophage marker developmental profile. These are much more likely to support our understanding of the regional heterogeneity of microglial cells during development and the dynamics of this population. Furthermore, the relevant mathematical models applied to these data would allow for covariate correction such as the post-mortem interval, brain weight, gestational age, gender, and anatomical location in order to reveal main effects on microglial numbers as well interactions between these independent variables. Significant advances in the field would be made with tightly controlled tissue samples from elective terminations and well-powered sample sizes in addition to emphasis on the previously mentioned points.

Finally, microglia feature as part of the pathophysiological signature of various neurodevelopmental disorders such as schizophrenia (5), intellectual disability (103), autism (104), and Rett's syndrome (105). Therefore, understanding their precise dynamics and how they contribute to altering brain networks is likely to help us design therapeutic strategies to enhance their activity, limit the damage that could be caused by their activation, or use them as diagnostic biomarkers in disease.

CONCLUSION

In the past decade, the field of study of microglia has rapidly expanded, and although rodent studies provide the opportunity to investigate how these cells populate the brain, develop, and proliferate, studies on the developing human brain are needed, particularly as the human and rodent adult brains have different microglial signatures and regional distributions. It remains unclear how these differences contribute to function and further studies correlating anatomy with function are needed.

AUTHOR CONTRIBUTIONS

DAM and DG-N conceived the structure and content of the manuscript. DAM wrote the manuscript and prepared the figures. DAM and DG-N drafted the manuscript.

REFERENCES

Chen Z, Trapp BD. Microglia and neuroprotection. *J Neurochem* (2016) 136(Suppl 1):10–7. doi:10.1111/jnc.13062

Lindholm D, Castrén E, Kiefer R, Zafra F, Thoenen H. Transforming growth factor-beta 1 in the rat brain: increase after injury and inhibition of astrocyte proliferation. *J Cell Biol* (1992) 117(2):395–400. doi:10.1083/jcb.117.2.395

Schafer DP, Lehrman EK, Kautzman AG, Koyama R, Mardinly AR, Yamasaki R, et al. Microglia sculpt postnatal neural circuits in an activity and complement-dependent manner. *Neuron* (2012) 74(4):691–705. doi:10.1016/j.neuron.2012.03.026

Neumann H, Kotter MR, Franklin RJ. Debris clearance by microglia: an essential link between degeneration and regeneration. *Brain* (2009) 132(Pt 2): 288–95. doi:10.1093/brain/awn109

Sekar A, Bialas AR, de Rivera H, Davis A, Hammond TR, Kamitaki N, et al. Schizophrenia risk from complex variation of complement component 4. *Nature* (2016) 530(7589):177–83. doi:10.1038/nature16549

Hong S, Beja-Glasser VF, Nfonoyim BM, Frouin A, Li S, Ramakrishnan S, et al. Complement and microglia mediate early synapse loss in Alzheimer mouse models. *Science* (2016) 352(6286):712–6. doi:10.1126/science.aad8373

Streit WJ, Mrak RE, Griffi WS. Microglia and neuroinflammation: a pathological perspective. *J Neuroinflammation* (2004) 1(1):14. doi:10.1186/1742-2094-1-14

Gomez-Nicola D, Perry VH. Microglial dynamics and role in the healthy and diseased brain: a paradigm of functional plasticity. *Neuroscientist* (2015) 21(2):169–84. doi:10.1177/1073858414530512

Gosselin D, Skola D, Coufal NG, Holtman IR, Schlachetzki JCM, Sajti E, et al. An environment-dependent transcriptional network specifies human microglia identity. *Science* (2017) 356(6344):eaal3222. doi:10.1126/science. aal3222

Galatro TF, Holtman IR, Lerario AM, Vainchtein ID, Brouwer N, Sola PR, et al. Transcriptomic analysis of purified human cortical microglia reveals age-associated changes. *Nat Neurosci* (2017) 20(8):1162–71. doi:10.1038/ nn.4597

Mittelbronn M, Dietz K, Schluesener HJ, Meyermann R. Local distribution of microglia in the normal adult human central nervous system differs by up to one order of magnitude. *Acta Neuropathol* (2001) 101(3):249–55.

Wierzba-Bobrowicz T, Kosno-Kruszewska E, Gwiazda E, Lechowicz W. The comparison of microglia maturation in different structures of the human nervous system. *Folia Neuropathol* (1998) 36(3):152–60.

Réu P, Khosravi A, Bernard S, Mold JE, Salehpour M, Alkass K, et al. The lifespan and turnover of microglia in the human brain. *Cell Rep* (2017) 20(4):779–84. doi:10.1016/j.celrep.2017.07.004

Askew K, Li K, Olmos-Alonso A, Garcia-Moreno F, Liang Y, Richardson P, et al. Coupled proliferation and apoptosis maintain the rapid turnover of microglia in the adult brain. *Cell Rep* (2017) 18(2):391–405. doi:10.1016/j. celrep.2016.12.041

Nissl F. Über einige Beziehungen zwischen Nervenzellerkrankungen und gliösen Erscheinungen bei verschiedenen Psychosen. *Arch Psychiatr* (1899) 32:656–76.

Kreutzberg GW. Autoradiographische Untersuchung über die Beteiligung von Gliazellen an der axonalen Reaktion im Facialiskern der Ratte. *Acta Neuropathol* (1966) 7(2):149–61. doi:10.1007/BF00686781

Rezaie P, Male D. Mesoglia & microglia – a historical review of the concept of mononuclear phagocytes within the central nervous system. *J Hist Neurosci* (2002) 11(4):325–74. doi:10.1076/jhin.11.4.325.8531

Cajal S, Ramon Y. Sobre un nuevo proceder de impregnacion de la neuroglia y sus resultados en las centros nerviosos del hombre y de los animales. *Trab Lab Invest Biol Univ Madr* (1913) 11:219.

Del Río Hortega P. Lo que debe entenderse por "tercer elemento" de los centros nerviosos. *Bol Soc Esp Biol* (1924) (11):33–5.

Del Riò Hortega P. *Microglia, in Cytology and Cellular Pathology of the Nervous System*. New York: P.B. Hoerber (1932). p. 482–534.

Del Río Hortega P. El "tercer elemento de los centros nerviosos". II. Intervención de la microglía en los procesos patológicos (células en baston- cito y cuerpos gránuloadiposos). *Bol Soc Esp Biol* (1919) (VIII):91–103.

Sierra A, de Castro F, Del Río-Hortega J, Rafael Iglesias-Rozas J, Garrosa M, Kettenmann H. The "Big-Bang" for modern glial biology: translation and comments on Pío del Río-Hortega 1919 series of papers on microglia. *Glia* (2016) 64(11):1801–40. doi:10.1002/glia.23046

von Santha K. Untersuchungen über die Entwicklung der Hortegaschen Mikroglia. *Arch Psychiatr* (1932) 96:36–50. doi:10.1007/BF02064322

Rydberg E. Cerebral injury in new-born children consequent on birth trauma, with an inquiry into the normal and pathological anatomy of the neuroglia. *Acta Path Microbiol Scand* (1932):247–97.

Juba A. Untersuchungen über die Entwicklung der Hortegaschen Mikroglia des Menschen. *Arch Psychiatr Nervenkr* (1934) 101:577–92.

Perry VH, Hume DA, Gordon S. Immunohistochemical localization of macrophages and microglia in the adult and developing mouse brain. *Neuroscience* (1985) 15(2):313–26. doi:10.1016/0306-4522(85)90215-5

Hickey WF, Kimura H. Perivascular microglial cells of the CNS are bone marrow-derived and present antigen in vivo. *Science* (1988) 239(4837):290–2. doi:10.1126/science.3276004

Janigro D. *Mammalian Brain Development*. Totowa, NJ: Humana Press (2009). Available from: http://dx.doi.org/10.1007/978-1-60761-287-2

Stiles J, Jernigan TL. The basics of brain development. *Neuropsychol Rev* (2010) 20(4):327–48. doi:10.1007/s11065-010-9148-4

Fujita S, Kitamura T. Origin of brain macrophages and the nature of the so-called microglia. *Acta Neuropathol* (1975) (Suppl 6):291–6.

Rezaie P, Male D. Colonisation of the developing human brain and spinal cord by microglia: a review. *Microsc Res Tech* (1999) 45(6):359–82. doi:10.1002/ (SICI)1097-0029(19990615)45:6<359::AID-JEMT4>3.0.CO;2-D

Hao C, Richardson A, Fedoroff S. Macrophage-like cells originate from neu- roepithelium in culture: characterization and properties of the macrophage- like cells. *Int J Dev Neurosci* (1991) 9(1):1–14. doi:10.1016/0736-5748(91) 90067-V

de Groot CJ, Huppes W, Sminia T, Kraal G, Dijkstra CD. Determination of the origin and nature of brain macrophages and microglial cells in mouse central nervous system, using non-radioactive in situ hybridization and immunoperoxidase techniques. *Glia* (1992) 6(4):301–9. doi:10.1002/glia. 440060408

Nogales FF. *The Human Yolk Sac and Yolk Sac Tumors*. Berlin, Heidelberg: Springer Berlin Heidelberg (1993). XVI,367 p.

Tavian M, Hallais MF, Péault B. Emergence of intraembryonic hemato- poietic precursors in the pre-liver human embryo. *Development* (1999) 126(4):793–803.

Tavian M, Péault B. Embryonic development of the human hematopoietic system. *Int J Dev Biol* (2005) 49(2–3):243–50. doi:10.1387/ijdb.041957mt

Bloom W, Bartelmez GW. Hematopoiesis in young human embryos. *Am J Anat* (1940) (67):21–53. doi:10.1002/aja.1000670103

Migliaccio G, Migliaccio AR, Petti S, Mavilio F, Russo G, Lazzaro D, et al. Human embryonic hemopoiesis. Kinetics of progenitors and precursors underlying the yolk sac-liver transition. *J Clin Invest* (1986) 78(1):51–60. doi:10.1172/JCI112572

Chen LT, Weiss L. The development of vertebral bone marrow of human fetuses. *Blood* (1975) 46(3):389–408.

Bystron I, Blakemore C, Rakic P. Development of the human cerebral cortex: Boulder Committee revisited. *Nat Rev Neurosci* (2008) 9(2):110–22. doi:10.1038/ nrn2252

Jakovcevski I, Filipovic R, Mo Z, Rakic S, Zecevic N. Oligodendrocyte development and the onset of myelination in the human fetal brain. *Front Neuroanat* (2009) 3:5. doi:10.3389/neuro.05.005.2009

Marín-Padilla M. The human brain intracerebral microvascular system: development and structure. *Front Neuroanat* (2012) 6:38. doi:10.3389/ fnana.2012.00038

Wierzba-Bobrowicz T, Gwiazda E, Poszwińska Z. Morphological study of microglia in human mesencephalon during the development and aging. *Folia Neuropathol* (1995) 33(2):77–83.

Saunders NR, Dreifuss JJ, Dziegielewska KM, Johansson PA, Habgood MD, Møllgård K, et al. The rights and wrongs of blood-brain barrier permeability studies: a walk through 100 years of history. *Front Neurosci* (2014) 8:404. doi:10.3389/fnins.2014.00404

Møllgård K, Saunders NR. The development of the human blood-brain and blood-CSF barriers. *Neuropathol Appl Neurobiol* (1986) 12(4):337–58. doi:10.1111/j.1365-2990.1986.tb00146.x

Antony JM, Paquin A, Nutt SL, Kaplan DR, Miller FD. Endogenous microglia regulate development of embryonic cortical precursor cells. *J Neurosci Res* (2011) 89(3):286–98. doi:10.1002/jnr.22533

Petanjek Z, Judaš M, Šimic G, Rasin MR, Uylings HB, Rakic P, et al. Extraordinary neoteny of synaptic spines in the human prefrontal cortex. *Proc Natl Acad Sci U S A* (2011) 108(32):13281–6. doi:10.1073/pnas. 1105108108

Huttenlocher PR. Synaptic density in human frontal cortex – developmental changes and effects of aging. *Brain Res* (1979) 163(2):195–205. doi:10.1016/ 0006-8993(79)90349-4

Akiyama H, McGeer PL. Brain microglia constitutively express beta-2 integrins. *J Neuroimmunol* (1990) 30(1):81–93. doi:10.1016/0165-5728(90)90055-R

Ginhoux F, Greter M, Leboeuf M, Nandi S, See P, Gokhan S, et al. Fate map- ping analysis reveals that adult microglia derive from primitive macrophages. *Science* (2010) 330(6005):841–5. doi:10.1126/science.1194637

Kierdorf K, Erny D, Goldmann T, Sander V, Schulz C, Perdiguero EG, et al. Microglia emerge from erythromyeloid precursors via Pu.1- and Irf8-dependent pathways. *Nat Neurosci* (2013) 16(3):273–80. doi:10.1038/ nn.3318

Erblich B, Zhu L, Etgen AM, Dobrenis K, Pollard JW. Absence of colony stimulation factor-1 receptor results in loss of microglia, disrupted brain devel- opment and olfactory deficits. *PLoS One* (2011) 6(10):e26317. doi:10.1371/ journal. pone.0026317

Janossy G, Bofill M, Poulter LW, Rawlings E, Burford GD, Navarrete C, et al. Separate ontogeny of two macrophage-like accessory cell populations in the human fetus. *J Immunol* (1986) 136(12):4354–61.

Morrison SJ, Scadden DT. The bone marrow niche for haematopoietic stem cells. *Nature* (2014) 505(7483):327–34. doi:10.1038/nature12984

Askew K, Gomez-Nicola D. A story of birth and death: insights into the forma- tion and dynamics of the microglial population. *Brain Behav Immun* (2017) 69:9–17. doi:10.1016/j.bbi.2017.03.009

Skidmore FD. An analysis of the age and size of 483 human embryos. *Teratology* (1977) 15(1):97–102. doi:10.1002/tera.1420150113

Müller F, O'Rahilly R. The human brain at stages 21–23, with particular refer- ence to the cerebral cortical plate and to the development of the cerebellum. *Anat Embryol (Berl)* (1990) 182(4):375–400. doi:10.1007/BF02433497

O'Rahilly R, Müller F. Developmental stages in human embryos: revised and new measurements. *Cells Tissues Organs* (2010) 192(2):73–84. doi:10.1159/ 000289817

O'Rahilly R, Bossy J, Müller F. [Introduction to the study of embryonic stages in man]. *Bull Assoc Anat (Nancy)* (1981) 65(189):141–236.

Monier A, Adle-Biassette H, Delezoide AL, Evrard P, Gressens P, Verney C. Entry and distribution of microglial cells in human embryonic and fetal cerebral cortex. *J Neuropathol Exp Neurol* (2007) 66(5):372–82. doi:10.1097/ nen.0b013e3180517b46

Monier A, Evrard P, Gressens P, Verney C. Distribution and differentiation of microglia in the human encephalon during the first two trimesters of gestation. *J Comp Neurol* (2006) 499(4):565–82. doi:10.1002/cne.21123

Choi BH. Hematogenous cells in the central nervous system of develop- ing human embryos and fetuses. *J Comp Neurol* (1981) 196(4):683–94. doi:10.1002/cne.901960412

Andjelkovic AV, Nikolic B, Pachter JS, Zecevic N. Macrophages/microglial cells in human central nervous system during development: an immuno- histochemical study. *Brain Res* (1998) 814(1–2):13–25. doi:10.1016/S0006- 8993(98)00830-0

Rakic P. Specification of cerebral cortical areas. *Science* (1988) 241(4862): 170–6. doi:10.1126/science.3291116

Marín O. Developmental timing and critical windows for the treatment of psychiatric disorders. *Nat Med* (2016) 22(11):1229–38. doi:10.1038/nm.4225

Fujimoto E, Miki A, Mizoguti H. Histochemical study of the differentiation of microglial cells in the developing human cerebral hemispheres. *J Anat* (1989) 166:253–64.

Rakic P. Mode of cell migration to the superficial layers of fetal monkey neocortex. *J Comp Neurol* (1972) 145(1):61–83. doi:10.1002/cne.901450105

Wonders CP, Anderson SA. The origin and specification of cortical interneu- rons. *Nat Rev Neurosci* (2006) 7(9):687–96. doi:10.1038/nrn1954

Anderson SA, Kaznowski CE, Horn C, Rubenstein JL, McConnell SK. Distinct origins of neocortical projection neurons and interneurons in vivo. *Cereb Cortex* (2002) 12(7):702–9. doi:10.1093/cercor/12.7.702

Petanjek Z, Berger B, Esclapez M. Origins of cortical GABAergic neurons in the cynomolgus monkey. *Cereb Cortex* (2009) 19(2):249–62. doi:10.1093/ cercor/bhn078

Jakovcevski I, Mayer N, Zecevic N. Multiple origins of human neocortical interneurons are supported by distinct expression of transcription factors. *Cereb Cortex* (2011) 21(8):1771–82. doi:10.1093/cercor/bhq245

Tau GZ, Peterson BS. Normal development of brain circuits. *Neuropsycho pharmacology* (2010) 35(1):147–68. doi:10.1038/npp.2009.115

Verney C, Monier A, Fallet-Bianco C, Gressens P. Early microglial colo- nization of the human forebrain and possible involvement in periventric- ular white-matter injury of preterm infants. *J Anat* (2010) 217(4):436–48. doi:10.1111/j.1469-7580.2010.01245.x

Wierzba-Bobrowicz T, Lewandowska E. Morphological study of endothelial cells in the human fetus during early period of gestation. *Folia Neuropathol* (1995) 33(4):241–5.

Rezaie P, Dean A, Male D, Ulfig N. Microglia in the cerebral wall of the human telencephalon at second trimester. *Cereb Cortex* (2005) 15(7):938–49. doi:10.1093/cercor/bhh194

Kershman J. Genesis of microglia in the human brain. *Arch NeurPsych* (1939) 41(1):24–50. doi:10.1001/archneurpsyc.1939.02270130034002

Hutchins KD, Dickson DW, Rashbaum WK, Lyman WD. Localization of morphologically distinct microglial populations in the developing human fetal brain: implications for ontogeny. *Brain Res Dev Brain Res* (1990) 55(1):95–102. doi:10.1016/0165-3806(90)90109-C

Kostovic I, Rakic P. Developmental history of the transient subplate zone in the visual and somatosensory cortex of the macaque monkey and human brain. *J Comp Neurol* (1990) 297(3):441–70. doi:10.1002/cne. 902970309

Allendoerfer KL, Shatz CJ. The subplate, a transient neocortical structure: its role in the development of connections between thalamus and cortex. *Annu Rev Neurosci* (1994) 17:185–218. doi:10.1146/annurev.ne.17.030194.001153

Gould SJ, Howard S. An immunohistological study of macrophages in the human fetal brain. *Neuropathol Appl Neurobiol* (1991) 17(5):383–90. doi:10.1111/j.1365-2990.1991.tb00738.x

Navascués J, Moujahid A, Almendros A, Marin-Teva JL, Cuadros MA. Origin of microglia in the quail retina: central-to-peripheral and vitreal-to-scleral migration of microglial precursors during development. *J Comp Neurol* (1995) 354(2):209–28. doi:10.1002/cne.903540205

Rezaie P, Cairns NJ, Male DK. Expression of adhesion molecules on human fetal cerebral vessels: relationship to microglial colonisation during develop- ment. *Brain Res Dev Brain Res* (1997) 104(1–2):175–89.

Rezaie P, Bohl J, Ulfig N. Anomalous alterations affecting microglia in the central nervous system of a fetus at 12 weeks of gestation: case report. *Acta Neuropathol* (2004) 107(2):176–80. doi:10.1007/s00401-003-0779-x

Billiards SS, Haynes RL, Folkerth RD, Trachtenberg FL, Liu LG, Volpe JJ, et al. Development of microglia in the cerebral white matter of the human fetus and infant. *J Comp Neurol* (2006) 497(2):199–208. doi:10.1002/cne.20991

Rezaie P, Dean A. Periventricular leukomalacia, inflammation and white matter lesions within the developing nervous system. *Neuropathology* (2002) 22(3):106–32. doi:10.1046/j.1440-1789.2002.00438.x

Esiri MM, al Izzi MS, Reading MC. Macrophages, microglial cells, and HLA-DR antigens in fetal and infant brain. *J Clin Pathol* (1991) 44(2):102–6. doi:10.1136/jcp.44.2.102

Nieuwenhuys R, Voogd J, Huijzen CV. *The Human Central Nervous System*. 4th ed. Berlin: Springer (2008). xiv,967 p.

Mildner A, Huang H, Radke J, Stenzel W, Priller J. P2Y. *Glia* (2017) 65(2): 375–87. doi:10.1002/glia.23097

Cho KH, Cheong JS, Kim JH, Abe H, Murakami G, Cho BH. Site-specific distribution of CD68-positive microglial cells in the brains of human midterm fetuses: a topographical relationship with growing axons. *Biomed Res Int* (2013) 2013:762303. doi:10.1155/2013/762303

Hutchins KD, Dickson DW, Rashbaum WK, Lyman WD. Localization of microglia in the human fetal cervical spinal cord. *Brain Res Dev Brain Res* (1992) 66(2):270–3. doi:10.1016/0165-3806(92)90091-A

Choi BH, Lapham LW. Radial glia in the human fetal cerebrum: a combined Golgi, immunofluorescent and electron microscopic study. *Brain Res* (1978) 148(2):295–311. doi:10.1016/0006-8993(78)90721-7

Rakic S, Zecevic N. Programmed cell death in the developing human telen- cephalon. *Eur J Neurosci* (2000) 12(8):2721–34. doi:10.1046/j.1460-9568.2000. 00153.x

Whish S, Dziegielewska KM, Møllgård K, Noor NM, Liddelow SA, Habgood MD, et al. The inner CSF-brain barrier: developmentally controlled access to the brain via intercellular junctions. *Front Neurosci* (2015) 9:16. doi:10.3389/fnins.2015.00016

Møllgård K, Saunders NR. Complex tight junctions of epithelial and of endothelial cells in early foetal brain. *J Neurocytol* (1975) 4(4):453–68. doi:10.1007/BF01261375

Gröntoft O. Intracranial haemorrhage and blood-brain barrier problems in the new-born; a pathologico-anatomical and experimental investigation. *Acta Pathol Microbiol Scand Suppl* (1954) 100:8–109.

Cunningham CL, Martínez-Cerdeño V, Noctor SC. Microglia regulate the number of neural precursor cells in the developing cerebral cortex. *J Neurosci* (2013) 33(10):4216–33. doi:10.1523/JNEUROSCI.3441-12.2013

Marín-Teva JL, Dusart I, Colin C, Gervais A, van Rooijen N, Mallat M. Microglia promote the death of developing Purkinje cells. *Neuron* (2004) 41(4):535–47. doi:10.1016/S0896-6273(04)00069-8

Dalmau I, Finsen B, Tønder N, Zimmer J, González B, Castellano B. Development of microglia in the prenatal rat hippocampus. *J Comp Neurol* (1997) 377(1):70–84. doi:10.1002/(SICI)1096-9861(19970106)377: 1<70::AID-CNE7>3.0.CO;2-G

Squarzoni P, Oller G, Hoeffel G, Pont-Lezica L, Rostaing P, Low D, et al. Microglia modulate wiring of the embryonic forebrain. *Cell Rep* (2014) 8(5):1271–9. doi:10.1016/j.celrep.2014.07.042

Pont-Lezica L, Beumer W, Colasse S, Drexhage H, Versnel M, Bessis A. Microglia shape corpus callosum axon tract fasciculation: functional impact of prenatal inflammation. *Eur J Neurosci* (2014) 39(10):1551–7. doi:10.1111/ ejn.12508

Paolicelli RC, Bolasco G, Pagani F, Maggi L, Scianni M, Panzanelli P, et al. Synaptic pruning by microglia is necessary for normal brain development. *Science* (2011) 333(6048):1456–8. doi:10.1126/science.1202529

Thion MS, Low D, Silvin A, Chen J, Grisel P, Schulte-Schrepping J, et al. Microbiome influences prenatal and adult microglia in a sex-specifi manner. *Cell* (2018) 172(3):500–16.e16. doi:10.1016/j.cell.2017.11.042

Coutinho E, Menassa DA, Jacobson L, West SJ, Domingos J, Moloney TC, et al. Persistent microglial activation and synaptic loss with behavioral abnormalities in mouse off pring exposed to CASPR2-antibodies in utero. *Acta Neuropathol* (2017) 134(4):567–83. doi:10.1007/s00401-017- 1751-5

Tetreault NA, Hakeem AY, Jiang S, Williams BA, Allman E, Wold BJ, et al. Microglia in the cerebral cortex in autism. *J Autism Dev Disord* (2012) 42(12):2569–84. doi:10.1007/s10803-012-1513-0

Derecki NC, Cronk JC, Lu Z, Xu E, Abbott SB, Guyenet PG, et al. Wild- type microglia arrest pathology in a mouse model of Rett syndrome. *Nature* (2012) 484(7392):105–9. doi:10.1038/nature10907

Silbereis JC, Pochareddy S, Zhu Y, Li M, Sestan N. The cellular and molecular landscapes of the developing human central nervous system. *Neuron* (2016) 89(2):248–68. doi:10.1016/j.neuron.2015.12.008

9

Lipoprotein Lipase is a Feature of Alternatively-Activated Microglia and may Facilitate Lipid Uptake in the CNS during Demyelination

Kimberley D. Bruce[1], Sachi Gorkhali[1], Katherine Given[2], Alison M. Coates[3], Kristen E. Boyle[4], Wendy B. Macklin[2] and Robert H. Eckel[1]*

[1]Division of Endocrinology, Metabolism, & Diabetes, Department of Medicine, University of Colorado Denver Anschutz Medical Campus, Aurora, CO, United States, [2]Department of Cell and Developmental Biology, University of Colorado School of Medicine, Aurora, CO, United States, [3]School of Health Sciences, Sansom Institute for Health Research, University of South Australia, Adelaide, SA, Australia, [4]Department of Pediatrics, Section of Nutrition, University of Colorado School of Medicine, Aurora, CO, United States

***Correspondence:**
Kimberley D. Bruce
kimberley.bruce@ucdenver.edu

Severe demyelinating disorders of the central nervous system (CNS) such as multiple sclerosis (MS), can be devastating for many young lives. To date, the factors resulting in poor remyelination and repair are not well understood, and reparative therapies that benefit MS patients have yet to be developed. We have previously shown that the activity and abundance of Lipoprotein Lipase (LPL)—the rate-limiting enzyme in the hydrolysis of triglyceride-rich lipoproteins—is increased in Schwann cells and macrophages following nerve crush injury in the peripheral nervous system (PNS), suggesting that LPL may help scavenge myelin-derived lipids. We hypothesized that LPL may play a similar role in the CNS. To test this, mice were immunized with MOG_{35-55} peptide to induce experimental allergic encephalomyelitis (EAE). LPL activity was increased ($p < 0.05$) in the brain at 30 days post-injection, coinciding with partial remission of clinical symptoms. Furthermore, LPL abundance and activity was up-regulated ($p < 0.05$) at the transition between de- and re-myelination in lysolecithin-treated *ex vivo* cerebellar slices. Since microglia are the key immune effector cells of the CNS we determined the role of LPL in microglia. Lipid uptake was decreased ($p < 0.001$) in LPL-deficient BV-2 microglial cells compared to WT. In addition, LPL-deficient cells showed dramatically reduced expression of anti-inflammatory markers, YM1 (−22 fold, $p < 0.001$), and arginase 1 (Arg1; −265 fold, $p < 0.001$) and increased expression of pro-inflammatory markers, such as iNOS compared to WT cells (+53 fold, $p < 0.001$). This suggests that LPL is a feature of reparative microglia, further supported by the metabolic and inflammatory profile of LPL-deficient microglia. Taken together, our data strongly suggest that LPL expression is a novel feature of a microglial phenotype that supports remyelination and repair through the clearance of lipid debris. This mechanism may be exploited to develop future reparative therapies for MS and primary neurodegenerative disorders (Alzheimer's disease (AD) and Parkinson's disease).

Keywords: lipoprotein lipase (LPL), microglia, multiple sclerosis, myelination, lipid metabolism

INTRODUCTION

Multiple sclerosis (MS) is the most common cause of neurological disability in young adults, affecting approximately 2.3 million people worldwide. The majority of MS patients show a relapsing-remitting (RR-MS) clinical phenotype, in which oligodendrocyte precursor cells (OPCs) attempt to remyelinate areas of myelin damage. However, as the disease progresses these attempts fail, leading to severe secondary progressive MS (P-MS; Antel et al., 2012). Despite recent advances in our understanding of the pathophysiology of MS, the factors resulting in poor remyelination and repair are not well understood. Consequently, much needed reparative therapies that benefit MS patients have yet to be developed.

Microglia are the key immune effector cells of the central nervous system (CNS), where they have essential functions such as synaptic pruning by engulfment of neuronal terminals (Hong and Stevens, 2016), and phagocytosis of dying cells and myelin debris (Ransohoff and Perry, 2009). CNS microglia have been repeatedly implicated in the pathophysiology of MS (Ransohoff and El Khoury, 2015). On one hand, CNS microglia may be considered "classically activated" and be associated with the production of neurotoxic molecules and pro-inflammatory cytokines that are detrimental to the repair processes (Becher et al., 2001; Heppner et al., 2005). In contrast, "alternatively activated" reparative microglia provide neurotrophic and immunosuppressive factors that counteract pathological processes and may help facilitate remyelination (Schwartz and Moalem, 2001; Hohlfeld, 2008). Although the plethora of intermediate microglial phenotypes associated within a disease process make it difficult to categorize microglial sub-populations into defined activation states, it is a useful starting point to investigate the basic mechanisms by which microglia participate in demyelination and remyelination. In MS, appropriately activated microglia may help clear debris after myelin damage (Lampron et al., 2015), expedite extracellular matrix deposition, facilitate regeneration, and offer trophic support to OPCs (Varin and Gordon, 2009). In support of this notion, when debris clearance by microglia is impeded, regeneration is delayed (Neumann et al., 2009). However, the microglial factors that actively facilitate these processes are unknown.

In peripheral tissues such as adipose tissue and skeletal/cardiac muscle the canonical function of lipoprotein lipase (LPL)—hydrolysis of triglyceride-rich lipoproteins—is well defined. However, the role of LPL in cells of the CNS is not well understood. Nonetheless, previous work from our laboratory and others have shown that LPL is expressed in several regions of the nervous system including the brain, spinal cord and peripheral nerves, with the highest expression observed in the hippocampus and spinal cord (Eckel and Robbins, 1984; Goldberg et al., 1989; Bessesen et al., 1993). Our studies show that in the peripheral nervous system (PNS), LPL is expressed in Schwann cells, macrophages, and dorsal root ganglia neurons, and its abundance and activity is increased following nerve crush injury (Huey et al., 2002; Nunes and Sousa, 2008). This increase in LPL is associated with macrophage infiltration, suggesting that LPL may be part of an acute response to scavenge and reutilize myelin-derived lipids in degenerating peripheral nerves (Huey et al., 2002). However, this has yet to be empirically determined.

A number of recent reports have highlighted LPL as a key feature of reparative CNS microglia. Single cell analysis of disease-associated-microglia (DAM) in a murine model of Alzheimer's disease (AD), shows that LPL is markedly increased in a unique microglial sub-type associated with phagocytosis and protection in advanced stages of AD (Keren-Shaul et al., 2017). In addition, in a cuprizone model of demyelination, LPL gene expression was elevated during demyelination, and tailed-off once remyelination was complete, suggesting that LPL is novel feature of a microglial phenotype that actively supports remyelination and repair (Olah et al., 2012). Despite these important observations the mechanisms underlying the role of LPL in CNS microglia have not been determined.

Here, we hypothesize that LPL may support repair through the clearance of myelin-derived lipids. In the current study we utilize *in vivo*, *ex vivo* and *in vitro* systems to demonstrate that LPL is increased during the onset of remyelination, is associated with an anti-inflammatory reparative microglial phenotype, and may facilitate the uptake of myelin-derived lipids in the CNS.

MATERIALS AND METHODS

Animals

This study was carried out in accordance with University of Colorado Institutional Animal Care and Use Committee (IACUC) guidelines IACUC for animal use, which are in agreement with the NIH Guide for the Care and Use of Laboratory Animals. Animal protocols were approved by the University of Colorado IACUC. Male C57Bl/6J wild-type mice 8–10 weeks-old were obtained from Jackson Labs (Bar Harbor, Maine, ME, USA). Mice were individually housed and maintained at ~20°C with a 12-h light/dark cycle and given unrestricted access to standard laboratory diet (Diet 8640; Harlan Teklad, Madison, WI, USA) and water. Paralyzed mice were afforded easier access to food and water to prevent dehydration.

Antigens

The peptide used in this study was the immunodominant MOG_{35-55} peptide (MEVGWYRSPFSRVVHLYRNGK; Mendel et al., 1995). The purity was assessed by HPLC (>97% pure) and amino acid composition was verified by mass spectrometry (Peptides International, Louisville, KY, USA). MOG_{35-55} peptide batches for *in vivo* work were all from one preparation. They were stored insoluble until required and then dissolved in saline to a concentration of 150 μg/μl. The unused portions were stored at 4°C for a maximum of 3 months.

Induction of Experimental Autoimmune Encephalitis (EAE)

Mice were injected subcutaneously at two sites on the femoral region with 200 μl of a mixture of MOG $_{35-55}$ peptide emulsified 1:1 with Freund's complete adjuvant (Difco Laboratories, Detroit, MI, USA). MOG treated mice were also boosted with pertussis toxin (List Laboratories, Campbell, CA, USA; 200 ng) I.P. on both the day of injection and also 48 h later. Control mice were injected with equal volumes of CFA and given saline instead of pertussis toxin. Tissues were harvested at days 10, 20, 30 and 60 post injection after the animals had been fasted for 4 h before being anesthetized with Avertin (2,2,2-tribromoethanol, 32 mg; Aldrich, Milwaukee, WI, USA).

Clinical Evaluation

Individual mice were observed daily for clinical signs of disease for up to 60 days after immunization. Mice were scored according to the following scale: 0, no detectable signs of experimental allergic encephalomyelitis (EAE); 0.5, incomplete tail paralysis; 1, complete limp tail; 2, hind limb weakness and unsteady gait; 3, complete hind limb paralysis; 4, total paralysis of both forelimbs and hind limbs; 5, moribund; 6, death (Okuda et al., 2002). Body weights and food intake was also measured daily. The data were plotted as daily mean clinical score for all animals in a particular treatment group.

General Cell Culture Conditions and Reagents

Immortalized BV-2 murine microglia cells were kindly donated by Dr. Peter Grace (University of Colorado, Boulder). BV-2 cells were grown in Dulbecco's modified Eagle's medium (DMEM) containing 10% fetal bovine serum (FBS) and 1% penicillin/streptomycin (P/S) and maintained at 37°C with 5% CO_2 (unless otherwise stated). Cells were grown to 70%–90% confluency for all experiments (unless otherwise specified).

LPL Targeting by shRNA

LPL was knocked down (KD) in BV-2 microglia using shRNA targeting, as previously described (Libby et al., 2015). In brief, for stable KD of LPL, 12 μg of control (202) or LPL shRNA (553) lentiviral vector was transfected into HEK 293FT cells along with 10.8 μg pΔ8.9 and 1.2 μg VSV-G using 45 μL Lipofectamine 2000. Virus was packaged for 48 h in 6 mL media, and two rounds of BV-2 transduction were performed. Stably-transduced cells were then selected for 3 days by growth in medium containing 5 μg/mL puromycin.

Real-Time RT PCR

Total RNA was isolated from BV-2 cells using RNeasy Plus Mini Kit (Qiagen, Valencia, CA, USA) according to the manufacturer's instructions. cDNA was prepared from 1 μg of total RNA using verso cDNA synthesis kit (Thermo Scientific, Lithuania). Relative gene expression levels quantified by real-time PCR using SYBR Select Master Mix (Applied Biosystems, Foster City, CA, USA). Primers were designed using the NCBI primer-blast tool and specific primers were designed to cross exon boundaries where available. The sequences for all primers used in the study are shown in **Table 1**. All PCR was performed with the following cycling conditions: 50°C for 2 min, 95°C for 10 min, followed by 40 cycles of 95°C for 15 s and 60 for 1 min using the StepOnePlus instrument (Applied Biosystem, Foster City, CA, USA) with complementary StepOne™ Software v2.3. Comparative Ct method was used to determine relative target gene expression and data were normalized to an average of the references genes Ubc and GAPDH.

Western Blots

BV-2 cells were grown to 90% confluency and total protein was harvested using RIPA buffer (150 nM NaCl, 1% Triton X-100, 0.5% Sodium Deoxycholate, 0.1% SDS, 50 mM Tris pH 8.0, and Complete Protease Inhibitor Cocktail). Insoluble cell debris were removed by centrifugation, and protein containing supernatants were retained. Protein quantification was performed using the Peirce™ BCA protein Assay kit (Thermo). Twenty-five microgram of total protein extracts were run on 12% Separating Gel and transferred to nitrocellulose membrane. The membrane was blocked in skim milk and incubated with a proprietary anti-mouse anti-LPL primary antibody raised in rabbit at a concentration of 3 μg/ml 397-1 (397-1, NeoBiolab, Cambridge, MA, USA) and 1 μg/ml mouse anti-GAPDH (MAB374, Millipore, Billerica, MA, USA) overnight at 4°C. Membranes were then incubated with secondary antibodies IRDye 800RD goat anti-rabbit and IRDye 680RD donkey anti-mouse (Li-Cor, Nebraska, NE, USA) at 1:10,000 dilution. Protein bands were visualized using the Li-COR Odyssey with complementary software (Image studio).

LPL Enzymatic Activity

The enzymatic activity of cell-surface heparin-released LPL was measured using a phospholipid (PL)/^{3}H-triolein substrate with human serum as a source of ApoC2 as described previously (Jensen et al., 2008).

LPS Stimulation

BV-2-553 and BV-2-202 and primary adult microglia were stimulated with 1 μg/ml lipopolysaccharide (LPS) in DMEM containing 1% FBS and 1% P/S media for 24 h. Gene expression and LPL activity assays were then conducted.

Lipid Uptake Experiments

Lipid Uptake Experiments were performed as previously described (Libby et al., 2015), with limited modifications. The assay was performed by preparing a synthetic PL/TG emulsion containing a ^{3}H triolein tracer. The emulsion was prepared by combining and sonicating the following: 5 mg ^{1}H triolein, 0.25 mg L-phosphatidylcholine, a trace amount of ^{3}H triolein (12.5 μCi total) tracer (NET431001MC, Perkin Elmer, Waltham, MA, USA), 0.9 mL water, 2 mL 1 M tris–HCl (pH 8.0), 800 μL FFA-free BSA (MP Biomedical), and 300 μL KRP. After sonication, the substrate was diluted into DMEM to a final concentration of 50 μM, and 100 μM triolein. Purified human ApoC2 (MyBioSource, SanDiego, CA, USA) was added to a final concentration of 5 μg/mL. Substrate was applied to over-night serum-starved BV-2-553 and BV-2-202 cells for 2 h. Cells were washed with 0.1% FFA-free BSA and lysed in 1 mL of RIPA buffer. Eight-hundred microliter of each lysis fraction was subjected to scintillation counting, while the rest was used for protein normalization.

TABLE 1 | Primer Sequences.

Gene	Accession	Forward	Reverse
LPL	NM_008509	ATGGATGGACGGTAACGGGAATGT	TGGATAATGTTGCTGGGCCCGATA
Arg1	NM_007482.3	GGTCTGTGGGGAAAGCCAAT	AACTGCCAGACTGTGGTCTC
YM1	NM_009892.3	GTACCCTGGGTCTCGAGGAA	CCTTGGAATGTCTTTCTCCACAG
iNOS	NM_010927.4	GGTGAAGGGACTGAGCTGTT	ACGTTCTCCGTTCTCTTGCAG
IGF-1	NM_010512.5	CTCAGAAGTCCCCGTCCCTA	ATTTTCTGCTCCGTGGGAGG
CCL3	NM_011337.2	CCATATGGAGCTGACACCCC	TCAGGAAAATGACACCTGGCT
IL-6	NM_031168.2	CTCTGCAAGAGACTTCCATCCA	GACAGGTCTGTTGGGAGTGG
IL-12	NM_001159424.2	TTCTCACCGTGCACATCCAA	GAGGAGGTAGCGTGATTGACA
MSR1	NM_031195.2	GGGAGTGTAGGCGGATCAAG	ATAGTAGGGTGCTCTGCCCA
CD36	NM_001159558.1	AGAATTCTCAGCTGCTCCGC	ACACATTTCAGAAGGCAGCAAC
VLDLR	NM_013703.2	TCAACTGCCCTTCTCGAACC	AGCCATCAACACAGTCTCGG
LDLR	NM_010700.3	CCAATCGACTCACGGGTTCA	TCACACCAGTTCACCCCTCT

Uptake of DiI Labeled Liposomes

Synthetic PL/TG chylomicrons were prepared as described above in the absence of radiolabeled triolein. PL containing vesicles were prepared by drying an 80:20 molar ratio of phosphatidylcholine (PC, Avanti) and phosphatidylserine (PS, Avanti) under nitrogen. HBS was added to the dried lipid for 1 h prior to sonication. Ten microliter of a 3 mg/ml solution of 1′-dioctadecyl-3,3,3′3′-tetramethylindocarbocyanine perchlorate (DiI) was added to 1 ml of either SC's or PL vesicles and incubated in the dark for 10 min at 37°C. Free label was removed with ultracentrifugation at 24,000 *g* for 10 min. After resuspension in sterile PBS, labeled liposomes were added to cells for 16 h (overnight) at a final concentration of 250 μg/ml.

Conditioned Media Experiments

Conditioned media (CM) was harvested from either BV-2-553 or BV-2-202 cells grown in DMEM containing 1% FBS for 72 h reaching 90% confluency. Following aspiration, the media was briefly centrifuged to remove cellular debris, and media was either used immediately for experiments or flash frozen and stored at −80°C.

Neuro-Discovery Array

Semi-quantitative analysis of cytokines and cell signaling proteins from BV-2-CM was performed using the RayBio® C-series Mouse Neuro Discovery Array C1 (RayBiotech, Norcross, GA, USA) according to manufacturer's instructions. In brief, membranes were incubated with 1 ml of CM (described above) overnight at 4°C, while the biotinylated antibody and streptavidin incubation steps were performed at 2 h at room temperature (RT). The array was then visualized using chemiluminescence detection.

Cerebellar Slice Culture, Lysolecithin and Conditioned Media Treatment

Sagittal cerebellar slices (300 μm) were prepared from PLP-eGFP mice (Mallon et al., 2002) at P10–P12 and cultured on MilliCell 0.4 μm membrane inserts (Millipore, Billerica, MA, USA) for 7 days in slice media (25% HBSS, 25% heat-inactivated horse serum, 50% MEM, 125 mM HEPES, 28 mM D-Glucose, 2 mM L-Glutamine; Sheridan and Dev, 2012). Slices were incubated at 37°C at 5% CO_2 and slice media was changed within 24 h after preparation, then every 3–4 days following. Lysolecithin (Lyso; Sigma) was prepared as a 5 mg/ml stock and stored at −20°C until use. To prepare a working solution, lysolecithin stock was diluted in slice media at 0.5 mg/ml and sterile filtered. Treatments (Slice media or lysolecithin) were applied both on top (50 μl) and below (250 μl) the membrane insert that contained the slice for 17 h at 37°C at 5% CO_2. After 17 h, slices were washed twice with slice media and then incubated in control media or CM from BV-2 microglia.

Slice Fixation and Immunostaining

After treatment, cerebellar slices were rinsed twice with PBS and fixed in 4% paraformaldehyde in PBS for 20 min. For immunohistochemistry, slices were rinsed in PBS and permeabilized in 1.5% or 10% (for myelin proteins) Triton X-100 in PBS for 20 min. Slices were rinsed, blocked with 5% normal donkey serum (NDS) in PBSTx (0.3%) for 1 h, and incubated with primary antibodies overnight at RT. Following three washes in PBS, secondary antibodies (Jackson ImmunoResearch) were applied 1:800 overnight at RT, slices were washed three times in PBS and mounted in Fluoromount G (Southern Biotech). The following primary antibodies were used: Rb α glial fibrillary acidic protein, GFAP (Sigma, G3893), Rb α Calbindin (Millipore, AB1778), Ms α myelin-associated glycoprotein, Myelin associated glycoprotein (MAG; Millipore, MAB1567), Rb α myelin-oligodendrocyte glycoprotein, MOG (Abcam, ab32760), Ms α myelin basic protein, MBP (Covance, SMI 94), Ckn α Neurofilament-H (Neuromics, CH22104), Gt α Iba1 (Abcam, ab5076), Rb α NG2 proteoglycan, GP α NG2 proteoglycan (gifts from Dr. William Stallcup, Burnham Institute), and Rb α Olig2 (a gift from Dr. Charles Stiles, Harvard University).

Immunocytochemistry of BV-2 Cells

BV-2-202 and BV-2-553 cells were grown to about 40% confluency on a coverslip. Media was then replaced with CM from BV-2-202 cells, BV-2-553 cells or regular control media for 24 h. Cells were fixed in 4% paraformaldehyde for 20 min at RT and incubated with 1 μg/ml of rabbit anti-Iba1 primary antibody (Wako, Osaka, Japan) overnight at 4°C. Cells were then incubated in Alexa Fluor® goat anti-rabbit 488 (Molecular Probes, Eugene, OR, USA) secondary antibody at 1:400 for

1 h at RT. Vectashield mounting media with DAPI (Vector Laboratories, Inc., Burlingame, CA, USA) was used to mount coverslip onto slide and stain nuclei. Confocal laser scanning microscopy (Olympus FV1000, Olympus, Tokyo, Japan, using Olympus Fluoview Software), was used to image the cells. Exposure parameters were identical in all slides.

For DiI labeled lipid uptake experiments the average fluorescence per cell was determined by measuring the mean fluoresce in $N = 10$ induvial cells from each experimental group using the cell magic wand plugin for ImageJ.

Primary Microglia Isolation

Primary microglia were isolated from adult brains of $Cx3CR1^{CreER}$ (Jackson Labs) which express a Cre-ERT2 and EYFP in microglia as previously described with very limited modifications (Lee and Tansey, 2013). Care was taken to dissociate cells with large, medium and small-sized hole polished Pasteur pipettes.

Immunohistochemistry of Primary Microglial Cells

Primary microglia cells were plated on poly-D-lysine coated glass coverslips and cultured for 2 days. Cells were washed with HBSS 1× and fixed in 4% paraformaldehyde at RT for 20 min. Cells were blocked for 1 h with 10% normal goat serum (NGS) in PBST. Cells were incubated with 1 μl Anti-Iba1, rabbit (Wako) for 16 h at 4°C. The cells were washed 3× with PBS. Secondary goat anti-rabbit 488 was diluted 1:400, cells were incubated for 1 h at RT. Coverslips were mounted with Vectashield Anti Fade Mounting Media with DAPI, and slides were imaged using confocal Olympus FV1000.

Metabolomics

Cell pellets (10^6 cells) were stored at −80°C and were extracted in ice-cold lysis/extraction buffer (methanol:acetonitrile:water 5:3:2 v/v). Amino acids (Nemkov et al., 2017), and other metabolites (Nemkov et al., 2015), were quantified in cell extracts as previously described.

Fatty Acid Oxidation (FAO) Assays

Fatty Acid Oxidation (FAO) assays were performed as described with limited modifications (Boyle et al., 2017). Control and LPL KD BV-2 microglia were incubated at 37°C in sealed 24-well plates containing differentiation media containing 0.25 μCi/ml [^{14}C]oleate and 0.25 μCi/ml [^{14}C]palmitate (PerkinElmer Life Sciences, Waltham, MA, USA), with 200 μM, 1:1 "cold" oleate:palmitate. After a 2 h incubation, the rate of FAO was determined by measuring the $^{14}CO_2$ released from the media following acidification. Measures were performed in quintuplet and data were corrected for total protein content, determined by BCA assay (Pierce Biotechnology, Inc., Waltham, MA, USA).

Statistical Analysis

The difference between mean values for each experimental group were determined by either an ANOVA (for three or more experimental groups) followed by *post hoc* analysis with Bonferroni correction for multiple comparisons, or unpaired *t*-test (for experiments with two experimental groups), since data showed Gaussian distribution.

RESULTS

LPL Activity Is Increased at the Interphase of De- and Remyelination

EAE is a CNS autoimmune disease that has been used to investigate human demyelinating diseases such as MS (Johns et al., 1995). EAE has been well characterized in rodents using both recombinant and synthetic myelin proteins including MOG or MOG peptides (Amor et al., 1994; Johns et al., 1995; Frausto et al., 2007). To induce EAE we chose the MOG_{35-55} peptide as a suitable antigen since MS patients predominantly generate anti-MOG antibodies (Goudonnet et al., 1990). Following immunization with MOG_{35-55} or vehicle, mice were scored daily for severity of clinical symptoms. By 20 days post immunization MOG_{35-55} mice had symptoms ranging from a limp tail (score of 1) to completely paralyzed hind limbs (score of 3) while the control mice showed no symptoms of paralysis. The control mice continued to show no signs of paralysis from 1 day through to 60 days after treatment while the MOG-treated mice had relapsing remitting symptoms (**Figure 1A**). At 20 days (when clinical symptoms were at their peak in the MOG group) and 30 days (when clinical symptoms reached a nadir in the MOG group), LPL activity was measured in whole brain tissue (**Figure 1B**). Interestingly, while there was no difference in LPL activity in the brain at 20 days, LPL activity was increased in the brain by 30 days ($P < 0.01$), suggesting that increased LPL activity was associated with the initiation of improved symptoms.

To further understand the dynamics of LPL function during active demyelination and subsequent remyelination, we measured LPL enzymatic activity and expression in lysolecithin-treated *ex vivo* organotypic slice cultures of mouse cerebellum. The *ex vivo* slice culture system closely resembles the *in vivo* environment of the CNS, with the added benefit of ease of high throughput experimental treatments (Harlow et al., 2015). In this system, short term treatment with lysolecithin produces dramatic demyelination in brain slices *ex vivo* (Birgbauer et al., 2004). Twenty-four hours after lysolecithin removal there is continued demyelination and a significant reduction of the myelin markers, MBP, MOG and cyclic nucleotide phosphodiesterase (CNPase; Birgbauer et al., 2004). Several days after lysolecithin treatment these markers began to increase, concomitant with the initiation of remyelination. Here we show that heparin-releasable LPL enzymatic activity from the surface of cerebellar slices was significantly increased after 2 days of recovery in normal medium (**Figure 1C**), suggesting that LPL is most active and abundant at the end of demyelination and the initiation of remyelination. In addition, LPL expression was significantly increased by 7 days post-lysolecithin treatment (**Figure 1D**), which is consistent with the recovery of myelin markers, MAG and proteolipid protein1 (PLP1) by 7 days (**Figures 1E–J**).

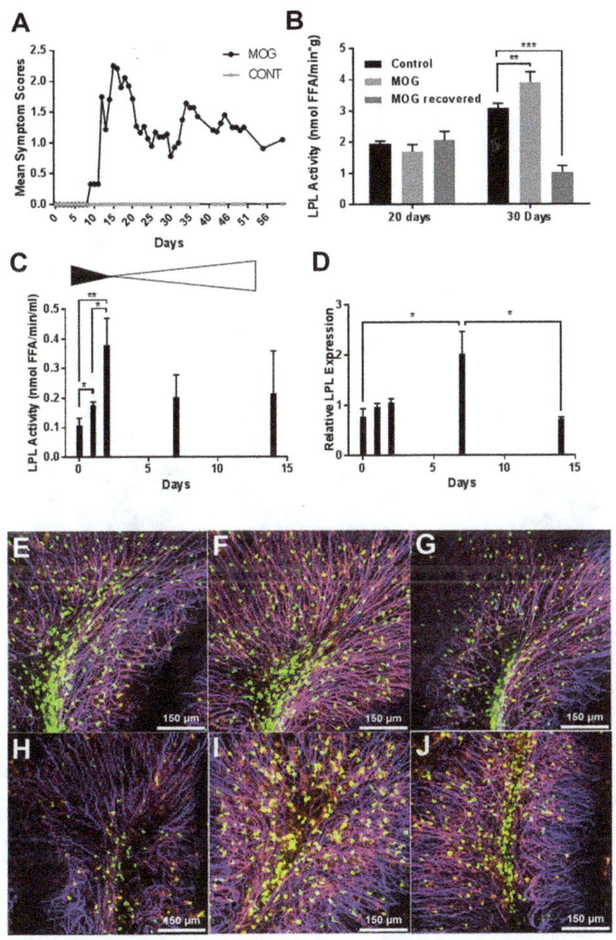

FIGURE 1 | Lipoprotein lipase (LPL) is increased during de- and re-myelination. **(A)** Mean symptom scores of MOG treated ($N = 10$) and Control ($N = 20$) mice**.** **(B)** LPL activity in brain tissue at 20 and 30 days MOG post injection from Control ($N = 20$), MOG-treated (symptomatic; $N = 10$), and MOG-recovered (MOG-treated but asymptomatic following an initial bout of symptoms, $N = 3$)**.** **(C)** LPL enzymatic activity in cerebellar slice cultures at 0, 1, 2, 7 and 14 days post LPC treatment ($N = 3$ per group, per time point). **(D)** LPL gene expression in cerebellar slice cultures at 0, 1, 2, 7 and 14 days post LPC treatment ($N = 3$ per group, per time point). **(E–J)** Lysolecithin-mediated demyelination of *ex vivo* cerebella brain slices ($N = 3$ per group, per time point). Blue—neurofilament (NFH), Green—PLPeGFP, Red—myelin associated glycoprotein (MAG). **(E–G)** No Lysolecithin control, 1, 4 and 7 days. **(H–J)** 1, 4 and 7 days post Lysolecithin treatment. $^{*}P < 0.05$ vs. CONT (at corresponding time point). $^{**}P < 0.01$ vs. CONT (at corresponding time point). $^{***}P < 0.001$ vs. CONT (at corresponding time point).

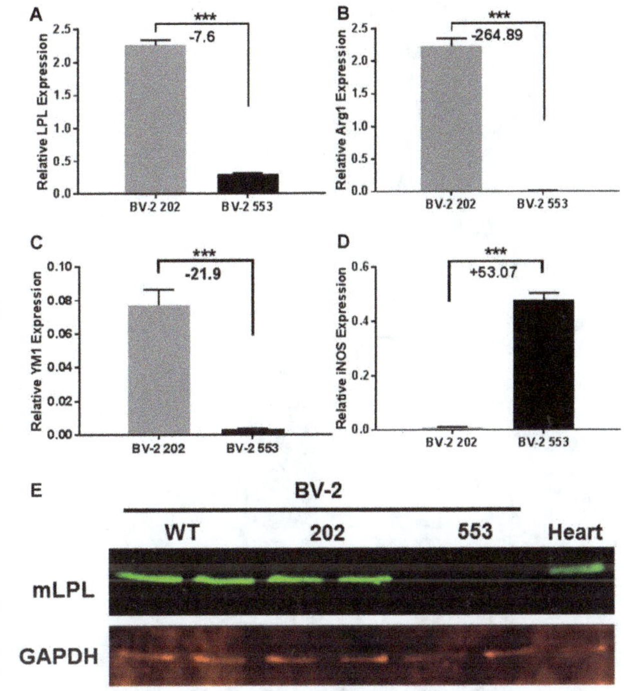

FIGURE 2 | Characterization of LPL knock down (KD) cells. **(A–D)** Gene expression in microglial BV-2 KD (553) vs. control (202) cells ($N = 4$ per group). **(E)** Western blot analysis showing reduced LPL protein expression in BV-2 553 cells. $^{***}P < 0.001$ vs. CONT.

The Loss of LPL in BV-2 Microglia Polarizes BV-2 Microglia to a Pro-inflammatory Phenotype

Since CNS microglia have been repeatedly implicated in myelination dynamics (Sun et al., 2017), and recent reports have highlighted the pivotal role of microglial LPL in remyelination and primary neurodegenerative disorders (Olah et al., 2012; Keren-Shaul et al., 2017), we utilized BV-2 microglia to investigate the role of LPL within these cells. BV-2 cells are an immortalized mouse microglial cell line that retains many of the morphological and functional characteristics of microglia, with the ability to stably alter the levels of LPL expression and activity (Blasi et al., 1990; Bocchini et al., 1992; Wen et al., 2006; Fu et al., 2017). We stably KD LPL expression in BV-2 cells using an LPL shRNA lentiviral vector, previously validated in our laboratory, to generate BV-2 553 (LPL KD) cells (Libby et al., 2015). An empty lentiviral vector was used to generate BV-2 202 (control) cells. Comparative gene expression between LPL KD and control cells showed a dramatic decrease in LPL KD cells in the expression of genes classically associated with an "M2" or "alternatively activated" phenotype. For example, the enzyme arginase 1 (Argl), which converts arginine to polyamines and contributes to wound healing and repair (Munder, 2009; Cherry et al., 2014), showed a −265-fold reduction in LPL KD cells (**Figure 2B**). Similarly, chitinase-like 3 (YM1), a secretory lectin that binds heparin sulfate and prevents degradation of extracellular matrix proteins showed a −22-fold reduction in gene expression in LPL KD cells (**Figure 2C**). In contrast, iNOS expression, commonly associated with an M1-like microglial phenotype was markedly elevated in LPL KD cells (+53 fold; **Figure 2D**). Taken together, these data strongly suggest that microglial LPL expression is a feature of alternatively activated microglia.

To further understand the inflammatory phenotype of BV-2 cells lacking LPL, the relative abundance of cytokines and chemokines in CM from LPL KD and control cells was determined. Monocyte chemoattractant protein (MCP-1), which

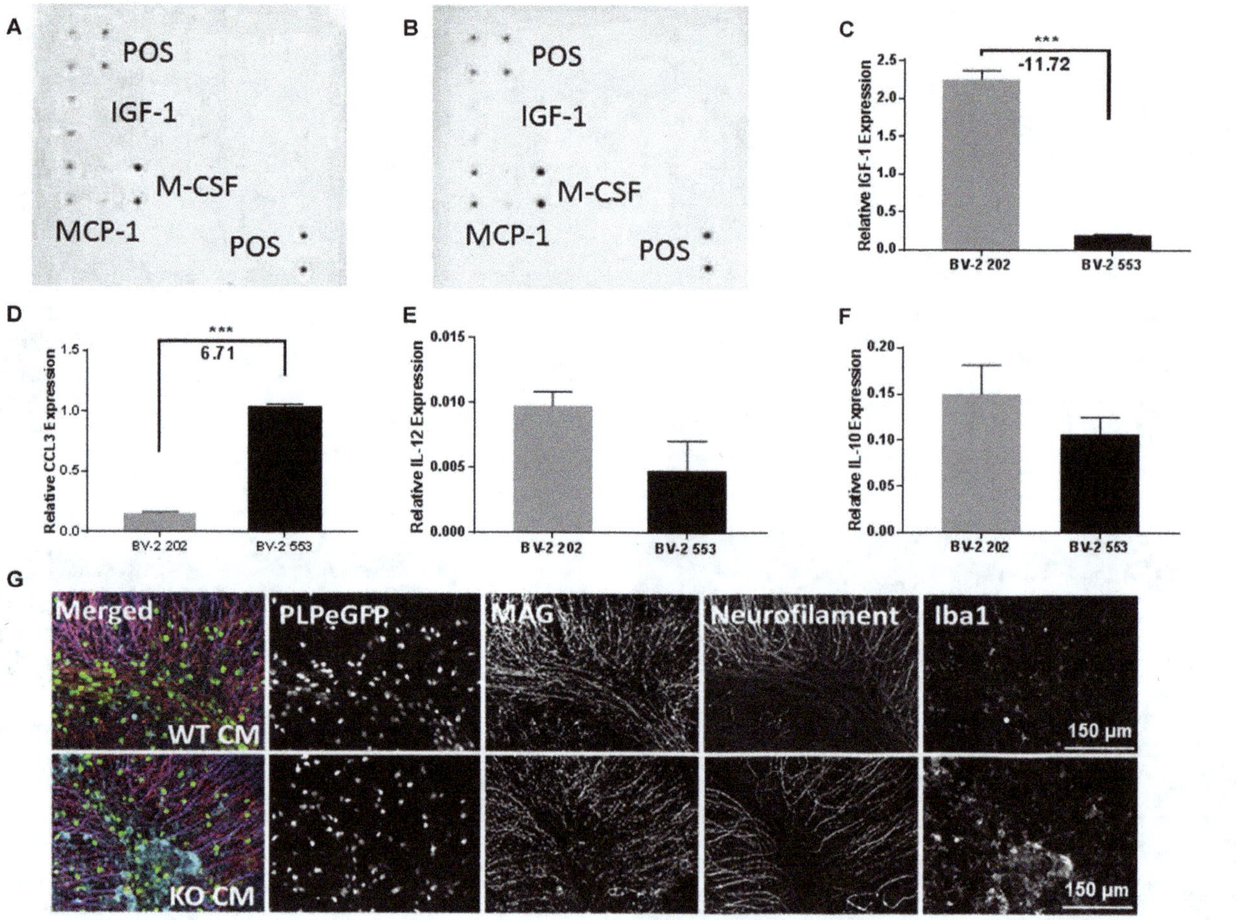

FIGURE 3 | Chemokine and cytokine production in LPL KD microglia. Ray biotech mouse neuro array was used to determine differences in chemokine expression in media from **(A)** BV-2 202 or **(B)** BV-2 553 cells. **(C–F)** Comparative gene expression between control (BV-2 202) and LPL KD (BV-2 553) cells (N = 4 per group). **(G)** *Ex vivo* brain slices following incubation with conditioned media (CM) from either control (WT CM) or LPL KD (KD CM) for 24 h, characterizing Plp-eGFP MAG, neurofilament or Iba1 expression (N = 3 per group). ***P < 0.001 vs. CONT.

is associated with microglial migration and proliferation, but not pro-inflammation (Hinojosa et al., 2011), was detected at a similar abundance in both control (**Figure 3A**), and LPL KD (**Figure 3B**) cells. Similarly, macrophage colony stimulating factor (M-CSF) was secreted at similar levels by both cell lines (**Figures 3A,B**). In contrast, insulin-like growth factor 1 (IGF-1), which is thought to play a role in the trophic support and development of neurons, was only detected in CM from control cells (**Figures 3A,B**). Since these findings were only semi-quantitative, we also used real-time PCR to quantify the level of IGF-1 gene expression in LPL KD vs. control cells. IGF-1 mRNA was significantly reduced (−12 fold) in LPL KD cells (**Figure 3C**). The expression of additional genes associated with a pro-inflammatory phenotype was also measured. Macrophage inflammatory protein (MIP)-1alpha (CCL3), associated with neuro-inflammation in the CNS (Zhu et al., 2016), was increased (+6.7 fold) in LPL KD cells (**Figure 3D**) compared to control. However, the expression of IL-12 and IL-10 were not significantly different between LPL KD and control cells (**Figures 3E,F**). To establish whether microglia-CM from either control cells or cells lacking LPL could elicit an inflammatory response *ex vivo*, brain slices were cultured with CM for 24 h. Histological analysis revealed that while myelin markers remained unchanged between groups, the expression of Iba1, associated with classically activated microglia, was increased in slices treated with CM from LPL KD cells (**Figure 3G**).

It is well established that LPS can stimulate an inflammatory response in BV-2 microglia (Orihuela et al., 2016; Sun et al., 2017). Since our data suggest that LPL is a feature of reparative microglia, we hypothesized that LPS-mediated microglial polarization would influence LPL dynamics. Following LPS treatment, both the enzymatic activity (**Figure 4A**) and mRNA expression (**Figure 4B**) of LPL were significantly reduced. Although the profile of the BV-2 microglia resembles that of primary microglia, it is important to assess the role of LPL in primary microglial cultures. Therefore, to test the dynamics of LPL following microglial polarization, primary microglia were isolated from adult brain (**Figure 4C**) and incubated with LPS for 24 h. Following LPS stimulation LPL gene expression was significantly reduced (**Figure 4D**).

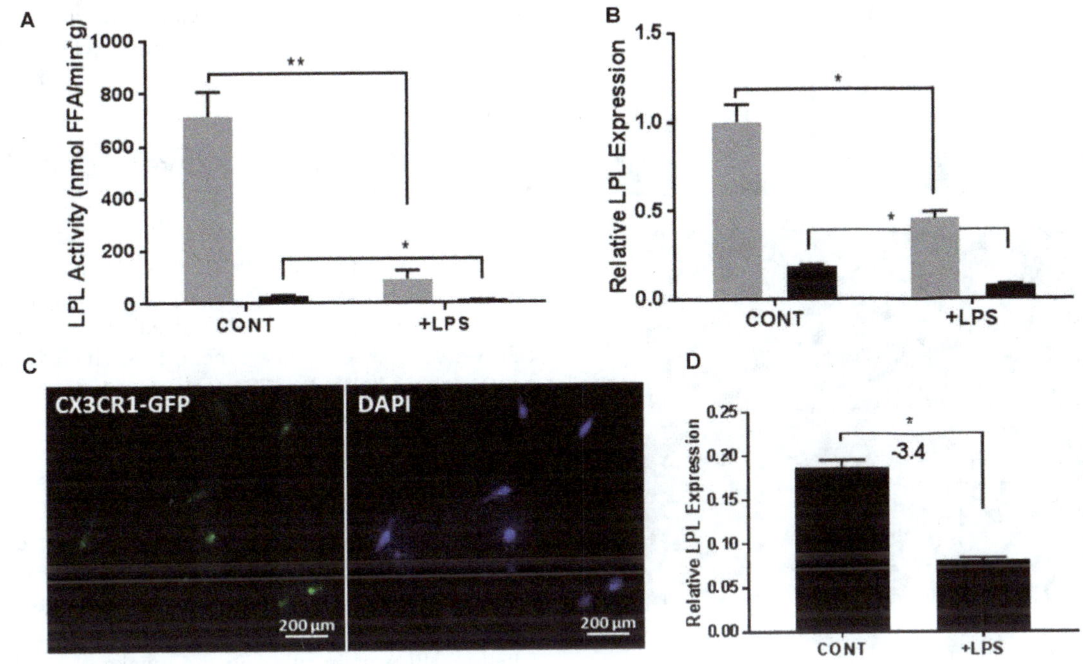

FIGURE 4 | Lipopolysaccharide (LPS) activation reduces LPL expression and activity in both immortalized cells (*N* = 4 per group; **A,B**) and primary microglial cells from adult brain (**C,D**; *N* = 3 per group). **P* < 0.05 vs. CONT. ***P* < 0.01 vs. CONT.

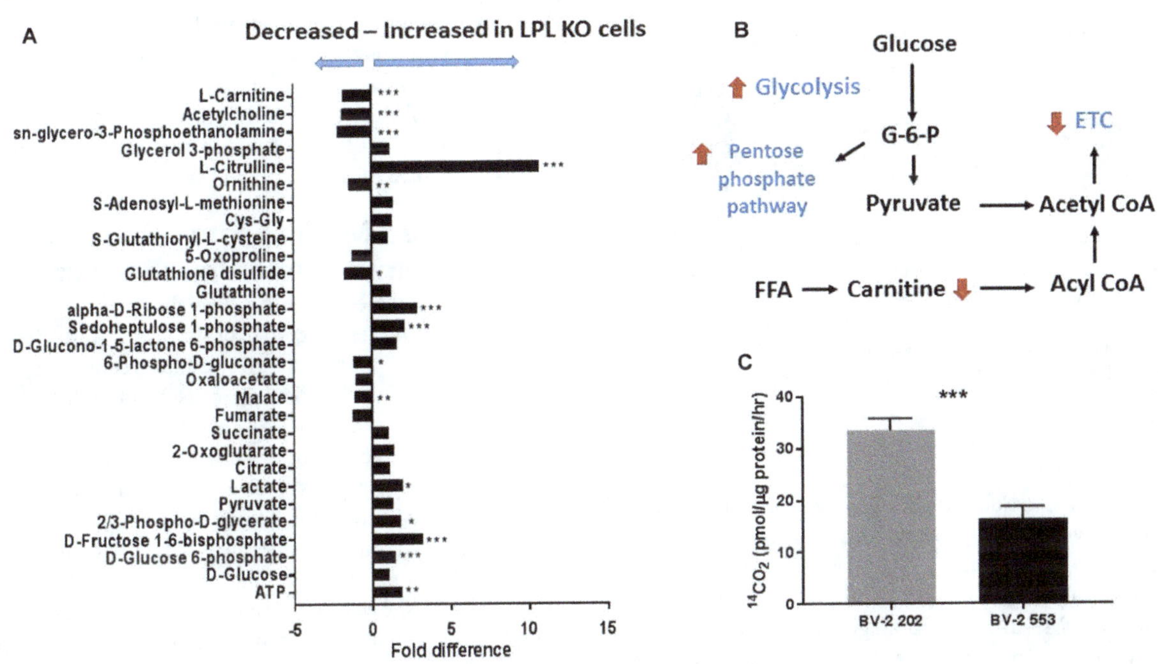

FIGURE 5 | Glycolysis is increased and fatty acid oxidation (FAO) is decreased in BV-2 microglial cells lacking LPL. **(A)** Global metabolomics in control and LPL KD BV-2 microglia (*N* = 3 per group). **(B)** Schematic representation of metabolic reprogramming following depletion of LPL in microglia. **(C)** FAO in control vs. LPL KD BV-2 microglia (*N* = 6 per group). **P* < 0.05 vs. CONT. ***P* < 0.01 vs. CONT. ****P* < 0.001 vs. CONT.

Although microglial activation states are complex, it is likely that microglial polarization resembles that of M1/M2 macrophages (Orihuela et al., 2016). In brief, metabolic functions in reparative (M2-like) macrophages is

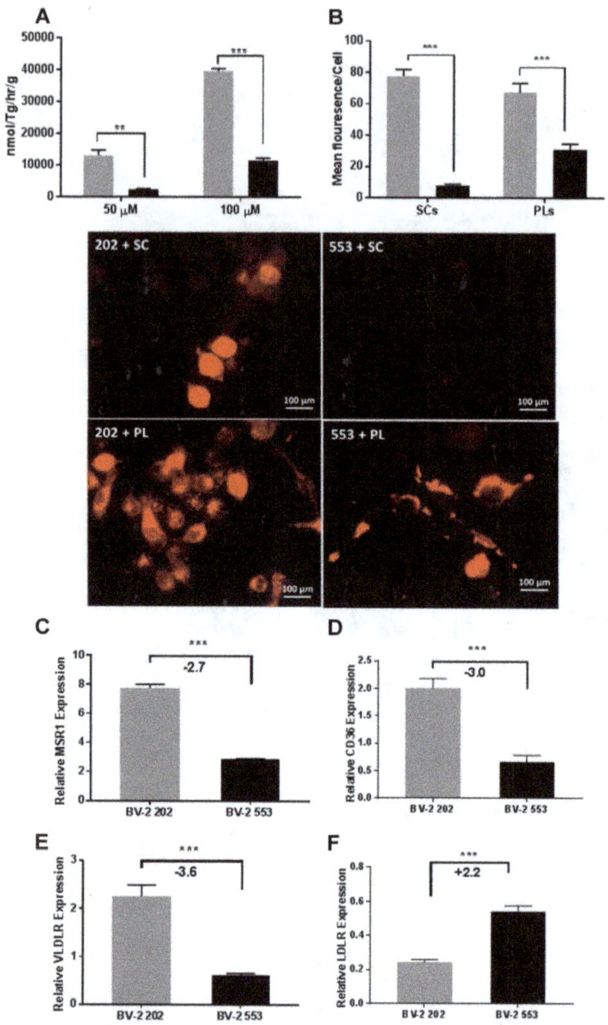

FIGURE 6 | LPL is involved in lipid and lipoprotein uptake in microglia. **(A)** Intracellular uptake of radiolabeled synthetic lipid vesicles containing 50 or 100 μM Triolein Tg. **(B)** Uptake/phagocytosis of 1′-dioctadecyl-3,3,3′3′-tetramethylindocarbocyanine perchlorate (DiI) labeled synthetic chylomicrons (SC) and phospholipid liposomes (PLs; $N = 3$ per group). **(C–F)** Expression of scavenger and lipoprotein receptors in BV-2 202 vs. 553 cells ($N = 4$ per group). $^{**}P < 0.01$ vs. CONT. $^{***}P < 0.001$ vs. CONT.

characterized by FA oxidation (FAO; Odegaard and Chawla, 2011), comparatively lower glucose consumption as compared to M1 macrophages, and arginine metabolism is shifted towards ornithine and polyamines (Mills et al., 2000). By contrast, acutely activated M1-like macrophages are shifted towards increased glycolysis (Rodríguez-Prados et al., 2010), activation of the pentose phosphate pathway (PPP; Haschemi et al., 2012), and the conversion of L-arginine to L-citrulline by iNOS. To further understand the involvement of microglial LPL in the metabolic reprogramming associated with polarization we performed global metabolomics in control and LPL KD BV-2 cells (**Figure 5A**). We found that glucose metabolism, PPP metabolism and L-citrulline were significantly increased in LPL KD cells (**Figure 5A**), consistent with the metabolic changes associated with M1-like polarization (**Figure 5B**). We also found that FAO was increased in control cells vs. LPL KD cells, consistent with the metabolic profile of M2-like macrophages (**Figure 5C**).

Lipoprotein Lipase Is Involved in Microglial Lipid Uptake

Several studies have implicated LPL in phagocytosis and repair. However, to date the precise mechanisms have not been determined. To test whether microglial LPL is involved in the uptake of free fatty acids (FFA), either control or LPL KD cells were incubated with synthetic TG-rich chylomicrons (SCs) containing radiolabeled FFAs. After incubation, the amount of intracellular radiolabeled FFA was quantified. FFA uptake was significantly increased in LPL expressing control microglia (**Figure 6A**). To address whether uptake was dependent on enzymatic hydrolysis by LPL, SCs were labeled with DiI and the internalization of intact (non-hydrolyzed) DiI labeled SC was visualized by fluorescence microscopy. SC uptake was undetectable in cells lacking LPL (**Figure 6B**). Since low-density lipoproteins are scarce in the brain we next asked whether LPL was involved in the uptake of myelin associated phospholipids (PLs). BV-2 cells were incubated with DiI labeled liposomes containing PC and PS. Although BV-2 cells lacking LPL internalized PLs, this was much greater in the control cells (**Figure 6B**). Taken together these data suggest that microglia LPL is involved in the uptake, and phagocytosis of lipids. Therefore, we measured the expression of factors associated with phagocytosis. We found that MSR1, CD36, and VLDLR expression was markedly reduced in LPL KD cells (**Figures 6C–E**); however, LDLR, which is the major importer of ApoE containing lipoproteins in the brain, was significantly increased (**Figure 6F**).

Conditioned Media From LPL KD Cells Inhibits Remyelination and Repair

To determine whether LPL directly plays a role in the remyelination process, *ex vivo* brain slices were demyelinated with lysolecithin and then allowed to recover in CM from LPL-secreting control cells and cells lacking LPL (**Figure 7**). After a 7-day recovery period, cells treated with lysolecithin, but then incubated in optimal slice media showed normal tissue morphology and expression of myelin markers. In slices treated with CM from control cells, the slice morphology was comparable to control, and there was no change in myelin marker expression. However, brain slices treated with CM from LPL KD cells showed a somewhat necrotic morphology with diffuse myelin marker staining and a significant reduction in neurofilament expression, suggesting that re-myelination was incomplete in this condition.

DISCUSSION

MS is a relatively common (1:1000 in susceptible populations) debilitating disease that can affect an individual for much of

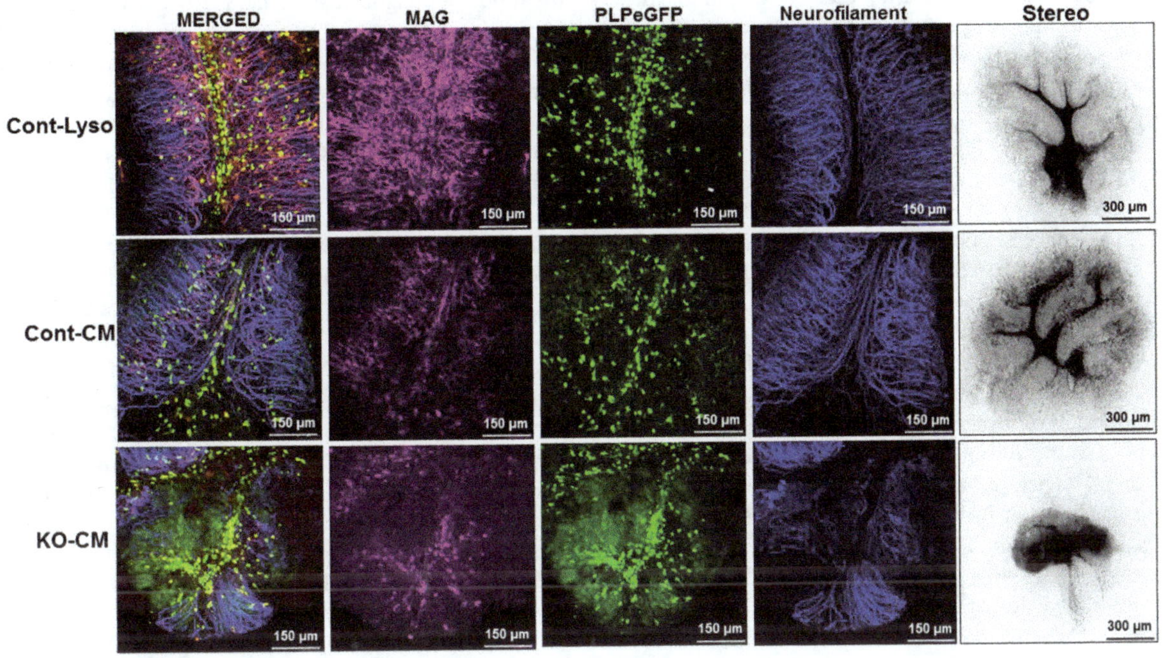

FIGURE 7 | CM from LPL KD cells prevents normal remyelination processes. *Ex vivo* cerebellar slices were demyelinated by overnight treatment (17 h) with lysolecithin. The next day slices were incubated with either standard slice media, or CM and allowed to recover for 7 days.

their life. During the RR-MS stage of the disease there is a discrete balance between de- and remyelination. In P-MS, these attempts to remyelinate fail leading to neurodegeneration and progressive disability. Understanding the dynamics of de- and remyelination may be key to identifying potential targets for MS treatments. Microglial function plays a key role in the remyelination process. The production of cytokines/chemokines from the appropriate microglial phenotype can stimulate the proliferation of OPCs to promote remyelination (Voss et al., 2012). In addition, microglial-phagocytosis of myelin debris plays in important role in the initiation of remyelination (Neumann et al., 2009). However, to date the factors involved in these processes remain elusive. Here we show for the first time that LPL, a lipid-processing enzyme with known roles in hydrolytic and non-hydrolytic lipid uptake, was a feature of an "alternatively activated" microglial phenotype associated with phagocytosis and remyelination.

In the present study we have demonstrated that LPL is increased at the initiation of remyelination in both *in vivo* and *ex vivo* model systems (**Figure 1**). Using the EAE model of demyelination we showed that LPL activity in the brain is significantly increased at 30 days following immunization. This time point correlates with a nadir in clinical symptoms, suggesting that LPL expression is associated with the partial remyelination observed in RR-MS. In further support of this notion, we have shown that LPL abundance and enzymatic activity is increased at the interphase of de- and remyelination in an *ex vivo* model of demyelination. These observations are consistent with findings from a cuprizone model of de- and remyelination in which LPL was identified as a key feature of a microglial phenotype that actively supports remyelination (Olah et al., 2012). Since cuprizone models of de- and remyelination have been used to identify the central role of debris clearance in these processes (Voss et al., 2012), these data suggest that LPL is a feature of a scavenging microglial phenotype that promotes regeneration.

Microglia play an important role in remyelination and regeneration process through their ability to clear debris, and their expression and secretion of cytokines and growth factors. Our data robustly show that cells lacking LPL polarize to a pro-inflammatory phenotype characterized by the expression of iNOS, inflammatory cytokines, and a metabolic shift towards glycolysis—similar to that of M1-like macrophages. In contrast, cells that express LPL maintain an alternately activated phenotype that is characterized by elevated Arg1 expression, IGF-1 production and increased FAO. The switch in expression between Arg1 and iNOS is particularly indicative of microglial polarization, since both Arg1 and iNOS use the same substrate (arginine) to either produce polyamines for wound healing and matrix deposition (Arg1; Munder, 2009), or nitric oxidase production contributing to inflammation (iNOS; Morris, 2007). Thus, Arg1 vs. iNOS represents a useful switch to an M2 vs. M1 phenotype. In the present study we show that the loss of LPL markedly increases iNOS expression while reducing Arg1 expression, suggesting a polarization towards an M1-like phenotype (**Figure 2**). In further support we show that L-citrulline, a product in the synthesis of NO by iNOS (Chaturvedi et al., 2007), was markedly increased in cells lacking LPL (**Figure 5**). Although defined phenotypic boundaries likely do not represent the full microglial spectra observed in complex disease states such as MS, the expression of these genes serve as useful categorical markers. In addition, the expression of chemokines and growth factors are also useful markers of microglial status. Therefore, it is of interest that microglial cells lacking LPL have significantly reduced IGF-1 secretion and expression (**Figure 3**), which has been classically associated with M2-like macrophages. Recent studies have shown that IGF-1 plays a key role in the CNS, where it exerts a trophic effect on neurogenesis and neuronal survival (Russo et al., 2005). Moreover, microglia are an important source of IGF-1, and its production is linked to microglial inflammatory status (Suh et al., 2013). Interestingly, reduced IGF-1 has been observed in AD (Alvarez et al., 2007), and in patients with MS (Shahbazi et al., 2017). Recent studies using rodent models of local lysolecithin induced demyelination have shown that intrathecal IGF-1 administration promotes remyelination (Hlavica et al., 2017). Taken together, these observations suggest that LPL expression promotes a microglial polarization towards an anti-inflammatory and reparative phenotype associated with IGF-1 production, a key factor that links microglial function to repair and remyelination in MS.

Our findings prompt the question; how does modulated LPL expression mediate microglial polarization? Since LPL depletion is associated with a shift towards glycolysis and away from FAO, it is likely that LPL loss causes a shift in lipid availability, substrate metabolism, and polarization towards an M1-like classically activated phenotype. In contrast, cells harboring endogenous LPL levels may have sufficient lipid-uptake, facilitating the shift towards FAO that is associated with M2-like alternatively activated and reparative microglia. Precisely why increased FAO is important for reparative microglial remains to be tested, but one may speculate that FAO provides the additional energy supply required for long-term reparative processes.

In support of the notion that LPL facilitates microglial lipid uptake, we have also shown that microglia lacking LPL are less able to take up FFAs from synthetic chylomicrons than cells expressing endogenous levels of LPL (**Figure 6**). These data are consistent with the canonical function of LPL—the hydrolysis of TG-rich lipoproteins—and suggests that microglial LPL may play a similar role in the CNS. Since TG-rich lipoproteins are scarce in the CNS, we hypothesized that LPL may facilitate the up-take of myelin-associated lipids. The phospholipids PC and PS are major components of myelin with pleotropic roles in the pathogenesis of MS (Ho et al., 2012). In the present study we show that LPL is involved in the uptake of PC and PS containing liposomes (**Figure 6**), highlighting the role of microglial LPL in myelin phagocytosis. It is well established that the phagocytosis and clearance of myelin debris is important for the remyelination process (Neumann et al., 2009). For example, ineffective microglial function in CXC3CR1 knockout mice has been shown to impair debris clearance and subsequent remyelination (Lampron et al., 2015). In addition, many recent reports have highlighted the potential role of microglial-LPL in phagocytosis and myelin clearance (Olah et al., 2012; Ma et al., 2013; Keren-Shaul et al., 2017). Although the precise mechanism regarding the mode of phagocytosis remains to be determined, recent data from mice deficient in Triggering receptor expressed on myeloid cells 2 (TREM 2)—a microglial immunoreceptor associated with phagocytosis and whose loss-of-function mutations in humans cause presenile dementia—have shown that LPL is markedly reduced in this model, suggesting that LPL is involved in TREM 2-mediated phagocytosis (Cantoni et al., 2015). In addition, the mechanism underlying the role of LPL in phagocytosis may be similar to its non-hydrolytic, or endocytotic bridging function. LPL can facilitate lipid-uptake through simultaneous interaction with lipoproteins, heparin sulfate proteoglycans (HSPG), or cell surface receptors such as lipoprotein receptors (Makoveichuk et al., 2004). Conversely, it is well established that lipoprotein receptors have a duel function in immune cells, and may also serve as scavenging receptors to facilitate phagocytosis. For example, the low density lipoprotein receptor-related protein-1 (LRP1) functions as a scavenger receptors for myelin debris (Fernandez-Castaneda et al., 2013). It is thus plausible to speculate that LPL may simultaneously bind to lipid-rich myelin debris and cell surface lipoprotein receptors to expedite myelin clearance and remyelination.

Collectively, our observations suggest that microglial LPL exerts simultaneous benefits for repair and remyelination. On one hand the presence of LPL is associated with a reparative microglial phenotype, the production of important neurotropic factors, and the maintenance of appropriate substrate availability for metabolic polarization towards and M2-like phenotype. In addition, microglial LPL is directly involved in the internalization of myelin associated lipids. In summary, our findings highlight the importance of LPL in demyelinating disorders and are the first to demonstrate that microglial-LPL is directly involved in reparative processes needed for remyelination. We anticipate that the role of LPL in debris clearance and uptake may be exploited to develop future therapies for demyelinating disorders such as MS, and also primary neurodegenerative disease.

AUTHOR CONTRIBUTIONS

KDB, WBM and RHE designed the experiments. KDB, SG, KG and AMC performed the experiments; analyzed the data. KDB, KEB, WBM and RHE interpreted data. KDB wrote the manuscript and all authors edited the manuscript.

ACKNOWLEDGMENTS

The funding for the animal work was provided by a Pilot Award from the National Multiple Sclerosis Society. *In vitro* and *ex vivo* work was funded by the Colorado Clinical and Translational Sciences Institute (CCTSI: CNS-Pilot Basic Research Team Award, CNSBT-15-92). We would like to acknowledge Ms. Leslie Knaub and Mr. Dalan Jensen for technical assistance with these studies. We would also like to acknowledge and thank Ms. Sarah Gehrke in the University of Colorado School of Medicine Metabolomics Core for technical assistance.

REFERENCES

Alvarez, A., Cacabelos, R., Sanpedro, C., García-Fantini, M., and Aleixandre, M. (2007). Serum TNF-α levels are increased and correlate negatively with free IGF-I in Alzheimer disease. *Neurobiol. Aging* 28, 533–536. doi: 10.1016/j.neurobiolaging.2006.02.012

Amor, S., Groome, N., Linington, C., Morris, M. M., Dornmair, K., Gardinier, M. V., et al. (1994). Identification of epitopes of myelin oligodendrocyte glycoprotein for the induction of experimental allergic encephalomyelitis in SJL and Biozzi AB/H mice. *J. Immunol.* 153, 4349–4356.

Antel, J., Antel, S., Caramanos, Z., Arnold, D. L., and Kuhlmann, T. (2012). Primary progressive multiple sclerosis: part of the MS disease spectrum or separate disease entity? *Acta Neuropathol.* 123, 627–638. doi: 10.1007/s00401-012-0953-0

Becher, B., Durell, B. G., Miga, A. V., Hickey, W. F., and Noelle, R. J. (2001). The clinical course of experimental autoimmune encephalomyelitis and inflammation is controlled by the expression of CD40 within the central nervous system. *J. Exp. Med.* 193, 967–974. doi: 10.1084/jem.193.8.967

Bessesen, D. H., Richards, C. L., Etienne, J., Goers, J. W., and Eckel, R. H. (1993). Spinal cord of the rat contains more lipoprotein lipase than other brain regions. *J. Lipid Res.* 34, 229–238.

Birgbauer, E., Rao, T. S., and Webb, M. (2004). Lysolecithin induces demyelination *in vitro* in a cerebellar slice culture system. *J. Neurosci. Res.* 78, 157–166. doi: 10.1002/jnr.20248

Blasi, E., Barluzzi, R., Bocchini, V., Mazzolla, R., and Bistoni, F. (1990). Immortalization of murine microglial cells by a v-raf/v-myc carrying retrovirus. *J. Neuroimmunol.* 27, 229–237. doi: 10.1016/0165-5728(90)90073-v

Bocchini, V., Mazzolla, R., Barluzzi, R., Blasi, E., Sick, P., and Kettenmann, H. (1992). An immortalized cell line expresses properties of activated microglial cells. *J. Neurosci. Res.* 31, 616–621. doi: 10.1002/jnr.490310405

Boyle, K. E., Friedman, J. E., Janssen, R. C., Underkofler, C., Houmard, J. A., and Rasouli, N. (2017). Metabolic inflexibility with obesity and the effects of fenofibrate on skeletal muscle fatty acid oxidation. *Horm. Metab. Res.* 49, 50–57. doi: 10.1055/s-0042-111517

Cantoni, C., Bollman, B., Licastro, D., Xie, M., Mikesell, R., Schmidt, R., et al. (2015). TREM2 regulates microglial cell activation in response to demyelination *in vivo*. *Acta Neuropathol.* 129, 429–447. doi: 10.1007/s00401-015-1388-1

Chaturvedi, R., Asim, M., Lewis, N. D., Algood, H. M., Cover, T. L., Kim, P. Y., et al. (2007). L-arginine availability regulates inducible nitric oxide synthase-dependent host defense against Helicobacter pylori. *Infect. Immun.* 75, 4305–4315. doi: 10.1128/iai.00578-07

Cherry, J. D., Olschowka, J. A., and O'Banion, M. K. (2014). Neuroinflammation and M2 microglia: the good, the bad, and the inflamed. *J. Neuroinflammation* 11:98. doi: 10.1186/1742-2094-11-98

Eckel, R. H., and Robbins, R. J. (1984). Lipoprotein lipase is produced, regulated and functional in rat brain. *Proc. Natl. Acad. Sci. U S A* 81, 7604–7607. doi: 10.1073/pnas.81.23.7604

Fernandez-Castaneda, A., Arandjelovic, S., Stiles, T. L., Schlobach, R. K., Mowen, K. A., Gonias, S. L., et al. (2013). Identification of the low density lipoprotein (LDL) receptor-related protein-1 interactome in central nervous system myelin suggests a role in the clearance of necrotic cell debris. *J. Biol. Chem.* 288, 4538–4548. doi: 10.1074/jbc.M112.384693

Frausto, R. F., Crocker, S. J., Eam, B., Whitmire, J. K., and Whitton, J. L. (2007). Myelin oligodendrocyte glycoprotein peptide-induced experimental allergic encephalomyelitis and T cell responses are unaffected by immunoproteasome deficiency. *J. Neuroimmunol.* 192, 124–133. doi: 10.1016/j.jneuroim.2007.09.024

Fu, W., Vukojevic, V., Patel, A., Soudy, R., MacTavish, D., Westaway, D., et al. (2017). Role of microglial amylin receptors in mediating β amyloid (Aβ)-induced inflammation. *J. Neuroinflammation* 14:199. doi: 10.1186/s12974-017-0972-9

Goldberg, I. J., Soprano, D. R., Wyatt, M. L., Vanni, T. M., Kirchgessner, T. G., and Schotz, M. C. (1989). Localization of lipoprotein lipase mRNA in selected rat tissues. *J. Lipid Res.* 30, 1569–1577.

Goudonnet, H., Magdalou, J., Mounie, J., Naoumi, A., Viriot, M. L., Escousse, A., et al. (1990). Differential action of thyroid hormones and chemically related compounds on the activity of UDP-glucuronosyltransferases and cytochrome P-450 isozymes in rat liver. *Biochim. Biophys. Acta* 1035, 12–19. doi: 10.1016/0304-4165(90)90167-u

Harlow, D. E., Saul, K. E., Komuro, H., and Macklin, W. B. (2015). Myelin proteolipid protein complexes with αv integrin and AMPA receptors *in vivo* and regulates AMPA-dependent oligodendrocyte progenitor cell migration through the modulation of cell-surface GluR2 expression. *J. Neurosci.* 35, 12018–12032. doi: 10.1523/JNEUROSCI.5151-14.2015

Haschemi, A., Kosma, P., Gille, L., Evans, C. R., Burant, C. F., Starkl, P., et al. (2012). The sedoheptulose kinase CARKL directs macrophage polarization through control of glucose metabolism. *Cell Metab.* 15, 813–826. doi: 10.1016/j.cmet.2012.04.023

Heppner, F. L., Greter, M., Marino, D., Falsig, J., Raivich, G., Hövelmeyer, N., et al. (2005). Experimental autoimmune encephalomyelitis repressed by microglial paralysis. *Nat. Med.* 11, 146–152. doi: 10.1038/nm1177

Hinojosa, A. E., Garcia-Bueno, B., Leza, J. C., and Madrigal, J. L. (2011). CCL2/MCP-1 modulation of microglial activation and proliferation. *J. Neuroinflammation* 8:77. doi: 10.1186/1742-2094-8-77

Hlavica, M., Delparente, A., Good, A., Good, N., Plattner, P. S., Seyedsadr, M. S., et al. (2017). Intrathecal insulin-like growth factor 1 but not insulin enhances myelin repair in young and aged rats. *Neurosci. Lett.* 648, 41–46. doi: 10.1016/j.neulet.2017.03.047

Ho, P. P., Kanter, J. L., Johnson, A. M., Srinagesh, H. K., Chang, E. J., Purdy, T. M., et al. (2012). Identification of naturally occurring fatty acids of the myelin sheath that resolve neuroinflammation. *Sci. Transl. Med.* 4:137ra73. doi: 10.1126/scitranslmed.3003831

Hohlfeld, R. (2008). Neurotrophic cross-talk between the nervous and immune systems: relevance for repair strategies in multiple sclerosis? *J. Neurol. Sci.* 265, 93–96. doi: 10.1016/j.jns.2007.03.012

Hong, S., and Stevens, B. (2016). Microglia: phagocytosing to clear, sculpt, and eliminate. *Dev. Cell* 38, 126–128. doi: 10.1016/j.devcel.2016.07.006

Huey, P. U., Waugh, K. C., Etienne, J., and Eckel, R. H. (2002). Lipoprotein lipase is expressed in rat sciatic nerve and regulated in response to crush injury. *J. Lipid Res.* 43, 19–25.

Jensen, D. R., Knaub, L. A., Konhilas, J. P., Leinwand, L. A., MacLean, P. S., and Eckel, R. H. (2008). Increased thermoregulation in cold-exposed transgenic mice overexpressing lipoprotein lipase in skeletal muscle: an avian phenotype? *J. Lipid Res.* 49, 870–879. doi: 10.1194/jlr.M700519-jlr200

Johns, T. G., Kerlero de Rosbo, N., Menon, K. K., Abo, S., Gonzales, M. F., and Bernard, C. C. (1995). Myelin oligodendrocyte glycoprotein induces a demyelinating encephalomyelitis resembling multiple sclerosis. *J. Immunol.* 154, 5536–5541.

Keren-Shaul, H., Spinrad, A., Weiner, A., Matcovitch-Natan, O., Dvir-Szternfeld, R., Ulland, T. K., et al. (2017). A unique microglia type associated with restricting development of Alzheimer's disease. *Cell* 169, 1276.e17–1290.e17. doi: 10.1016/j.cell.2017.05.018

Lampron, A., Larochelle, A., Laflamme, N., Préfontaine, P., Plante, M. M., Sanchez, M. G., et al. (2015). Inefficient clearance of myelin debris by microglia impairs remyelinating processes. *J. Exp. Med.* 212, 481–495. doi: 10.1084/jem.20141656

Lee, J. K., and Tansey, M. G. (2013). Microglia isolation from adult mouse brain. *Methods Mol. Biol.* 1041, 17–23. doi: 10.1007/978-1-62703-520-0_3

Libby, A. E., Wang, H., Mittal, R., Sungelo, M., Potma, E., and Eckel, R. H. (2015). Lipoprotein lipase is an important modulator of lipid uptake and storage in hypothalamic neurons. *Biochem. Biophys. Res. Commun.* 465, 287–292. doi: 10.1016/j.bbrc.2015.08.026

Ma, Y., Bao, J., Zhao, X., Shen, H., Lv, J., Ma, S., et al. (2013). Activated cyclin-dependent kinase 5 promotes microglial phagocytosis of fibrillar β-amyloid by up-regulating lipoprotein lipase expression. *Mol. Cell. Proteomics* 12, 2833–2844. doi: 10.1074/mcp.M112.026864

Makoveichuk, E., Castel, S., Vilaró, S., and Olivecrona, G. (2004). Lipoprotein lipase-dependent binding and uptake of low density lipoproteins by THP-1 monocytes and macrophages: possible involvement of lipid rafts. *Biochim. Biophys. Acta* 1686, 37–49. doi: 10.1016/j.bbalip.2004.08.015

Mallon, B. S., Shick, H. E., Kidd, G. J., and Macklin, W. B. (2002). Proteolipid promoter activity distinguishes two populations of NG2-positive cells throughout neonatal cortical development. *J. Neurosci.* 22, 876–885.

Mendel, I., Kerlero de Rosbo, N., and Ben-Nun, A. (1995). A myelin oligodendrocyte glycoprotein peptide induces typical chronic experimental autoimmune encephalomyelitis in H-2b mice: fine specificity and T cell receptor V β expression of encephalitogenic T cells. *Eur. J. Immunol.* 25, 1951–1959. doi: 10.1002/eji.1830250723

Mills, C. D., Kincaid, K., Alt, J. M., Heilman, M. J., and Hill, A. M. (2000). M-1/M-2 macrophages and the Th1/Th2 paradigm. *J. Immunol.* 164, 6166–6173. doi: 10.4049/jimmunol.164.12.6166

Morris, S. M. Jr. (2007). Arginine metabolism: boundaries of our knowledge. *J. Nutr.* 137, 1602S–1609S. doi: 10.1093/jn/137.6.1602s

Munder, M. (2009). Arginase: an emerging key player in the mammalian immune system. *Br. J. Pharmacol.* 158, 638–651. doi: 10.1111/j.1476-5381.2009.00291.x

Nemkov, T., D'Alessandro, A., and Hansen, K. C. (2015). Three-minute method for amino acid analysis by UHPLC and high-resolution quadrupole orbitrap mass spectrometry. *Amino Acids* 47, 2345–2357. doi: 10.1007/s00726-015-2019-9

Nemkov, T., Hansen, K. C., and D'Alessandro, A. (2017). A three-minute method for high-throughput quantitative metabolomics and quantitative tracing experiments of central carbon and nitrogen pathways. *Rapid Commun. Mass Spectrom.* 31, 663–673. doi: 10.1002/rcm.7834

Neumann, H., Kotter, M. R., and Franklin, R. J. (2009). Debris clearance by microglia: an essential link between degeneration and regeneration. *Brain* 132, 288–295. doi: 10.1093/brain/awn109

Nunes, A. F., and Sousa, M. M. (2008). Transthyretin knockout mouse nerves have increased lipoprotein lipase and sphingolipid content following crush. *Neurosci. Lett.* 446, 83–87. doi: 10.1016/j.neulet.2008.09.054

Odegaard, J. I., and Chawla, A. (2011). Alternative macrophage activation and metabolism. *Annu. Rev. Pathol.* 6, 275–297. doi: 10.1146/annurev-pathol-011110-130138

Okuda, Y., Okuda, M., and Bernard, C. C. (2002). Gender does not influence the susceptibility of C57BL/6 mice to develop chronic experimental autoimmune encephalomyelitis induced by myelin oligodendrocyte glycoprotein. *Immunol. Lett.* 81, 25–29. doi: 10.1016/s0165-2478(01)00339-x

Olah, M., Amor, S., Brouwer, N., Vinet, J., Eggen, B., Biber, K., et al. (2012). Identification of a microglia phenotype supportive of remyelination. *Glia* 60, 306–321. doi: 10.1002/glia.21266

Orihuela, R., McPherson, C. A., and Harry, G. J. (2016). Microglial M1/M2 polarization and metabolic states. *Br. J. Pharmacol.* 173, 649–665. doi: 10.1111/bph.13139

Ransohoff, R. M., and El Khoury, J. (2015). Microglia in health and disease. *Cold Spring Harb. Perspect. Biol.* 8:a020560. doi: 10.1101/cshperspect.a020560

Ransohoff, R. M., and Perry, V. H. (2009). Microglial physiology: unique stimuli, specialized responses. *Annu. Rev. Immunol.* 27, 119–145. doi: 10.1146/annurev.immunol.021908.132528

Rodríguez-Prados, J. C., Través, P. G., Cuenca, J., Rico, D., Aragonés, J., Martín-Sanz, P., et al. (2010). Substrate fate in activated macrophages: a comparison between innate, classic and alternative activation. *J. Immunol.* 185, 605–614. doi: 10.4049/jimmunol.0901698

Russo, V. C., Gluckman, P. D., Feldman, E. L., and Werther, G. A. (2005). The insulin-like growth factor system and its pleiotropic functions in brain. *Endocr. Rev.* 26, 916–943. doi: 10.1210/er.2004-0024

Schwartz, M., and Moalem, G. (2001). Beneficial immune activity after CNS injury: prospects for vaccination. *J. Neuroimmunol.* 113, 185–192. doi: 10.1016/s0165-5728(00)00447-1

Shahbazi, M., Abdolmohammadi, R., Ebadi, H., and Farazmandfar, T. (2017). Novel functional polymorphism in IGF-1 gene associated with multiple sclerosis: a new insight to MS. *Mult. Scler. Relat. Disord.* 13, 33–37. doi: 10.1016/j.msard.2017.02.002

Sheridan, G. K., and Dev, K. K. (2012). S1P1 receptor subtype inhibits demyelination and regulates chemokine release in cerebellar slice cultures. *Glia* 60, 382–392. doi: 10.1002/glia.22272

Suh, H. S., Zhao, M. L., Derico, L., Choi, N., and Lee, S. C. (2013). Insulin-like growth factor 1 and 2 (IGF1, IGF2) expression in human microglia: differential regulation by inflammatory mediators. *J. Neuroinflammation* 10:37. doi: 10.1186/1742-2094-10-37

Sun, D., Yu, Z., Fang, X., Liu, M., Pu, Y., Shao, Q., et al. (2017). LncRNA GAS5 inhibits microglial M2 polarization and exacerbates demyelination. *EMBO Rep.* 18, 1801–1816. doi: 10.15252/embr.2016 43668

Varin, A., and Gordon, S. (2009). Alternative activation of macrophages: immune function and cellular biology. *Immunobiology* 214, 630–641. doi: 10.1016/j.imbio.2008.11.009

Voss, E. V., Skuljec, J., Gudi, V., Skripuletz, T., Pul, R., Trebst, C., et al. (2012). Characterisation of microglia during de- and remyelination: can they create a repair promoting environment? *Neurobiol. Dis.* 45, 519–528. doi: 10.1016/j.nbd.2011.09.008

Wen, W., Sanelli, T., Ge, W., Strong, W., and Strong, M. J. (2006). Activated microglial supernatant induced motor neuron cytotoxicity is associated with upregulation of the TNFR1 receptor. *Neurosci. Res.* 55, 87–95. doi: 10.1016/j.neures.2006.02.004

Zhu, X., Wei, D., Chen, O., Zhang, Z., Xue, J., Huang, S., et al. (2016). Upregulation of CCL3/MIP-1α regulated by MAPKs and NF-κB mediates microglial inflammatory response in LPS-induced brain injury. *Acta Neurobiol. Exp.* 76, 304–317. doi: 10.21307/ane-2017-029

10

CHL1 is Expressed and Functions as a Malignancy Promoter in Glioma Cells

Zhai Yang[1†], Qing Xie[1†], Cheng-Liang Hu[1], Qiong Jiang[1], Hui-Fan Shen[1], Melitta Schachner[1,2] and Wei-Jiang Zhao[1*]*

[1]Center for Neuroscience, Shantou University Medical College, Shantou, China, [2]Keck Center for Collaborative Neuroscience and Department of Cell Biology and Neuroscience, Rutgers University, Piscataway, NJ, United States

***Correspondence:**
Melitta Schachner
schachner@stu.edu.cn
Wei-Jiang Zhao
neuromancn@aliyun.com

†These authors have contributed equally to this work.

The cell adhesion molecule with homology to L1CAM (close homolog of L1) (CHL1) is a member of the cell adhesion molecule L1 (L1CAM) gene family. Although CHL1 expression and function have been reported in several tumors, the roles of CHL1 in the development of glioma remain unclear. In the present study, we investigated the effects of CHL1 on proliferation indexes and activation of Akt1 and Erk signaling by siRNA in U-87 MG human glioblastoma and human U251 and SHG-44 glioma cells. We found that siRNA targeting CHL1 significantly down-regulated the expression of CHL1 mRNA and protein accompanied by reduced cell proliferation and transmigration invasion in all three cell lines. Down-regulating CHL1 expression also reduced cell survival, as measured by the Bax/Bcl-2 ratio, and increased activation of caspase-3. In subcutaneous U-87 MG cell xenograft tumors in nude mice, intratumoral administration of siRNA targeting CHL1 treatment significantly down-regulated CHL1 expression *in vivo*, accompanied by increased levels of activated caspase-3. Our combined results confirmed for the first time that in contrast to findings about CHL1 in most other cancer types, CHL1 functions in promoting cell proliferation, metastasis and migration in human glioma cells both *in vitro* and *in vivo*. These results indicate that CHL1 is a therapeutic target in the clinical management of glioma/glioblastoma.

Keywords: glioma, glioma cell lines, close homolog of L1 cell adhesion molecule (CHL1), siRNA, malignancy

INTRODUCTION

Gliomas are a set of highly invasive glial cell-derived tumors that originate in the central nervous system, and they account for 40% to 50% of all intracranial tumors (Gabriel et al., 2014). Despite advancements in surgery, chemotherapy and radiation oncology technology, the average survival time is 9.7 months. In addition, patients with lowly differentiated astroglioma and glioblastoma exhibit high recurrence, high mortality and low cure rates, and the 5-year survival rate remains less than 5% (Sathornsumetee et al., 2007; Nakazato, 2008; Jemal et al., 2010). All these clinical data suggest that the development of glioma is a multifactor-based process in which a series of molecules are involved.

We previously reported that cell adhesion molecule L1 (L1CAM) is involved in regulating tumor progression and invasion under the modulation of Neuregulin-1 (Nrg1), suggesting the roles of L1 family members in glioma development (Zhao and Schachner, 2013). The close homolog of cell adhesion molecule L1 (CHL1) belongs to the transmembrane adhesion molecule of the immunoglobulin superfamily and exhibits biological functions similar to that of L1. CHL1 contains an N-terminal signal sequence, six immunoglobulin (Ig)-like domains, five fibronectin type III repeat (FN III) sequences, a transmembrane domain and a conserved intracellular domain containing the sequence that functions in the cell skeleton protein anykrin recognition sequence (FIGAY; Holm et al., 1996; Zhang et al., 1998; Maness and Schachner, 2007). CHL1 was identical to melanoma cell adhesion molecule (MCAM), which was previously reported as one of the extravillous trophoblasts (EVT) markers (Higuchi et al., 2003). CHL1 is located within or near the uveal melanoma susceptibility locus UVM2 at 3p25 (Tschentscher et al., 2003). In addition, CHL1 interacts genetically with both CTF7/ECO1 and CTF18/CHL12 to modulate sister-chromatid cohesion (Skibbens, 2004).

The cleavage and release of the CHL1 extracellular domain initiates autocrine signaling and reduces cellular adhesion to promote cell motility (Katic et al., 2014). In the central nervous system, CHL1 promotes Purkinje and granule cell survival and granule cell migration during cerebellar development (Jakovcevski et al., 2009). In *Caenorhabditis elegans*, the chl-1 gene is required for normal development and fertility, whereas CHL1 mutations can lead to lineage-independent cell proliferation defects (Chung et al., 2011). CHL1 dysfunction has been implicated in abnormal thalamocortical circuitry, schizophrenia and autism (Morag et al., 2011). Mutations in the coding region of CHL1 are involved in the etiology of schizophrenia in both Chinese and Japanese populations (Sakurai et al., 2002; Chen et al., 2005). Patients with heterozygous deletion of CHL1 gene can suffer cognitive impairment (Tassano et al., 2014). Research on the roles of CHL1 in tumorigenesis has gradually attracted attention (He et al., 2013). Overexpression of CHL1 was also observed in serous epithelial ovarian cancers (EOCs; Manderson et al., 2009). However, single-nucleotide polymorphism (SNP)-mass array demonstrated the absence and down-regulation of CHL1 expression in primary esophageal squamous cell carcinoma (ESCC) tumors and ESCC cell lines (Qin et al., 2008). Down-regulation/silencing of CHL1 is present in a majority of primary tumors, and its up-regulation is associated with invasive/metastatic growth. In one study, frequent down-regulation of CHL1 was detected in 11 types of cancer, mainly including breast, kidney, colon, thyroid, and stomach. In contrast, only five types (lung, ovary, uterus, liver and trachea) of cancer exhibited frequent up-regulation (Senchenko et al., 2011). These combined data indicated that the functional expression of CHL1 is tumor phenotype dependent. However, CHL1 expression in the development, metastasis, and progression of gliomas both *in vitro* and *in vivo* remains unclear.

To address this issue, we systematically investigated the roles of CHL1 in glioma behaviors mainly using siRNA targeting CHL1 in glioma cells. We evaluated the roles of CHL1 in cell proliferation, metastasis, colony formation, and AKT1 and ERK signaling in these cells. Finally, siRNA targeting CHL1 was intratumorally administered to U-87 MG cell-derived subcutaneous xenografts to further confirm the observations *in vitro*. In summary, CHL1 is vitally involved in the regulation of the occurrence and development of glioma. Targeting the role of CHL1 may represent a promising therapeutic means for the management of glioma.

MATERIALS AND METHODS

Animals

Nude mice were purchased from Beijing Vital River Animal Center (Beijing, China). All the procedures related to handling, care, and treatment in the present research were performed according to the guidelines approved by Institutional Animal Care and Use Committees (IACUC) of Shantou University Medical College.

Cell Culture and CHL1 siRNA Transfection

Normal human astro glia cell HEB cell line and human glioma cell lines U251 and SHG44 and human glioblastoma U-87 MG cell line (Chinese Type Culture Collection, Shanghai, China) were cultured in Dulbecco's modified Eagle's medium (DMEM, Thermo Scientific HyClone, Beijing, China) supplemented with 50 U/ml of a penicillin/streptomycin mixture (Solarbio Biotech Corp. Beijing, China) and 10% fetal bovine serum (Sijiqing Biotech Corp, Hangzhou, China). All cells were routinely grown in 75-cm^2 cell culture plates (Corning Inc., Corning, NY, USA) at 37°C with 5% CO_2 in a humidified atmosphere. The cells were collected in logarithmic phase for the following experiments. On the day before transfection, cells were digested by trypsin (0.25%, Solarbio Biotech Corp., Beijing, China), counted and seeded in a six-well plate at an optimal concentration. When the cells achieved 80% confluence, the medium was changed with serum-free DMEM, and cells were incubated overnight. Control siRNA or siRNA targeting CHL1 (10 nM for both; **Table 1**) complexed with Entranster^{TM}-R4000 (Cat. No. 4000-3, Engreen, Beijing, China) was transfected into three cell lines. In the vehicle control group, cells were treated with the same volume of transfection reagent. The efficiency of CHL1 siRNA was confirmed by RT-PCR and Western blot.

Cell Viability Assay

U251, SHG44 and U-87 MG cells were seeded onto the 96-well plate at 5000 cells/well in 200 μl of DMEM supplemented with 10% FBS. These cells were transfected with control siRNA or siRNA targeting CHL1 or treated with transfection reagent as mentioned above. Transfected and non-transfected cells were incubated under the conditions of 5% CO_2, saturated humidity, and 37°C for 24, 48, 72, and 96 h. Then, 20 μl

TABLE 1 | Sequences for random control siRNA and siRNAs against CHL1.

siRNA	Sequence	
	Sense (5′-3′)	Anti-sense (5′-3′)
Random control	UUCUCCGAACG UGUCACGUtt	ACGUGACACGUU CGGAGAAtt
CHL1	GGAGCUAAUUU GACCAUAUtt	AUAUGGUCAAAUU AGCUCCtt

of 5 mg/ml MTT (Beyotime, Jiangsu, China) was added to each well, and cells were further cultured for 4 h. Then, the culture medium was removed, and 150 μl of dimethyl sulfoxide (DMSO, Sigma) was added. The optical density was measured at 570 nm using a multiwell spectrophotometer (Infinite M1000, Tecan, Switzerland). Cell growth curves were plotted using the average absorbance at 570 nm from triplicate samples of three independent experiments.

Colony Formation Assay

Forty-eight hours after transfection, U251, SHG44 and U-87 MG cells were seeded onto six-well plates at a density of 500 cells/well in triplicate. After 14 days of culture, cells were fixed with methanol, stained with 0.5% crystal violet, and visualized under a phase-contrast light microscope (Olympus, IX51, Japan). Cells were then lysed in 1% SDS, and the colony formation was indexed by the optical density measured at 564 nm using a multiwell spectrophotometer (Infinite M1000, Tecan, Switzerland).

Cell Senescence Assay

CHL1 siRNA was tested in U251, SHG44 and U-87 MG cells to assess the effect of CHL1 on cell senescence. Cells (1×10^5 cells/well) in culture medium were allowed to adhere overnight to 24-well plates. When 80% confluence was achieved, the medium was aspirated and replaced with fresh medium containing RNA-siRNA-mate complexes (10 nM of either the control or CHL1 siRNA per well). The cells were further cultured for 48 h. Cells treated with the same volume of siRNA-mate were used as the vehicle control. After 48 h of treatment, the cells were fixed. β-galactosidase/X-Gal complex was added to each well and incubated overnight at 37°C according to the manufacturer's protocol (Beyotime, Jiangsu, China). Cell senescence was indexed by the activation of β-galactosidase reflected by the development of deep blue color of X-Gal (Dimri et al., 1995). Photographs were obtained from at least five random bright-field areas. The percentage of deep blue-stained cells was counted to indicate the senescence response of cells to CHL1 down-regulation.

Transwell Migration Assay

Given that L1 potentiates the migration of glioma cells, we hypothesized that CHL1 may also possess similar functions. We then tested the potential role of CHL1 for inducing glioma cell migration using the Transwell migration assay (Zhao and Schachner, 2013). For the Transwell migration assay, U-87 MG, SHG44, and U251 cells were pretreated individually with the vehicle control, CHL1 siRNA, and random control siRNA in DMEM for 48 h. The culture medium was then aspirated, and cells were resuspended in DMEM and seeded onto the upper chamber (1×10^5 per well) of each Transwell insert consisting of a filter (Becton Dickinson Labware, Franklin Lakes, NJ, USA) with 8-μm pores. The underside of the filter was pretreated with 100 μg/mL fibronectin in PBS (Millipore) to ensure the attachment of the migrated cells to this side of the filter. The lower chamber was loaded with 500 μl of DMEM only. At 18 h after plating, the cells that had failed to move to the underside of the filter were removed using a cotton-tipped applicator. The cells retained on the underside of the filter were rinsed 3× with PBS and fixed in 4% paraformaldehyde in PBS. The migrated cells were stained with 0.1% crystal violet in acetic acid to determine their morphology. The crystal violet staining was evaluated at 200× magnification with bright field microscopy. The number of cells in each field was counted using Image Tool II software.

Xenograft Studies

For subcutaneous implantation, 10 4-week-old female BALB/c nude mice (Vital River, Beijing, China) were randomly divided into control siRNA group (n = 5) and siRNA targeting CHL1 group (n = 5). Mice were anesthetized with 100 $mg \cdot kg^{-1}$ ketamine, and 5×10^5 U-87 MG cells were injected into the right flank near the upper extremity. After 4 weeks, tumor length and width were measured with calipers in cephalad-to-caudad and left-to-right dimensions, and measurements continued at one-day intervals. Tumor volume was calculated each day using the formula: volume = length × width^2 × 0.5 and expressed in mm^3. When the tumor volume reached approximately 150 mm^3, control siRNA or CHL1-siRNA complexed with Entranster^{TM}-*in vivo* were intratumorally injected at 2 mg/kg for the 1st time and at 4 mg/kg 7 days after the 1st injection (EngreenBiosystem Co. Ltd., Beijing, China) was undertaken. The second intratumoral injection was performed on the 7th day after the 1st intratumoral injection. After the 16th day of measurements, mice were anesthetized and euthanized by decapitation to remove the tumors.

RNA Isolation and Reverse Transcriptase PCR (RT-PCR) Analysis

Total RNA from glioma cells was extracted using RNAiso extraction kit (Tiangen, Beijing, China) according to the manufacturer's protocol, and reverse transcription was performed using StarScrip II First-strand cDNA Synthesis Mix (GenStar, Beijing, China). A 35-cycle PCR using the following conditions was performed (except for the GAPDH primer set): 94°C for 2 min, 94°C for 30 s, 59°C for 20 s, 74°C for

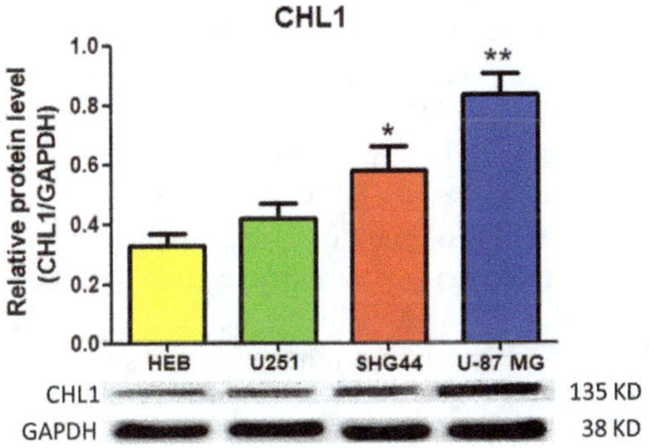

FIGURE 1 | Western blot analysis of the protein levels of CHL1 detected in normal human glial HEB cells and 3 glioma/glioblastoma cell lines. CHL1 was weakly expressed in normal human HEB glial cells. Its levels in all the 3 glioma/glioblastoma cells were higher than that in normal human HEB glial cells, with the statistical significance detected in SHG44 cells (*$p < 0.05$ vs. HEB cells) and U-87 MG cells (**$p < 0.01$ vs. HEB cells). n = 3 for each group. Student's t-test for independent samples was used.

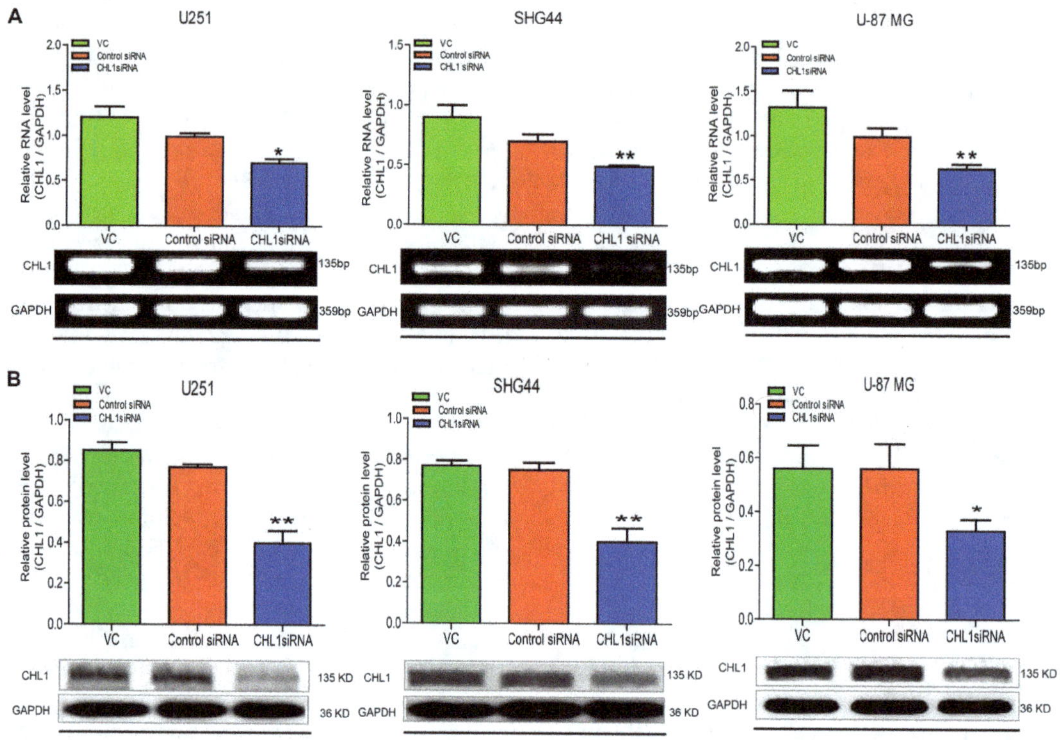

FIGURE 2 | Treatment of siRNA targeting CHL1 in three human glioma cell lines. Total RNA was isolated from U251, SHG44 and U-87 MG cells treated with vehicle control (vc), control siRNA (control siRNA) or siRNA targeting CHL1 (CHL1 siRNA). RT-PCR and Western blot analysis were then used to measure both relative mRNA and protein levels of CHL1. **(A)** RT-PCR analysis of the mRNA levels of CHL1 in U251, SHG44 and U-87 MG cells treated with vehicle control (vc), control siRNA and siRNA targeting CHL1, and **(B)** Western blot analysis of the protein levels of CHL1 detected in U251, SHG44 and U-87 MG cells treated with vehicle control (vc), control siRNA and siRNA targeting CHL1. Data are presented as means ± standard error of the mean (SEM) ($n = 3$, *$p < 0.05$; **$p < 0.01$, independent Student's *t*-test).

40 s and a final extension step at 72°C for 10 min. For GAPDH cDNA amplification, 28 reaction cycles were employed. We subjected 5 μl of the PCR products to gel electrophoresis using a 2.0% agarose gel (Gene Choice) containing Gelred (1:10,000; Biotium, Hayward, CA, USA). The bands were identified under UV light. The primers used for PCR for detecting the mRNA expression were listed as follows: hCHL1-forward primer: 5′-TCAAAGGAAGCCTTCGGTCC-3′ and hCHL1-reverse primer: 5′-TAGATCCAGCGTAGGCACCA-3′; GAPDH forward primer: 5′-TATAAATTGAGCCCGCAGCC-3′ and GAPDH reverse primer: 5′-TTCCCGTTCTCAGCCTTGAC-3′. The signal intensity was quantified using Image Tool II software via average densitometry multiplied by the number of pixel (National Institutes of Health, Bethesda, MD, USA). The relative mRNA level of a protein was indexed by its signal intensity to that of GAPDH.

Western Blot Analysis

The cells and tumor samples were lysed in a RIPA buffer mixture (Solarbio Biotech, Beijing, China) supplemented with PMSF (1:200, Solarbio Biotech). The cell lysates were centrifuged at 14,000× *g* for 15 min at 4°C, and the supernatants were collected for Western blot analysis (Zhao W. and Ren, 2011; Zhao W. J. and Ren, 2011). Equivalent quantities of the lysates from the cells were heated at 95°C in 20% sample loading buffer (0.125 M Tris-HCl, pH 6.8, 20% glycerol, 10% SDS, 0.1% bromophenol blue and 5% β-mercaptoethanol), resolved using an 8% SDS-PAGE and electroblotted onto polyvinylidenedifluoride membranes (PVDF, Millipore, Billerica, MA, USA). Non-specific protein binding sites were blocked with 5% BSA diluted in Tris-buffered saline (TBS, pH 7.3) buffer containing 0.05% Tween-20 (TBST). Membranes were incubated with a rat anti-human CHL1 antibody that specifically targets the extracellular domain of CHL1 (1:500, R&D Systems, cat. no. MAB2126, Minneapolis, MN, USA), mouse monoclonal anti-Bcl-2 antibody (1:1000, cat. no. sc-7382, Santa Cruz, CA, USA), rabbit polyclonal anti-Bax antibody (1:1000, cat. no. sc-526, Santa Cruz), rabbit polyclonal anti-PCNA antibody (1:1000, cat. no. sc-7907, Santa Cruz), rabbit polyclonal anti-caspase-3 antibody (1:1000, cat. no. sc-7148, Santa Cruz), mouse monoclonal anti phosphorylated extracellular signal regulated kinase 1/2 (anti-pErk1/2) antibody (1:1000, cat. no. sc-7383, Santa Cruz), mouse monoclonal anti-Erk1/2 antibody (1:1000, cat. no. sc-135900, Santa Cruz), mouse monoclonal anti-pAkt1 antibody (1:1000, cat. no. sc-81433, Santa Cruz), mouse monoclonal anti-Akt1 antibody (1:1000, cat. no. sc-55523, Santa Cruz) and mouse monoclonal anti-glyceraldehyde-3-phosphate dehydrogenase (GADPH) antibody (1:1000, cat. no. sc-365062, Santa Cruz) overnight at 4°C.

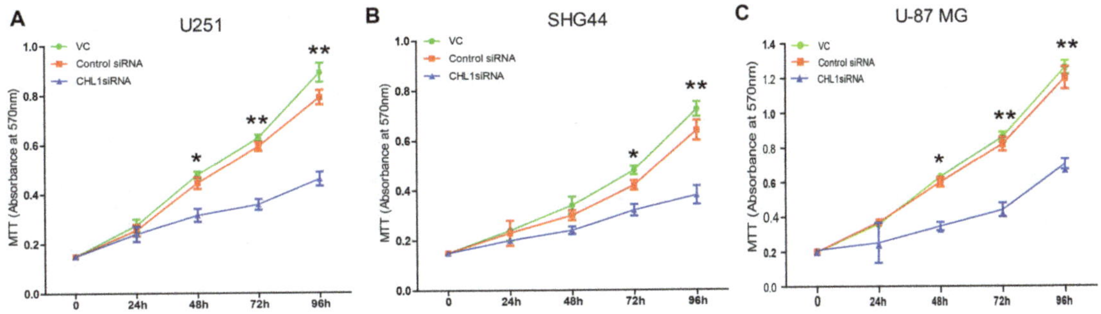

FIGURE 3 | Knockdown of CHL1 affects the proliferation and survival of U251, SHG44 and U-87 MG glioma cells. Cells were seeded on 96-well plates in triplicate, and proliferation rates were measured by MTT assay to evaluate the effect of CHL1 on the proliferation of U251, SHG44 and U-87 MG glioma/glioblastoma cells. **(A–C)** Changes of the proliferation rate in U251 **(A)**, SHG44 **(B)**, and U-87 MG **(C)** cells treated with vehicle control (vc), control siRNA (control siRNA) or siRNA targeting CHL1 (CHL1 siRNA). The data were expressed as the means ± SEM of three independent experiments (*$p < 0.05$ and **$p < 0.01$ vs. either vehicle control or control siRNA; independent Student's *t*-test).

After washing the membrane with TBST three times at room temperature (5 min each wash), the membranes were further incubated with horseradish peroxidase conjugated goat anti mouse secondary antibody (1:1000, cat. no. BA1051, Boster Biological Technology, Wuhan, China), anti rabbit secondary antibody (1:1000, cat. no. BA1055, Boster Biological Technology), or rabbit anti-rat secondary antibody (1:1000, cat. no. BA1058, Boster Biological Technology) for 1 h. Subsequently, the membranes were washed with TBST three times (5 min each wash) at room temperature. The immunoreactive bands were visualized using an enhanced chemiluminescence kit (Bio Rad Laboratories, Richmond, CA, USA) and an imaging system (Alpha Innotech, San Leandro, CA, USA). The signal intensity was quantified using Image Tool II software via average densitometry multiplied by the number of pixels (National Institutes of Health, Bethesda, MD, USA). The relative expression level of the protein under study was indicated by the ratio of its signal intensity to that of GAPDH.

Immunohistochemical Analysis

Immunohistochemical staining of paraffin sections was performed as described (Zhao et al., 2009). The human U-87 MG glioma cell xenograft tissues were cryosectioned at 8 μm thickness. Antigen retrieval was performed using 10 mM citrate buffer (pH 6.0), and endogenous peroxidase clearance was performed by incubation in 3% H_2O_2. Then, sections were blocked with 10% normal goat serum in PBS at room temperature for 30 min, and samples were subjected to incubation with the following primary antibodies: rat anti-human CHL1 antibody (1:100, cat. no. MAB2126, R&D Systems), rabbit polyclonal anti-PCNA antibody (1:200, cat. no. sc-7907, Santa Cruz), rabbit polyclonal anti-caspase-3 antibody (1:200, cat. no. sc-7148, Santa Cruz), rabbit polyclonal anti-GFAP antibody (1:500, cat. no. BA0056, Boster Biological Technology) at 4°C overnight. Bound antibody was visualized using the AEC method. Counterstaining was performed with Mayer's hematoxylin. H&E (Zhongshan Goldbridge Biotechnology Co., LTD, Beijing, China) and immunohistochemical stainings were analyzed using a Jiangnan light microscope (DN-10B, Jiangnan, Nanjing, Jiangsu).

STATISTICAL ANALYSIS

In vitro experiments were repeated at least three times using independent culture preparations. All data are presented as group mean values with standard error of the mean (SEM). Statistical analyses were performed using SPSS (Statistical Package for the Social Sciences) 10.0 software (SPSS, Chicago, IL, USA). The data from CHL1 siRNA group were compared with those from either vehicle control or control siRNA group by using Student's *t*-test for independent samples. *p*-values < 0.05 were considered statistically significant.

RESULTS

Comparison of the Protein Levels of CHL1 in 3 Glioma/Glioblastoma Cell Lines with that in Normal Human Glia HEB Cells

We first used Western blot to evaluate and compare the expression of CHL1 protein in HEB and three human glioma/glioblastoma cell lines. As demonstrated in **Figure 1**, CHL1 was weakly expressed in normal human HEB glial cells. Its levels in all the 3 glioma/glioblastoma cells were higher than that in normal human HEB glial cells, with the statistical significance detected in SHG44 cells ($p < 0.05$ vs. HEB cells) and U-87 MG cells ($p < 0.01$ vs. HEB cells). Western blot of CHL1 using samples from all the four cell lines on the uncut PVDF membrane suggests that the CHL1 antibody used in the present study is highly specific (**Supplementary Figure S1**).

Down-Regulation of CHL1 Expression Affects Glioma Cell Proliferation and Survival

We then explored whether glioma cell oncogenicity is dependent upon CHL1. We reduced CHL1 expression using Entranster^{TM}-

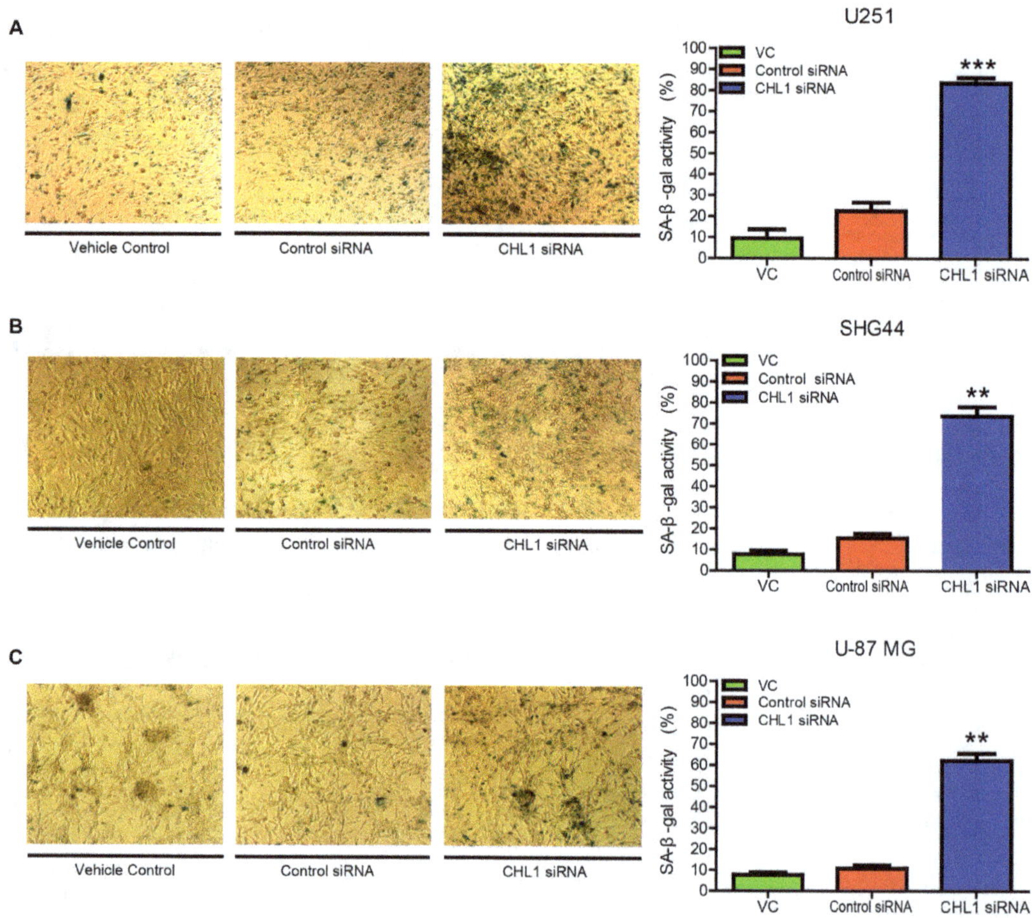

FIGURE 4 | Knockdown of CHL1 affects the senescence of glioma/glioblastoma cells *in vitro*. **(A–C)** U251 **(A)**, SHG44 **(B)** and U-87 MG **(C)** cells were seeded onto 24-well plates and treated with vehicle control, control siRNA and siRNA targeting CHL1, and senescent cells were then detected by senescence-associated β-galactosidase staining (200×). The data were expressed as the means ± SEM from four independent experiments (**$p < 0.01$; ***$p < 0.001$ vs. either vehicle control or control siRNA; Independent Student's *t*-test).

R4000-mediated CHL1-siRNA transfection inU251, SHG44 and U-87 MG glioma cells to specifically knock down endogenous CHL1 expression. CHL1 expression was significantly suppressed by siRNA targeting CHL1 mRNA and protein levels in all three cell lines investigated (**Figures 2A,B**). As seen from the MTT assay, compared with the vehicle control group and negative control group, knocking down CHL1 significantly reduced the viability of U251 cells at 48, 72 and 96 h ($p < 0.05$ vs. controls at 48 h, and $p < 0.01$ vs. controls at 72 h and 96 h; **Figure 3**). A similar pattern of cell viability inhibition in response to CHL1 knockdown was observed in both SHG44 glioma and U-87 MG glioblastoma cells, with remarkable viability inhibition observed at both 72 h and 96 h time points ($p < 0.01$ vs. controls at both time points; **Figure 3**). These results suggested that CHL1 is involved in the viability of glioma/glioblastoma cells.

Reducing CHL1 Expression Promotes Cell Senescence

To investigate whether down-regulation of CHL1 leads to cell senescence, we performed senescence staining via a β-galactosidase activity assay. After glioma cells were transfected with CHL1 siRNA or negative control siRNA, cells were fixed and stained for β-galactosidase activity using the X-Gal substrate at 37°C for 24 h. In the vehicle control with U251 cells, 10.16% of the total cells were X-Gal-positive, whereas the average percentage of X-Gal-positive cells upon treatment with negative control siRNA and CHL1 siRNA were 21.67% and 82.46%, respectively. Thus, knocking down CHL1 expression induced a significant increase in SA-β-gal-positive senescent cells compared with the vehicle control and negative control siRNA ($p < 0.001$ for both; **Figure 4A**). Similarly, SHG44 and U-87 MG cells were also sensitive to CHL1 knockdown-induced senescence. Both the CHL1 knockdown groups (74.89% and 61.27% for SHG44 and U-87 MG cells, respectively; **Figures 4B,C**) exhibited a remarkable increase in the number of SA-β-gal-positive cells compared with vehicle control (9.87% and 8.36% for SHG44 and U-87 MG cells, respectively; $p < 0.01$ vs. vehicle control for both cell lines; **Figures 4B,C**) and negative control siRNA (17.56% and 10.53% for SHG44 and U-87 MG

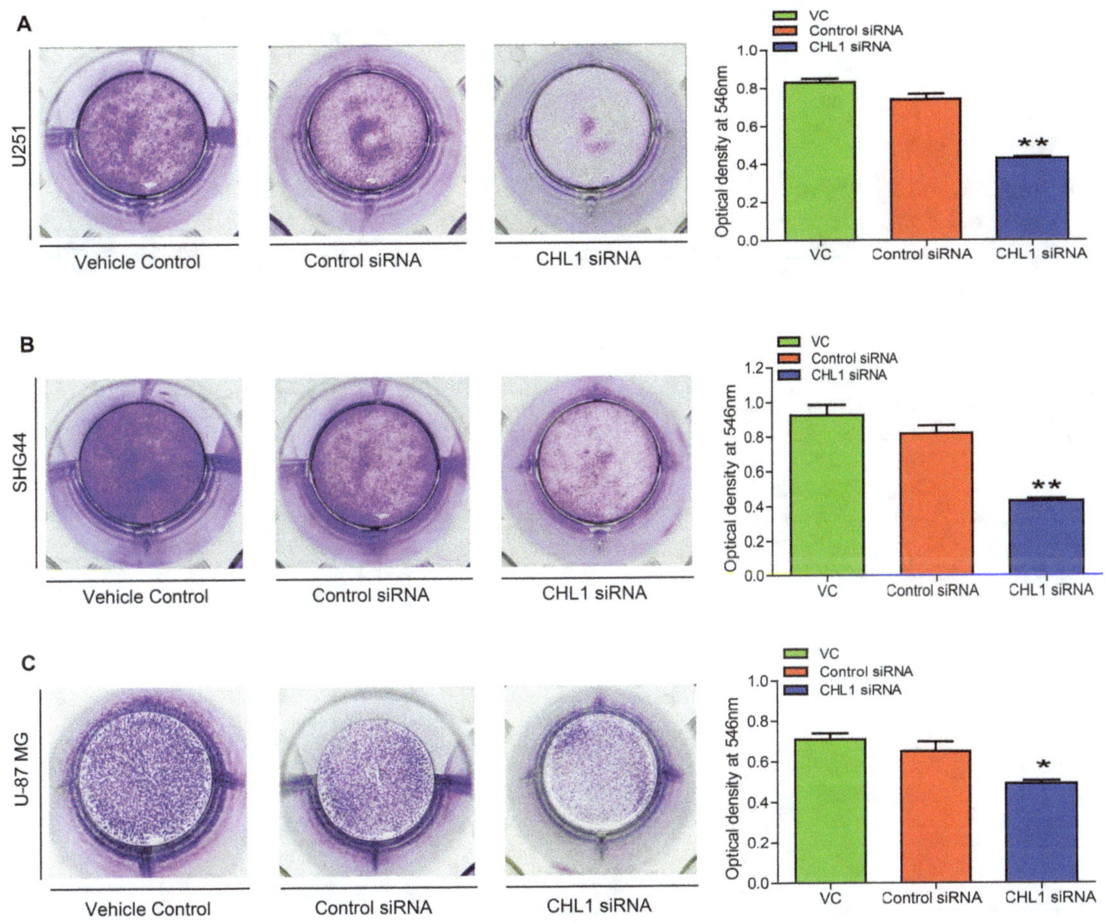

FIGURE 5 | Knockdown of CHL1 reduced colony formation capacity of glioma/glioblastoma cells *in vitro*. Cell colony was stained by crystal violet, which was then dissolved in 1% SDS and the optical density was measured at 546 nm under a microplate reader. **(A)** The colony formation assay revealed that knockdown of CHL1 reduced the colony formation of U251 cells, as was revealed by the optical density detected at 546 nm from three independent experiments (*$p < 0.05$; **$p < 0.01$ vs. either vehicle control or control siRNA). **(B,C)** Similar results were found in SHG44 **(B)** and U-87 MG **(C)** cells for the colony formation experiment. The data were expressed as the means ± SEM from 4 independent experiments (*$p < 0.05$ and **$p < 0.01$ vs. both vehicle control and control siRNA; independent Student's *t*-test).

cells, respectively; $p < 0.01$ vs. negative control for both cell lines; **Figures 4B,C**). These results indicated that knocking down CHL1 expression significantly promoted glioma cell senescence.

Knockdown of CHL1 Expression Inhibits Glioma Cell Colony Formation *in Vitro*

To test whether knocking down CHL1 expression suppresses colony formation, we performed a colony formation assay using U251, SHG44 and U-87 MG cells. After 14 days, the cells were stained with crystal violet and imaged to analyze colony formation rates. Then, crystal violet was dissolved in 1% SDS, and the optical density was measured at 546 nm using a microplate reader. The results demonstrated that CHL1 down-regulation induced a significant reduction in the colony formation rate compared with the negative control siRNA in U251 cell line ($p < 0.05$, **Figure 5A**). Similarly, SHG44 and U-87 MG cell lines with CHL1 knocked down also exhibited a remarkable decrease in the colony formation compared with the negative control siRNA group ($p < 0.01$, **Figures 5B,C**).

Knockdown of Expression Inhibits Cell Invasion *in Vitro*

The number of invading U251 cells in the CHL1-siRNA-transfected group was 100 ± 9, which was significantly lower than that in the groups of cell streated with either the vehicle control or the negative control siRNA (312 ± 11 and 276 ± 8, respectively; $p < 0.05$ for both comparisons (**Figure 6A**). In addition, the numbers of invading SHG44 and U-87 MG cells in the CHL1-siRNA-transfected groups were also significantly decreased compared with cells treated with vehicle or negative control siRNA (84 ± 6 vs. 239 ± 7 and 224 ± 4 for SHG44 cells and 54 ± 5 vs. 187 ± 6 and 205 ± 3 for U-87 MG cells, respectively; $p < 0.001$ for comparisons in SHG44 cells, and $p < 0.01$ for both comparisons in U-87 MG cells; **Figures 6B,C**).

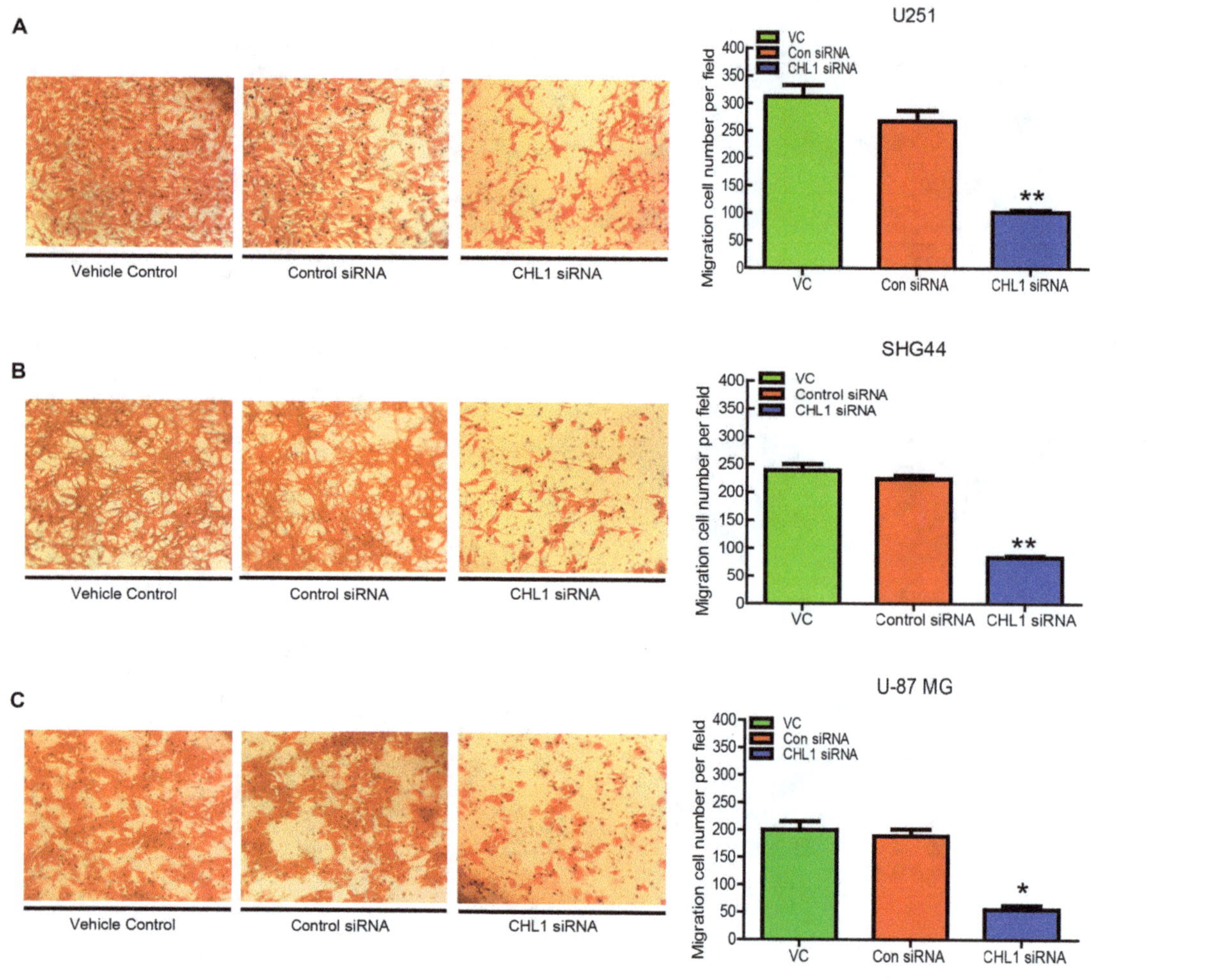

FIGURE 6 | Knockdown of CHL1 suppressed the migration of glioma/glioblastoma cells *in vitro*. Transwell migration assays were carried out using U251 **(A)**, SHG44 **(B)** and U-87 MG **(C)** cells transfected with CHL siRNA. Representative fields containing migrated cells attached to the underside of the membrane were presented. The migration ability was indexed by the relative number of migrated cells from three independent experiments. The data were expressed as the means ± SEM from 3 independent experiments (*$p < 0.05$; **$p < 0.01$ vs. both vehicle control and control siRNA (independent Student's *t*-test).

Knockdown of Expression Affects Cell Survival/Apoptosis Signaling Pathways in Glioma Cells

To confirm the results of the MTT assay, Western blot was performed to determine the expression of the apoptosis-related proteins Bcl-2, Bax, active caspase-3 and PCNA in the three human glioma cell lines. We found that 48 h post transfection, the ratio of Bax to Bcl-2 was up-regulated in three siRNA-CHL1-transfected cell lines compared with cells treated with vehicle control or the negative control siRNA ($p < 0.001$ for both comparisons in U251 cells, and $p < 0.01$ for both comparisons in regarding SHG44 and U-87 MG cells; **Figure 7A**). In addition, active caspase-3 protein levels were significantly increased in all 3 siRNA-CHL1-transfected cell lines compared with cells treated with the negative control or vehicle control ($p < 0.05$ for both comparisons in both SHG44 and U-87 MG cells, and $p < 0.01$ for both comparisons in U251 cells; **Figure 7B**). In addition, PCNA was examined as a possible proliferation indicator in glioma cells. The results demonstrated that PCNA protein levels were significantly reduced in three siRNA-CHL1-transfected cell lines compared with cells treated with vehicle control or negative control siRNA ($p < 0.05$ for both comparisons in SHG44 cells, and $P < 0.01$ for both comparisons in both U251 and U-87 MG cells; **Figure 7C**). Taken together, these results indicated that CHL1-dependent anti-apoptosis in glioma cells may be partially mediated by regulation of the death receptor signaling pathway composed of Bax, Bcl-2 and active caspase-3.

Effects of siRNA Targeting CHL1 on Main Signaling Pathways in Human Glioma Cells

Accumulating evidence indicates that the Ras/MAPK/ERK and PI3/AKT signaling pathways contribute to cell growth, proliferation, and survival (Asati et al., 2016). Western blot analysis was performed to determine pErk and pAkt

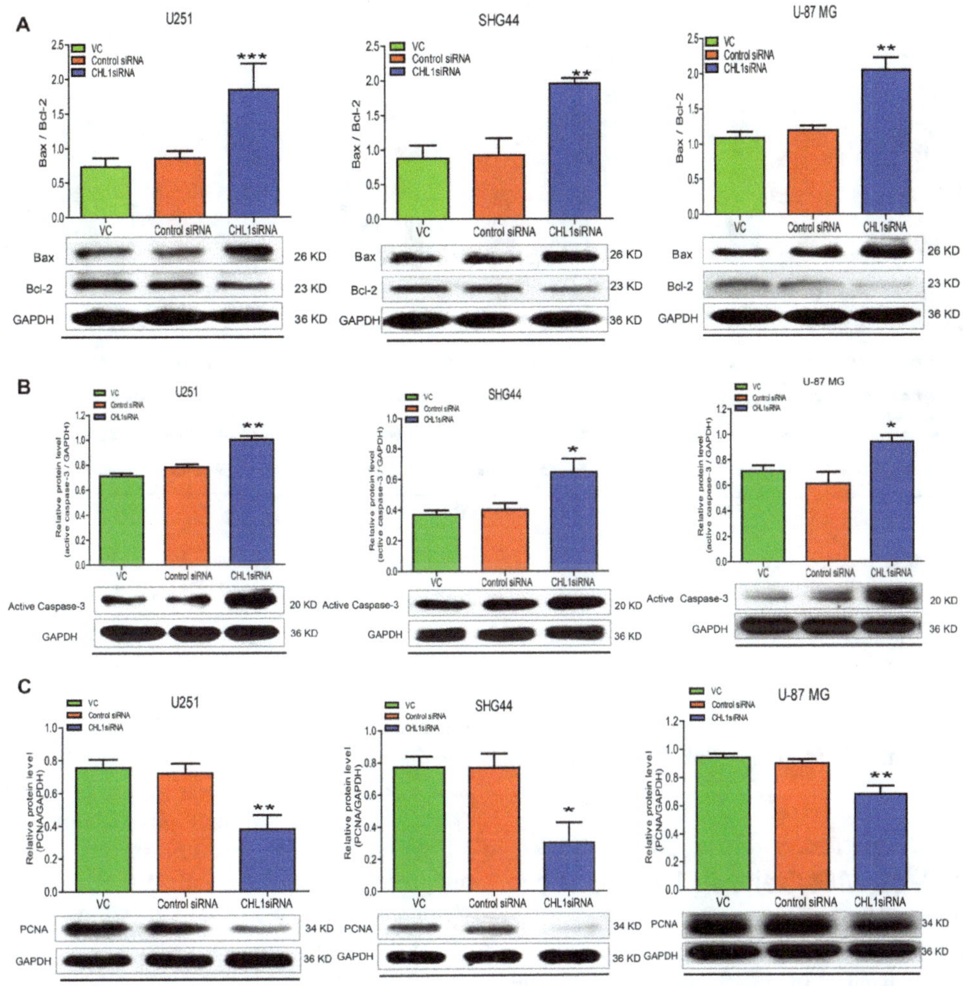

FIGURE 7 | Knockdown of CHL1 affects apoptosis signaling molecules in glioma/glioblastoma cells. U251, SHG44 and U-87 MG cells were seeded onto 48-well plates and treated with vehicle control, control siRNA and siRNA targeting CHL1, respectively. Western blot analysis was performed to determine the levels the apoptosis-related proteins, including changes of the ratio of Bax to Bcl-2 (Bax/Bcl-2) **(A)**, active caspase-3 **(B)** and PCNA **(C)** in glioma/glioblastoma cells. GAPDH was used as the loading control. The data were expressed as the means ± SEM from three independent experiments (*$p < 0.05$; **$p < 0.01$; ***$p < 0.001$ vs. either vehicle control or control siRNA; independent Student's *t*-test).

levels in U251, SHG44 and U-87 MG cells in response to CHL1 siRNA treatment for 48 h. We found that 48 h post transfection, pAkt/AKT levels were significantly decreased in three CHL1 siRNA-transfected human glioma cell lines compared with cells transfected with negative control siRNA ($p < 0.05$ for both comparisons in both SHG44 and U-87 MG cells and $p < 0.01$ for both comparisons in U251 cells (**Figure 8**). However, CHL1 knockdown exhibited no apparent effect on pErk/ERK protein levels at 48 h compared with treatment with either the vehicle control or the negative control siRNA ($p > 0.05$ for both comparisons in all three cell lines).

CHL1 Regulates U-87 MG Glioma Cell Growth *in Vivo*

We established subcutaneous xenografts of U-87 MG glioblastoma cells in nude mice. To determine whether the tumor formation of glioblastoma oncogenicity is dependent upon CHL1, we reduced CHL1 expression using CHL1-siRNA combined with Entranster^{TM}-*in vivo*, which was intratumorally injected to specifically knockdown endogenous CHL1 expression *in vivo*. Intratumoral injection of siRNA targeting CHL1 significantly reduces the fold increases in tumor volume at most time points post siRNA transfection ($p < 0.05$ vs. control siRNA at 3, 5, 7, 10, 11, 12 and 13 day points post 1st siRNA injection; $p < 0.01$ vs. control siRNA at 4, 6, and 9 day points post the 1st siRNA injection; **Figures 9A–C**). The final tumor volume in mice treated with siRNA targeting CHL1 was lower than that in mice treated with control siRNA, although no statistical significance was detected (1689.41 ± 239.46 vs. 1877.50 ± 325.11 mm^3, p = 0.2768 vs. control siRNA, **Figure 9D**). However, the total increased tumor volume in mice treated with siRNA targeting CHL1 was significantly lower than that in mice treated with control siRNA ($p < 0.05$ vs.

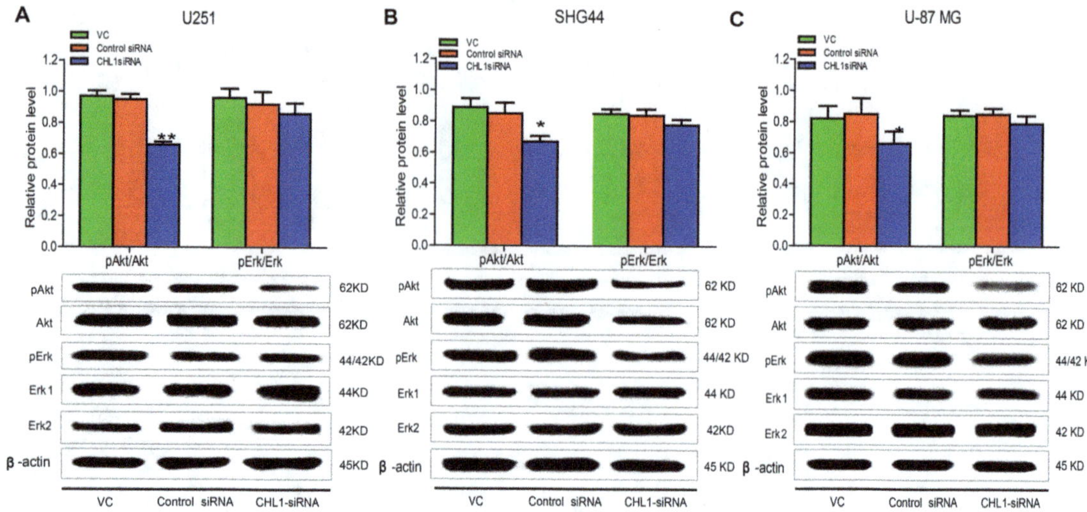

FIGURE 8 | Knockdown of CHL1 reduced the phosphorylation levels of ERK and AKT. Western blot was used to analyze the levels of pAkt and pErk in three cell lines after treatment with vehicle control, negative control and CHL1 siRNA for 48 h. pAkt and pErk protein levels in U251 **(A)**, SHG44 **(B)** and U-87 MG cells **(C)** were presented. GAPDH was used as a loading control. The data were expressed as the means ± SEM from 3 independent experiments (*$p < 0.05$ and **$p < 0.01$ vs. both vehicle control and control siRNA; independent Students *t*-test).

control siRNA, data not shown). These results demonstrate that reducing CHL1 expression can partially inhibit the growth of glioma *in vivo*.

Morphological Changes in Subcutaneous Xenografts of U-87 MG Cells in Response to CHL1 Transfection

H&E staining and immunohistochemical staining were performed to analyze changes in the xenograft tumor in response to CHL1 down-regulation. The results demonstrated that the staining intensities of CHL1, PCNA and GFAP were apparently reduced, whereas active caspase-3 intensity was apparently increased (**Figure 10**).

DISCUSSION

Despite recent progress in the treatment of melanoma and other tumors, treatment of glioma remains disappointing. Traditional therapeutic means, including radiotherapy and chemotherapy, and neurosurgical removal do not completely eradicate the tumor, and recurrence often occurs (Sathornsumetee et al., 2007). L1 is among the most intensively studied cell adhesion molecules in glioma investigation. Inhibition of L1 expression by siRNA and administration of L1 ectodomain-binding antibodies reduced the migration of glioblastoma cells *in vitro* (Izumoto et al., 1996) and disrupted glioma stem cell proliferation, leading to apoptosis (Bao et al., 2008). Reducing L1 expression *in vivo* also suppressed tumor growth and increased the survival of tumor-bearing animals (Cheng et al., 2011).

The overexpression of cell adhesion molecules, including L1, and specific inhibition of the function of these molecules suggests target-directed and promising therapeutic methods for glioma (Skibbens, 2004; Wolterink et al., 2010). The mAbs specifically reacting with L1CAM efficiently prolong survival and reduce tumor burden in a model of SKOV3ip cells in CD1 nude mice. Antibody-based functional impairment of L1 also altered the expression of cellular genes associated with apoptosis and tumor growth (Wolterink et al., 2010). Additional reports demonstrated that cell adhesion molecule VCAM-1-positive glioblastoma tumor stem cells (GTSC) possess a high rate of proliferation as measured by PCNA expression (Zarnescu et al., 2011). However, neural cell adhesion molecule (NCAM) levels were down-regulated as the malignancy of astrocytomas increased, and this effect is inversely correlated with PCNA (Huang et al., 2001). These data suggested that different cell adhesion molecule members may function differentially in the development of glioma and promoted us to explore the potential roles of CHL1 in promoting glioma cell migration, proliferation and metastasis.

Similar to L1, CHL1 also plays major roles in axonal guidance in the developing brain and mediates the maintenance and remodeling of neural circuits in the adult brain (Sakurai et al., 2002; Hitt et al., 2012). In the present work, CHL1 was lowly expressed in human astroglia cells, but was highly expressed in human glioma cell lines, suggesting its potential roles in the development of glioma. To date, paradoxical data about the roles of CHL1 under the tumorigenic conditions were reported in non-neuronal cell-derived tumors. He et al. (2013) reported that CHL1 is down-regulated in human breast cancer and is related to lower grade. Down-regulating CHL1 expression results in increased proliferation and invasion, and CHL1 deficiency also promotes tumor formation *in vivo*. Knocking down CHL1 expression by miR-10a increased colony formation activity, migration and invasion of human cervical cancer cells, whereas over-expression of CHL1 abolished the effects of

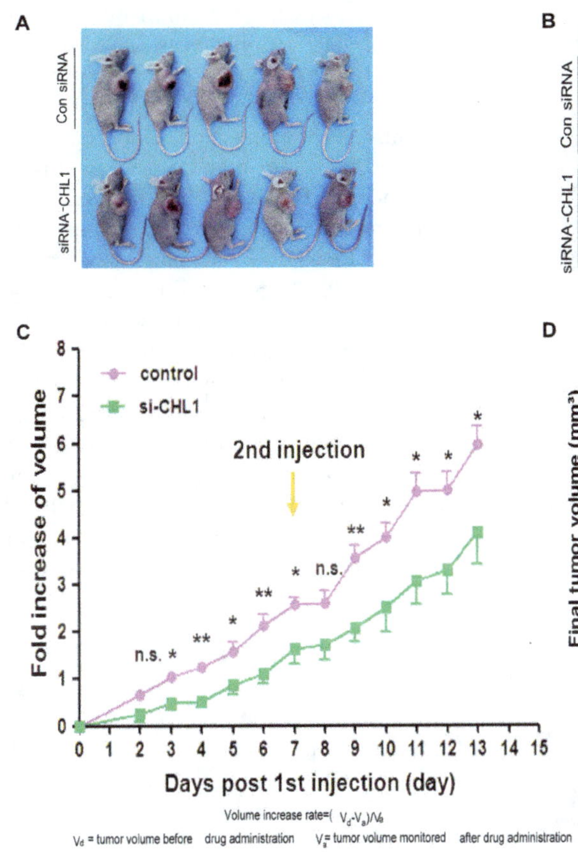

FIGURE 9 | CHL1 regulates growth of U-87 MG glioma cells *in vivo*. **(A,B)** Two weeks after the 1st intratumoral injection, all mice were killed by cervical dislocation. The *in situ* tumors and the dissected tumor tissues were photographed. **(C)** The fold increase of volume at each day points post the 1st intratumoral injection of either control siRNA or CHL1 siRNA complexed with the Entranster^{TM}-*in vivo*. **(D)** Column diagram showing the final average tumor volumes from both control siRNA and CHL1 siRNA-treated group ($n = 5$, $p = 0.2768$ vs. the control siRNA group) ($^{*}p < 0.05$; $^{**}p < 0.01$ vs. control siRNA; Independent Student's t-test).

miR-10a (Long et al., 2012). In addition, miR-182 promotes cell proliferation and invasion through direct suppression of CHL1 in papillary thyroid carcinoma (PTC; Zhu et al., 2014). miR-590-5p is up-regulated in human cervical cancer and promotes cervical cancer cell growth, cell invasion and colony formation by negatively regulating CHL1 at the posttranscriptional level (Chu et al., 2014). In addition, CHL1 negatively modulates the proliferation and neuronal differentiation of neural progenitor cells (NPCs) by CHL1/ERK1/2 MAPK signaling (Huang et al., 2011). Although CHL1 functions as a putative tumor suppressor during primary tumor growth and is silenced to facilitate *in situ* tumor growth, re-expression of CHL1 on the edge of the tumor mass may promote local invasive growth and enable further metastatic spread in ovary, colon and breast cancers (Senchenko et al., 2011). In addition, CHL1 facilitates the identification of two major histological types of renal cancer as a potentially novel specific biomarker in early pathogenesis (Senchenko et al., 2011). In pituitary adenoma (PA), differential expression of CHL1 may potentially predict a recurrence phenotype (Marko et al., 2012). Senescence is a stress response that stably blocks proliferation and functions as a tumor suppressor in aging and precancerous cells (Collado and Serrano, 2010; Wagner et al., 2015). The lysosomal-β-galactosidase gene is the source of senescence associated-β-galactosidase activity, and that level of lysosomal-β-galactosidase protein increase during senescence (Lee et al., 2006). The presence of senescence in response to CHL1 knockdown indicates the potential role of CHL1 in promoting cell survival during the development of glioma.

The discrepancy of our findings with those observed in non-glia-derived tumors may be due to the fact that glioma cells express a variety of molecules that interact with CHL1. CHL1 also potentiates integrin-dependent haptotactic cell migration toward the extracellular matrix via a potential integrin interaction motif Asp-Gly-Glu-Ala (DGEA) in the sixth immunoglobulin domain (Buhusi et al., 2003). CHL1–CHL1 homophilic interactions inhibit the neurite outgrowth-promoting functions of CHL1, whereas integrins in cis- and trans-configurations are conducive to heterophilic CHL1 interactions (Buhusi et al., 2003; Demyanenko et al., 2004; Jakovcevski et al., 2007, 2009). Observations indicate that homophilic CHL1 trans-interactions regulate differentiation of neuronal progenitor cells at early postnatal stages, whereas heterophilic trans-interactions of CHL1 with vitronectin, integrins and the plasminogen activator system regulate neuritogenesis and neuronal cell migration at a

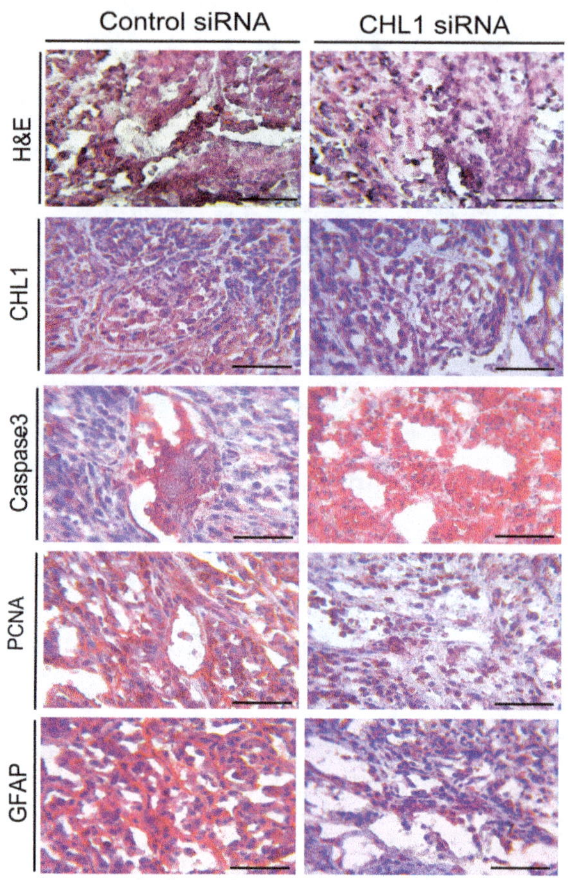

FIGURE 10 | H&E staining and immunohistochemical staining analyses for the CHL1, caspase-3, PCNA and GFAP molecules in glioblastoma xenograft tissues from both control siRNA and CHL1 siRNA-treated groups. Scale bars represent 25 μm.

later stage of cerebellar morphogenesis (Hillenbrand et al., 1999; Katic et al., 2014).

The activation of the AKT and ERK signaling pathways plays an important role in the regulation of cell apoptosis, invasion and metastasis of gliomas. Feng et al. (2010) reported that TAM was present in the rat C6 glioma cell line through activation of PI3K/Akt, JNK and ERK signaling pathways to mediate the physiological processes of glioma cell apoptosis. Studies by Li et al. (2013) reported that a DC electric field can activate the AKT and ERK signaling pathways in U251 glioma cells, thus affecting the direction of tumor migration. Wu et al. (2010) found that PI3K-dependent activation of Erk1/2 signaling pathway up-regulates CHL1 expression in the primary culture of astrocytes. Whether the effect of CHL1 on glioma is closely related to the ERK and AKT signaling pathways has not been reported. We demonstrate that downregulation of CHL1 reduces the activation of Akt1 with no apparent effect on Erk1/2 activation. This finding suggests that the modulating role of CHL1 on glioma behaviors may be partially mediated by Akt1 signaling.

Intratumoral injection represents a promising therapeutic method for tumor management via directly introducing siRNA or other forms of RNA into the tumor tissue. Unlike traditional xenografts of cells with a targeting molecule knocked down, *in vivo* knockdown is more similar to the clinical setting. Zheng et al. (2015) introduced shRNA against LunX into lung cancer tissue *in vivo*, which successfully disrupted local invasion, micrometastasis formation and metastatic colonization of lung cancer cells, thus inhibiting the initial and final steps of the invasion-metastasis cascade. Using an *in vivo* tumorigenicity assay, Wang et al. (2012) reported that daily intratumoral injection of let-7a-mimics in nude mice suppresses NIRF expression and reduces tumor growth. Although intratumoral injection of siRNA targeting CHL1 in the present study reduces the volume increase of glioma xenografts, the growth inhibition efficacy is limited. This effect may be partially attributed to low siRNA transfection and expansion due to the high intensity of cell proliferation in the glioma tissue, which is different from that of lung cancer. Increases in the siRNA transfection efficiency may lead to a more promising therapeutic result.

In summary, our work preliminarily demonstrated the roles of CHL1 in the development of gliomas, which may provide a scientific basis for the molecular targeting treatment of gliomas.

AUTHOR CONTRIBUTIONS

ZY and QX performed most of the experiments, analyzed the data, and drafted the sections of Materials and Methods, and Results. C-LH was responsible for figure processing and adjusting the format of the draft. QJ and H-FS assisted the performance of some experiments. MS provided suggestions and comments for the project and contributed to writing of the manuscript. W-JZ conceived and designed the experiments, undertook some of the experiments, wrote the manuscript and revised the article critically and approved the final version to be submitted.

ACKNOWLEDGMENTS

The authors are grateful to the National Natural Science Foundation of China (Grant Nos. 81471279 and 81171138 to W-JZ) for support. The authors are also grateful to the Li Kashing Foundation (Grant No. LD030601 to MS). This work was also supported by a Talent grant of Shantou University Medical College (Grant No. 250122 0118 to W-JZ).

REFERENCES

Asati, V., Mahapatra, D. K., and Bharti, S. K. (2016). PI3K/Akt/mTOR and Ras/Raf/MEK/ERK signaling pathways inhibitors as anticancer agents: structural and pharmacological perspectives. *Eur. J. Med. Chem.* 109, 314–341. doi: 10.1016/j.ejmech.2016.01.012

Bao, S., Wu, Q., Li, Z., Sathornsumetee, S., Wang, H., McLendon, R. E., et al. (2008). Targeting cancer stem cells through L1CAM suppresses glioma growth. *Cancer Res.* 68, 6043–6048. doi: 10.1158/0008-5472.CAN-08-1079

Buhusi, M., Midkiff, B. R., Gates, A. M., Richter, M., Schachner, M., and Maness, P. F. (2003). Close homolog of L1 is an enhancer of integrin-mediated cell migration. *J. Biol. Chem.* 278, 25024–25031. doi: 10.1074/jbc.M303084200

Chen, Q. Y., Chen, Q., Feng, G. Y., Lindpaintner, K., Chen, Y., Sun, X., et al. (2005). Case-control association study of the close homologue of L1 (CHL1) gene and schizophrenia in the Chinese population. *Schizophr. Res.* 73, 269–274. doi: 10.1016/j.schres.2004.06.001

Cheng, L., Wu, Q., Guryanova, O. A., Huang, Z., Huang, Q., Rich, J. N., et al. (2011). Elevated invasive potential of glioblastoma stem cells. *Biochem. Biophys. Res. Commun.* 406, 643–648. doi: 10.1016/j.bbrc.2011.02.123

Chu, Y., Ouyang, Y., Wang, F., Zheng, A., Bai, L., Han, L., et al. (2014). MicroRNA-590 promotes cervical cancer cell growth and invasion by targeting CHL1. *J. Cell. Biochem.* 115, 847–853. doi: 10.1002/jcb.24726

Chung, G., O'Neil, N. J., and Rose, A. M. (2011). CHL-1 provides an essential function affecting cell proliferation and chromosome stability in *Caenorhabditis elegans*. *DNA Repair* 10, 1174–1182. doi: 10.1016/j.dnarep.2011.09.011

Collado, M., and Serrano, M. (2010). Senescence in tumours: evidence from mice and humans. *Nat. Rev. Cancer* 10, 51–57. doi: 10.1038/nrc2772

Demyanenko, G. P., Schachner, M., Anton, E., Schmid, R., Feng, G., Sanes, J., et al. (2004). Close homolog of L1 modulates area-specific neuronal positioning and dendrite orientation in the cerebral cortex. *Neuron* 44, 423–437. doi: 10.1016/j.neuron.2004.10.016

Dimri, G. P., Lee, X., Basile, G., Acosta, M., Scott, G., Roskelley, C., et al. (1995). A biomarker that identifiessenescenthumancells in culture and in aging skin *in vivo*. *Proc. Natl. Acad. Sci. U S A* 92, 9363–9367. doi: 10.1073/pnas.92.20.9363

Feng, Y., Huang, J., Ding, Y., Xie, F., and Shen, X. (2010). Tamoxifen-induced apoptosis of rat C6 glioma cells via PI3K/Akt, JNK and ERK activation. *Oncol. Rep.* 24, 1561–1567. doi: 10.3892/or_00001018

Gabriel, A., Batey, J., Capogreco, J., Kimball, D., Walters, A., Tubbs, R. S., et al. (2014). Adult brain cancer in the U.S. black population: a surveillance, epidemiology, and end results (SEER) analysis of incidence, survival and trends. *Med. Sci. Monit.* 20, 1510–1517. doi: 10.12659/MSM.890762

He, L. H., Ma, Q., Shi, Y. H., Ge, J., Zhao, H. M., Li, S. F., et al. (2013). CHL1 is involved in human breast tumorigenesis and progression. *Biochem. Biophys. Res. Commun.* 438, 433–438. doi: 10.1016/j.bbrc.2013.07.093

Higuchi, T., Fujiwara, H., Egawa, H., Sato, Y., Yoshioka, S., Tatsumi, K., et al. (2003). Cyclic AMP enhances the expression of an extravillous trophoblast marker, melanoma cell adhesion molecule, in choriocarcinoma cell JEG3 and human chorionic villous explant cultures. *Mol. Hum. Reprod.* 9, 359–366. doi: 10.1093/molehr/gag044

Hillenbrand, R., Molthagen, M., Montag, D., and Schachner, M. (1999). The close homologue of the neural adhesion molecule L1 (CHL1): patterns of expression and promotion of neurite outgrowth by heterophilic interactions. *Eur. J. Neurosci.* 11, 813–826. doi: 10.1046/j.1460-9568.1999.00496.x

Hitt, B., Riordan, S. M., Kukreja, L., Eimer, W. A., Rajapaksha, T. W., and Vassar, R. (2012). β-Site amyloid precursor protein (APP)-cleaving enzyme 1 (BACE1)-deficient mice exhibit a close homolog of L1 (CHL1) loss-of-function phenotype involving axon guidance defects. *J. Biol. Chem.* 287, 38408–38425. doi: 10.1074/jbc.M112.415505

Holm, J., Hillenbrand, R., Steuber, V., Bartsch, U., Moos, M., Lübbert, H., et al. (1996). Structuralfeatures of a close homologue of L1 (CHL1) in the mouse: a new member of the L1 family of neural recognition molecules. *Eur. J. Neurosci.* 8, 1613–1629. doi: 10.1111/j.1460-9568.1996.tb01306.x

Huang, C. X., Hu, S. X., and Chen, B. (2001). Expression of the neural cell adhesion molecule in human astrocytomas. *Hunan Yi Ke Da Xue Xue Bao* 26, 543–545.

Huang, X., Zhu, L. L., Zhao, T., Wu, L. Y., Wu, K. W., Schachner, M., et al. (2011). CHL1 negatively regulates the proliferation and neuronal differentiation of neural progenitor cells through activation of the ERK1/2 MAPK pathway. *Mol. Cell. Neurosci.* 46, 296–307. doi: 10.1016/j.mcn.2010.09.013

Izumoto, S., Ohnishi, T., Arita, N., Hiraga, S., Taki, T., and Hayakawa, T. (1996). Gene expression of neural cell adhesion molecule L1 in malignant gliomas and biological significance of L1 in glioma invasion. *Cancer Res.* 56, 1440–1444.

Jakovcevski, I., Siering, J., Hargus, G., Karl, N., Hoelters, L., Djogo, N., et al. (2009). Close homologue of adhesion molecule L1 promotes survival of Purkinje and granule cells and granule cell migration during murine cerebellar development. *J. Comp. Neurol.* 513, 496–510. doi: 10.1002/cne.21981

Jakovcevski, I., Wu, J., Karl, N., Leshchyns'ka, I., Sytnyk, V., Chen, J., et al. (2007). Glial scar expression of CHL1, the close homolog of the adhesion molecule L1, limits recovery after spinal cord injury. *J. Neurosci.* 27, 7222–7233. doi: 10.1523/JNEUROSCI.0739-07.2007

Jemal, A., Siegel, R., Xu, J., and Ward, E. (2010). Cancer statistics, 2010. *CA Cancer J. Clin.* 60, 277–300. doi: 10.3322/caac.20073

Katic, J., Loers, G., Kleene, R., Karl, N., Schmidt, C., Buck, F., et al. (2014). Interaction of the cell adhesion molecule CHL1 with vitronectin, integrins and the plasminogen activator inhibitor-2 promotes CHL1-induced neurite outgrowth and neuronal migration. *J. Neurosci.* 34, 14606–14623. doi: 10.1523/JNEUROSCI.3280-13.2014

Lee, B. Y., Han, J. A., Im, J. S., Morrone, A., Johung, K., Goodwin, E. C., et al. (2006). Senescence-associated β-galactosidase is lysosomal β-galactosidase. *Aging Cell* 5, 187–195. doi: 10.1111/j.1474-9726.2006.00199.x

Li, F., Chen, T., Hu, S., Lin, J., Hu, R., and Feng, H. (2013). Superoxide mediates direct current electric field-induced directional migration of glioma cells through the activation of AKT and ERK. *PLoS One* 8:e61195. doi: 10.1371/journal.pone.0061195

Long, M. J., Wu, F. X., Li, P., Liu, M., Li, X., and Tang, H. (2012). MicroRNA-10a targets CHL1 and promotes cell growth, migration and invasion in human cervical cancer cells. *Cancer Lett.* 324, 186–196. doi: 10.1016/j.canlet.2012.05.022

Manderson, E. N., Birch, A. H., Shen, Z., Mes-Masson, A. M., Provencher, D., and Tonin, P. N. (2009). Molecular genetic analysis of a cell adhesion molecule with homology to L1CAM, contactin 6 and contactin 4 candidate chromosome 3p26pter tumor suppressor genes in ovarian cancer. *Int. J. Gynecol. Cancer* 19, 513–525. doi: 10.1111/IGC.0b013e3181a3cd38

Maness, P. F., and Schachner, M. (2007). Neural recognition molecules of the immunoglobulin superfamily: signaling transducers of axon guidance and neuronal migration. *Nat. Neurosci.* 10, 19–26. doi: 10.1038/nn0207-263b

Marko, N. F., Coughlan, C., and Weil, R. J. (2012). Towards an integrated molecular and clinical strategy to predict early recurrence in surgically resected non-functional pituitary adenomas. *J. Clin. Neurosci.* 19, 1535–1540. doi: 10.1016/j.jocn.2012.01.038

Morag, A., Pasmanik-Chor, M., Oron-Karni, V., Rehavi, M., Stingl, J. C., and Gurwitz, D. (2011). Genome-wide expression profiling of human lymphoblastoid cell lines identifies CHL1 as a putative SSRI antidepressant response biomarker. *Pharmacogenomics* 12, 171–184. doi: 10.2217/pgs.10.185

Nakazato, Y. (2008). The 4th edition of WHO classification of tumours of the central nervous system published in 2007. *No Shinkei Geka* 36, 473–491.

Qin, Y. R., Fu, L., Sham, P. C., Kwong, D. L., Zhu, C. L., Chu, K. K., et al. (2008). Single-nucleotide polymorphism-mass array reveals commonly deleted regions at 3p22 and 3p14.2 associate with poor clinical outcome in esophageal squamous cell carcinoma. *Int. J. Cancer* 123, 826–830. doi: 10.1002/ijc.23577

Sakurai, K., Migita, O., Toru, M., and Arinami, T. (2002). An association between a missense polymorphism in the close homologue of L1 (CHL1, CALL) gene and schizophrenia. *Mol. Psychiatry* 7, 412–415. doi: 10.1038/sj.mp.4000973

Sathornsumetee, S., Rich, J. N., and Reardon, D. A. (2007). Diagnosis and treatment of high-grade astrocytoma. *Neurol. Clin.* 25, 1111–1139. doi: 10.1016/j.ncl.2007.07.004

Senchenko, V. N., Krasnov, G. S., Dmitriev, A. A., Kudryavtseva, A. V., Anedchenko, E. A., Braga, E. A., et al. (2011). Differential expression of CHL1 gene during development of major human cancers. *PLoS One* 6:e15612. doi: 10.1371/journal.pone.0015612

Skibbens, R. V. (2004). Chl1p, a DNA helicase-like protein in budding yeast, functions in sister-chromatid cohesion. *Genetics* 166, 33–42. doi: 10.1534/genetics.166.1.33

Tassano, E., Biancheri, R., Denegri, L., Porta, S., Novara, F., Zuffardi, O., et al. (2014). Heterozygous deletion of CHL1 gene: detailed array-CGH and clinical characterization of a new case and review of the literature. *Eur. J. Med. Genet.* 57, 626–629. doi: 10.1016/j.ejmg.2014.09.007

Tschentscher, F., Hüsing, J., Hölter, T., Kruse, E., Dresen, I. G., Jöckel, K. H., et al. (2003). Tumor classification based on gene expression profiling shows that uveal melanomas with and without monosomy 3 represent two distinct entities. *Cancer Res.* 10, 2578–2584.

Wagner, J., Damaschke, N., Yang, B., Truong, M., Guenther, C., McCormick, J., et al. (2015). Overexpression of the novel senescence marker β-galactosidase (GLB1) in prostate cancer predicts reduced PSA recurrence. *PLoS One* 10:e0124366. doi: 10.1371/journal.pone.0124366

Wang, F., Zhang, P., Ma, Y., Yang, J., Moyer, M. P., Shi, C., et al. (2012). NIRF is frequently upregulated in colorectal cancer and its oncogenicity can be suppressed by let-7a microRNA. *Cancer Lett.* 314, 223–231. doi: 10.1016/j.canlet.2011.09.033

Wolterink, S., Moldenhauer, G., Fogel, M., Kiefel, H., Pfeifer, M., Lüttgau, S., et al. (2010). Therapeutic antibodies to human L1CAM: functional characterization and application in a mouse model for ovarian carcinoma. *Cancer Res.* 70, 2504–2515. doi: 10.1158/0008-5472.CAN-09-3730

Wu, J., Wrathall, J. R., and Schachner, M. (2010). Phosphatidylinositol 3-kinase/protein kinase Cdelta activation induces close homolog of adhesion molecule L1 (CHL1) expression in cultured astrocytes. *Glia* 58, 315–328. doi: 10.1002/glia.20925

Zarnescu, O., Brehar, F. M., Bleotu, C., and Gorgan, R. M. (2011). Co-localization of PCNA, VCAM-1 and caspase-3 with nestin in xenografts derived from human anaplastic astrocytoma and glioblastoma multiforme tumor spheres. *Micron* 42, 793–800. doi: 10.1016/j.micron.2011.04.005

Zhang, X., Davis, J. Q., Carpenter, S., and Bennett, V. (1998). Structural requirements for association of neurofascin with ankyrin. *J. Biol. Chem.* 273, 30785–30794. doi: 10.1074/jbc.273.46.30785

Zhao, W. J., and Ren, S. G. (2011). Endogenous neuregulin-1 expression in the anterior pituitary of female Wistar-Furth rats during the estrous cycle. *Nan Fang Yi Ke Da Xue Xue Bao* 31, 921–927.

Zhao, W., and Ren, S. G. (2011). Neuregulin-1 (Nrg1) is mainly expressed in rat pituitarygonadotroph cells and possibly regulates prolactin (PRL) secretionin a juxtacrine manner. *J. Neuroendocrinol.* 23, 1252–1262. doi: 10.1111/j.1365-2826.2011.02223.x

Zhao, W. J., and Schachner, M. (2013). Neuregulin 1 enhances cell adhesion molecule l1 expression in human glioma cells and promotes their migration as a function of malignancy. *J. Neuropathol. Exp. Neurol.* 72, 244–255. doi: 10.1097/NEN.0b013e3182863dc5

Zhao, W., Shen, H., Yuan, F., Li, G., Sun, Y., Shi, Z., et al. (2009). Induction stage-dependent expression of vascular endothelial growth factor and aquaporin-1 in diethylstilbestrol-treated rat pituitary. *Eur. J. Histochem.* 53, 53–60. doi: 10.4081/ejh.2009.53

Zheng, X., Cheng, M., Fu, B., Fan, X., Wang, Q., Yu, X., et al. (2015). Targeting LUNX inhibits non-small cell lung cancer growth and metastasis. *Cancer Res.* 75, 1080–1090. doi: 10.1158/0008-5472.CAN-14-1831

Zhu, H., Fang, J., Zhang, J., Zhao, Z., Liu, L., Wang, J., et al. (2014). miR-182 targets CHL1 and controls tumor growth and invasion in papillary thyroid carcinoma. *Biochem. Biophys. Res. Commun.* 450, 857–862. doi: 10.1016/j.bbrc.2014.06.073

11

Molecular Mechanisms Involved in Schwann Cell Plasticity

Angélique Boerboom[1], *Valérie Dion*[1], *Alain Chariot*[2,3] *and Rachelle Franzen*[1*]

[1] GIGA-Neurosciences, University of Liège, Liège, Belgium, [2] GIGA-Molecular Biology of Diseases, University of Liège, Liège, Belgium, [3] Walloon Excellence in Lifesciences and Biotechnology (WELBIO), Wavre, Belgium

***Correspondence:**
Rachelle Franzen
rfranzen@ulg.ac.be

Schwann cell incredible plasticity is a hallmark of the utmost importance following nerve damage or in demyelinating neuropathies. After injury, Schwann cells undergo dedifferentiation before redifferentiating to promote nerve regeneration and complete functional recovery. This review updates and discusses the molecular mechanisms involved in the negative regulation of myelination as well as in the reprogramming of Schwann cells taking place early following nerve lesion to support repair. Significant advance has been made on signaling pathways and molecular components that regulate SC regenerative properties. These include for instance transcriptional regulators such as c-Jun or Notch, the MAPK and the Nrg1/ErbB2/3 pathways. This comprehensive overview ends with some therapeutical applications targeting factors that control Schwann cell plasticity and highlights the need to carefully modulate and balance this capacity to drive nerve repair.

Keywords: Schwann cell, plasticity, molecular mechanisms, nerve injury, peripheral neuropathy

INTRODUCTION

The peripheral nervous system (PNS) shows a surprising capacity of regeneration compared to the central nervous system (CNS). This ability of peripheral nerves to recover quickly following damage is to a large extent due to the remarkable plasticity of Schwann cells (SCs), the glial cells of the PNS (Jessen et al., 2015). SCs are derived from neural crest cells that differentiate into SC precursors

---

Abbreviations: ALS, Amyotrophic lateral sclerosis; BACE1, Beta-Site amyloid precursor protein cleaving enzyme 1; BDNF, Brain-derived neurotrophic factor; BMP7, Bone morphogenetic protein 7; Cadm3, Cell adhesion molecule 3; cAMP, Cyclic adenosine monophosphate; Cdkn1c, Cyclin-dependent kinase inhibitor 1C; CIDP, Chronic inflammatory demyelinating polyneuropathy; CMT, Charcot-Marie-Tooth; Cthrc1, Collagen triple helix repeat containing 1; EZH2, Enhancer of zeste homolog 2; Dixdc1, DIX domain containing-1; DDIT4, DNA damage-inducible transcript 4 protein; Dlg1, Disks large homolog 1; Dock7, Dedicator of cytokinesis protein 7; DRG, Dorsal root ganglion; Dusp15, Dual specificity phosphatase 15; Egr-1, Early growth response protein 1; Egr-2/Krox-20Early growth response 2; Egr-3, Early growth response protein 3; Erk, Extracellular signal regulated kinase; GDNF, Glial cell-derived neurotrophic factor; GFAP, Glial fibrillary acidic protein; GPCR, G protein-coupled receptor; Id2, DNA-binding protein inhibitor 2; IVIG, Immunoglobulins; JNK, c-jun N-terminal kinase; LKB1/STK11, Liver kinase B1/ Serine/threonine-protein kinase 11; LPA1, Lysophosphatidic acid receptor 1; LXR, Liver X receptor; MAPK, Mitogen activated protein kinase; MBP, Myelin basic protein; MPZ, Myelin protein zero; MyD88, Myeloid differentiation primary response gene 88; NF-κB, Nuclear factor-kappa B; NMJ, Neuromuscular junction; Notch, Neurogenic locus notch homolog protein; Nrg1, Neuregulin 1; NuRD, Nucleosome remodeling and deacetylase complex; p75NTR, Low affinity neurotrophin receptor; Pax-3, Paired box protein 3; PI 3-kinase, Phosphatidylinositol 3-kinase; PMP22, Peripheral myelin protein 22; Pou3f1/Oct6, POU domain class 3 transcription factor / Octamer-binding factor 6; PTEN, Phosphatase and tensin homolog deleted on chromosome 10; S1PR, Sphingosine-1-phosphate receptor; Sam68, Src-associated in mitosis of 68 kD; Sox2, SRY-related HMG-box gene 2; SSeCKS, Src-suppressed protein kinace C substrate; STK11, Serine/threonine protein kinase; TACE, Tumor necrosis factor-α–converting enzyme; TGFβ, Transforming growth factor beta; Zeb2, Zinc-finger E-box-binding homeobox 2.

and then into immature SCs between E12 and E15 in mice. Around birth, axonal sorting and myelination start in the peripheral nerves. Some SCs establish a 1:1 relationship with large-diameter axons, wrap them multiple times to form a thick and compact myelin sheath. Myelin-forming SCs allow the fast conduction of action potentials by insulating the axons. The non-myelinating SCs typically associate with several small-diameter axons to form Remak bundles. Nerve homeostasis, trophic support and myelin maintenance are other important functions of SCs in adulthood (Jessen and Mirsky, 2005).

Besides their major roles in normal nerve physiology, SCs play a key function for repair in many pathological conditions thanks to their striking plasticity (Zochodne, 2008). For example, after a peripheral nerve injury, they are capable of switching into a SC immature-like phenotype that drives nerve repair. Over the last decades, major progress has been made in unraveling molecular mechanisms and signaling pathways that drive SC dedifferentiation and regulate their plasticity. In this review, we will update and discuss the recent studies that identified molecular components involved in SC plasticity and their possible therapeutic implications.

REMARKABLE PLASTICITY AND REGENERATIVE PROPERTIES OF SCHWANN CELLS AFTER PERIPHERAL NERVE INJURY

The unique plasticity of SCs has been extensively studied in transgenic animals after cut or crush injuries. Indeed, following a nerve lesion, quiescent, highly specialized myelinating and non-myelinating SCs reprogram into proliferative progenitor-like repair SCs that drive the entire regeneration process (Chen et al., 2007; Jessen and Mirsky, 2016). The reprogramming of SCs in physiopathological conditions is most of the time defined as dedifferentiation. However, transdifferentiation seems more appropriate, as repair SCs, besides re-expressing immature SC markers, exhibit completely different features (Arthur-Farraj et al., 2012). Indeed, the injury-induced conversion of mature SCs in regeneration-promoting cells is an active phenomenon. It involves a down-regulation of pro-myelinating genes including early growth response 2 *(Egr-2,* more frequently named *Krox-20*), POU domain class 3 transcription factor 1 *(Pou3f1* or *Oct-6),* myelin protein zero *(MPZ) or* myelin basic protein *(MBP)* as well as an up-regulation of markers of immature, de-differentiated SCs such as c-Jun, low affinity neurotrophin receptor (p75NTR) or glial fibrillary acidic protein (GFAP) but also specific repair-supportive genes (Jessen and Mirsky, 2008). Following damage, the nerve undergoes a series of complex multicellular and molecular events in which SCs play a role of orchestrator (**Figure 1**).

Quickly after nerve injury, damaged axons in the distal stump degenerate in an active process called Wallerian degeneration (Waller, 1850). Yet unidentified signals from damaged nerves induce the reprogramming of SCs. These downregulate pro-myelinating genes and start clearing their myelin sheaths through a mechanism of autophagy called myelinophagy (Gomez-Sanchez et al., 2015). Axonal and myelin debris are also phagocyted by resident and blood-derived macrophages recruited by SCs (Hirata and Kawabuchi, 2002; Lee et al., 2006; Barrette et al., 2008). An inflammatory reaction occurs: many blood cells invade the lesion site and secrete numerous cytokines and chemokines (Martini et al., 2008; Gaudet et al., 2011; Rotshenker, 2011). Following nerve axotomy, particularly, the basal lamina of SCs and the connective tissue are interrupted (Zochodne, 2008). A tissue bridge is formed between the two stumps of the nerve over the lesion site. Fibroblasts play a major role in building this bridge by interacting with SCs (Parrinello et al., 2010). Newly formed vasculature is also crucial to guide the SCs and the growing axons through the lesion site (Cattin et al., 2015). Many chemical and physical interactions happen between the actors present in the injured nerve, creating a permissive and favorable environment for regeneration (Cattin and Lloyd, 2016).

Irrespective of whether the injury is a crush or a cut, repair SCs in the distal stump proliferate, secrete several trophic factors that support glial and neuronal survival and regrowth including artemin, brain-derived neurotrophic factor (BDNF) or glial cell line-derived neurotrophic factor (GDNF) (Fernandez-Valle et al., 1995; Kim et al., 2000; Boyd and Gordon, 2003; Fontana et al., 2012). They also align in tracts named bands of Büngner and provide a trophic and physical support for axons to regrow and reinnervate correctly their targets (Weinberg and Spencer, 1978; Stoll and Müller, 1999). At the neuromuscular junction, specialized terminal SCs direct reinnervation by helping the axons to find their paths toward their appropriate sites (Son and Thompson, 1995). After axonal regeneration, repair SCs readily exit the cell cycle and differentiate again into myelinating and non-myelinating SCs to support the complete functional recovery. Nevertheless, most of the time, the newly-formed myelin sheaths are shorter and thinner than expected based on axonal diameter (Schröder, 1972).

MOLECULAR MECHANISMS THAT MODULATE SCHWANN CELL PLASTICITY

The adaptive reprogramming of SCs after a nerve injury is a prerequisite for regeneration. Therefore, understanding the molecular components that regulate the phenotypic change of adult specialized SCs into repair-supportive SCs can lead to the development of new therapeutics that can boost repair in various peripheral neuropathies. In this section, we will review the most relevant molecular components or pathways that have been identified for controlling SC plasticity, the negative regulation of myelination and their response to nerve injury (**Tables 1, 2; Figure 2**).

Transcriptional Regulators

The Transcription Factor c-Jun

The transcription factor c-Jun does not appear to be essential during SC development but multiple arguments demonstrate its central function in SC reprogramming. Firstly, the time course of c-Jun expression supports its role in SC plasticity. Indeed, c-Jun is down-regulated post-natally during SC differentiation

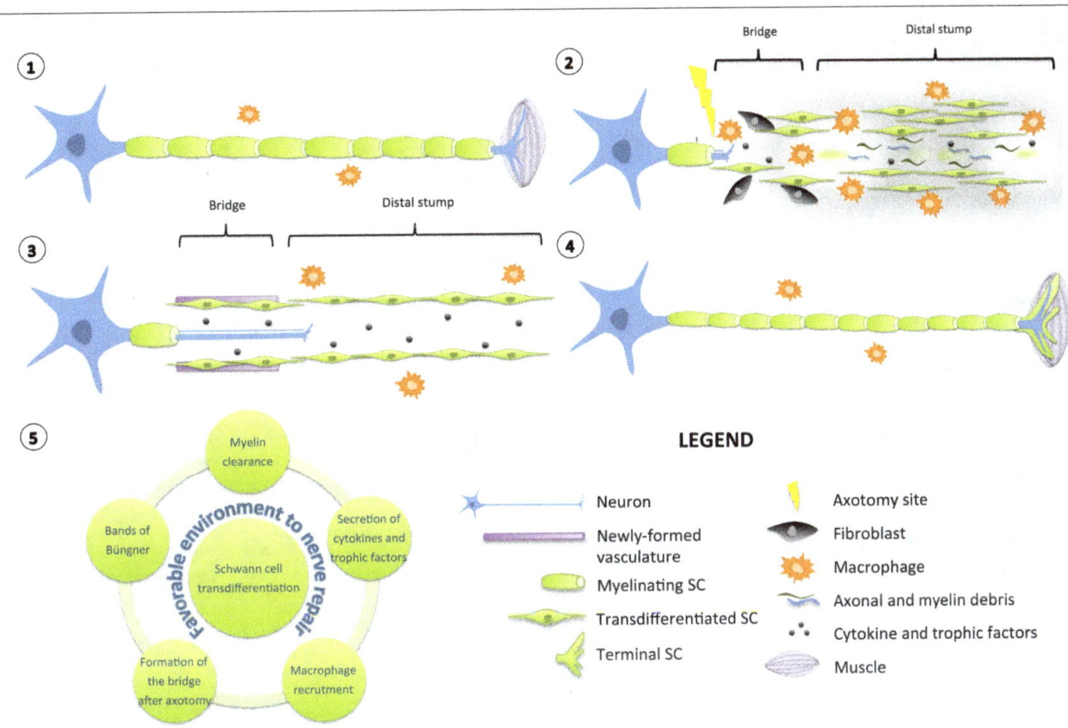

FIGURE 1 | Diagram of Schwann cell response to nerve injury. (1) Schematic representation of a single neuron with myelinating SCs and resident macrophages. For simplification, the basal lamina around SCs is not shown. (2) After injury, the nerve distal to the injury site degenerates and undergoes a series of complex multicellular and molecular events in which SCs play a role of orchestrator. SCs transdifferentiate into repair-promoting cells, creating a permissive and favorable environment for nerve regeneration. SCs downregulate pro-myelinating genes and clear their myelin sheaths. They proliferate, secrete several pro-inflammatory cytokines and trophic factors that support glial and neuronal survival and regrowth. Axonal and myelin debris are also phagocyted by resident and blood-derived macrophages recruited by SCs. SCs interact with fibroblasts to build a bridge between the two stumps of the nerve over the lesion site. (3) Newly formed vasculature guides the SCs and the growing axons through the lesion site. In the distal stump, SCs align in tracts named bands of Büngner to provide a trophic and physical support for axons to regrow. (4) After axonal regeneration, transdifferentiated SCs readily exit the cell cycle, differentiate again and remyelinate the axons to support the complete functional recovery. However, the newly-formed myelin sheaths are most of the time shorter and thinner than expected based on axonal diameter. Specialized terminal SCs direct reinnervation by helping the axons to find their paths toward their initial targets. (5) Diagram displaying the various roles played by a transdifferentiated SC to create a favorable environment for nerve repair.

and myelination and is highly up-regulated under pathological conditions such as peripheral nerve injuries, demyelinating diseases or other peripheral neuropathies (Stewart, 1995; Parkinson et al., 2008; Hutton et al., 2011; Hantke et al., 2014; Klein et al., 2014). Secondly, c-Jun is a negative regulator of myelination and a cross-antagonist of Krox-20, one of the main pro-myelinating transcription factors. Deletion of c-Jun in SCs *in vitro* strikingly facilitates Krox-20-induced expression of MBP or MPZ. Conversely, forced expression of c-Jun inhibits the expression of myelin proteins. Moreover, the expression of c-Jun and Krox-20 is mutually exclusive: while immature SCs express relatively high levels of c-Jun and low levels of Krox-20, myelinating cells switch to high levels of Krox-20 and low levels of c-Jun (Parkinson et al., 2004, 2008). Thirdly, strong evidence *in vivo* exhibits the involvement of c-Jun in major aspects of the SC response following nerve injury. Myelination during development occurs normally in the c-Jun conditional knockout mice (Parkinson et al., 2008). However, the absence of c-Jun following a nerve lesion results in a failure of generating the repair SCs, leading to impaired axonal regeneration, target reinnervation and functional recovery. The myelin sheath degradation and clearance are also strikingly slowed in the absence of c-Jun *in vitro* and *in vivo*. Moreover, mutant SCs fail to form the Büngner regenerative tracks and to express the adhesion molecules and trophic factors critical for nerve repair (Parkinson et al., 2008; Arthur-Farraj et al., 2012; Fontana et al., 2012).

The role of c-Jun in SCs after nerve injury is to activate a repair program to support regeneration. The high levels of c-Jun found in different peripheral neuropathies that do not involve nerve injuries like Charcot-Marie Tooth (CMT) raises the question on the precise function of c-Jun in these diseases. Does c-Jun counteract the pathology, attenuate its symptoms and promote repair or does its up-regulation cause the disease? In two models of demyelinating diseases, CMT1A and CMT1X, c-Jun is increased in SCs that retain myelin differentiation (Hantke et al., 2014; Klein et al., 2014). The myelin sheaths are also relatively normal in a mouse mutant showing a 5–8 fold c-Jun overexpression in SCs (levels that are lower than those seen after a nerve cut though) (Jessen and Mirsky, 2016). Moreover, c-Jun and repair-associated genes are up-regulated in a mouse model in which the serine/threonine-protein kinase (LKB1, also

TABLE 1 | Most relevant molecular components involved in Schwann cell plasticity, negative regulation of myelination and nerve repair.

Molecular component	*In vivo* model(s)	Phenotype(s)	Reference(s)
TRANSCRIPTIONAL REGULATORS			
c-Jun	• $MPZ^{Cre}/cJun^{fl/fl}$	• Injury: lack of trophic factor expression, no alignment in bands of Büngner, no axon regeneration and no functional recovery	Parkinson et al., 2008; Arthur-Farraj et al., 2012; Fontana et al., 2012
Notch	• $MPZ^{Cre}/RBPJ^{fl/fl};Notch^{fl/fl}$; $NCID^{CASL-STOPfl/fl}$ • Rats treated with Jagged1 (Notch signaling activator)	• Development: acceleration of myelination and thicker myelin. Injury: delayed myelin breakdown • Injury: improved axonal regeneration and functional recovery	Woodhoo et al., 2009; Wang et al., 2015
Zeb2	• $PLP^{CreERT2}/Zeb2^{fl/fl}$	• Injury: disturbed regeneration and lack of remyelination	Quintes et al., 2016; Wu L. M. N. et al., 2016
NF-κB	• GFAP-IκBα-dn	• Injury: delayed axonal regeneration and disturbed remyelination	Morton et al., 2012
MAPK			
Raf/Erk	• MPZ^{RafTR} (tamoxifen-inducible Raf activation)	• Normal adult nerve: transient activation leads to demyelination, SC proliferation and increased P75NTR expression. Sustained activation leads to inflammation.	Napoli et al., 2012
Rac/JNK	• Mice transfected with dominant negative Rac1/Rac1-RNAi • Mice injected with Rac inhibitor • Dock7 shRNA transgenic mice (decrease of Rac/JNK signaling)	• Injury: reduced myelin sheath fragmentation • Injury: decreased c-Jun and p75NTR expression • Development: enhanced myelin thickness	Jung et al., 2011; Yamauchi et al., 2011; Shin et al., 2013
p38	• Mice injected with p38 inhibitor • $MPZ^{Cre}/p38a^{fl/fl}$ • New-born rats treated with BMP7 (activation of p38MAPK signaling)	• Injury: reduced SC demyelination and dedifferentiation • Development: acceleration of myelination Injury: delay in myelin clearance, small increase of re-myelination • Development: delayed peripheral myelination	Yang et al., 2012; Liu et al., 2016b; Roberts et al., 2016
PI3K/Akt/mTOR			
PTEN	• $PLP^{CreERT2}/PTEN^{fl/fl}$ (hyper-activation of PI3K/Akt signaling) • $PLP^{CreERT2}/PTEN^{fl/fl}$treated with rapamycin (mTOR inhibitor)	• Development: hypermyelination, myelin outfoldings • Amelioration of the myelin pathology	Goebbels et al., 2010, 2012
Dlg1	• Mice injected with Dlg1 shRNA (activation of PI3K/Akt signaling) • $MPZ^{Cre}Dlg1^{fl/f}$ (activation of PI3K/Akt signaling)	• Development: myelin outfoldings and peripheral nerve overmyelination • Development: transient hypermyelination	Cotter et al., 2010; Noseda et al., 2013
DDIT4	• DDIT4 KO mouse (over-activation of PI3K/AKT/mTOR signaling)	• Development: sustained hypermyelination	Noseda et al., 2013
WNT SIGNALING AND LXR			
LXR	• LXR KO mouse • Mice treated with paraquat (LXR activation)	• Development: altered myelination • Development: severe myelin sheath disorganization	Makoukji et al., 2011; Hichor et al., 2016
TLR SIGNALING			
TLRs	• TLR2, TLR4, MyD88 KO mice	• Injury: impaired expression of inflammatory modulators, macrophage recruitment and activation, axonal regeneration and functional recovery	Boivin et al., 2007
GPCR SIGNALING			
Gpr126	• $PLP^{CreERT2}/Gpr126^{fl/fl}$	• Injury: defects in remyelination, macrophage recruitment and axon regeneration	Mogha et al., 2016b

(Continued)

TABLE 1 | Continued

Molecular component	*In vivo* model(s)	Phenotype(s)	Reference(s)
Nrg1/ErbB2-B3			
Nrg1	• Animals treated with Nrg1 isoforms • Inactivation in axons: SLICK-A^{Cre}; CAG^{CreERT2}/Nrg1$^{fl/fl}$ • Inactivation in SCs: Dhh^{Cre}/Nrg1$^{fl/fl}$	• Injury: improved nerve regeneration and functional recovery • Injury: slower axon regeneration, target reinnervation, remyelination, functional recovery • Injury: impaired remyelination	Chen et al., 1998; Joung et al., 2010; Fricker et al., 2011, 2013; Yildiz et al., 2011; Stassart et al., 2013
BACE1	• BACE1 KO mouse	• Injury: impaired remyelination but enhanced axon regeneration	Hu et al., 2008; Farah et al., 2011
TACE	• HB9^{Cre}/TACE$^{fl/fl}$	• Development: hypermyelination	La Marca et al., 2011
ErbB2	• Mice treated with ErbB2 inhibitor • Plp1^{CreERT2}ErbB2$^{fl/fl}$	• Injury: decreased demyelination • Injury: no effect on SC proliferation and survival	Guertin et al., 2005; Atanasoski et al., 2006

TABLE 2 | Summary of the regulators and their roles in SCs.

Negative regulation of myelination during development	Positive regulation of myelination during development
c-Jun, Notch, Sox2, Pax-3, Id2, Egr-1, Egr-3, Cdkn1c, Hes5, Dock7, Cthrc1, p38MAPK, BMP7, SSeCKS, Cadm3, Dlg1, PTEN, DDIT4, AMPK, LXR?	Zeb2, NF-κB?, PI3K/Akt/mTOR, LXR?, Wnt, Gpr126, Gpr44, LPA1, Nrg1/ErbB2/3
Transdifferentiation after injury	**Remyelination after injury**
c-Jun, Notch, NF-κB?, Sox2, Ras/Raf/Erk, Rac/JNK, p38 MAPK, Sam68, Dixdc1, TLRs, Gpr126, Nrg1/ErbB2/3?	Zeb2, NF-κB?, Nrg1/ErbB2/3

called STK11) is inactivated in SCs and in which axons are damaged without overt demyelination (Beirowski et al., 2014). This suggests that moderated c-Jun up-regulation is compatible with myelination, does not necessarily provoke demyelination and that it is sufficient to activate the transcription of injury-related genes (Jessen and Mirsky, 2016). In this context, it is likely that the role of c-Jun in those diseases is to promote neuroprotection and repair. However, if c-Jun is activated in an uncontrolled and exacerbated manner, c-Jun could also be detrimental. For example, the increase of c-Jun expression in some human schwannomas may directly or indirectly be associated with the abnormal proliferation and the absence of myelin markers (Shivane et al., 2013).

Notch Signaling

Neurogenic locus notch homolog protein (Notch) is a transmembrane receptor protein that, upon ligand binding, is cleaved and generates an intracellular domain (NICD), which acts as a transcriptional regulator. Notch controls SC proliferation and promotes the generation of immature SCs from Schwann cell precursors (SCPs) *in vivo* but also acts as a negative regulator of myelination. It is down-regulated as myelination proceeds and its inactivation or over-activation leads to premature or delayed myelin formation, respectively. Moreover, the inhibition of Notch signaling in adult mice decelerates myelin breakdown following a nerve lesion (Woodhoo et al., 2009). The addition of recombinant jagged1, an activator of Notch, in rat nerves after injury improves nerve regeneration and functional recovery (Wang et al., 2015). This indicates that stimulation of Notch signaling in SCs could represent an interesting therapeutic strategy to efficiently promote nerve repair. However, aberrant activation of Notch signaling in SCs may also participate to the development of malignant peripheral nerve sheath tumors (Li et al., 2004).

The Transcriptional Repressor Zeb2

More recently, the function of a transcriptional repressor, Zeb2, has been investigated in SC myelination and nerve repair by two different teams (Quintes et al., 2016; Wu L. M. N. et al., 2016). They both showed that mice lacking Zeb2 in SCs develop a severe peripheral neuropathy, caused by a failure of SC maturation. Zeb2 transcriptional repression of negative regulators of myelination such as Notch and SRY-related HMG-box gene 2 (Sox2) is necessary for SC lineage progression. Although Zeb2 seems dispensable for the establishment of the repair SCs following nerve injury, its absence disrupts nerve regeneration and remyelination (Quintes et al., 2016). Mechanistically, Zeb2 controls SC differentiation and remyelination by recruiting histone deacetylases 1 and 2 (HDAC1/2) and NuRD complexes and by inhibiting a Notch–Hey2 signaling axis (Wu L. M. N. et al., 2016).

NF-κB

The nuclear factor κB (NF-κB) transcription factor regulates a lot of physiological processes and mediates the physiopathological inflammatory response in many diseases. Several studies suggest that NF-κB activation is essential for the differentiation and myelination of SCs *in vitro* (Nickols et al., 2003; Yoon et al., 2008; Limpert and Carter, 2010). Furthermore, it has been shown that the deacetylation of NF-κB mediated by HDAC 1/2 is necessary to control myelination in the PNS (Chen et al., 2011). Paradoxically, by using transgenic mice in which NF-κB activation is inhibited in SCs, Morton and colleagues demonstrated that NF-κB is dispensable for developmental myelination *in vivo* (Morton

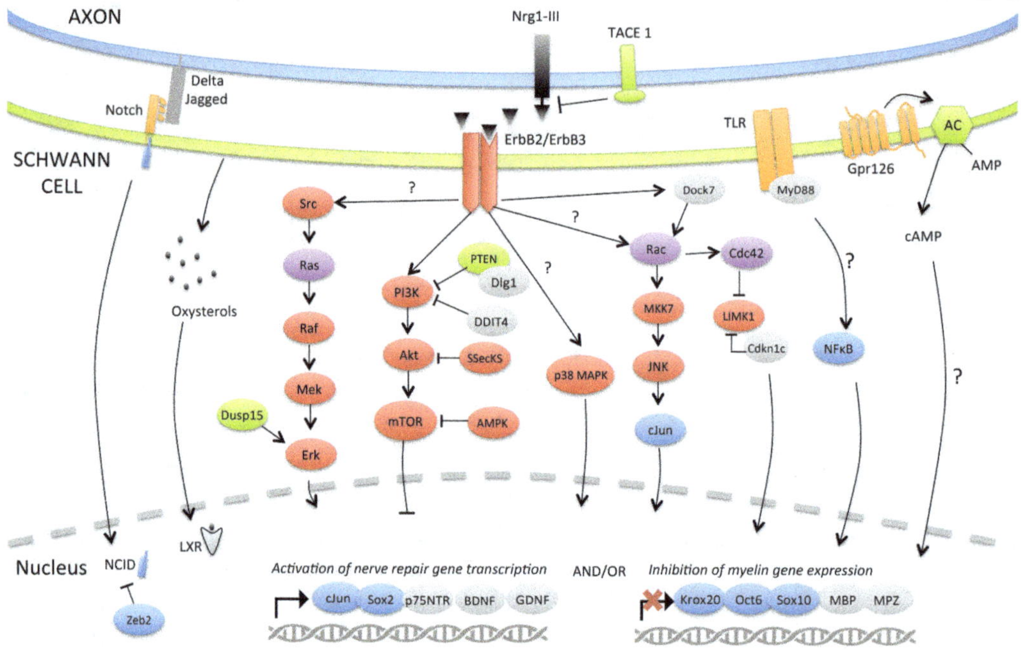

FIGURE 2 | Molecular signaling pathways involved in SC plasticity, negative regulation of myelination and nerve repair. The injury-induced reprogramming of SCs in regeneration-promoting cells involves a down-regulation of pro-myelinating genes including *Krox-20, Oct-6, Sox10, MPZ, or MBP* as well as an up-regulation of markers characteristic of immature, de-differentiated SCs such as c-Jun and p75NTR but also specific repair-supportive proteins like BDNF or GDNF. The signaling pathways or molecular components activating (c-Jun, Notch, MAPKs, TLRs, Gpr126, NF-κB) or inhibiting (PI3K/Akt/mTOR) this reprogramming are represented. Some negative regulators of myelination during development (TACE1, LXRs, Dusp15, PTEN, Dlg1, DDIT4, Dock7, Zeb2) and possibly playing a role after injury are also shown.

et al., 2013). However, they showed that the inhibition of NF-κB signaling in SCs following sciatic nerve crush transiently delays axonal regeneration and remyelination (Morton et al., 2012). This highlights again the differences between immature SCs during development and transdifferentiated repair-SCs. Importantly, NF-κB activity is robustly enhanced after peripheral nerve damage and in inflammatory neuropathies (Andorfer et al., 2001; Laura et al., 2006; Smith et al., 2009; Fu et al., 2010). The different NF-κB-induced transcriptional targets responsible for SC beneficial effects after nerve injury still have to be determined. Of note, the placental growth factor (PlGF) which is up-regulated by NF-κB in SCs following injury, is critical for SC proliferation, myelin sheath degradation, macrophage invasion and axonal regeneration (Chaballe et al., 2011). Targeting NF-κB or its downstream targets in SCs may offer new therapeutic strategies for PNS regeneration. Nevertheless, it is essential to keep in mind that aberrant activation of NF-κB may also lead to schwannoma (Ammoun et al., 2014; Dilwali et al., 2015).

Sox-2, Pax-3, Id2, Egr-1, Egr-3

Other factors that regulate gene transcription have been studied in SC dedifferentiation. These are most of the time negative regulators of myelination that functionally counteract positive regulators of myelination including Krox-20 or Oct-6 and are likely to have major roles after nerve damage (Jessen and Mirsky, 2008). For instance, the transcription regulators Sox-2, paired box protein 3 (Pax-3), DNA-binding protein inhibitor 2 (Id2) and early growth response proteins 1 and 3 (Egr-1 and Egr-3) could represent good candidates for regulating SC dedifferentiation in damaged nerves. As for c-Jun or Notch, they are inactive or expressed at low levels in myelinating cells and re-expressed at high levels under conditions that lead to SC dedifferentiation such as nerve injury (Kioussi et al., 1995; Nikam et al., 1995; Stewart et al., 1997; Le et al., 2005; Parkinson et al., 2008; Doddrell et al., 2012). *In vitro*, Sox2, Pax3, and Id2 decrease or increase myelin gene expression when overexpressed or depleted, respectively (Kioussi et al., 1995; Mager et al., 2008; Parkinson et al., 2008; Doddrell et al., 2012). Egrl and Egr3 are both necessary to up-regulate p75NTR in SC primary cultures, known to be re-expressed in SCs following a nerve injury (Nikam et al., 1995; Topilko et al., 1997; Gao et al., 2007). It remains to be unraveled by *in vivo* evidence whether those different factors are required for SC response after nerve injury.

Besides its likely role in SC plasticity, it has been shown that Sox-2 is necessary in SCs for the formation of a nerve bridge after a complete transection. This process is mediated by an ephrin-B/EphB2 signaling between fibroblasts and SCs that initiates cell sorting, followed by directional collective cell migration of SCs out of the nerve stumps to guide axons across the injury wound (Parrinello et al., 2010).

MAPK Signaling

A growing number of studies provide evidence on the functions of mitogen activated protein kinase (MAPK) family proteins in the regulation of SC plasticity. Extracellular signal regulated kinase (Erk), c-jun N-terminal kinase (JNK) and p38 MAP kinase are rapidly and highly activated in the SCs of the distal stump following a nerve lesion and play specific, overlapping or complementary roles in the SC response after injury (Sheu et al., 2000; Harrisingh et al., 2004; Zrouri et al., 2004; Agthong et al., 2006; Parkinson et al., 2008; Lee et al., 2014; Ronchi et al., 2016).

Ras/Raf/Erk

Harrisingh and coworkers were the first ones to demonstrate the role of the Ras/Raf/Erk signaling in SC dedifferentiation (Harrisingh et al., 2004). They showed that the ectopic activation of Raf, an upstream activator of Erk, suppresses the differentiation of primary SCs induced by cyclic adenosine monophosphate (cAMP). Raf activation in cultured differentiated SCs also drives their dedifferentiation. Importantly, they demonstrated that sustained activation of the Ras/Raf/Erk pathway in myelinated dorsal root ganglion (DRG) neuron-SC co-cultures induces demyelination even in the presence of normal axon signaling. Consistent with this idea, another paper reported that the dual specificity phosphatase 15 (Dusp15) is necessary for full activation of Erk and that its activation represses the expression of several myelin genes including *MBP* (Rodríguez-Molina et al., 2017). While high and sustained activation of Ras/Raf/Erk signaling seems essential for SC plasticity, Ras over-activation might also be deleterious and lead to the development of SC-derived tumors in neurofibromatosis type 1 patients (Harrisingh et al., 2004). A strong activation of Erk has actually been reported in human primary schwannoma cells (Ammoun et al., 2008).

More recently, the strategy of Raf high activation has been used *in vivo* to determine whether this signaling was central in controlling SC plasticity (Napoli et al., 2012). Napoli et al. demonstrated that a tamoxifen-inducible Raf transgene in SCs is sufficient to induce the down-regulation of myelin proteins and the expression of dedifferentiation markers such as p75NTR, even in the absence of axonal damage. The duration of Raf activation, controlled by the period of tamoxifen treatment, determines the time frame of SC demyelination and proliferation. A prolonged tamoxifen treatment and activation of Raf in SCs also induces an inflammatory response, which includes the breakdown of the blood-nerve barrier and the infiltration of macrophages, which promote nerve regeneration.

Although the repression of Erk signaling with specific inhibitors blocks SC mitosis, cytokine expression and demyelination, it only prevents partially the induction of p75NTR or GDNF and the shutdown of myelin gene expression (Napoli et al., 2012; Shin et al., 2013). Thus, even if the sustained and pronounced activation of the Ras/Raf/Erk pathway is fundamental in regulating SC plasticity, this supports the idea that several independent mechanisms control multiple aspects of SC plasticity. Moreover, different studies showed that Erk activation is actually pro-myelinating and that its inhibition blocks SC differentiation and myelination *in vivo* (Grossmann et al., 2009; He et al., 2010; Newbern et al., 2011). Therefore, Erk signaling is necessary both for the differentiation of SCPs and for the dedifferentiation of adult SCs following a nerve injury. The reconciling explanation could be that distinct levels of Erk activity would define the state of SC differentiation. Low or basal activity would be required for SC differentiation while high Erk levels would drive dedifferentiation and proliferation (Napoli et al., 2012; Newbern and Snider, 2012).

Finally, the role of c-Jun as a downstream target of Erk signaling is an attractive idea but it is actually quite controversial and further *in vivo* studies are necessary to better understand the signaling mechanisms downstream of Erk in SC plasticity. Indeed, Syed et al. showed that Erk inhibition abolishes both the demyelination and the expression of c-Jun induced by Neuregulin 1 (Nrg1) treatment in SC-DRG neuron co-cultures (Syed et al., 2010). Additionally, the Lloyd group indicated that c-Jun is strongly up-regulated after Raf over-activation in SCs *in vivo* (Napoli et al., 2012). However, others reported that the induction of c-Jun and SC dedifferentiation triggered by cAMP removal *in vitro* depends on JNK rather than on Erk activity (Monje et al., 2010; Shin et al., 2013). Actually, other data revealed that Rac/JNK and Raf/Erk might be responsible for different complementary functions in SC plasticity. For example, microarray experiments showed that the expression of regeneration-associated genes, including *p75NTR* or *GDNF* is dependent on JNK but not on Erk signaling. In addition, it has been demonstrated that most Erk-dependent genes in dedifferentiated SCs do not require Rac activity (Shin et al., 2013).

Rac/JNK

JNK is another MAPK that plays important functions in SCs. It has been demonstrated that c-Jun activation by JNK is essential for SC migration and proliferation *in vitro* (Parkinson et al., 2001, 2004; Yamauchi et al., 2003). Parkinson and colleagues showed that overexpressing the MAPK kinase 7 (MKK7), a protein upstream of JNK, in SC primary cultures, inhibits myelin gene expression by elevating c-Jun, suggesting a role for JNK signaling in SC plasticity (Parkinson et al., 2008). This was confirmed by Monje et al. who revealed that a reduction of cAMP *in vitro* is sufficient to induce SC dedifferentiation through a JNK-dependent mechanism (Monje et al., 2010). More recently, it has also been demonstrated that the active Rac1 GTPase (Rac), an upstream regulator of MKK7 and JNK, inhibits SC differentiation. In SC cultures, Rac up-regulates c-Jun and down-regulates Krox-20 through the MKK7/JNK pathway, but not through the Raf/Erk pathway. Furthermore, MKK7 activation and the induction of c-Jun observed in sciatic nerves after axotomy were blocked by Rac inhibition (Shin et al., 2013). In addition, Rac activation has been observed in the distal stump of injured nerves, and specific Rac inhibition with a dominant negative Rac1 and Rac1-RNAi decelerates myelin sheath fragmentation. Rac would actually promote myelin sheath fragmentation by controlling actin polymerization in the Schmidt-Lantermann incisures of SCs (Jung et al., 2011).

The cyclin-dependent kinase inhibitor 1 (Cdkn1c), also called p57Kip2, is a protein first known to block G_1/S transition during cell cycle that also regulates actin filament dynamics through the translocation of LIM domain kinase 1 (LIMK1), a downstream effector of Rac/Cdc42/Rho (Yokoo et al., 2003; Heinen et al., 2013). Knocking down Cdkn1c in cultured SCs leads to cell cycle exit, morphological changes and up-regulation of myelin genes. In SC-DRG neuron co-cultures, Cdkn1c depletion accelerates myelination. A global gene expression analysis also revealed a shift of the transcriptional expression program toward the pattern of differentiating SCs, strongly suggesting a role for Cdkn1c as an important negative regulator of myelination (Heinen et al., 2008). The same team also showed that Cdkn1c is a direct target of the enhancer of zeste homolog 2 (EZH2) and identify Hes5 as a new transcriptional repressor of myelin genes (Heinen et al., 2012). Indeed, EZH2 binds Cdkn1c promoter and its suppression results in a decrease of histone H3K27 trimethylation, an induction of Cdkn1c expression and a down-regulation of myelin genes in cultured primary SCs.

Yamauchi and co-workers reported that the atypical guanine-nucleotide exchange factor dedicator of cytokinesis protein 7 (Dock7), activates Rac/Cdc42/JNK signaling in SCs to positively regulate migration and negatively regulate differentiation and myelination (Yamauchi et al., 2008, 2011). Indeed, knocking down Dock7 in primary SCs results in an increase of cAMP-mediated differentiation and in a shorter duration of activation of Rac/Cdc42 and JNK while silencing Dock7 in mice increases myelin thickness.

Finally, collagen triple helix repeat containing 1 (Cthrc1) is another protein that has been linked to Rac activation in SCs. Apra and colleagues investigated Cthrc1 function in SCs both *in vitro* and *in vivo* (Apra et al., 2012). They showed that its silencing inhibits SC proliferation and promotes their migration and myelination *in vitro*. By using a transgenic mouse line, they demonstrated that the overexpression of Cthrc1 in SCs leads to a delayed myelination with SCs maintaining a proliferative state. More recently, Zhou et al identified miR-9 as an important negative regulator of SC migration targeting Cthrc1 after nerve injury (Zhou et al., 2014). The up-regulation of miR-9 decreases SC migration in culture and following nerve injury. miR-9 has also been shown to directly target Cthrc1, which in turn inactivates downstream Rac. Finally, miR-9 is down-regulated at a speed comparable to the rate of SC migration during nerve regeneration, suggesting that miR-9 down-regulation could be linked to the phenotypic modulation of SCs after nerve injury.

All these data confirm that Rac/JNK signaling is essential for SC plasticity, proliferation and migration following nerve injury. However, stimulating this pathway for nerve repair therapy must be cautioned as its upregulation has also been observed in primary human schwannoma cells (Kaempchen et al., 2003).

p38 MAPK

Yang and co-workers provided the most relevant data that suggest a role for p38 MAPK in SC plasticity and their response to injury (Yang et al., 2012). They reported that inhibition of p38 MAPK activity in mice blocks SC demyelination and dedifferentiation following nerve injury, despite ongoing axon degeneration. In myelinating co-cultures, they also showed that p38 MAPK mediates myelin breakdown induced by Nrg1. Finally, while the ectopic activation of p38 MAPK is sufficient to provoke myelin breakdown and c-Jun induction, its inhibition in SC-DRG neuron co-cultures promotes myelination. Roberts and collaborators recently investigated the *in vivo* function of the major p38 isoform in mouse conditional knockouts. In agreement with previous findings, the lack of p38alpha in SCS accelerates myelination during development, delays myelin clearance and slightly increases remyelination following injury. However, axonal regeneration and functional repair were not affected in the absence of p38alpha (Roberts et al., 2016).

A recent study on bone morphogenetic protein 7 (BMP7) in SCs revealed a role for this protein as a negative regulator of myelination through the activation of p38 MAPK signaling (Liu et al., 2016b). Liu and collaborators showed that BMP7 expression is inversely correlated with myelin gene expression during normal peripheral myelination. In addition, BMP7 treatment decreases cAMP-induced myelin gene expression and up-regulates c-Jun in primary SC cultures. By using p38 MAPK inhibitor, gene silencing and rescue experiments, they reported that p38 MAPK is responsible for BMP7-mediated suppression of myelin gene expression. Finally, when injected in newborn rats, BMP7 retards peripheral myelination.

PI3K/Akt/mTOR Signaling

Phosphatidylinositol 3-kinase (PI3K) signaling has been implicated in the regulation of proliferation, survival and differentiation of different cell types (Rameh and Cantley, 1999). The pathway is strongly activated in primary cultured SCs following treatment with soluble Nrg1 or contact with neurite membranes. Moreover, PI3K signaling is required for SC proliferation and survival *in vitro* (Maurel and Salzer, 2000; Li et al., 2001). Its inhibition affects the initiation of myelination but is not necessary for myelin maintenance in SC-neuron co-cultures (Maurel and Salzer, 2000). In addition, the selective activation of Akt in SCs by gene transfer results in an increased myelination in SC-DRG neuron co-cultures and allogenic nerve graft experiments. This last paper also demonstrated opposing functions for PI3K/Akt and Erk pathways in controlling SC myelination (Ogata et al., 2004). The imbalanced activation of the PI3K/Akt and Erk pathways has also been revealed in transgenic rodent models of CMT1A in which SCs develop persistent differentiation defects during early postnatal development (Fledrich et al., 2014). The transgenic expression of constitutively active Akt in SCs, however does not seem to disrupt PNS myelination *in vivo* (Flores et al., 2008).

Studies investigating the role of specific proteins indirectly indicated a role for PI3K/Akt signaling in the regulation of SC proliferation and differentiation after nerve injury. For example, Src-associated in mitosis of 68 kD (Sam68) has been suggested to promote SC proliferation by enhancing the PI3K/Akt pathway and to act on regeneration following sciatic nerve crush (Wu et al., 2016b). DIX domain containing-1 (Dixdc1), which is also linked to PI3K/Akt activation, promotes SC proliferation after sciatic nerve crush. Dixdc1 is up-regulated following nerve damage and its depletion leads to reduced PI3K/Akt activation and SC proliferation (Wu et al., 2016a). Consistently, 17β-estradiol has been shown to promote SC proliferation and early remyelination in the bridge after nerve transection through Akt activation (Chen et al., 2016a). Src-suppressed protein kinase C substrate (SSeCKS) has been indicated as a suppressor of SC differentiation and myelination by inhibiting Akt activation.

SSeCKS is effectively down-regulated during SC differentiation and its depletion promotes Akt phosphorylation and myelin gene expression in cAMP-treated primary cultures or myelination when co-cultured with DRG neurons (Ji et al., 2010). SSeCKS is also essential for SC adhesion, spreading and migration (Yan et al., 2009). In addition, Chen and collaborators showed that cell adhesion molecule 3 (Cadm3) might act as a negative modulator of PNS myelination. Cadm3 interferes with PI3K/Akt activation and inhibits Schwann cell myelination *in vitro* (Chen M. S. et al., 2016).

The mammalian disks large homolog 1 (Dlg1) inhibits PI3K/Akt signaling via its interaction with the phosphatase and tensin homolog deleted on chromosome 10 (PTEN). This mechanism is necessary to transiently inhibit the axonal stimulation of myelination and to limit myelin sheath thickness. Myelin outfoldings and overmyelination followed later by pathologic demyelination are actually observed after Dlg1 silencing via RNA interference in mouse sciatic nerves (Cotter et al., 2010). Dlg1/PTEN functions and consequently Akt/mTOR signaling are also disturbed in different mouse models of demyelinating neuropathies and peripheral nerve sheath tumors (Bolino et al., 2004; Cotter et al., 2010; Goebbels et al., 2010, 2012; Keng et al., 2012). More specifically, Goebbels et al showed that mice lacking PTEN develop a tomaculous neuropathy characterized by focal hypermyelination and myelin outfoldings. Moreover, treating these mice with rapamycin, a mTOR inhibitor, significantly ameliorates the pathology (Goebbels et al., 2010, 2012). Another study on PTEN signaling showed that the hypermyelination observed in mice lacking Dlg1 is actually transitory. This study also identified DNA damage-inducible transcript 4 protein (DDIT4) as a sustained negative regulator of myelination. High DDIT4 expression in SCs precedes the peak of Dlg1 and Akt activity in peripheral nerves. Loss of DDIT4 expression both *in vitro* and *in vivo* results in sustained hypermyelination and enhanced PI3K/Akt/mTOR activation (Noseda et al., 2013). Therefore, it is likely that PI3K/Akt signaling has to be precisely regulated and balanced to induce SC differentiation or dedifferentiation.

mTOR is a signaling pathway that integrates a lot of signals (growth factors, nutrients, energy and oxygen) necessary for growth and proliferation (Norrmén and Suter, 2013). Sherman and collaborators demonstrated that the inactivation of mTOR in SCs provokes an arrest of myelination after axonal sorting and reduces axon growth (Sherman et al., 2012). More recently, it has been reported that the conditional ablation of raptor, an essential component of the mTORC1 complex, in SCs leads to hypomyelination and abnormal lipid biosynthesis (Norrmén et al., 2014). The pathways that converge to mTOR are numerous and include Nrg1/ErbB, PI3K/Akt, MAPKS or AMP-activated kinase (AMPK) signaling (Norrmén and Suter, 2013). AMPK is a heterotrimeric serine/threonine protein kinase acting as an inhibitor of mTOR and described as a crucial energy sensor important to maintain energy homeostasis (Hardie et al., 2012). Recently, Liu and colleagues revealed that AMPK is a critical negative regulator of SC myelination in the PNS (Liu et al., 2016a). They first showed that AMPK is gradually decreased as myelination proceeds. The high activation of AMPK decreases myelin gene expression and stimulates c-Jun expression in SCs whereas its inhibition or depletion induces myelin gene expression. The injection of an inhibitor of AMPK in newborn rats downregulates c-Jun expression, enhances lipid and protein synthesis and increases myelin gene expression and myelin sheath thickness.

Wnt Signaling and LXR

Wnt/β-catenin pathway has been demonstrated to be crucial for PNS development. Cell cultures and genetic experiments revealed that Wnt signaling is required for SC lineage progression, proliferation and myelination (Hari et al., 2002; Lee et al., 2004; Gess et al., 2008; Grigoryan et al., 2013). Recently, it has been proposed that interplay between Wnt pathway and liver X receptors (LXR) induces the fine-tuning of myelin gene expression (Makoukji et al., 2011). Oxysterols acting as ligands for LXR, are natural compounds that originate from the enzymatic oxidation of cholesterol. They are involved in cholesterol homeostasis and in the progression of different neurodegenerative disorders (Lütjohann et al., 2000; Russell, 2000; Leoni et al., 2002). Makoukji and co-workers provided evidence for the implication of LXR and oxysterols in the negative regulation of myelin gene expression (Makoukji et al., 2011). They showed that oxysterols are expressed in a SC line and in sciatic nerves and that they repress the gene expression of *MPZ* and *PMP22* by a mechanism involving LXR α and β and the inhibition of the Wnt/β-catenin-dependent pathway. Unexpectedly, although myelin gene transcripts are up-regulated in mice lacking LXR α and β, presumably due to increased β-catenin activity, myelin protein expression is decreased and myelin sheath thickness is reduced. They conclude that this hypomyelination in the knockout animals may be the consequence of an altered cholesterol homeostasis and an inefficient myelin protein trafficking from the endoplasmic reticulum.

Recently, paraquat, a redox-active herbicide, has been shown to elicit an oxidative stress in the normal sciatic nerves of adult mice and a dramatic disorganization of myelin sheaths, causing severe locomotor and sensory deficits through a stimulation of the LXR pathway leading to reduced Wnt signaling (Hichor et al., 2016). Indeed, the authors provided evidence that paraquat alters myelin gene expression by activating LXR signaling and by preventing β-catenin recruitment on the myelin gene promoter. They also indicated that treating paraquat-exposed mice with lithium, a Wnt/β-catenin modulator acting as a GSK3 inhibitor, prevents the deleterious defects of paraquat on sciatic nerves. The effect of lithium chloride has also been studied after nerve injury. It accelerates axon and myelin debris clearance and remyelination by inhibiting GSK3 and increasing β-catenin levels (Chen et al., 2016b).

Toll-like Receptor Signaling

The Toll-like receptors (TLRs) are a class of proteins that play a key role in initiating the immune and inflammatory response. After nerve injury, damage-associated molecular patterns and ligands released from or expressed by activated or necrotic damaged cells result in the activation of TLRs in different cell types including SCs (Thakur et al., 2017). TLR3, TLR4, and TLR7 are constitutively expressed by SCs and TLR1 is upregulated following nerve injury, strongly suggesting a role for TLR signaling in SC-driven nerve repair (Goethals et al., 2010). *In vitro*, the addition of necrotic neurons to SCs activates their inflammatory response through TLR signaling. It increases their gene expression of inflammatory mediators such as tumor necrosis factor-alpha (TNF-α) or monocyte chemoattractant protein-1 (MCP-1) but this effect is reduced in SCs from mice lacking TLR3 (Lee et al., 2006). Boivin and collaborators provided *in vivo* evidence that TLR signaling is involved in WD and nerve regeneration following a nerve injury, possibly through NF-κB activation (Boivin et al., 2007). The early expression of inflammatory modulators such as MCP-1, macrophage recruitment and activation, axonal regeneration and functional recovery are impaired in the mice deficient in TLR signaling. The results are indeed similar in mice lacking TLR2, TLR4 or their adaptor, myeloid differentiation primary response gene 88 (MyD88) known to activate NF-κB. Interestingly, a single injection of TLR2 and TLR4 ligands at the site of the lesion speeds up myelin clearance and functional recovery. Besides its likely role in SC response to nerve injury, TLR signaling has also been linked to the development of neuropathic pain, making TLRs promising therapeutic targets (Thakur et al., 2017).

GPCR Signaling

To date, three different G protein-coupled receptors (GPCRs) have been demonstrated to be implicated in SC differentiation. These are Gpr126, Gpr44, and lysophosphatidic acid receptor 1 (LPA1) (Mogha et al., 2016a). Monk and colleagues were the first ones to show that a GPCR, Gpr126, is crucial for the initiation of SC myelination (Monk et al., 2009). The analysis of mice lacking LPA1 then revealed that this receptor is required for SC migration, axonal sorting and proper myelination (Anliker et al., 2013). Finally, Trimarco et al. provided *in vitro* and *in vivo* evidence for the involvement of Gpr44 and its ligand the prostaglandin D2 in PNS myelination (Trimarco et al., 2014). Very recently, Mogha and collaborators investigated the potentiel functions of Gpr126 in peripheral nerve injury and repair as its expression is maintained in adult SCs (Mogha et al., 2016b). By using an inducible conditional Gpr126 knockout model, they observed defects in remyelination, macrophage recruitment and axon regeneration following a nerve crush. Since approximately 30% of all approved drugs target GPCRs, Gpr126 may represent an attractive potential target to stimulate repair in myelinating disease or after nerve damage.

Nrg1/ErbB2/3 Signaling

Currently, the signals from damaged nerves that induce the reprogramming of SCs are not identified. Some argued that these can be linked to the Nrg1/ErbB pathway, one of the best described interdependent relationship between axons and SCs. Neuregulins form a large family of epidermal growth factor-like proteins that signal through the ErbB tyrosine kinase receptors (Falls, 2003). The Nrg1/ErbB signaling is first of all known for its critical functions in SC development (for review, Newbern and Birchmeier, 2010; Grigoryan and Birchmeier, 2015; Willem, 2016). Indeed, numerous genetic models have been generated to unravel fundamental functions of the Nrg1/ErbB signaling axis. Differential expression of Nrg1 isoforms, their proteolytic processing or ErbB receptor localization and trafficking are important parameters that control the Nrg1/ErbB pathway and its underlying functions. Depending on the developmental stage, the pathway significantly regulates SC survival, proliferation, migration, differentiation and myelination (Syroid et al., 1996; Morris et al., 1999; Leimeroth et al., 2002; Taveggia et al., 2005; Chen et al., 2006; Freidin et al., 2009; Newbern and Birchmeier, 2010). This functional diversity is presumably accomplished through the interaction of Nrg1/ErbB signaling with other pathways. For instance, Notch signaling, which is necessary for SCP proliferation and survival, increases the expression of ErbB2 in SCPs and their sensitivity to Nrg1 (Woodhoo et al., 2009). The tyrosine phosphatase Shp2 is another essential component of the Nrg-1/ErbB pathway that promotes SC proliferation and migration (Grossmann et al., 2009). Later in SC development, critical interactions between Nrg1/ErbB and cAMP or NF-κB signaling are necessary for SC differentiation and myelination (Limpert and Carter, 2010; Arthur-Farraj et al., 2011). Although the importance of Nrg1/ErbB signaling in SC development is well established, its role in adult SC plasticity is more controversial. In adulthood, the pathway seems dispensable for the maintenance of the myelin sheaths, as the absence of Nrg1 in peripheral axons or of ErbB2 in SCs has no effect on axon or myelin sheath integrity (Atanasoski et al., 2006; Fricker et al., 2011, 2013). However, increasing evidence indicates that Nrg1 is necessary for the SC response to nerve injury.

The fact that the Nrg1/ErbB system is selectively and highly regulated during peripheral nerve degeneration and regeneration represents a strong argument in favor of a role for this signaling pathway in SC plasticity (Carroll et al., 1997; Kwon et al., 1997; Ronchi et al., 2016). Recently, a study using a rat surgical model of delayed nerve repair showed that Nrg1 and some SC markers are highly down-regulated after chronic degeneration. Since long-term denervated Schwann cells are known to be partly responsible for delayed nerve repair, it suggests that Nrg1 plays an important function in SC activity after denervation (Ronchi et al., 2017). Guertin and colleagues showed that ErbB2 is activated in myelinating SCs after sciatic nerve axotomy and that its transient activation is sufficient to initiate SC demyelination in compartmentalized cell culture chambers. In addition, the treatment of rats with an ErbB2 inhibitor results in a reduction of demyelination after nerve transection (Guertin et al., 2005). Consistently, Syed et al demonstrated that a robust activation of ErbB2/3 via high concentration of soluble Nrg1 induces SC dedifferentiation and demyelination as well as increased c-Jun expression in primary cultures (Syed et al., 2010). Furthermore, *in vitro* experiments recapitulating nerve injuries provide evidence that Nrg1 treatment shows beneficial effects on SC migration and proliferation (Mahanthappa et al., 1996; Li et al., 1998). Finally, Lee and collaborators showed that Nrg1 on motor axons controls the terminal SC-mediated synapse elimination at the neuromuscular junctions (NMJs) during development. They also revealed that NMJs of adult transgenic mice overexpressing Nrg1 in motor axons exhibited continued remodeling, identifying Nrg1 as a molecular determinant for SC-driven neuromuscular synaptic plasticity (Lee et al., 2016).

Nrg1/ErbB signaling has also been linked to different pathways mentioned earlier for regulating SC plasticity. For example, Nrg1 has been proposed as the signal driving high Erk activation after injury (Napoli et al., 2012). High concentrations of Nrg1 indeed highly activates Erk signaling and induces SC dedifferentiation and demyelination (Syed et al., 2010; Newbern and Snider, 2012). In agreement with this, Tapinos and collaborators also showed that direct binding of Mycobacterium leprae onto ErbB2 receptors on SCs activates Erk signaling and promotes demyelination (Tapinos et al., 2006). Another possibility is that Nrg1 could trigger Rac/JNK activation. Indeed, the inhibition of ErbB2 signaling has been shown to prevent MKK7 activation, c-jun expression, and Rac-dependent gene transcription in cultures of sciatic nerve explants (Shin et al., 2013). ErbB2 also directly binds and activates the negative regulator of myelination, Dock7, upstream of Rac/JNK while the depletion of Dock7 attenuates Nrg1 effects in primary SC cultures (Yamauchi et al., 2008, 2011). Although these data suggest a role for Nrg1/ErbB signaling in SC dedifferentiation, *in vivo* studies revealed contradictory results. Atanasoski et al. showed that lack of ErbB2 in adult SCs does not affect their proliferation and survival after nerve injury, despite reduced levels of phosphorylated MAPK (Atanasoski et al., 2006). In addition, the absence of the axonal Nrg1 following a nerve crush does not disturb SC proliferation associated with nerve degeneration or the clearance of myelin debris by macrophages. Therefore, adult SCs would not require major Nrg1/ErbB signaling for their proliferation and survival after nerve damage, in contrast to what is observed *in vitro* and during development.

While the implication of Nrg1 signaling in the dedifferentiation of SCs is unclear, its involvement in myelination but also in remyelination after nerve injury is now well established. The transmembrane axonal isoforms Nrg1 type III signals in a juxtacrine manner through ErbB2/3 receptors on SCs and determines their myelin thickness, both during development and following a nerve lesion (Michailov et al., 2004; Taveggia et al., 2005; Fricker et al., 2011, 2013). The axonal Nrg1 is rate limiting but not essential for remyelination since compensations are observed at later stages (Fricker et al., 2013). All the same, myelin sheaths are often thinner after injury, suggesting an insufficent simulation by axonal signals (Schröder, 1972). Another mechanism is actually induced in SCs to supplement these poor neuronal cues. Nerve damage triggers the expression of soluble Nrg1 type I by SCs themselves. This particular autocrine and paracrine signaling in SCs is only required after nerve injury and promotes SC survival, redifferentiation and remyelination (Fricker and Bennett, 2011; Stassart et al., 2013).

Besides the importance of Nrg1/ErbB signaling in SC remyelination, it also plays a key role in repair processes after nerve injury (Fricker and Bennett, 2011). Fricker and colleagues showed that the absence of Nrg1 in adult axons after sciatic nerve crush results in defects of remyelination but also in slower axon regeneration, impaired muscle reinnervation and delayed functional recovery (Fricker et al., 2011, 2013). In a mouse model lacking β-Site amyloid precursor protein cleaving enzyme 1 (BACE1), a protein known to cleave and activate Nrg1 type III, SC remyelination is impaired and nerve repair is surprisingly accelerated (Hu et al., 2008). Indeed, knockout mice show an accelerated clearance of axonal and myelin debris from degenerated fibers, faster axonal regeneration and earlier reinnervation of neuromuscular junctions, compared to littermate controls (Farah et al., 2011). The enhanced myelin clearance might clearly be due to the initial hypomyelination observed in adult BACE1 KO sciatic nerves. However, how BACE1 deletion enhances peripheral regeneration still needs to be clarified. Another enzyme capable of cleaving the axonal Nrg1, counteracting BACE1 and interfering with SC differentiation is the tumor necrosis factor-α–converting enzyme (TACE, also named ADAM17). TACE depletion in SC-DRG neuron co-cultures increases myelination. The absence of TACE in motoneurons *in vivo* results in hypermyelination. The function of TACE is neuron autonomous since its knockdown in SCs does not affect myelination. Mechanistically, TACE cleaves Nrg1-III, which is no longer capable of activating PI3K signaling and inducing myelination (La Marca et al., 2011). Therefore, TACE can be considered as a modulator of NRG1 type III activity and as a negative regulator of SC myelination. The role of Nrg1 in peripheral regeneration has also been demonstrated in different studies examining the effects of Nrg1 treatment following nerve injury. The application of exogenous Nrg1 or strategies that elevates Nrg1 levels after nerve damage improve myelin clearance, axon regeneration, remyelination and functional recovery (Chen et al., 1998; Joung et al., 2010; Fricker and Bennett, 2011; Yildiz et al., 2011). Therefore, treatment with recombinant forms of Nrg1 may represent interesting therapeutic avenues to improve nerve repair following injury. It is however necessary to better understand the mechanisms by which Nrg1 treatment improves nerve regeneration.

Finally, it is essential to remember that the inappropriate overactivation of Nrg1/ErbB signaling pathway may be harmful in some circumstances and lead to demyelinating neuropathies or peripheral tumors. For instance, constitutive activation of the Nrg1/ErbB pathway promotes the proliferation of peripheral neuroepithelioma and neoplastic SC line (Frohnert et al., 2003; Fallon et al., 2004). Similarly, transgenic mice over-expressing Nrg1 in myelinating SCs develop hypertrophic neuropathies and malignant peripheral nerve sheath tumors (Huijbregts et al., 2003). In addition, leprosy, which is an important cause of demyelinating neuropathy in the world, is initiated by an excessive activation of ErbB2 by *M. leprae* (Tapinos et al., 2006).

PATHOPHYSIOLOGICAL IMPLICATIONS AND THERAPEUTIC STRATEGIES

Peripheral neuropathies are responsible for significant morbidity and decreased quality of life due to weakness, sensory loss and neuropathic pain. The causes are multiple: peripheral nerve traumatic injuries, inherited genetic diseases such as CMT, metabolic disorders including diabetes, infectious, toxic and inflammatory disorders (Zhou and Notterpek, 2016). Peripheral nerves have an impressive ability to regenerate compared to the CNS. However, the clinical recovery of patients suffering from peripheral neuropathies is often incomplete due to slow regeneration rate, target mis-reinnervation and lack of a long-term regeneration-supportive environment. After injury, peripheral axons regenerate at a very slow rate of approximately 1 mm per day, depending on the lesion site. Moreover, the regenerative capacity of the PNS decreases over time. Indeed, SCs become unsupportive to regeneration after long periods and they also exhibit diminished plasticity with age (Fu and Gordon, 1995; Zochodne, 2012; Painter et al., 2014). Of note, Joshi and coworkers recently demonstrated that SCs show an altered expression of c-Jun and Cdkn1c and that their pro-regenerative functions are diminished in a model of chronic inflammatory demyelinating polyneuropathy (CIDP) (Joshi et al., 2016). A better understanding of the signaling pathways that drive SC reprogramming and plasticity is of great interest and may offer new opportunities to enhance nerve repair. SCs are the main effectors of regeneration in many peripheral diseases and recent studies clearly positioned SCs at the hub for organizing the environment to clear myelin debris, promote axonal regrowth, remyelination and allow complete functional repair following injury (Zochodne, 2012; Kim et al., 2013).

The different signaling pathways and molecular components described in the present review represent interesting targets to boost the endogenous regeneration potential of SCs. For instance, the treatment with different Nrg1 isoforms or strategies to stimulate ErbB signaling have actually been shown to improve myelin clearance, axon regeneration, remyelination and functional recovery after nerve injury in different studies (Chen et al., 1998; Joung et al., 2010; Fricker and Bennett, 2011; Yildiz et al., 2011). In a CMT1A rat model, soluble Nrg1 therapy during early postnatal life prevents abnormal developmental demyelination and subsequent axon loss observed in adulthood (Fledrich et al., 2014). The overexpression of Nrg1 in mouse models of amyotrophic lateral sclerosis (ALS) and partial muscle denervation improves functional recovery by enhancing collateral reinnervation sprouting through both Akt and Erk pathways (Mancuso et al., 2016). Notch activation through the addition of a recombinant jagged1 after injury also promotes nerve regeneration and functional recovery (Wang et al., 2015). The inhibition of mTOR signaling with rapamycin in mice that lack PTEN and develop a tomaculous neuropathy significantly ameliorates the pathology (Goebbels et al., 2012). Finally, stimulating TLR signaling at the site of the lesion accelerates myelin clearance and functional recovery (Boivin et al., 2007).

Additional molecules, not directly linked to pathways analyzed in this review, have been shown to regulate SC differentiation/dedifferentiation and improve nerve repair. For example, Fingolimod, an agonist of the sphingosine-1-phosphate receptor (S1PR), previously known to affect SC migration and myelination *in vitro* and currently in clinical trial for the treatment of CIDP, has been demonstrated to modulate SC plasticity (Köhne et al., 2012; Heinen et al., 2015). Fingolimod activates different dedifferentiation markers such as c-Jun, increases the expression of growth factors, slows myelination and improves axon regeneration. Therefore, Fingolimod supports the generation of a repair promoting cellular phenotype and could be used for the treatment of peripheral nerve damages and diseases (Heinen et al., 2015). Its mechanism of action, however, still remains to be elucidated. Another recent example is the addition of TGF-β1 and forskolin directly on chronically injured nerves that can reactivate the growth-supportive environment through the reprogramming of chronically denervated SCs, the re-induction of their proliferation, and the re-expression of regeneration-associated proteins (Sulaiman and Nguyen, 2016). Immunoglobulins (IVIG) are another therapeutic strategy widely used to treat immune-mediated neuropathies such as CIDP. Tzekova and colleagues demonstrated that IVIG positively influence SC differentiation and maturation and increase their potential to induce axonal outgrowth (Tzekova et al., 2015).

Other approaches that are based on SC regenerative properties and improve nerve repair can also be addressed. First, exercise and electrical stimulation have been demonstrated to promote SC-driven peripheral nerve regeneration in both animal models and in human patients (Gordon and English, 2016). Second, transplantations of SCs from healthy nerves or engineered from different tissues thanks to the emergence of the induced pluripotent stem cell technology are another interesting method to support nerve healing (Zhou and Notterpek, 2016). Actually, numerous SC-based therapies also exist to promote repair in the lesioned CNS (Matsas et al., 2008). Third, there has been major progress in developing biomaterials that improve regeneration by providing trophic factors or by supporting the regenerative functions of transplanted cells *in situ* (Marquardt and Sakiyama-Elbert, 2013). For example, Liu and colleagues recently showed that Salidroside and tissue engineering using SCs and poly lactic-co-glycolic acid promote peripheral nerve regeneration and functional recovery after sciatic nerve transection in rats. Salidroside increases the proliferation and regenerative function of SCs, probably through the modulation of neurotrophic factors (Liu et al., 2017). Fourth, gene therapy that allows to deliver genetic material directly into somatic cells is another option to treat neuropathies that involve gene mutations or to improve nerve regeneration after injuries (Zhou and Notterpek, 2016). In hereditary diseases including CMTs, loss or gain of function mutations could be restored through the re-expression of the absent or non-functional protein or through the down-regulation of the mutant mRNA via RNA interference strategy, respectively. For example, Kagiava and coworkers recently demonstrated that the intrathecal injection of a lentiviral vector with a myelin-specific promoter restored the expression of a neuropathy-associated gene and rescued a model of demyelinating peripheral neuropathy (Kagiava et al., 2016). Also, other studies based on gene therapy in order to improve nerve regeneration used lentiviral vectors that potentiate the therapeutic functions of SCs transplanted in artificial nerves, in autografts or present in damaged nerves (Hoyng et al., 2015). Even if SCs are known to mainly support nerve repair through an enhanced expression of neurotrophic factors that stimulate axon regeneration and myelination, this review highlights the numerous additional roles and regenerative functions of SCs in nerve injuries and in demyelinating diseases. Therefore, targeting the molecular mechanisms that regulate SC reprogramming by gene therapy can be highly promising to stimulate nerve healing. In the future, the challenges will be to optimize gene therapy in order to specifically target SCs and to create safe and regulatable vectors to potentiate their regenerative functions in the appropriate timings. Lipid-based nanoparticles such as liposomes or extracellular vesicles are other valuable tools for the delivery of small molecules or genes that can modulate signaling pathways and promote nerve repair (Takeda and Xu, 2015). In order to selectively introduce material into SCs, Lee and colleagues developed phospholipid-based liposomes that can enter SCs when injected in tail vein and showed that the nanoparticles escape the endogenous degradative mechanisms *in vitro* (Lee et al., 2013). In the following years, it is expected that the improvements in cell and tissue engineering and in approaches that target SC capacity to promote regeneration will lead to the development of innovative and efficient PNS repair strategies.

The different therapeutic options that are built on or that stimulate SC plasticity represent new avenues for nerve regeneration but should be considered with prudence for different reasons. First, a prolonged stimulation of SC dedifferentiation could disrupt the remyelination of the regenerated axons, necessary for rapid nerve conduction and complete functional recovery. Also, the aberrant activation of some pathways including Notch, NF-κB, Ras/Raf/Erk, Rac/JNK, or Nrg1/ErbB can lead to abnormal cell proliferation and the development of peripheral nerve tumors. The better understanding of the molecular mechanisms regulating SC plasticity is thus necessary to set up therapies that modulate both temporally and quantitatively SC capacity to drive nerve repair without affecting remyelination and increasing the chances of triggering tumor development.

CONCLUSIONS

The remarkable capacity of regeneration of the PNS compared to the CNS is mainly due to the fantastic plasticity of SCs. The molecular components involved in the negative regulation of myelination and in the reprogramming of SCs in response to nerve injury provide important mechanistic insights and new therapeutic approaches into peripheral neuropathies as well as in peripheral nerve regeneration. However, this review demonstrates the complexity of the molecular signaling pathways and reveals that little is known about the interactions between them. Moreover, the temporal and quantitative activation of the different signals identified is of great importance but is not completely elucidated. Future studies are thus required to better understand the cross-talks, networks and activation patterns and timings between the different molecular mechanisms controlling SC plasticity.

AUTHOR CONTRIBUTIONS

Manuscript writing: AB; Manuscript revision: VD, AC, and RF.

ACKNOWLEDGMENTS

This work was supported by Grants from the FSR-FNRS, the Fonds Léon Fredericq, the Special Funds of the University of Liège, the Fondation Charcot and the Association Belge contre les Maladies Neuro-Musculaires.

REFERENCES

Agthong, S., Kaewsema, A., Tanomsridejchai, N., and Chentanez, V. (2006). Activation of MAPK ERK in peripheral nerve after injury. *BMC Neurosci.* 7:45. doi: 10.1186/1471-2202-7-45

Ammoun, S., Flaiz, C., Ristic, N., Schuldt, J., and Hanemann, C. O. (2008). Dissecting and targeting the growth factor-dependent and growth factor-independent extracellular signal-regulated kinase pathway in human schwannoma. *Cancer Res.* 68, 5236–5245. doi: 10.1158/0008-5472.CAN-07-5849

Ammoun, S., Provenzano, L., Zhou, L., Barczyk, M., Evans, K., Hilton, D. A., et al. (2014). Axl/Gas6/NFκB signalling in schwannoma pathological proliferation, adhesion and survival. *Oncogene* 33, 336–346. doi: 10.1038/onc.2012.587

Andorfer, B., Kieseier, B. C., Mathey, E., Armati, P., Pollard, J., Oka, N., et al. (2001). Expression and distribution of transcription factor NF-κB and inhibitor IκB in the inflamed peripheral nervous system. *J. Neuroimmunol.* 116, 226–232. doi: 10.1016/S0165-5728(01)00306-X

Anliker, B., Choi, J. W., Lin, M.-E., Gardell, S. E., Rivera, R. R., Kennedy, G., et al. (2013). Lysophosphatidic acid (LPA) and its receptor, LPA1, influence embryonic schwann cell migration, myelination, and cell-to-axon segregation. *Glia* 61, 2009–2022. doi: 10.1002/glia.22572

Apra, C., Richard, L., Coulpier, F., Blugeon, C., Gilardi-Hebenstreit, P., Vallat, J., et al. (2012). Cthrc1 is a negative regulator of myelination in Schwann cells. *Glia* 60, 393–403. doi: 10.1002/glia.22273

Arthur-Farraj, P. J., Latouche, M., Wilton, D. K., Quintes, S., Chabrol, E., Banerjee, A., et al. (2012). c-Jun reprograms Schwann cells of injured nerves to generate a repair cell essential for regeneration. *Neuron* 75, 633–647. doi: 10.1016/j.neuron.2012.06.021

Arthur-Farraj, P., Wanek, K., Hantke, J., Davis, C. M., Jayakar, A., Parkinson, D. B., et al. (2011). Mouse schwann cells need both NRG1 and cyclic AMP to myelinate. *Glia* 59, 720–733. doi: 10.1002/glia.21144

Atanasoski, S., Scherer, S. S., Sirkowski, E., Leone, D., Garratt, A. N., Birchmeier, C., et al. (2006). ErbB2 signaling in Schwann cells is mostly dispensable for maintenance of myelinated peripheral nerves and proliferation of adult Schwann cells after injury. *J. Neurosci.* 26, 2124–2131. doi: 10.1523/JNEUROSCI.4594-05.2006

Barrette, B., Hébert, M.-A., Filali, M., Lafortune, K., Vallières, N., Gowing, G., et al. (2008). Requirement of myeloid cells for axon regeneration. *J. Neurosci.* 28, 9363–9376. doi: 10.1523/JNEUROSCI.1447-08.2008

Beirowski, B., Babetto, E., Golden, J. P., Chen, Y.-J., Yang, K., Gross, R. W., et al. (2014). Metabolic regulator LKB1 plays a crucial role in Schwann cell-mediated axon maintenance. *Nat. Neurosci.* 17, 1351–1361. doi: 10.1038/nn.3809

Boivin, A., Pineau, I., Barrette, B., Filali, M., Vallières, N., Rivest, S., et al. (2007). Toll-like receptor signaling is critical for wallerian degeneration and functional recovery after peripheral nerve injury. *J. Neurosci.* 27, 12565–12576. doi: 10.1523/JNEUROSCI.3027-07.2007

Bolino, A., Bolis, A., Previtali, S. C., Dina, G., Bussini, S., Dati, G., et al. (2004). Disruption of Mtmr2 produces CMT4B1-like neuropathy with myelin outfolding and impaired spermatogenesis. *J. Cell Biol.* 167, 711–721. doi: 10.1083/jcb.200407010

Boyd, J. G., and Gordon, T. (2003). Glial cell line-derived neurotrophic factor and brain-derived neurotrophic factor sustain the axonal regeneration of chronically axotomized motoneurons *in vivo*. *Exp. Neurol.* 183, 610–619. doi: 10.1016/S0014-4886(03)00183-3

Carroll, S. L., Miller, M. L., Frohnert, P. W., Kim, S. S., and Corbett, J. A. (1997). Expression of neuregulins and their putative receptors, ErbB2 and ErbB3, is induced during Wallerian degeneration. *J. Neurosci.* 17, 1642–1659.

Cattin, A.-L., Burden, J. J., Van Emmenis, L., Mackenzie, F. E., Hoving, J. J. A., Calavia, N. G., et al. (2015). Macrophage-induced blood vessels guide Schwann cell-mediated regeneration of peripheral nerves. *Cell* 162, 1127–1139. doi: 10.1016/j.cell.2015.07.021

Cattin, A.-L., and Lloyd, A. C. (2016). The multicellular complexity of peripheral nerve regeneration. *Curr. Opin. Neurobiol.* 39, 38–46. doi: 10.1016/j.conb.2016.04.005

Chaballe, L., Close, P., Sempels, M., Delstanche, S., Fanielle, J., Moons, L., et al. (2011). Involvement of placental growth factor in Wallerian degeneration. *Glia* 59, 379–396. doi: 10.1002/glia.21108

Chen, L. E., Liu, K., Seaber, A. V., Katragadda, S., Kirk, C., and Urbaniak, J. R. (1998). Recombinant human glial growth factor 2 (rhGGF2) improves functional recovery of crushed peripheral nerve (a double-blind study). *Neurochem. Int.* 33, 341–351. doi: 10.1016/S0197-0186(98)00037-0

Chen, M.-S., Kim, H., Jagot-Lacoussiere, L., and Maurel, P. (2016). Cadm3 (Necl-1) interferes with the activation of the PI3 kinase/Akt signaling cascade and inhibits Schwann cell myelination *in vitro*. *Glia* 64, 2247–2262. doi: 10.1002/glia.23072

Chen, S., Velardez, M. O., Warot, X., Yu, Z.-X., Miller, S. J., Cros, D., et al. (2006). Neuregulin 1-erbB signaling is necessary for normal myelination and sensory function. *J. Neurosci.* 26, 3079–3086. doi: 10.1523/JNEUROSCI.3785-05.2006

Chen, Y., Guo, W., Xu, L., Li, W., Cheng, M., Hu, Y., et al. (2016a). 17β-Estradiol promotes Schwann cell proliferation and differentiation, accelerating early remyelination in a mouse peripheral nerve injury model. *BioMed Res. Int.* 2016:7891202. doi: 10.1155/2016/7891202

Chen, Y., Wang, H., Yoon, S. O., Xu, X., Hottiger, M. O., Svaren, J., et al. (2011). HDAC-mediated deacetylation of NF-[κ] B is critical for Schwann cell myelination. *Nat. Neurosci.* 14, 437–441. doi: 10.1038/nn.2780

Chen, Y., Weng, J., Han, D., Chen, B., Ma, M., Yu, Y., et al. (2016b). GSK3β inhibition accelerates axon debris clearance and new axon remyelination. *Am. J. Transl. Res.* 8, 5410–5420.

Chen, Z.-L., Yu, W.-M., and Strickland, S. (2007). Peripheral Regeneration. *Annu. Rev. Neurosci.* 30, 209–233. doi: 10.1146/annurev.neuro.30.051606.094337

Cotter, L., Ozcelik, M., Jacob, C., Pereira, J. A., Locher, V., Baumann, R., et al. (2010). Dlg1-PTEN interaction regulates myelin thickness to prevent damaging peripheral nerve overmyelination. *Science* 328, 1415–1418. doi: 10.1126/science.1187735

Dilwali, S., Briët, M. C., Kao, S.-Y., Fujita, T., Landegger, L. D., Platt, M. P., et al. (2015). Preclinical validation of anti-nuclear factor-κ B therapy to inhibit human vestibular schwannoma growth. *Mol. Oncol.* 9, 1359–1370. doi: 10.1016/j.molonc.2015.03.009

Doddrell, R. D. S., Dun, X., Moate, R. M., Jessen, K. R., Mirsky, R., and Parkinson, D. B. (2012). Regulation of Schwann cell differentiation and proliferation by the Pax-3 transcription factor. *Glia* 60, 1269–1278. doi: 10.1002/glia.22346

Fallon, K. B., Havlioglu, N., Hamilton, L. H., Cheng, T. P. H., and Carroll, S. L. (2004). Constitutive activation of the neuregulin-1/erbB signaling pathway promotes the proliferation of a human peripheral neuroepithelioma cell line. *J. Neurooncol.* 66, 273–284. doi: 10.1023/B:NEON.0000014521.28294.84

Falls, D. L. (2003). Neuregulins: functions, forms, and signaling strategies. *Exp. Cell Res.* 284, 14–30. doi: 10.1016/S0014-4827(02)00102-7

Farah, M. H., Pan, B. H., Hoffman, P. N., Ferraris, D., Tsukamoto, T., Nguyen, T., et al. (2011). Reduced BACE1 activity enhances clearance of myelin debris and regeneration of axons in the injured peripheral nervous system. *J. Neurosci.* 31, 5744–5754. doi: 10.1523/JNEUROSCI.6810-10.2011

Fernandez-Valle, C., Bunge, R. P., and Bunge, M. B. (1995). Schwann cells degrade myelin and proliferate in the absence of macrophages: evidence from *in vitro* studies of Wallerian degeneration. *J. Neurocytol.* 24, 667–679. doi: 10.1007/BF01179817

Fledrich, R., Stassart, R. M., Klink, A., Rasch, L. M., Prukop, T., Haag, L., et al. (2014). Soluble neuregulin-1 modulates disease pathogenesis in rodent models of Charcot-Marie-Tooth disease 1A. *Nat. Med.* 20, 1055–1061. doi: 10.1038/nm.3664

Flores, A. I., Narayanan, S. P., Morse, E. N., Shick, H. E., Yin, X., Kidd, G., et al. (2008). Constitutively active Akt induces enhanced myelination in the CNS. *J. Neurosci.* 28, 7174–7183. doi: 10.1523/JNEUROSCI.0150-08.2008

Fontana, X., Hristova, M., Da Costa, C., Patodia, S., Thei, L., Makwana, M., et al. (2012). c-Jun in Schwann cells promotes axonal regeneration and motoneuron survival via paracrine signaling. *J. Cell Biol.* 198, 127–141. doi: 10.1083/jcb.201205025

Freidin, M., Asche, S., Bargiello, T. A., Bennett, M. V. L., and Abrams, C. K. (2009). Connexin 32 increases the proliferative response of Schwann cells to neuregulin-1 (Nrg1). *Proc. Natl. Acad. Sci. U.S.A.* 106, 3567–3572. doi: 10.1073/pnas.0813413106

Fricker, F. R., Antunes-Martins, A., Galino, J., Paramsothy, R., La Russa, F., Perkins, J., et al. (2013). Axonal neuregulin 1 is a rate limiting but not essential factor for nerve remyelination. *Brain* 136, 2279–2297. doi: 10.1093/brain/awt148

Fricker, F. R., and Bennett, D. L. (2011). The role of neuregulin-1 in the response to nerve injury. *Future Neurol.* 6, 809–822. doi: 10.2217/fnl.11.45

Fricker, F. R., Lago, N., Balarajah, S., Tsantoulas, C., Tanna, S., Zhu, N., et al. (2011). Axonally derived neuregulin-1 is required for remyelination and regeneration after nerve injury in adulthood. *J. Neurosci.* 31, 3225–3233. doi: 10.1523/JNEUROSCI.2568-10.2011

Frohnert, P. W., Stonecypher, M. S., and Carroll, S. L. (2003). Constitutive activation of the neuregulin-1/ErbB receptor signaling pathway is essential for the proliferation of a neoplastic Schwann cell line. *Glia* 43, 104–118. doi: 10.1002/glia.10232

Fu, E. S., Zhang, Y. P., Sagen, J., Candiotti, K. A., Morton, P. D., Liebl, D. J., et al. (2010). Transgenic inhibition of glial NF-κB reduces pain behavior and inflammation after peripheral nerve injury. *Pain* 148, 509–518. doi: 10.1016/j.pain.2010.01.001

Fu, S. Y., and Gordon, T. (1995). Contributing factors to poor functional recovery after delayed nerve repair: prolonged axotomy. *J. Neurosci.* 15, 3876–3885.

Gao, X., Daugherty, R. L., and Tourtellotte, W. G. (2007). Regulation of low affinity neurotrophin receptor (p75NTR) by early growth response (Egr) transcriptional regulators. *Mol. Cell. Neurosci.* 36, 501–514. doi: 10.1016/j.mcn.2007.08.013

Gaudet, A. D., Popovich, P. G., and Ramer, M. S. (2011). Wallerian degeneration: gaining perspective on inflammatory events after peripheral nerve injury. *J. Neuroinflammation* 8:110. doi: 10.1186/1742-2094-8-110

Gess, B., Halfter, H., Kleffner, I., Monje, P., Athauda, G., Wood, P. M., et al. (2008). Inhibition of N-cadherin and beta-catenin function reduces axon-induced Schwann cell proliferation. *J. Neurosci. Res.* 86, 797–812. doi: 10.1002/jnr.21528

Goebbels, S., Oltrogge, J. H., Kemper, R., Heilmann, I., Bormuth, I., Wolfer, S., et al. (2010). Elevated phosphatidylinositol 3,4,5-trisphosphate in glia triggers cell-autonomous membrane wrapping and myelination. *J. Neurosci.* 30, 8953–8964. doi: 10.1523/JNEUROSCI.0219-10.2010

Goebbels, S., Oltrogge, J. H., Wolfer, S., Wieser, G. L., Nientiedt, T., Pieper, A., et al. (2012). Genetic disruption of Pten in a novel mouse model of tomaculous neuropathy. *EMBO Mol. Med.* 4, 486–499. doi: 10.1002/emmm.201200227

Goethals, S., Ydens, E., Timmerman, V., and Janssens, S. (2010). Toll-like receptor expression in the peripheral nerve. *Glia* 58, 1701–1709. doi: 10.1002/glia.21041

Gomez-Sanchez, J. A., Carty, L., Iruarrizaga-Lejarreta, M., Palomo-Irigoyen, M., Varela-Rey, M., Griffith, M., et al. (2015). Schwann cell autophagy, myelinophagy, initiates myelin clearance from injured nerves. *J. Cell Biol.* 210, 153–168. doi: 10.1083/jcb.201503019

Gordon, T., and English, A. W. (2016). Strategies to promote peripheral nerve regeneration: electrical stimulation and/or exercise. *Eur. J. Neurosci.* 43, 336–350. doi: 10.1111/ejn.13005

Grigoryan, T., and Birchmeier, W. (2015). Molecular signaling mechanisms of axon-glia communication in the peripheral nervous system. *Bioessays* 37, 502–513. doi: 10.1002/bies.201400172

Grigoryan, T., Stein, S., Qi, J., Wende, H., Garratt, A. N., Nave, K.-A., et al. (2013). Wnt/Rspondin/beta-catenin signals control axonal sorting and lineage progression in Schwann cell development. *Proc. Natl. Acad. Sci. U.S.A.* 110, 18174–18179. doi: 10.1073/pnas.1310490110

Grossmann, K. S., Wende, H., Paul, F. E., Cheret, C., Garratt, A. N., Zurborg, S., et al. (2009). The tyrosine phosphatase Shp2 (PTPN11) directs Neuregulin-1/ErbB signaling throughout Schwann cell development. *Proc. Natl. Acad. Sci. U.S.A.* 106, 16704–16709. doi: 10.1073/pnas.0904336106

Guertin, A. D., Zhang, D. P., Mak, K. S., Alberta, J. A., and Kim, H. A. (2005). Microanatomy of axon/glial signaling during Wallerian degeneration. *J. Neurosci.* 25, 3478–3487. doi: 10.1523/JNEUROSCI.3766-04.2005

Hantke, J., Carty, L., Wagstaff, L. J., Turmaine, M., Wilton, D. K., Quintes, S., et al. (2014). c-Jun activation in Schwann cells protects against loss of sensory axons in inherited neuropathy. *Brain* 137, 2922–2937. doi: 10.1093/brain/awu257

Hardie, D. G., Ross, F. A., and Hawley, S. A. (2012). AMPK: a nutrient and energy sensor that maintains energy homeostasis. *Nat. Rev. Mol. Cell Biol.* 13, 251–262. doi: 10.1038/nrm3311

Hari, L., Brault, V., Kleber, M., Lee, H.-Y., Ille, F., Leimeroth, R., et al. (2002). Lineage-specific requirements of beta-catenin in neural crest development. *J. Cell Biol.* 159, 867–880. doi: 10.1083/jcb.200209039

Harrisingh, M. C., Perez-Nadales, E., Parkinson, D. B., Malcolm, D. S., Mudge, A. W., and Lloyd, A. C. (2004). The Ras/Raf/ERK signalling pathway drives Schwann cell dedifferentiation. *EMBO J.* 23, 3061–3071. doi: 10.1038/sj.emboj.7600309

He, Y., Kim, J. Y., Dupree, J., Tewari, A., Melendez-Vasquez, C., Svaren, J., et al. (2010). Yy1 as a molecular link between neuregulin and transcriptional modulation of peripheral myelination. *Nat. Neurosci.* 13, 1472–1480. doi: 10.1038/nn.2686

Heinen, A., Beyer, F., Tzekova, N., Hartung, H.-P., and Küry, P. (2015). Fingolimod induces the transition to a nerve regeneration promoting Schwann cell phenotype. *Exp. Neurol.* 271, 25–35. doi: 10.1016/j.expneurol.2015.05.002

Heinen, A., Kremer, D., Göttle, P., Kruse, F., Hasse, B., Lehmann, H., et al. (2008). The cyclin-dependent kinase inhibitor p57kip2 is a negative regulator of Schwann cell differentiation and *in vitro* myelination. *Proc. Natl. Acad. Sci. U.S.A.* 105, 8748–8753. doi: 10.1073/pnas.0802659105

Heinen, A., Lehmann, H. C., and Küry, P. (2013). Negative regulators of Schwann cell differentiation—novel targets for peripheral nerve therapies? *J. Clin. Immunol.* 33, 18–26. doi: 10.1007/s10875-012-9786-9

Heinen, A., Tzekova, N., Graffmann, N., Torres, K. J., Uhrberg, M., Hartung, H.-P., et al. (2012). Histone methyltransferase enhancer of zeste homolog 2 regulates Schwann cell differentiation. *Glia* 60, 1696–1708. doi: 10.1002/glia.22388

Hichor, M., Sampathkumar, N. K., Montanaro, J., Borderie, D., Petit, P. X., Gorgievski, V., et al. (2016). Paraquat induces peripheral myelin disruption and locomotor defects: crosstalk with LXR and Wnt pathways. *Antioxid. Redox Signal.* doi: 10.1089/ars.2016.6711. [Epub ahead of print].

Hirata, K., and Kawabuchi, M. (2002). Myelin phagocytosis by macrophages and nonmacrophages during Wallerian degeneration. *Microsc. Res. Tech.* 57, 541–547. doi: 10.1002/jemt.10108

Hoyng, S. A., de Winter, F., Tannemaat, M. R., Blits, B., Malessy, M. J. A., and Verhaagen, J. (2015). Gene therapy and peripheral nerve repair: a perspective. *Front. Mol. Neurosci.* 8:32. doi: 10.3389/fnmol.2015.00032

Hu, X., He, W., Diaconu, C., Tang, X., Kidd, G. J., Macklin, W. B., et al. (2008). Genetic deletion of BACE1 in mice affects remyelination of sciatic nerves. *FASEB J.* 22, 2970–2980. doi: 10.1096/fj.08-106666

Huijbregts, R. P. H., Roth, K. A., Schmidt, R. E., and Carroll, S. L. (2003). Hypertrophic neuropathies and malignant peripheral nerve sheath tumors in transgenic mice overexpressing glial growth factor beta3 in myelinating Schwann cells. *J. Neurosci.* 23, 7269–7280.

Hutton, E. J., Carty, L., Laurá, M., Houlden, H., Lunn, M., Brandner, S., et al. (2011). c-Jun expression in human neuropathies: a pilot study. *J. Peripher. Nerv. Syst.* 16, 295–303. doi: 10.1111/j.1529-8027.2011.00360.x

Jessen, K. R., and Mirsky, R. (2005). The origin and development of glial cells in peripheral nerves. *Nat. Rev. Neurosci.* 6, 671–682. doi: 10.1038/nrn1746

Jessen, K. R., and Mirsky, R. (2008). Negative regulation of myelination: relevance for development, injury, and demyelinating disease. *Glia* 56, 1552–1565. doi: 10.1002/glia.20761

Jessen, K. R., and Mirsky, R. (2016). The repair Schwann cell and its function in regenerating nerves. *J. Physiol.* 594, 3521–3531. doi: 10.1113/jp270874

Jessen, K. R., Mirsky, R., and Lloyd, A. C. (2015). Schwann cells: development and role in nerve repair. *Cold Spring Harb. Perspect. Biol.* 7:a020487. doi: 10.1101/cshperspect.a020487

Ji, Y., Tao, T., Cheng, C., Yang, H., Wang, Y., Yang, J., et al. (2010). SSeCKS is a suppressor in Schwann cell differentiation and myelination. *Neurochem. Res.* 35, 219–226. doi: 10.1007/s11064-009-0045-2

Joshi, A. R., Holtmann, L., Bobylev, I., Schneider, C., Ritter, C., Weis, J., et al. (2016). Loss of Schwann cell plasticity in chronic inflammatory demyelinating polyneuropathy (CIDP). *J. Neuroinflammation* 13, 255. doi: 10.1186/s12974-016-0711-7

Joung, I., Yoo, M., Woo, J. H., Chang, C. Y., Heo, H., and Kwon, Y. K. (2010). Secretion of EGF-like domain of heregulinbeta promotes axonal growth and functional recovery of injured sciatic nerve. *Mol. Cells* 30, 477–484. doi: 10.1007/s10059-010-0137-5

Jung, J., Cai, W., Lee, H. K., Pellegatta, M., Shin, Y. K., Jang, S. Y., et al. (2011). Actin polymerization is essential for myelin sheath fragmentation during Wallerian degeneration. *J. Neurosci.* 31, 2009–2015. doi: 10.1523/JNEUROSCI.4537-10.2011

Kaempchen, K., Mielke, K., Utermark, T., Langmesser, S., and Hanemann, C. O. (2003). Upregulation of the Rac1/JNK signaling pathway in primary human schwannoma cells. *Hum. Mol. Genet.* 12, 1211–1221. doi: 10.1093/hmg/ddg146

Kagiava, A., Sargiannidou, I., Theophilidis, G., Karaiskos, C., Richter, J., Bashiardes, S., et al. (2016). Intrathecal gene therapy rescues a model of demyelinating peripheral neuropathy. *Proc. Natl. Acad. Sci. U.S.A.* 113, E2421–E2429. doi: 10.1073/pnas.1522202113

Keng, V. W., Rahrmann, E. P., Watson, A. L., Tschida, B. R., Moertel, C. L., Jessen, W. J., et al. (2012). PTEN and NF1 inactivation in Schwann cells produces a severe phenotype in the peripheral nervous system that promotes the development and malignant progression of peripheral nerve sheath tumors. *Cancer Res.* 72, 3405–3413. doi: 10.1158/0008-5472.CAN-11-4092

Kim, H. A., Mindos, T., and Parkinson, D. B. (2013). Plastic fantastic: Schwann cells and repair of the peripheral nervous system. *Stem Cells Transl. Med.* 2, 553–557. doi: 10.5966/sctm.2013-0011

Kim, H. A., Pomeroy, S. L., Whoriskey, W., Pawlitzky, I., Benowitz, L. I., Sicinski, P., et al. (2000). A developmentally regulated switch directs regenerative growth of Schwann cells through cyclin D1. *Neuron* 26, 405–416. doi: 10.1016/S0896-6273(00)81173-3

Kioussi, C., Gross, M. K., and Gruss, P. (1995). Pax3: a paired domain gene as a regulator in PNS myelination. *Neuron* 15, 553–562. doi: 10.1016/0896-6273(95)90144-2

Klein, D., Groh, J., Wettmarshausen, J., and Martini, R. (2014). Nonuniform molecular features of myelinating Schwann cells in models for CMT1: distinct disease patterns are associated with NCAM and c-Jun upregulation. *Glia* 62, 736–750. doi: 10.1002/glia.22638

Köhne, A., Stettner, M., Jangouk, P., Dehmel, T., Hartung, H.-P., Lehmann, H. C., et al. (2012). Fingolimod impedes Schwann cell-mediated myelination: implications for the treatment of immune neuropathies? *Arch. Neurol.* 69, 1280–1289. doi: 10.1001/archneurol.2012.394

Kwon, Y. K., Bhattacharyya, A., Alberta, J. A., Giannobile, W. V., Cheon, K., Stiles, C. D., et al. (1997). Activation of ErbB2 during wallerian degeneration of sciatic nerve. *J. Neurosci.* 17, 8293–8299.

La Marca, R., Cerri, F., Horiuchi, K., Bachi, A., Feltri, M. L., Wrabetz, L., et al. (2011). TACE (ADAM17) inhibits Schwann cell myelination. *Nat. Neurosci.* 14, 857–865. doi: 10.1038/nn.2849

Laura, M., Mazzeo, A., Aguennouz, M., Santoro, M., Catania, M. A., Migliorato, A., et al. (2006). Immunolocalization and activation of nuclear factor-κB in the sciatic nerves of rats with experimental autoimmune neuritis. *J. Neuroimmunol.* 174, 32–38. doi: 10.1016/j.jneuroim.2006.01.004

Le, N., Nagarajan, R., Wang, J. Y. T., Araki, T., Schmidt, R. E., and Milbrandt, J. (2005). Analysis of congenital hypomyelinating Egr2Lo/Lo nerves identifies Sox2 as an inhibitor of Schwann cell differentiation and myelination. *Proc. Natl. Acad. Sci. U.S.A.* 102, 2596–2601. doi: 10.1073/pnas.0407836102

Lee, H., Jo, E.-K., Choi, S.-Y., Oh, S. B., Park, K., Soo Kim, J., et al. (2006). Necrotic neuronal cells induce inflammatory Schwann cell activation via TLR2 and TLR3: implication in Wallerian degeneration. *Biochem. Biophys. Res. Commun.* 350, 742–747. doi: 10.1016/j.bbrc.2006.09.108

Lee, H. J., Shin, Y. K., and Park, H. T. (2014). Mitogen activated protein kinase family proteins and c-jun signaling in injury-induced Schwann cell plasticity. *Exp. Neurobiol.* 23, 130–137. doi: 10.5607/en.2014.23.2.130

Lee, H.-Y., Kleber, M., Hari, L., Brault, V., Suter, U., Taketo, M. M., et al. (2004). Instructive role of Wnt/beta-catenin in sensory fate specification in neural crest stem cells. *Science* 303, 1020–1023. doi: 10.1126/science.1091611

Lee, S., Ashizawa, A. T., Kim, K. S., Falk, D. J., and Notterpek, L. (2013). Liposomes to target peripheral neurons and Schwann cells. *PLoS ONE* 8:e78724. doi: 10.1371/journal.pone.0078724

Lee, Y. I., Li, Y., Mikesh, M., Smith, I., Nave, K.-A., Schwab, M. H., et al. (2016). Neuregulin1 displayed on motor axons regulates terminal Schwann cell-mediated synapse elimination at developing neuromuscular junctions. *Proc. Natl. Acad. Sci. U.S.A.* 113, E479–E487. doi: 10.1073/pnas.1519156113

Leimeroth, R., Lobsiger, C., Lussi, A., Taylor, V., Suter, U., and Sommer, L. (2002). Membrane-bound neuregulin1 type III actively promotes Schwann cell differentiation of multipotent Progenitor cells. *Dev. Biol.* 246, 245–258. doi: 10.1006/dbio.2002.0670

Leoni, V., Masterman, T., Diczfalusy, U., De Luca, G., Hillert, J., and Björkhem, I. (2002). Changes in human plasma levels of the brain specific oxysterol 24S-hydroxycholesterol during progression of multiple sclerosis. *Neurosci. Lett.* 331, 163–166. doi: 10.1016/S0304-3940(02)00887-X

Li, H., Wigley, C., and Hall, S. M. (1998). Chronically denervated rat Schwann cells respond to GGF *in vitro*. *Glia* 24, 290–303.

Li, Y., Rao, P. K., Wen, R., Song, Y., Muir, D., Wallace, P., et al. (2004). Notch and Schwann cell transformation. *Oncogene* 23, 1146–1152. doi: 10.1038/sj.onc.1207068

Li, Y., Tennekoon, G. I., Birnbaum, M., Marchionni, M. A., and Rutkowski, J. L. (2001). Neuregulin signaling through a PI3K/Akt/Bad pathway in Schwann cell survival. *Mol. Cell. Neurosci.* 17, 761–767. doi: 10.1006/mcne.2000.0967

Limpert, A. S., and Carter, B. D. (2010). Axonal neuregulin 1 type III activates NF-κB in Schwann cells during myelin formation. *J. Biol. Chem.* 285, 16614–16622. doi: 10.1074/jbc.M109.098780

Liu, H., Lv, P., Zhu, Y., Wu, H., Zhang, K., Xu, F., et al. (2017). Salidroside promotes peripheral nerve regeneration based on tissue engineering strategy using Schwann cells and PLGA: *in vitro* and *in vivo*. *Sci. Rep.* 7:39869. doi: 10.1038/srep39869

Liu, X., Peng, S., Zhao, Y., Zhao, T., Wang, M., Luo, L., et al. (2016a). AMPK negatively regulates peripheral myelination via activation of c-Jun. *Mol. Neurobiol.* doi: 10.1007/s12035-016-9913-3. [Epub ahead of print].

Liu, X., Zhao, Y., Peng, S., Zhang, S., Wang, M., Chen, Y., et al. (2016b). BMP7 retards peripheral myelination by activating p38 MAPK in Schwann cells. *Sci. Rep.* 6:31049. doi: 10.1038/srep31049

Lütjohann, D., Papassotiropoulos, A., Björkhem, I., Locatelli, S., Bagli, M., Oehring, R. D., et al. (2000). Plasma 24S-hydroxycholesterol (cerebrosterol) is increased in Alzheimer and vascular demented patients. *J. Lipid Res.* 41, 195–198.

Mager, G. M., Ward, R. M., Srinivasan, R., Jang, S.-W., Wrabetz, L., and Svaren, J. (2008). Active gene repression by the Egr2· NAB complex during peripheral nerve myelination. *J. Biol. Chem.* 283, 18187–18197. doi: 10.1074/jbc.M803330200

Mahanthappa, N. K., Anton, E. S., and Matthew, W. D. (1996). Glial growth factor 2, a soluble neuregulin, directly increases Schwann cell motility and indirectly promotes neurite outgrowth. *J. Neurosci.* 16, 4673–4683.

Makoukji, J., Meffre, D., Grenier, J., Liere, P., Lobaccaro, J.-M. A., Schumacher, M., et al. (2011). Interplay between LXR and Wnt/β-catenin signaling in the negative regulation of peripheral myelin genes by oxysterols. *J. Neurosci.* 31, 9620–9629. doi: 10.1523/JNEUROSCI.0761-11.2011

Mancuso, R., Martínez-Muriana, A., Leiva, T., Gregorio, D., Ariza, L., Morell, M., et al. (2016). Neuregulin-1 promotes functional improvement by enhancing collateral sprouting in SOD1(G93A) ALS mice and after partial muscle denervation. *Neurobiol. Dis.* 95, 168–178. doi: 10.1016/j.nbd.2016.07.023

Marquardt, L. M., and Sakiyama-Elbert, S. E. (2013). Engineering peripheral nerve repair. *Curr. Opin. Biotechnol.* 24, 887–892. doi: 10.1016/j.copbio.2013.05.006

Martini, R., Fischer, S., López-Vales, R., and David, S. (2008). Interactions between Schwann cells and macrophages in injury and inherited demyelinating disease. *Glia* 56, 1566–1577. doi: 10.1002/glia.20766

Matsas, R., Lavdas, A. A., Papastefanaki, F., and Thomaidou, D. (2008). Schwann cell transplantation for CNS repair. *Curr. Med. Chem.* 15, 151–160. doi: 10.2174/092986708783330593

Maurel, P., and Salzer, J. L. (2000). Axonal regulation of Schwann cell proliferation and survival and the initial events of myelination requires PI 3-kinase activity. *J. Neurosci.* 20, 4635–4645.

Michailov, G. V., Sereda, M. W., Brinkmann, B. G., Fischer, T. M., Haug, B., Birchmeier, C., et al. (2004). Axonal neuregulin-1 regulates myelin sheath thickness. *Science* 304, 700–703. doi: 10.1126/science.1095862

Mogha, A., D'Rozario, M., and Monk, K. R. (2016a). G protein-coupled receptors in myelinating glia. *Trends Pharmacol. Sci.* 37, 977–987. doi: 10.1016/j.tips.2016.09.002

Mogha, A., Harty, B., Carlin, D., Joseph, J., Sanchez, N., Suter, U., et al. (2016b). Gpr126/Adgrg6 has Schwann cell autonomous and nonautonomous functions in peripheral nerve injury and repair. *J. Neurosci.* 36, 12351–12367. doi: 10.1523/JNEUROSCI.3854-15.2016

Monje, P. V., Soto, J., Bacallao, K., and Wood, P. M. (2010). Schwann cell dedifferentiation is independent of mitogenic signaling and uncoupled to proliferation: role of cAMP and JNK in the maintenance of the differentiated state. *J. Biol. Chem.* 285, 31024–31036. doi: 10.1074/jbc.M110.116970

Monk, K. R., Naylor, S. G., Glenn, T. D., Mercurio, S., Perlin, J. R., Dominguez, C., et al. (2009). A G protein–coupled receptor is essential for Schwann cells to initiate myelination. *Science* 325, 1402–1405. doi: 10.1126/science.1173474.

Morris, J. K., Lin, W., Hauser, C., Marchuk, Y., Getman, D., and Lee, K. F. (1999). Rescue of the cardiac defect in ErbB2 mutant mice reveals essential roles of ErbB2 in peripheral nervous system development. *Neuron* 23, 273–283. doi: 10.1016/S0896-6273(00)80779-5

Morton, P. D., Dellarole, A., Theus, M. H., Walters, W. M., Berge, S. S., and Bethea, J. R. (2013). Activation of NF-κB in Schwann cells is dispensable for myelination *in vivo*. *J. Neurosci.* 33, 9932–9936. doi: 10.1523/JNEUROSCI.2483-12.2013

Morton, P. D., Johnstone, J. T., Ramos, A. Y., Liebl, D. J., Bunge, M. B., and Bethea, J. R. (2012). Nuclear factor-κB activation in Schwann cells regulates regeneration and remyelination. *Glia* 60, 639–650. doi: 10.1002/glia.22297

Napoli, I., Noon, L. A., Ribeiro, S., Kerai, A. P., Parrinello, S., Rosenberg, L. H., et al. (2012). A central role for the ERK-signaling pathway in controlling Schwann cell plasticity and peripheral nerve regeneration *in vivo*. *Neuron* 73, 729–742. doi: 10.1016/j.neuron.2011.11.031

Newbern, J., and Birchmeier, C. (2010). Nrg1/ErbB signaling networks in Schwann cell development and myelination. *Semin. Cell Dev. Biol.* 21, 922–928. doi: 10.1016/j.semcdb.2010.08.008

Newbern, J. M., Li, X., Shoemaker, S. E., Zhou, J., Zhong, J., Wu, Y., et al. (2011). Specific functions for ERK/MAPK signaling during PNS development. *Neuron* 69, 91–105. doi: 10.1016/j.neuron.2010.12.003

Newbern, J. M., and Snider, W. D. (2012). Bers-ERK Schwann cells coordinate nerve regeneration. *Neuron* 73, 623–626. doi: 10.1016/j.neuron.2012.02.002

Nickols, J. C., Valentine, W., Kanwal, S., and Carter, B. D. (2003). Activation of the transcription factor NF-κB in Schwann cells is required for peripheral myelin formation. *Nat. Neurosci.* 6, 161–167. doi: 10.1038/nn995

Nikam, S. S., Tennekoon, G. I., Christy, B. A., Yoshino, J. E., and Rutkowski, J. L. (1995). The zinc finger transcription factor Zif268/Egr-1 is essential for Schwann cell expression of the p75 NGF receptor. *Mol. Cell. Neurosci.* 6, 337–348. doi: 10.1006/mcne.1995.1026

Norrmén, C., Figlia, G., Lebrun-Julien, F., Pereira, J. A., Trötzmüller, M., Köfeler, H. C., et al. (2014). mTORC1 controls PNS myelination along the mTORC1-RXRγ-SREBP-lipid biosynthesis axis in Schwann cells. *Cell Rep.* 9, 646–660. doi: 10.1016/j.celrep.2014.09.001

Norrmén, C., and Suter, U. (2013). Akt/mTOR signalling in myelination. *Biochem. Soc. Trans.* 41, 944LP–950LP. doi: 10.1042/BST20130046

Noseda, R., Belin, S., Piguet, F., Vaccari, I., Scarlino, S., Brambilla, P., et al. (2013). DDIT4/REDD1/RTP801 is a novel negative regulator of Schwann cell myelination. *J. Neurosci.* 33, 15295–15305. doi: 10.1523/JNEUROSCI.2408-13.2013

Ogata, T., Iijima, S., Hoshikawa, S., Miura, T., Yamamoto, S., Oda, H., et al. (2004). Opposing extracellular signal-regulated kinase and Akt pathways control Schwann cell myelination. *J. Neurosci.* 24, 6724–6732. doi: 10.1523/JNEUROSCI.5520-03.2004

Painter, M. W., Brosius Lutz, A., Cheng, Y.-C., Latremoliere, A., Duong, K., Miller, C. M., et al. (2014). Diminished schwann cell repair responses underlie age-associated impaired axonal regeneration. *Neuron* 83, 331–343. doi: 10.1016/j.neuron.2014.06.016

Parkinson, D. B., Bhaskaran, A., Arthur-Farraj, P., Noon, L. A., Woodhoo, A., Lloyd, A. C., et al. (2008). c-Jun is a negative regulator of myelination. *J. Cell Biol.* 181, 625–637. doi: 10.1083/jcb.200803013

Parkinson, D. B., Bhaskaran, A., Droggiti, A., Dickinson, S., D'Antonio, M., Mirsky, R., et al. (2004). Krox-20 inhibits Jun-NH2-terminal kinase/c-Jun to control Schwann cell proliferation and death. *J. Cell Biol.* 164, 385–394. doi: 10.1083/jcb.200307132

Parkinson, D. B., Dong, Z., Bunting, H., Whitfield, J., Meier, C., Marie, H., et al. (2001). Transforming growth factor beta (TGFbeta) mediates Schwann cell death *in vitro* and *in vivo*: examination of c-Jun activation, interactions with survival signals, and the relationship of TGFbeta-mediated death to Schwann cell differentiation. *J. Neurosci.* 21, 8572–8585.

Parrinello, S., Napoli, I., Ribeiro, S., Wingfield Digby, P., Fedorova, M., Parkinson, D. B., et al. (2010). EphB signaling directs peripheral nerve regeneration through Sox2-dependent Schwann cell sorting. *Cell* 143, 145–155. doi: 10.1016/j.cell.2010.08.039

Quintes, S., Brinkmann, B. G., Ebert, M., Frob, F., Kungl, T., Arlt, F. A., et al. (2016). Zeb2 is essential for Schwann cell differentiation, myelination and nerve repair. *Nat. Neurosci.* 19, 1050–1059. doi: 10.1038/nn.4321

Rameh, L. E., and Cantley, L. C. (1999). The role of phosphoinositide 3-kinase lipid products in cell function. *J. Biol. Chem.* 274, 8347–8350. doi: 10.1074/jbc.274.13.8347

Roberts, S. L., Dun, X.-p., Dee, G., Gray, B., Mindos, T., and Parkinson, D. B. (2016). The role of p38alpha in Schwann cells in regulating peripheral nerve myelination and repair. *J. Neurochem.* doi: 10.1111/jnc.13929. [Epub ahead of print].

Rodríguez-Molina, J. F., Lopez-Anido, C., Ma, K. H., Zhang, C., Olson, T., Muth, K. N., et al. (2017). Dual specificity phosphatase 15 regulates Erk activation in Schwann cells. *J. Neurochem.* 140, 368–382. doi: 10.1111/jnc.13911

Ronchi, G., Cillino, M., Gambarotta, G., Fornasari, B. E., Raimondo, S., Pugliese, P., et al. (2017). Irreversible changes occurring in long-term denervated Schwann cells affect delayed nerve repair. *J. Neurosurg.* doi: 10.3171/2016.9.JNS16140. [Epub ahead of print].

Ronchi, G., Haastert-Talini, K., Fornasari, B. E., Perroteau, I., Geuna, S., and Gambarotta, G. (2016). The Neuregulin1/ErbB system is selectively regulated during peripheral nerve degeneration and regeneration. *Eur. J. Neurosci.* 43, 351–364. doi: 10.1111/ejn.12974

Rotshenker, S. (2011). Wallerian degeneration: the innate-immune response to traumatic nerve injury. *J. Neuroinflammation* 8:109. doi: 10.1186/1742-2094-8-109

Russell, D. W. (2000). Oxysterol biosynthetic enzymes. *Biochim. Biophys. Acta* 1529, 126–135. doi: 10.1016/S1388-1981(00)00142-6

Schröder, J. M. (1972). Altered ratio between axon diameter and myelin sheath thickness in regenerated nerve fibers. *Brain Res.* 45, 49–65. doi: 10.1016/0006-8993(72)90215-6

Sherman, D. L., Krols, M., Wu, L.-M. N., Grove, M., Nave, K.-A., Gangloff, Y.-G., et al. (2012). Arrest of myelination and reduced axon growth when Schwann cells lack mTOR. *J. Neurosci.* 32, 1817–1825. doi: 10.1523/JNEUROSCI.4814-11.2012

Sheu, J. Y., Kulhanek, D. J., and Eckenstein, F. P. (2000). Differential patterns of ERK and STAT3 phosphorylation after sciatic nerve transection in the rat. *Exp. Neurol.* 166, 392–402. doi: 10.1006/exnr.2000.7508

Shin, Y. K., Jang, S. Y., Park, J. Y., Park, S. Y., Lee, H. J., Suh, D. J., et al. (2013). The Neuregulin-Rac-MKK7 pathway regulates antagonistic c-jun/Krox20 expression in Schwann cell dedifferentiation. *Glia* 61, 892–904. doi: 10.1002/glia.22482

Shivane, A., Parkinson, D. B., Ammoun, S., and Hanemann, C. O. (2013). Expression of c-Jun and Sox-2 in human Schwannomas and traumatic neuromas. *Histopathology* 62, 651–656. doi: 10.1111/his.12062

Smith, D., Tweed, C., Fernyhough, P., and Glazner, G. W. (2009). Nuclear factor-κB activation in axons and Schwann cells in experimental sciatic nerve injury and its role in modulating axon regeneration: studies with etanercept. *J. Neuropathol. Exp. Neurol.* 68, 691–700. doi: 10.1097/NEN.0b013e3181a7c14e

Son, Y.-J., and Thompson, W. J. (1995). Schwann cell processes guide regeneration of peripheral axons. *Neuron* 14, 125–132. doi: 10.1016/0896-6273(95)90246-5

Stassart, R. M., Fledrich, R., Velanac, V., Brinkmann, B. G., Schwab, M. H., Meijer, D., et al. (2013). A role for Schwann cell-derived neuregulin-1 in remyelination. *Nat. Neurosci.* 16, 48–54. doi: 10.1038/nn.3281

Stewart, H. J. S. (1995). Expression of c-Jun, Jun B, Jun D and cAMP response element binding protein by Schwann cells and their precursors *in vivo* and *in vitro*. *Eur. J. Neurosci.* 7, 1366–1375. doi: 10.1111/j.1460-9568.1995.tb01128.x

Stewart, H. J. S., Zoidl, G., Rossner, M., Brennan, A., Zoidl, C., Nave, K. A., et al. (1997). Helix-loop-helix proteins in Schwann cells: a study of regulation and subcellular localization of Ids, REB, and E12/47 during embryonic and postnatal development. *J. Neurosci. Res.* 50, 684–701. doi: 10.1002/(SICI)1097-4547(19971201)50:5<684::AID-JNR6>3.0.CO;2-D

Stoll, G., and Müller, H. W. (1999). Nerve injury, axonal degeneration and neural regeneration: basic insights. *Brain Pathol.* 9, 313–325. doi: 10.1111/j.1750-3639.1999.tb00229.x

Sulaiman, W., and Nguyen, D. H. (2016). Transforming growth factor beta 1, a cytokine with regenerative functions. *Neural Regen. Res.* 11, 1549–1552. doi: 10.4103/1673-5374.193223

Syed, N., Reddy, K., Yang, D. P., Taveggia, C., Salzer, J. L., Maurel, P., et al. (2010). Soluble neuregulin-1 has bifunctional, concentration-dependent effects on Schwann cell myelination. *J. Neurosci.* 30, 6122–6131. doi: 10.1523/JNEUROSCI.1681-09.2010

Syroid, D. E., Maycox, P. R., Burrola, P. G., Liu, N., Wen, D., Lee, K. F., et al. (1996). Cell death in the Schwann cell lineage and its regulation by neuregulin. *Proc. Natl. Acad. Sci. U.S.A.* 93, 9229–9234. doi: 10.1073/pnas.93.17.9229

Takeda, Y. S., and Xu, Q. (2015). Synthetic and nature-derived lipid nanoparticles for neural regeneration. *Neural Regen. Res.* 10, 689–690. doi: 10.4103/1673-5374.156946

Tapinos, N., Ohnishi, M., and Rambukkana, A. (2006). ErbB2 receptor tyrosine kinase signaling mediates early demyelination induced by leprosy bacilli. *Nat. Med.* 12, 961–966. doi: 10.1038/nm1433

Taveggia, C., Zanazzi, G., Petrylak, A., Yano, H., Rosenbluth, J., Einheber, S., et al. (2005). Neuregulin-1 type III determines the ensheathment fate of axons. *Neuron* 47, 681–694. doi: 10.1016/j.neuron.2005.08.017

Thakur, K. K., Saini, J., Mahajan, K., Singh, D., Jayswal, D. P., Mishra, S., et al. (2017). Therapeutic implications of toll-like receptors in peripheral neuropathic pain. *Pharmacol. Res.* 115, 224–232. doi: 10.1016/j.phrs.2016.11.019

Topilko, P., Levi, G., Merlo, G., Mantero, S., Desmarquet, C., Mancardi, G., et al. (1997). Differential regulation of the zinc finger genes *Krox-20* and *Krox-24 (Egr-1)* suggests antagonistic roles in Schwann cells. *J. Neurosci. Res.* 50, 702–712. doi: 10.1002/(SICI)1097-4547(19971201)50:5<702::AID-JNR7>3.0.CO;2-L

Trimarco, A., Forese, M. G., Alfieri, V., Lucente, A., Brambilla, P., Dina, G., et al. (2014). Prostaglandin D2 synthase/GPR44: a signaling axis in PNS myelination. *Nat. Neurosci.* 17, 1682–1692. doi: 10.1038/nn.3857

Tzekova, N., Heinen, A., Bunk, S., Hermann, C., Hartung, H.-P., Reipert, B., et al. (2015). Immunoglobulins stimulate cultured Schwann cell maturation and promote their potential to induce axonal outgrowth. *J. Neuroinflammation* 12, 107. doi: 10.1186/s12974-015-0331-7

Waller, A. (1850). Experiments on the section of the glossopharyngeal and hypoglossal nerves of the frog, and observations of the alterations produced thereby in the structure of their primitive fibres. *Philos. Trans. R. Soc. Lond.* 140, 423–429. doi: 10.1098/rstl.1850.0021

Wang, J., Ren, K.-Y., Wang, Y.-H., Kou, Y.-H., Zhang, P.-X., Peng, J.-P., et al. (2015). Effect of active Notch signaling system on the early repair of rat sciatic nerve injury. *Artif. Cells Nanomed. Biotechnol.* 43, 383–389. doi: 10.3109/21691401.2014.896372

Weinberg, H. J., and Spencer, P. S. (1978). The fate of Schwann cells isolated from axonal contact. *J. Neurocytol.* 7, 555–569. doi: 10.1007/BF01260889

Willem, M. (2016). Proteolytic processing of Neuregulin-1. *Brain Res. Bull.* 126, 178–182. doi: 10.1016/j.brainresbull.2016.07.003

Woodhoo, A., Alonso, M. B. D., Droggiti, A., Turmaine, M., D'Antonio, M., Parkinson, D. B., et al. (2009). Notch controls embryonic Schwann cell differentiation, postnatal myelination and adult plasticity. *Nat. Neurosci.* 12, 839–847. doi: 10.1038/nn.2323

Wu, L. M. N., Wang, J., Conidi, A., Zhao, C., Wang, H., Ford, Z., et al. (2016). Zeb2 recruits HDAC-NuRD to inhibit Notch and controls Schwann cell differentiation and remyelination. *Nat. Neurosci.* 19, 1060–1072. doi: 10.1038/nn.4322

Wu, W., Liu, Q., Liu, Y., Yu, Z., and Wang, Y. (2016a). Dixdc1 targets CyclinD1 and p21 via PI3K pathway activation to promote Schwann cell proliferation after sciatic nerve crush. *Biochem. Biophys. Res. Commun.* 478, 956–963. doi: 10.1016/j.bbrc.2016.08.058

Wu, W., Liu, Y., and Wang, Y. (2016b). Sam68 promotes Schwann cell proliferation by enhancing the PI3K/Akt pathway and acts on regeneration after sciatic nerve crush. *Biochem. Biophys. Res. Commun.* 473, 1045–1051. doi: 10.1016/j.bbrc.2016.04.013

Yamauchi, J., Chan, J. R., and Shooter, E. M. (2003). Neurotrophin 3 activation of TrkC induces Schwann cell migration through the c-Jun

N-terminal kinase pathway. *Proc. Natl. Acad. Sci. U.S.A.* 100, 14421–14426. doi: 10.1073/pnas.2336152100

Yamauchi, J., Miyamoto, Y., Chan, J. R., and Tanoue, A. (2008). ErbB2 directly activates the exchange factor Dock7 to promote Schwann cell migration. *J. Cell Biol.* 181, 351–365. doi: 10.1083/jcb.200709033

Yamauchi, J., Miyamoto, Y., Hamasaki, H., Sanbe, A., Kusakawa, S., Nakamura, A., et al. (2011). The atypical Guanine-nucleotide exchange factor, dock7, negatively regulates Schwann cell differentiation and myelination. *J. Neurosci.* 31, 12579–12592. doi: 10.1523/JNEUROSCI.2738-11.2011

Yan, M., Cheng, C., Jiang, J., Liu, Y., Gao, Y., Guo, Z., et al. (2009). Essential role of SRC suppressed C kinase substrates in Schwann cells adhesion, spreading and migration. *Neurochem. Res.* 34, 1002–1010. doi: 10.1007/s11064-008-9869-4

Yang, D. P., Kim, J., Syed, N., Tung, Y., Bhaskaran, A., Mindos, T., et al. (2012). p38 MAPK activation promotes denervated Schwann cell phenotype and functions as a negative regulator of Schwann cell differentiation and myelination. *J. Neurosci.* 32, 7158–7168. doi: 10.1523/JNEUROSCI.5812-11.2012

Yildiz, M., Karlidag, T., Yalcin, S., Ozogul, C., Keles, E., Alpay, H. C., et al. (2011). Efficacy of glial growth factor and nerve growth factor on the recovery of traumatic facial paralysis. *Eur. Arch. Otorhinolaryngol.* 268, 1127–1133. doi: 10.1007/s00405-011-1492-3

Yokoo, T., Toyoshima, H., Miura, M., Wang, Y., Iida, K. T., Suzuki, H., et al. (2003). p57Kip2 regulates actin dynamics by binding and translocating LIM-kinase 1 to the nucleus. *J. Biol. Chem.* 278, 52919–52923. doi: 10.1074/jbc.M309334200

Yoon, C., Korade, Z., and Carter, B. D. (2008). Protein kinase A-induced phosphorylation of the p65 subunit of nuclear factor-κB promotes Schwann cell differentiation into a myelinating phenotype. *J. Neurosci.* 28, 3738–3746. doi: 10.1523/JNEUROSCI.4439-07.2008

Zhou, S., Gao, R., Hu, W., Qian, T., Wang, N., Ding, G., et al. (2014). MiR-9 inhibits Schwann cell migration by targeting Cthrc1 following sciatic nerve injury. *J. Cell Sci.* 127(Pt 5), 967–976. doi: 10.1242/jcs.131672

Zhou, Y., and Notterpek, L. (2016). Promoting peripheral myelin repair. *Exp. Neurol.* 283, 573–580. doi: 10.1016/j.expneurol.2016.04.007

Zochodne, D. W. (2008). *Neurobiology of Peripheral Nerve Regeneration.* Cambridge: Cambridge University Press.

Zochodne, D. W. (2012). The challenges and beauty of peripheral nerve regrowth. *J. Peripher. Nerv. Syst.* 17, 1–18. doi: 10.1111/j.1529-8027.2012.00378.x

Zrouri, H., Le Goascogne, C., Li, W. W., Pierre, M., and Courtin, F. (2004). The role of MAP kinases in rapid gene induction after lesioning of the rat sciatic nerve. *Eur. J. Neurosci.* 20, 1811–1818. doi: 10.1111/j.1460-9568.2004.03641.x

12

Alpha-1 Antitrypsin Attenuates M1 Microglia-Mediated Neuroinflammation in Retinal Degeneration

Tian Zhou[†], Zijing Huang[†], Xiaowei Zhu[†], Xiaowei Sun, Yan Liu, Bing Cheng, Mei Li, Yizhi Liu, Chang He and Xialin Liu**

State Key Laboratory of Ophthalmology, Zhongshan Ophthalmic Center, Sun Yat-sen University, Guangzhou, China

***Correspondence:**
Chang He
hech33@mail.sysu.edu.cn;
Xialin Liu
liuxl28@mail.sysu.edu.cn

[†]These authors have contributed equally to this work.

Neurodegenerative diseases are a set of disorders characterized by progressive neuronal death and are associated with microglia-mediated neuroinflammation. Recently, neuroinflammation is proposed as a promising therapeutic target for many neurodegenerative diseases. Alpha-1 antitrypsin (AAT) is recognized as a novel immunomodulatory agent in autoimmune diseases and transplantation, however, its impact on neuroinflammation and neurodegeneration remains unknown. This study aims to explore the effects of AAT on microglia-mediated neuroinflammation and retinal degeneration in rd1 mouse model. We found reduced expression of AAT in rd1 retina, and AAT supplement exhibited certain protective effect on retinal degeneration, presenting with increased amount of photoreceptor nuclei, and amplified wave amplitudes in electroretinogram analysis. Of note, AAT shifted microglia phenotype from pro-inflammatory M1 ($CD16/CD32^+$, $iNOS^+$) to anti-inflammatory M2 ($CD206^+$, $Arg1^+$) both *in vivo* and *in vitro*, underscoring the concept of immunomodulation on microglia polarization by AAT during neurodegeneration. Furthermore, AAT suppressed the activation of STAT1, promoted the expression of IRF4 while inhibited IRF8 expression, indicating the involvement of these signaling pathways in AAT immunomodulation. Collectively, our data provided evidence for a novel protective role of AAT through immunomodulation on microglia polarization. Attenuating neuroinflammation by AAT may be beneficial to retard neurodegeneration in rd1 mice.

Keywords: alpha-1 antitrypsin, immunomodulation, microglia polarization, rd1 mice, neuroinflammation

INTRODUCTION

Neurodegenerative diseases are a heterogeneous group of disorders characterized by progressive loss of specific neurons and association with neuroinflammation. Retinitis pigmentosa (RP), one of the most common inherited retinal degeneration, is a major cause of incurable vision loss (1). It is a highly complex neurodegenerative disease defined by progressive degeneration of photoreceptors or retinal pigment epithelium (RPE) caused by various mutations (2). So far, more than 200 mutant genes have been identified (3, 4) and gene therapy therefore faces the challenge of genetic heterogeneity,

---

Abbreviations: AAT, alpha-1 antitrypsin; RP, retinitis pigmentosa; RPE, retinal pigment epithelium; ERG, electroretinogram; OCT, optical coherence tomography; GCL, ganglion cell layer; IPL, inner plexiform layer; INL, inner nuclear layer; OPL, outer plexiform layer; ONL, outer nuclear layer; CNS, central nervous system; IRF, interferon-regulatory factor.

which makes mutation-independent approaches highly desirable (5). Drugs that target on the broad biological processes which are common across various mutations in RP may show some efficacy in this complicated disorder (6).

Neuroinflammation is now recognized as a hallmark of many chronic neurodegenerative disorders and attenuating neuroinflammation has been proved to be a potential therapeutic strategy for these diseases (7, 8). It has been reported in RP patients with a substantial number of inflammatory cells found in the vitreous cavity and significantly elevated expression of a variety of pro-inflammatory cytokines and chemokines (9), highlighting that inflammation was implicated in retinal degeneration. Growing evidence suggested that microglia, the tissue-resident immune cells, would initiate immune responses with a chronic inflammation sustained in RP (10). The onset of microglia-mediated inflammation is identified as a common feature of RP (11).

Alpha-1 antitrypsin (AAT), an endogenous serine protease inhibitor for neutrophil elastase (NE), is emerging as a novel immunomodulatory agent to intervene immune diseases and transplantation (12). It is mainly produced by hepatocytes and maintains homeostasis with normal serum levels of 2–5 mg/mL (13). Patients with inherited AAT deficiency are predisposed to the development of lung diseases due to excessive destructive NE (14). And the AAT augmentation is well acknowledged in clinic for the treatment of these individuals (15). Interestingly, it was reported that a patient with RP was identified heterozygous for AAT deficiency (16). It is still unknown if there is a relationship between AAT deficiency and RP, and whether AAT augmentation is suitable for RP patients. Actually, besides its ability for elastase degradation, AAT was recently reported to suppress adverse immune responses under various conditions. For example, AAT application could avert type 1 diabetes and prolong the survival of islet allograft in NOD mice (12, 17). AAT administration reduced the susceptibility of islets to inflammation (18). These findings proposed AAT as a novel alternate for immunosuppression. The microglia-mediated neuroinflammation is an important pathological process in neurodegenerative diseases and immune modulation targeting on microglia is a promising treatment for these diseases. Therefore, we proposed that AAT may have the potential to attenuate microglia-mediated neuroinflammation and retinal degeneration.

In the present study, the impact of AAT immunotherapy on retinal inflammation and neurodegeneration was examined in rd1 mice, a typical mouse model of RP, as well as in cell culture. The results showed that AAT suppressed microglia-mediated neuroinflammtion and protected the degenerative retina. Our data proved that AAT may be beneficial to retinal degeneration by shifting retinal microglia from proinflammatory M1 to anti-inflammatory M2 phenotype.

MATERIALS AND METHODS

Rd1 Mice and AAT Treatment

Rd1 (FVB/N) mice were purchased from Beijing Vital River Laboratory Animal Technology, Co., Beijing, China and were kept in a specific pathogen-free facility in Animal Laboratories of Zhongshan Ophthalmic Center. This study was approved by the animal experimental ethics committee of Zhongshan Ophthalmic Center, Sun Yat-sen University (authorized number: 2014-039). All the experiments were carried out in accordance with the approved guidelines of Animal Care and Use Committee of Zhongshan Ophthalmic Center and the Association Research in Vision and Ophthalmology (ARVO) Statement for the Use of Animals in Ophthalmic and Vision Research. Rd1 mice and C57BL/6J mice at P4, P10, P14, and P20 were sacrificed for detecting AAT expression in retina. For treatment, rd1 mice received intraperitoneal injections of AAT (soluted in PBS 80 mg/kg, Sigma Chemical Co., St. Louis) every other day from P4 to P14. Mice were sacrificed at P16 and eyes were enucleated for further investigation, normal C57 mice of the same age were used as the control.

Primary Microglia and BV2 Cell Line Culture

Primary microglia were isolated and cultured as described previously (19). In brief, retinae were collected from 3-day-old Sprague–Dawley rats and crashed drastically into single cell suspensions. The cells were then resuspended in DMEM/F-12 medium containing 10% FBS, 100 U/mL penicillin, and 100 mg/mL streptomycin, and were seeded at a density of 1×10^6 cells/mL. The culture medium was changed 24 h later and twice a week thereafter. 2 weeks later, after shaking the flasks at 200 rpm for 1 h, the suspensions were collected and centrifuged at $300 \times g$ for 8 min at 4°C. The cells were harvested and identified as microglia using IBA1 staining. The BV2 murine microglial cell line was purchased from Kunming Institute of Zoology, Chinese Academy of Sciences, China. The microglia cells were pre-treated with AAT (1 mg/mL dissolved in PBS, A6105, Sigma) or PBS for 2 h and then stimulated with hydrogen peroxide (200 μM). Eighteen hours later, the cells were harvested for further analysis.

Electroretinogram (ERG) Recordings

In vivo ERG were performed on rd1 mice treated with AAT or PBS ($n = 9$ for each group) and C57 mice of the same age ($n = 6$). After dark adaptation overnight, mice were anesthetized with pentobarbital sodium diluted in saline. The pupils were dilated with 1% tropicamide and 2.5% phenylephrine and the corneas were anesthetized with 0.5% tetracaine HCl eye drops. ERGs were recorded with gold plated wire loop electrodes contacting the cornea surface as the active electrodes. Stainless steel needle electrodes were inserted into the skin near the eye and into the tail serving as reference and ground leads respectively. The mice were exposed to full-field scotopic flashes of 1.3 ms duration presented by a Ganzfeld (Roland Consult, Germany) with the intensities of 3.0 and 10.0 log cd•s/m^2. For each intensity, five responses were averaged to the luminance of flash stimuli. Amplitudes of the major ERG components (a- and b-wave) were measured (RETIsystem software) using automated and manual modes.

Imaging Mouse Retina by Optical Coherence Tomography (OCT)

Optical coherence tomography examination with SPECTRALIS-OCT (Heidelberg, Germany) was performed to detect the retinal structure in rd1 mice treated with AAT or PBS (n = 9 for each group) and normal C57 mice (n = 6). Briefly, the mice were anesthetized with pentobarbital sodium diluted in saline and the pupils were dilated with 1% tropicamide and 2.5% phenylephrine. Each scan was performed for at least two times, with realignment each time. Both the nasal and temporal retinae within 6 × 3 mm^2 area adjacent to optic nerve disk were scanned, consisted of 25 scan slices to form 3D acquisition. From the 3D imaging, average retinal thickness was measured within a circle area, with 1.5 mm in radius, centering at 3 mm away from the optic nerve disk.

Hematoxylin & Eosin Staining

Eyes were fixed in formalin overnight, embedded in paraffin, and were cut into 3 μm vertical slices. Sections were washed in deionized water for 5 min and incubated with hematoxylin buffer for 10 min at room temperature. Then these sections were rinsed in deionized water and dipped in 1% eosin solution for 15 s. After rehydrated in alcohol gradients, slices were washed again and mounted. Histological analyses of retinal tissues were observed under microscope (Leica DM4000, Germany). The amount of cell nuclei in outer nuclear layer (ONL) was counted and analyzed. Images were processed and analyzed with Image J software (Public Domain, imagej.nih.gov/ij/).

Immunofluorescence Staining on Retinal Cryosection and Whole Mounts

Eyes were enucleated and fixed in 4% paraformaldehyde for 60 min. For cryosection, eyes were embedded in O.C.T. compound (Tissue-Tek) with adjustment of the rim of corneal limbus in vertical direction. The frozen sample was then sliced transversely (8 μm) with a cryostat at −20°C and the cross-sections throughout optic nerve were used for staining and analysis. For retinal whole mounts, the retinae were dissected out. Both cryosections and retinal whole mounts were blocked with 0.5% Triton X-100/5% BSA for 2 h at room temperature and were incubated with primary antibodies overnight at 4°C. After washing with PBS, the slices were incubated with secondary antibodies for 1 h and counterstained with DAPI (Invitrogen) for 5 min before mounted. The primary antibodies included anti-AAT antibody (ab166610, Abcam, MA), anti-Brn3a antibody (ab81213, Abcam), anti-GFAP antibody (ab7260, Abcam), anti-CD11b antibody (ab8878, Abcam), anti-CD68 antibody (ab53444, Abcam), anti-IBA1 antibody (019-19471, Wako; ab15690, Abcam), anti-CD16/32 antibody (553141, BD Biosciences), anti-CD206 antibody (AF2535, R&D Systems), anti-iNOS antibody, anti-Argl antibody, and anti-Rhodopsin antibody (sc-7271, sc-18355, sc-57432, Santa Cruz, MA, USA). Antibodies used in this experiment were summarized in Table S1 in Supplementary Material. TUNEL staining (*In Situ* Cell Death Detection Kit, Fluorescein; Roche, IN, USA) was performed according to the manufacturer's instructions. The images were obtained using Zeiss Axiophot fluorescent microscope and LSCM (LSM710, Carl Zeiss).

Western Blotting

Retinal and cellular protein were harvested and homogenized in lysis buffer (RIPA, Biocolors, Shanghai, China) containing protease and phosphatase inhibitor mini tablets (Thermo Fisher Scientific, No. 88668; USA). The protein concentration was determined by bicinchoninic acid protein assay. Equal amount of protein was used and western blotting was performed as previously described (20). Primary antibodies included anti-AAT antibody (ab166610, Abcam, MA), anti-STAT1 antibody, anti-pSTAT1 antibody (14994S, 7649S, CST), anti-IRF4 antibody (PA5-21144, Thermo Fisher Scientific, USA), anti-IRF8 antibody, anti β-actin antibody (ab28696, ab28696, Abcam, MA, USA), anti-iNOS antibody, and anti-Argl antibody (sc-7271, sc-18355, Santa Cruz, CA, USA). The gray intensity of protein blots was measured using Image J software (US National Institutes of Health).

Quantitative PCR Analysis

The mRNA levels of AAT were detected by real time PCR. The total RNA of retinae were extracted with TRIzol (Invitrogen) and converted into first-strand cDNA using random hexamer primers and the Reverse Transcriptase Superscript II Kit (Invitrogen) according to the manufacturer's instructions. Real-time PCR was performed in a total volume of 20 μL containing 2 μL of cDNA, 10 μL of 2 × SYBR Premix Ex Taq, 7 μL ddH$_2$O, and 10 μmol/L of the primer pairs. The sequence of the used primers was: AAT forward: 5′-TCCCATGAGATCGCTACAAAC-3′; reverse: 5′-TGAT AATGGTTCTTGGCCTCT-3′; GAPDH forward: 5′-GCCAAGG CTGTGGGCAAGGT-3′; reverse: 5′-TCTCCAGGCGGCACGT CAGA-3′. The PCR amplification protocols consisted of 95°C for 30 s and up to 40 cycles of 95°C for 5 s and 60°C for 34 s according to the manufacturer's instructions.

Statistics

Statistical analysis was performed using GraphPad Prism (GraphPad Software, Version 6.0, La Jolla, CA, USA). For immunofluorescence on retinal whole mounts, three images were captured in the center, mid-periphery, and periphery areas of each retina respectively. Six retinae from six mice were used in each group for analysis. Representative images were shown in the according Figures. For immunofluorescence, Hematoxylin & eosin, and TUNEL staining on retinal section, at least three sections of each retina and three retinae of each group were used. In every section, three images were captured in the center, mid-periphery, and periphery area, respectively. In experiments including qPCR and western blot, samples were collected from three retinae of three individual mice. All *in vitro* experiments were performed in triplicate and repeated independently for at least three times. Unpaired student t-test was used to compare the means between two groups. Data were presented as mean ± SEM and P values <0.05 were considered statistically significant.

RESULTS

The Expression of AAT Was Reduced in the Degenerative Retinae of rd1 Mice

Although AAT is mainly produced by hepatocytes, it could also derive from macrophages, monocytes, and other cells (21, 22). We first detected the expression level of endogenic AAT in rd1 mice retinae. Strikingly, the RNA (**Figure 1A**) and protein (**Figure 1B**)

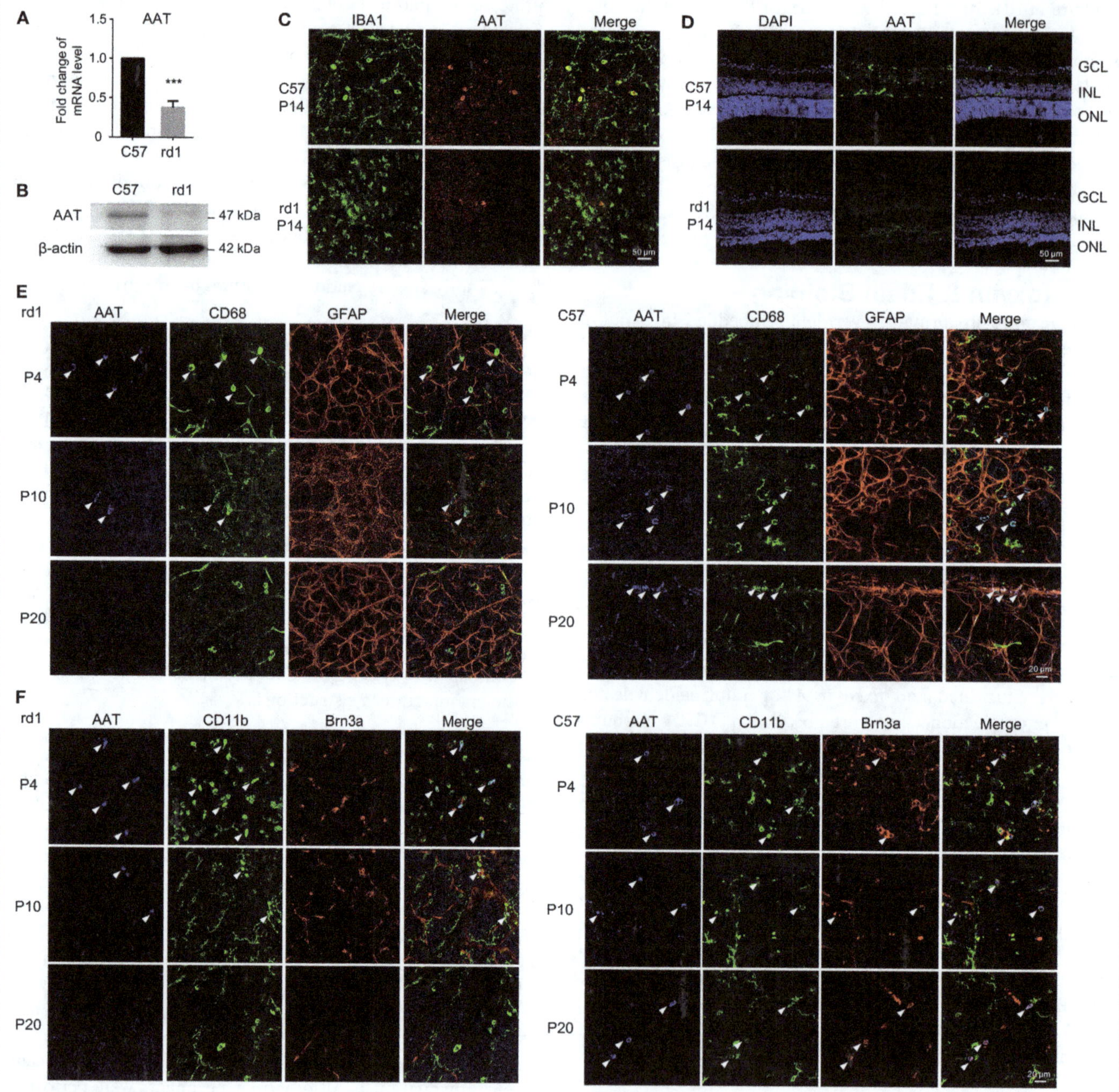

FIGURE 1 | Alpha-1 antitrypsin (AAT) expression decreased in rd1 mice. **(A,B)** Real time PCR **(A)** and western blot **(B)** results revealed that the expression of AAT decreased in retina from rd1 mice compared with C57 controls at P14. $^{***}p < 0.001$ (two-tailed unpaired *t*-test). **(C)**. There were some cells co-staining with AAT (red) and IBA1 (green) in C57-retina, whereas lack of AAT was observed in the retina from rd1 mice at P14. **(D)** Immunofluorescence on retinal section showed that AAT was mainly distributed in the inner nuclear layer in C57 mice, while the rd1 mice lacked AAT expression. **(E,F)** In retinal whole mounts, expression of AAT was decreased as the disease progressed from P4 to P20 in rd1 mice, whereas C57 mice presented stable expression of AAT during the same time. Of note, AAT was co-labeled mostly with CD68 **(E)** and CD11b **(F)**, also markers for microglia in the retina, slightly with Brn3a$^+$ retinal ganglion cell **(F)**, but not with GFAP$^+$ astrocytes/Müller cells **(E)**. Scare bar, 50 µm.

levels of AAT were reduced significantly in rd1 retina compared with the C57 controls at P14 when photoreceptors experienced severe loss (23, 24). Furthermore, the high expression of AAT in C57-retina was localized mainly on IBA1$^+$ microglia, whereas low level was detected on rd1-microglia at P14 (**Figure 1C**). The immunostaining on retinal cryosection showed that AAT was mostly distributed along the INL (**Figure 1D**). To further depict temporal expression of AAT on other retinal cell types, we performed co-staining of AAT and markers for several retinal cell types at different time-points. Strikingly, decreased expression of

AAT was observed in rd1 retina as the disease progressed from P4 to P20 and its expression also reduced markedly compared with the same age C57 controls, whereas both mice presented similar levels of AAT at P4 (a few days before the onset of rod photoreceptor degeneration at P9 in rd1) (**Figures 1E,F**). Of note, AAT co-labeled mostly with CD68 and CD11b, markers for microglia in the retina, while slightly with Brn3a$^+$ retinal ganglion cell (RGC), but not with GFAP$^+$ astrocytes/Mullers (**Figures 1E,F**), indicating that microglia was possibly the predominant cell type expressing AAT in the retina.

AAT Supplement Attenuated the Degenerative Retina of rd1 Mice

Given the decreased level of AAT in rd1 mice, we proposed that AAT supplement might be an alternative intervention for retinal degeneration. To test the possibility of this hypothesis, we injected 80 mg/kg AAT intraperitoneally in rd1 mice every other day from P4 to P14, and evaluated the retinal structure and thickness using SD-OCT and histologic analysis at P16. Twenty-five linear scans within 6 × 3 mm^2 area adjacent to optic nerve disk were obtained at nasal and temporal retinae (**Figure 2A**). As depicted in the 3D imaging of **Figure 2B**, the average retinal thickness was measured within a circle area, with 1.5 mm in radius, centering at 3 mm away from optic nerve disk. As expected, C57 mice presented normal retinal structures with multiple layers, including the nerve fiber layer, inner plexiform layer, inner nuclear layer, outer plexiform layer, ONL, external limiting membrane (ELM), inner segment/outer segment (IS/OS), and RPE. However, these retinal layers were hard to distinguish the PBS-treated rd1 mice with poorly visible ONL, ELM, and IS/OS junctions. Of note, the rd1 mice exhibited a low-reflecting ONL layer after AAT supplement (**Figure 2C**). The retinal volume and thickness were calculated and mild increment could be observed in rd1 mice after AAT supplement (**Figures 2D,E**). Histologic analysis was also performed and representative sections were shown in **Figure 3A**. A thicker retina was observed in AAT-treated retinae at P16. Despite the retinal thickness was slightly elevated in the rd1 mice treated with AAT compared to those under PBS treatment, the increased amount of photoreceptor nuclei in the ONL layer was more significant (**Figures 3B,C**), indicating the pronounced loss of photoreceptors in rd1 mice was ameliorated by AAT.

AAT Supplement Improved Retinal Function From Decrement in rd1 Mice at P16

Next, the retinal function was accessed by ERG at P16. The C57 control mice displayed typical a- and b-wave responses, whereas the PBS-treated rd1 mice exhibited virtually flat signal recordings under a variety of scotopic testing conditions. After AAT supplement, recognizable a- and b-waves were recovered, though the ERG responses were still diminished in comparison to C57 mice (**Figure 4**). Average amplitudes of a- and b-wave for each group of rd1 mice were analyzed. Both a- and b-wave amplitudes were found to be protected with increased amplitudes in the AAT-treated ones, confirming the extinct ERG responses in rd1 mice could be recovered after AAT supplement.

Inherited Photoreceptor Degeneration Was Protected by AAT

To further explore whether the AAT preserved the degenerative photoreceptors, rhodopsin and TUNEL staining was performed on retinal sections. More rhodopsin-positive rod photoreceptors were identified in the ONL of AAT-treated rd1 mice compared with the PBS control (**Figure 5A**). TUNEL-positive cells were

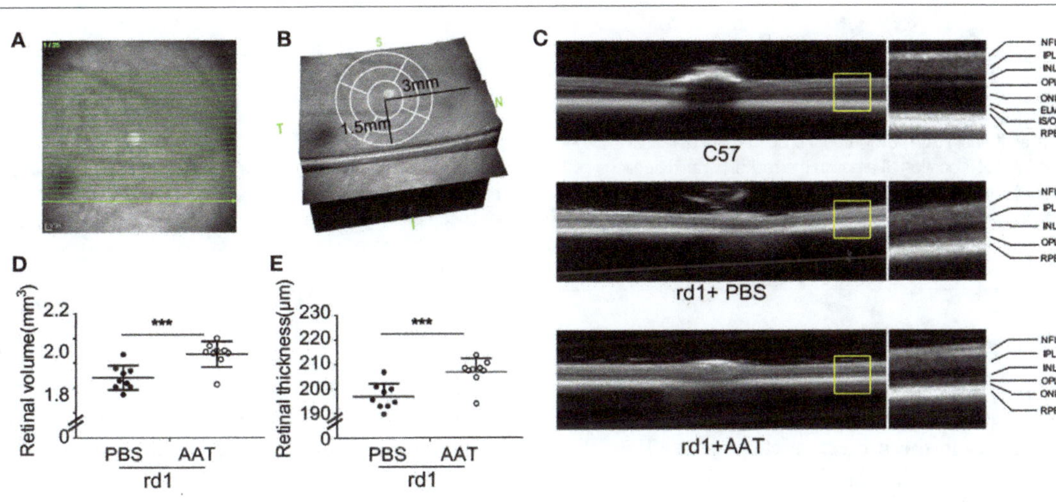

FIGURE 2 | Alpha-1 antitrypsin (AAT) reduced retinal degeneration in rd1 mice. **(A)** The scanning model used in the study. Twenty-five linear scans within 6 × 3 mm^2 area adjacent to optic nerve disk were obtained at nasal and temporal retinae. **(B)** The average retinal thickness was measured in circle area with a radius of 1.5 mm, centering at 3 mm away from optic nerve disk. **(C)** The retinal structure was well-organized with multiple layers in the C57 mice ($n = 6$), while the layers of ONL, ELM, and IS/OS junction were poorly visible in the PBS-treated rd1 mice at P16. However, the AAT-treated rd1 displayed a visible low-reflecting ONL layer. Abbreviations: NFL, nerve fiber layer; IPL, inner plexiform layer; INL, inner nuclear layer; OPL, outer plexiform layer; ONL, outer nuclear layer; ELM, external limiting membrane; IS/OS, inner segment/outer segment; RPE, retinal pigment epithelium. The retinal volume **(D)** and thickness **(E)** were increased in the AAT-treated rd1 mice compared to PBS-treated ones. $n = 9$ mice for each group. ***$p < 0.001$ (two-tailed unpaired t-test).

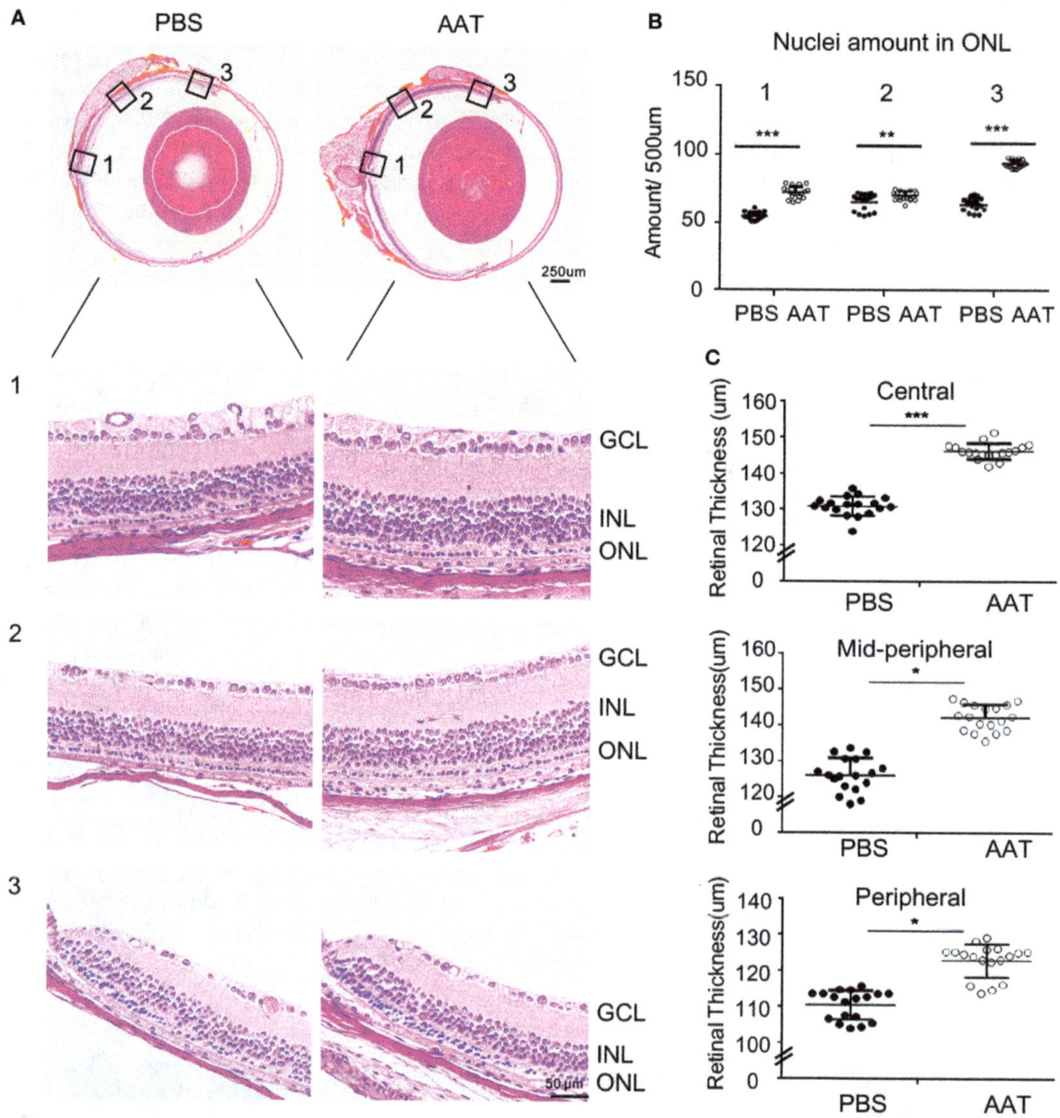

FIGURE 3 | Alpha-1 antitrypsin (AAT) alleviated the decrement of retinal thickness in rd1 mice. **(A)** Panoramic view of the central [Location 1], mid-peripheral [Location 2], and peripheral [Location 3] areas in the retinal paraffin sections by H&E staining. The representative images showed denser nuclei in the outer nuclear layer (ONL) and thicker retinal thickness of the rd1 mice treated with AAT compared to those with PBS treatment at P16. Scare bar, 50 μm. **(B)** The amounts of cellular nuclei in ONL at indicated locations were elevated significantly after AAT treatment in comparison with PBS controls. **(C)** The total thicknesses of retina were increased in the center, mid-periphery, and periphery areas in AAT-treated rd1 mice, compared with those in PBS-treated controls. Six sections of each retina and three retina of each group were used for analysis. $^*p < 0.05$, $^{**}p < 0.01$, $^{***}p < 0.001$ (two-tailed unpaired *t*-test).

evident in the ONL of PBS-treated rd1 mice at P16, while scarce TUNEL-positive cells were observed in the AAT-treated ones (**Figures 5A,B**), indicating the protective effect of AAT on the apoptotic rods.

AAT Suppressed Pro-Inflammatory M1 Microglial Polarization During Retinal Degeneration

Microglia was recognized as the main immune cell in retina and its activation contributed to neuroinflammation during retinal degeneration (10, 25). In this study, we noticed decreased number of IBA1$^+$ positive microglia in the AAT-treated rd1 retina, especially in the degenerative photoreceptor layer of ONL (**Figures 5A,C**). Since the activated microglia displayed different phenotypes at various conditions, we detected the pro-inflammatory M1 phenotype at the setting of degenerative retina in rd1 mice. CD16/32 is widely used as a classic marker for M1 microglia. As demonstrated in **Figure 6A**, AAT supplement obviously decreased the ratio of CD16/32$^+$IBA1$^+$ M1 microglia compared with that of the PBS-treated ones, prominently at the peripheral retina. In retinal sections, CD16/32$^+$IBA1$^+$ microglia

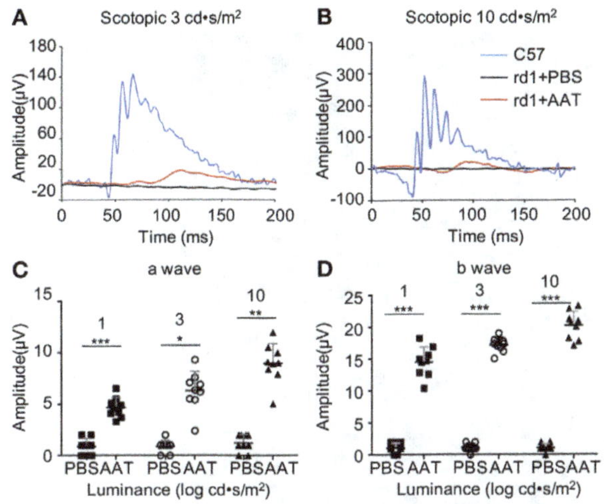

FIGURE 4 | Alpha-1 antitrypsin (AAT) supplement protected retinal function in rd1 mice. **(A,B)** Retinal function was measured at P16 by electroretinogram using single-flash recordings at light intensities of 3.0 and 10.0 log cd•s/m², respectively. C57 mice presented with typical a- and b-wave responses ($n = 6$), while the PBS-treated rd1 mice showed nearly undetectable amplitude of a- or b-waves under a variety of scotopic testing conditions. **(C,D)** AAT treatment induced mild increase in a-wave but significant elevation of b-wave amplitudes in the rd1 mice. $n = 9$ mice for each group. $^{*}p < 0.05$, $^{**}p < 0.01$, $^{***}p < 0.001$ (two-tailed unpaired t-test).

accumulated in the ONL of PBS-treated mice, whereas a lack of this pro-inflammatory microglia was observed in the AAT group (**Figure 6B**). In cultured BV2 microglia under oxidative stress by hydrogen peroxide stimulation, which mimic the general injury in the rd1 retina, AAT supplement also downregulated the expression of pro-inflammatory marker $CD16/32^{+}$ on microglia (**Figure 6C**). These results suggested that AAT suppressed the pro-inflammatory M1 microglia during retinal degeneration.

Microglia Skewed Toward Anti-Inflammatory M2 Phenotype in the Presence of AAT Supplement

The effect of AAT on the anti-inflammatory M2 microglia was also evaluated. CD206 and Arg1, two classic markers for M2 phenotype, were detected by immunofluorescence on retinal whole mounts and sections. We found significantly elevated ratio of $CD206^{+}IBA1^{+}$ microglia in retinal whole mounts from the AAT-treated group compared with the PBS-treated mice (**Figure 7A**). Similarly, although $Arg1^{+}IBA1^{+}$ M2 microglia were absent in the PBS-treated rd1 retinae, these M2 microglia appeared after AAT supplement (**Figure 7B**). From retinal sections, we found increased $CD206^{+}$ and $Arg1^{+}$ microglia locating mainly in the ganglion cell layer and ONL layers (Figures S1A,B in Supplementary Material). *In vitro*, immunostaining (**Figure 7C**) and western blot (**Figure 7D**) data revealed that AAT downregulated the expression of iNOS, another typical M1 marker, and upregulated Arg1 expression. CD206 also increased in the AAT-treated BV2 microglia (Figure S1C in Supplementary Material). These data suggested that AAT could induce microglia to adopt M2 phenotype.

AAT Modulated IRF4/8 Activation and Phosphorylation of STAT1 in Microglia *In Vivo* and *In Vitro*

The interferon-regulatory factor (IRF) members are transcriptional regulators of macrophage polarization, with IRF8 and IRF4 associated with the polarization toward the M1 state and M2 state, respectively (26, 27). We next detected the expressions of IRF4, IRF8, and relevant STAT1 signaling using western blot. The data revealed that in the AAT-treated retinae from rd1 mice, IRF8 expression was significantly suppressed while the level of IRF4 was slightly increased (**Figures 8A,B**). In addition, STAT1 signaling, of which phosphorylation could induce IRF8 expression, was also inactivated with the lack of p-STAT1 after AAT supplement (**Figures 8A,B**). Similarly, AAT stimulation downregulated the expression of IRF8 and upregulated IRF4 level in the cultured primary microglia. In addition, the phosphorylation of STAT1 was inhibited in the AAT-treated primary microglia cells (**Figures 8C,D**). Thus it is possibly that AAT exerted its modulatory effect through regulating the bias of IRF4 and IRF8 expression and STAT1 activation.

DISCUSSION

Neuroinflammation plays a key role in the process of RP (9, 10, 28). In the present study, the efficacy of AAT, an old drug showing new tricks of immunomodulatory property, was evaluated for preventing the neuroinflammation during the pathogenesis of retinal degeneration in rd1 mice. Our data provided evidence for the first time that AAT could suppress the neuroinflammation and attenuate neurodegeneration in the rd1 mice through the induction of anti-inflammatory M2 microglia, underlining that an immunomodulatory therapy may be an efficient strategy to retard retinal degeneration.

Alpha-1 antitrypsin, a 52-kDa glycoprotein encoded by the Serpina gene, is produced mainly by hepatocytes and maintained in circulation (22). Actually, the circulating AAT could be transported by vascular endothelial cells in an endocytosis pathway (29, 30). It is likely that AAT supplement by intraperitoneal injection enter retina through the blood–retinal barrier. Interestingly, AAT could be produced locally by mononuclear phagocytes in tissues and the expression is often reduced in a context of inflammation (21). For example, the amount of AAT in NOD diabetic islet was significantly less than that in C57BL/6 mice (12, 16, 17). In this study, we also found reduced expression of AAT in the rd1 retina, which was associated with neuroinflammation. So it is reasonable and promising of AAT supplement to treat RP due to the decreased level of AAT.

Accumulating evidence indicated that AAT exerted a novel immunomodulatory effect at the setting of inflammation. It is currently available and is used as medication in clinical trial for treating individuals with recent onset of type 1 diabetes and graft-versus-host disease, based on its potent immunomodulatory property (31, 32). Besides, AAT exhibits notable safety profile in clinical usage (33). Hence, it merits further investigation in other settings and our data of its efficacy in RP provide evidence to extend its usages in attenuating inflammation and protect the

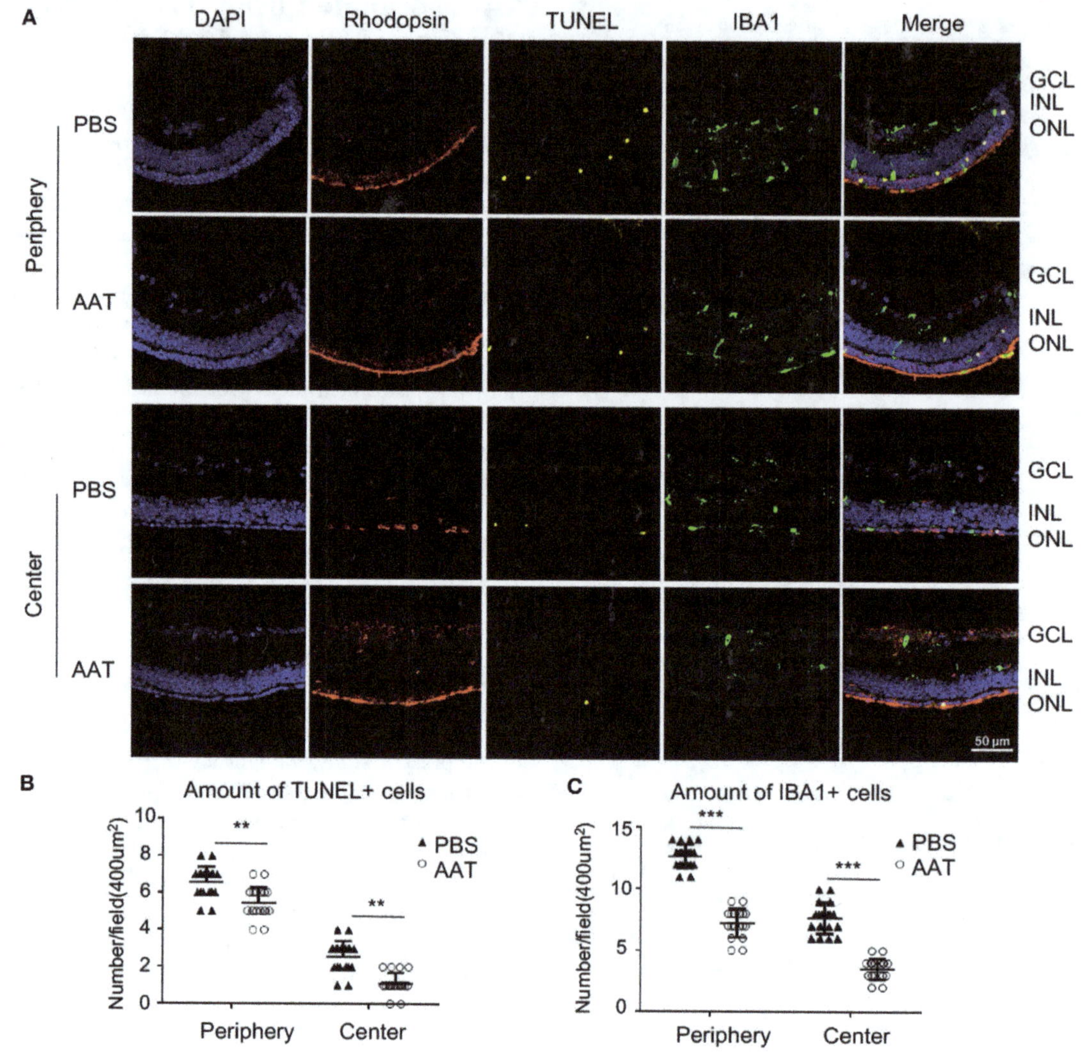

FIGURE 5 | Protective effect of alpha-1 antitrypsin (AAT) on degenerative rods. **(A)** The representative images of immunostaining on retinal sections revealed that rhodopsin-positive rod photoreceptors were scattered in the PBS-treated rd1 mice, but widely distributed and well-organized in the outer nuclear layer after AAT treatment. The TUNEL$^+$ apoptotic cells decreased in AAT supplement group, whereas the amount of IBA1$^+$ microglia also decreased comparing with the PBS-treated ones. The location "Center" refers to the panoramic view area 1 in **Figure 3A**, and "periphery" refers to the area 3. Scare bar, 50 μm. Statistical analysis of the TUNEL$^+$ **(B)** and IBA1$^+$ cells **(C)** revealed that all their amounts significantly decreased either in the center or periphery. Six sections of each retina and three retina of each group were used for analysis. $^{**}p < 0.01$, $^{***}p < 0.001$ (two-tailed unpaired *t*-test).

degenerative retina. In addition, it is well known that rd1 mice, an acute autosomal recessive form for human RP, carries a mutation affecting the expression of β subunit of phosphodiesterase 6, leading to the accumulation of cGMP that is thought to trigger photoreceptor degeneration in a short time. So these mice are hard to get complete recovery by intervention with biological agents. Despite the effect of immune therapy was not as remarkable as that of gene therapy in the rd1 mouse, suppression of neuroinflammation by AAT would provide a suitable microenvironment for cell survival and a stable treatment window for other therapies, considering the fact that neuroinflammation is associated with various type of RP, it serves as a hallmark of common pathological process of RP.

Alpha-1 antitrypsin exerted its immunomodulatory effect on many cellular targets (32, 34). Previous studies showed that neutrophils and monocyte/macrophage could produce more IL-10 and reduced the generation of TNF-α after AAT stimulation (35). AAT could also regulate the activities of NK cells through the interaction of NK cells and dendritic cells (DC) (34). It is worth noting that AAT modifies the phenotype of DC and B lymphocytes toward a tolerogenic pattern, exhibiting the immunomodulatory potential (36). In the present study, retinal microglia in rd1 mice experienced M2 polarization after AAT supplement, providing a novel immunological cellular mechanism by which AAT exerts its pluripotential anti-inflammatory effects. Despite the retina was believed to be immune privilege,

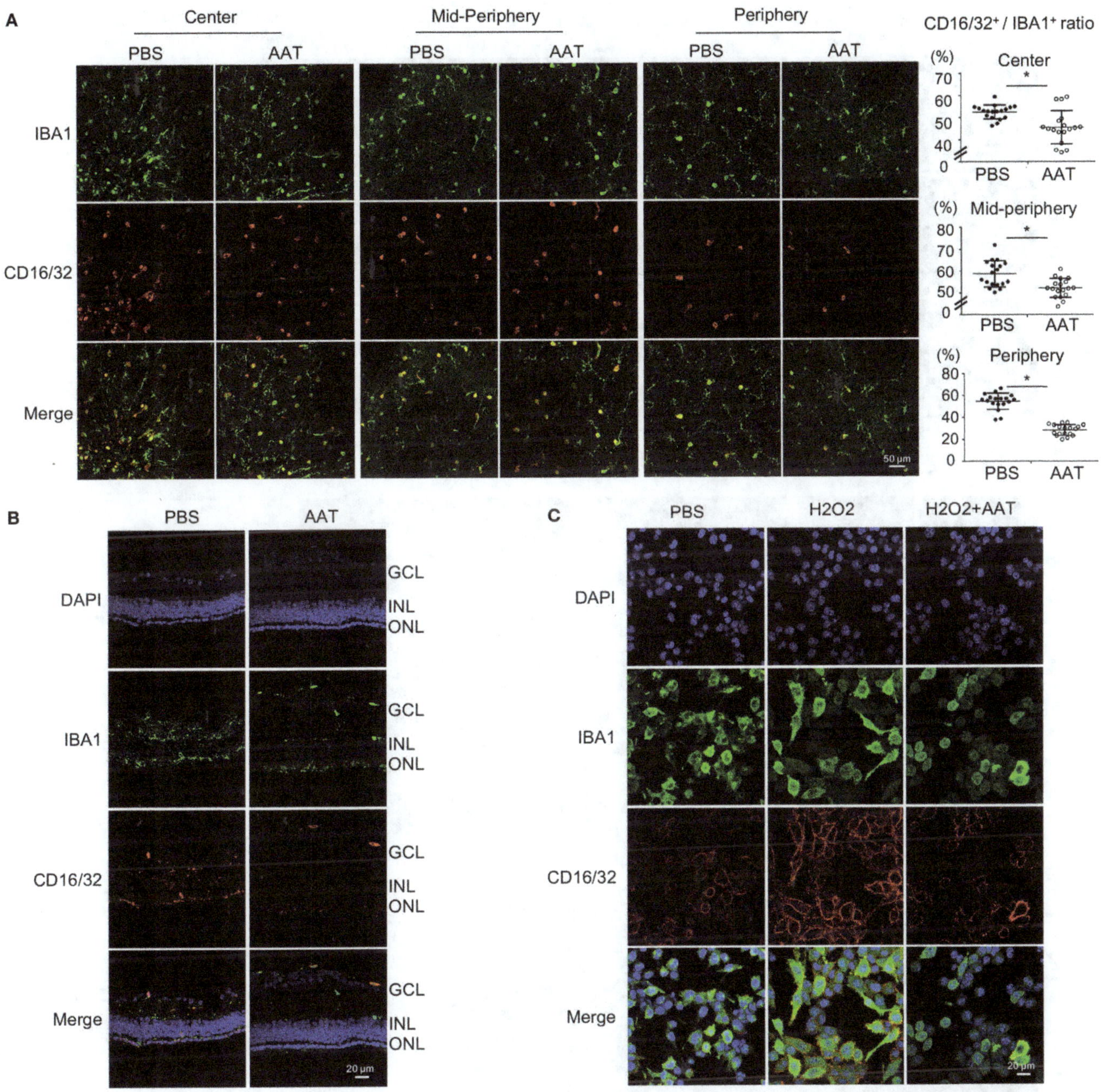

FIGURE 6 | Alpha-1 antitrypsin (AAT) suppressed pro-inflammatory M1 microglial polarization during retinal degeneration. **(A)**. In retinal whole mounts, the amount of CD16/32$^+$IBA1$^+$ M1 microglia in the central, mid-peripheral, and peripheral retina decreased significantly after AAT supplement. Scare bar, 50 μm. Depicted is mean ± SEM of three fields/eyes from six eyes. $^*p < 0.05$ (two-tailed unpaired *t*-test). **(B)** In the retinal section, CD16/32$^+$IBA1$^+$ microglia prominently accumulated in the outer nuclear layer in PBS treatment group whereas the amount of CD16/32$^+$IBA1$^+$ pro-inflammatory microglia significantly decreased in the AAT-treated group. Scare bar, 20 μm. **(C)** In the cultured microglia under oxidative stress by hydrogen peroxide stimulation, AAT supplement significantly suppressed the pro-inflammatory M1 phenotype of microglia, presenting with less CD16/32$^+$ cells co-stained with IBA1$^+$ cells. Scare bar, 20 μm.

its function was monitored by the active microglia, the main resident innate immune cells in CNS, who could initiate a cascade of cellular responses and orchestrate the immune condition in retinal diseases (10, 37). Similar to other immune cells like macrophages and DC, microglia adopt various phenotypes to adapt themselves to different insults (38). Our previous study has demonstrated that retinal microglia were activated and particularly polarized to a pro-inflammatory M1 phenotype at the rapid rods degenerative phase of rd1 mice (23), suggesting that M1 microglia contribute to retinal inflammation and the conversion to M2 anti-inflammatory statue seemed to be an efficient therapy for preventing retinal degeneration. Our

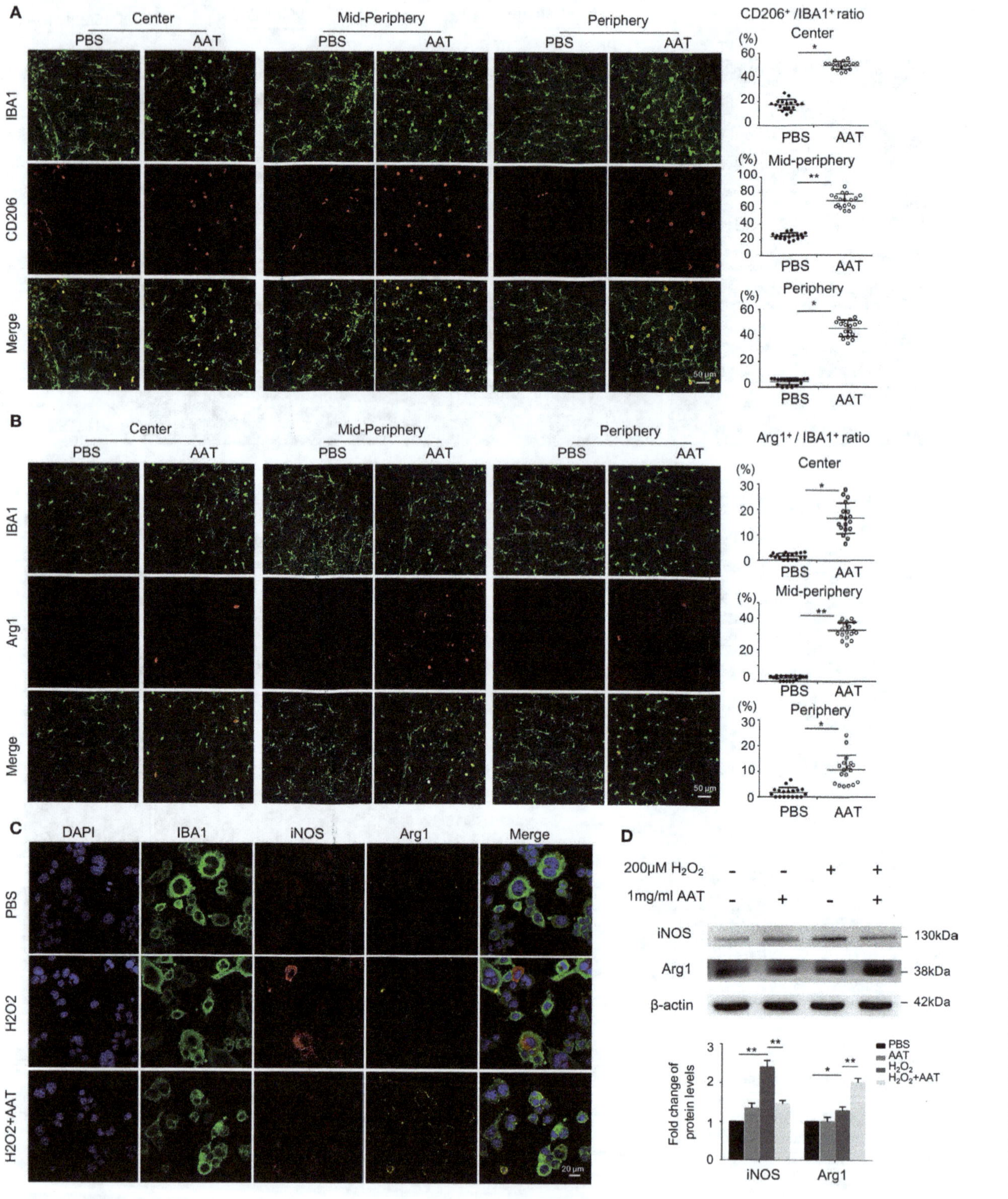

FIGURE 7 | Microglia skewed toward anti-inflammatory M2 phenotype in the presence of alpha-1 antitrypsin (AAT) supplement. **(A)** In the retinal whole mounts, the amount of CD206$^+$IBA1$^+$ microglia significantly increased in AAT-treated group compared with the PBS-treated mice, particularly in the central and mid-peripheral retina. **(B)** Arg1$^+$ cells, another M2 microglia marker, were absent in the PBS-treated rd1 retina, while after AAT supplement, these M2 microglia appeared and most concentrated in the mid-peripheral areas. Scare bar, 50 μm. Depicted is mean ± SEM of three fields/eyes from six eyes. $^*p < 0.05$, $^{**}p < 0.01$ (two-tailed unpaired *t*-test). **(C)** Immunostaining results showed that iNOS was elevated after hydrogen peroxide stimulation on cultured BV2 microglia, in particular expressed in the cytoplasm. In the AAT-treated microglia, not only the expression of iNOS significantly decreased, but also apparent upregulation of Arg1 expression was observed. Scare bar, 20 μm. **(D)** Western blot revealed that AAT downregulated the expression of iNOS while upregulated the Arg1 expression, the counteracting factor of iNOS.

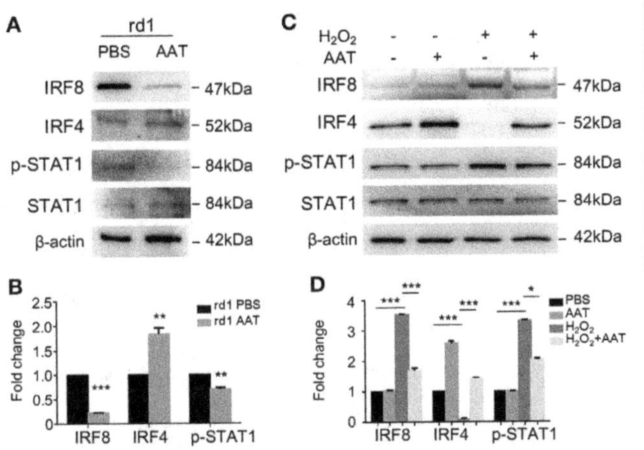

FIGURE 8 | Alpha-1 antitrypsin (AAT) modulated IRF4/8 activation and phosphorylation of STAT1 *in vivo* and *in vitro*. **(A,B)** Western blot revealed that in AAT-treated retina of rd1 mice, the IRF8 expression significantly decreased while the expression of IRF4 increased slightly. Moreover, the STAT1 signaling was suppressed after AAT supplement. **(C,D)** *In vitro*, hydrogen peroxide stimulation upregulated the expression level of IRF8 and promoted phosphorylation of STAT1 in the primary cultured microglia, while AAT treatment reversed the expression trend with decreased IRF8 and STAT1 and increased IRF4. $n = 3$. $^{*}p < 0.05$, $^{**}p < 0.01$, $^{***}p < 0.001$ (two-tailed unpaired t-test).

findings suggested that AAT upregulated M2 anti-inflammatory activity and ameliorated the retinal degeneration, offering a novel immunological mechanism of AAT, particularly in the neurodegenerative diseases, in which microglia polarization is of importance. Of course, the effects of AAT on other retinal cell types could not be excluded in the rd1 mice, and it would be interesting to explore in future study.

A number of studies have implicated that transcription factors cooperatively regulated the downstream M1/M2-associated genes of macrophage/microglia (39). Among them, the IRF family could regulate maturation and activation of immune cells (40). IRF4 and IRF8 are two homologs in the IRF family whose expressions are largely restricted to lymphoid and myeloid cells. In retina, IRF4 and IRF8 were reported to be expressed mainly in microglia (26), which shared features with myeloid/hematopoietic lineage cells. Therefore, their expression and alteration in the whole retina were believed, to some extent, to be associated with microglial activities. IRF8 could induce the expressions of a large panel of M1-related genes, such as IFN-β, IL-12, iNOS, and so on (40, 41), whereas IRF4 functions as a negative regulator of TLRs signaling, associated with the polarization to M2 state (41, 42). In this study, high IRF8 and low IRF4 were observed in rd1 retina. *In vitro*, we found hydrogen peroxide insults induced a marked elevation of IRF8 and reduction of IRF4 in primary microglia, associating with more $iNOS^{+}$ M1 and less $ARG1^{+}$ M2 microglia, further indicating the critical role of IRF8 on M1 polarization of microglia. Strikingly, AAT altered the expression profile with lower IRF8 and higher IRF4 and induced microglia polarization toward M2 and away from M1 *in vivo* and *in vitro*, supporting the concept that AAT modulated microglia polarization partly through regulating the balance of IRF4 and IRF8 expression. A further understanding of mechanisms involved in microglia polarization and its regulation by AAT would facilitate the intervention of neuroinflammation, as well as improve and broaden the usage of AAT in neurodegenerative diseases.

In summary, we demonstrated in the present study that treatment with AAT attenuated neuroinflammation through shifting M1–M2 microglia polarization and alleviated retinal degeneration in rd1 mice. Furthermore, AAT regulated the expression level of IRF8 and IRF4, which were associated with reduced STAT1 phosphorylation, indicating that these signaling pathways were required in AAT-induced M2 polarization. Our data provide evidences of the anti-inflammatory and immunomodulatory properties of AAT and support the possibility that immunomodulatory therapy through regulating microglia M1/M2 polarization may be benefit for treating retinal degeneration.

AUTHOR CONTRIBUTIONS

CH and XL designed and analyzed the experiments and wrote the manuscript. TZ, ZH, XZ, XS, YL, ML, and BC performed the experiments and analyzed the data. YL assisted in the experimental design and provided key research tools.

REFERENCES

Busskamp V, Duebel J, Balya D, Fradot M, Viney TJ, Siegert S, et al. Genetic reactivation of cone photoreceptors restores visual responses in retinitis pigmentosa. *Science* (2010) 329(5990):413–7. doi:10.1126/science. 1190897

Talib M, van Schooneveld MJ, van Genderen MM, Wijnholds J, Florijn RJ, Ten Brink JB, et al. Genotypic and phenotypic characteristics of CRB1- associated retinal dystrophies: a long-term follow-up study. *Ophthalmology* (2017) 124(6):884–95. doi:10.1016/j.ophtha.2017.01.047

Daiger SP, Bowne SJ, Sullivan LS. Genes and mutations causing autosomal dominant retinitis pigmentosa. *Cold Spring Harb Perspect Med* (2014) 5(10):a017129. doi:10.1101/cshperspect.a017129

Pierrache LHM, Kimchi A, Ratnapriya R, Roberts L, Astuti GDN, Obolensky A, et al. Whole-exome sequencing identifies biallelic IDH3A variants as a cause of retinitis pigmentosa accompanied by pseudocoloboma. *Ophthalmology* (2017) 124(7):992–1003. doi:10.1016/j.ophtha.2017.03.010

Sengillo JD, Justus S, Tsai YT, Cabral T, Tsang SH. Gene and cell-based thera- pies for inherited retinal disorders: an update. *Am J Med Genet C Semin Med Genet* (2016) 172(4):349–66. doi:10.1002/ajmg.c.31534

Athanasiou D, Aguila M, Bellingham J, Li W, McCulley C, Reeves PJ, et al. The molecular and cellular basis of rhodopsin retinitis pigmentosa reveals potential strategies for therapy. *Prog Retin Eye Res* (2017) 62:1–23. doi:10.1016/j. preteyeres.2017.10.002

Heneka MT, Kummer MP, Latz E. Innate immune activation in neurode- generative disease. *Nat Rev Immunol* (2014) 14(7):463–77. doi:10.1038/ nri3705

Blank T, Goldmann T, Koch M, Amann L, Schon C, Bonin M, et al. Early microglia activation precedes photoreceptor degeneration in a mouse model of CNGB1-linked retinitis pigmentosa. *Front Immunol* (2017) 8:1930. doi:10.3389/ fimmu.2017.01930

Yoshida N, Ikeda Y, Notomi S, Ishikawa K, Murakami Y, Hisatomi T, et al. Clinical evidence of sustained chronic inflammatory reaction in retinitis pigmentosa. *Ophthalmology* (2013) 120(1):100–5. doi:10.1016/j. ophtha.2012.07.006

Perry VH, Nicoll JA, Holmes C. Microglia in neurodegenerative disease. *Nat Rev Neurol* (2010) 6(4):193–201. doi:10.1038/nrneurol.2010.17

Peng B, Xiao J, Wang K, So KF, Tipoe GL, Lin B. Suppression of microglial activation is neuroprotective in a mouse model of human retinitis pigmen- tosa. *J Neurosci* (2014) 34(24):8139–50. doi:10.1523/JNEUROSCI.5200- 13.2014

Lewis EC, Mizrahi M, Toledano M, Defelice N, Wright JL, Churg A, et al. Alpha1-antitrypsin monotherapy induces immune tolerance during islet allograft transplantation in mice. *Proc Natl Acad Sci U S A* (2008) 105(42):16236–41. doi:10.1073/pnas.0807627105

Dhami R, Zay K, Gilks B, Porter S, Wright JL, Churg A. Pulmonary epithe- lial expression of human alpha1-antitrypsin in transgenic mice results in delivery of alpha1-antitrypsin protein to the interstitium. *J Mol Med* (1999) 77(4):377–85. doi:10.1007/s001090050364

Travis J, Salvesen GS. Human plasma proteinase inhibitors. *Annu Rev Biochem* (1983) 52:655–709. doi:10.1146/annurev.bi.52.070183.003255

Stockley RA, Turner AM. Alpha-1-antitrypsin deficiency: clinical vari- ability, assessment, and treatment. *Trends Mol Med* (2014) 20(2):105–15. doi:10.1016/j.molmed.2013.11.006

Rajagopalan B. Non-bullous ichthyosiform erythroderma associ- ated with retinitis pigmentosa. *Am J Med Genet* (2001) 99(3):181–4. doi:10.1002/1096-8628(2001)9999:9999<::AID-AJMG1164>3.0.CO;2-#

Lewis EC, Shapiro L, Bowers OJ, Dinarello CA. Alpha1-antitrypsin mono- therapy prolongs islet allograft survival in mice. *Proc Natl Acad Sci U S A* (2005) 102(34):12153–8. doi:10.1073/pnas.0505579102

Strom TB. Saving islets from allograft rejection. *Proc Natl Acad Sci U S A* (2005) 102(36):12651–2. doi:10.1073/pnas.0506079102

Dong N, Xu B, Shi H, Lu Y. miR-124 regulates amadori-glycated albumin- induced retinal microglial activation and inflammation by targeting Rac1. *Invest Ophthalmol Vis Sci* (2016) 57(6):2522–32. doi:10.1167/iovs.15-18224

Huang Z, Zhou T, Sun X, Zheng Y, Cheng B, Li M, et al. Necroptosis in microglia contributes to neuroinflammation and retinal degeneration through TLR4 activation. *Cell Death Differ* (2018) 25(1):180–9. doi:10.1038/ cdd.2017.141

van 't Wout EF, van Schadewijk A, Savage ND, Stolk J, Hiemstra PS. Alpha1-antitrypsin production by proinflammatory and antiinflammatory macro-phages and dendritic cells. *Am J Respir Cell Mol Biol* (2012) 46(5):607–13. doi:10.1165/rcmb.2011-0231OC

Lewis EC. Expanding the clinical indications for alpha(1)-antitrypsin therapy. *Mol Med* (2012) 18:957–70. doi:10.2119/molmed.2011.00196

Zhou T, Huang Z, Sun X, Zhu X, Zhou L, Li M, et al. Microglia polar- ization with M1/M2 phenotype changes in rd1 mouse model of retinal degeneration. *Front Neuroanat* (2017) 11:77. doi:10.3389/fnana.2017. 00077

Punzo C, Cepko C. Cellular responses to photoreceptor death in the rd1 mouse model of retinal degeneration. *Invest Ophthalmol Vis Sci* (2007) 48(2): 849–57. doi:10.1167/iovs.05-1555

Karlstetter M, Ebert S, Langmann T. Microglia in the healthy and degen- erating retina: insights from novel mouse models. *Immunobiology* (2010) 215(9–10):685–91. doi:10.1016/j.imbio.2010.05.010

Xu H, Zhu J, Smith S, Foldi J, Zhao B, Chung AY, et al. Notch-RBP-J signaling regulates the transcription factor IRF8 to promote inflammatory macrophage polarization. *Nat Immunol* (2012) 13(7):642–50. doi:10.1038/ ni.2304

Ivashkiv LB. Epigenetic regulation of macrophage polarization and function. *Trends Immunol* (2013) 34(5):216–23. doi:10.1016/j.it.2012.11.001

Viringipurampeer IA, Bashar AE, Gregory-Evans CY, Moritz OL, Gregory-Evans K. Targeting inflammation in emerging therapies for genetic retinal disease. *Int J Inflam* (2013) 2013:581751. doi:10.1155/2013/ 581751

Sohrab S, Petrusca DN, Lockett AD, Schweitzer KS, Rush NI, Gu Y, et al. Mechanism of alpha-1 antitrypsin endocytosis by lung endothelium. *FASEB J* (2009) 23(9):3149–58. doi:10.1096/fj.09-129304

Lockett AD. Alpha-1 antitrypsin transcytosis and secretion. *Methods Mol Biol* (2017) 1639:173–84. doi:10.1007/978-1-4939-7163-3_17 de Serres F, Blanco I. Role of alpha-1 antitrypsin in human health and disease. *J Intern Med* (2014) 276(4):311–35. doi:10.1111/joim.12239

Ehlers MR. Immune-modulating effects of alpha-1 antitrypsin. *Biol Chem* (2014) 395(10):1187–93. doi:10.1515/hsz-2014-0161

Ehlers MR, Nepom GT. Immune-directed therapy for type 1 diabetes at the clinical level: the Immune Tolerance Network (ITN) experience. *Rev Diabet Stud* (2012) 9(4):359–71. doi:10.1900/RDS.2012.9.359

Guttman O, Yossef R, Freixo-Lima G, Rider P, Porgador A, Lewis EC. Alpha1-antitrypsin modifies general NK cell interactions with dendritic cells and specific interactions with islet beta-cells in favor of protection from autoimmune diabetes. *Immunology* (2015) 144(3):530–9. doi:10.1111/imm.12403

Jonigk D, Al-Omari M, Maegel L, Muller M, Izykowski N, Hong J, et al. Anti-inflammatory and immunomodulatory properties of alpha1-antitrypsin without inhibition of elastase. *Proc Natl Acad Sci U S A* (2013) 110(37): 15007–12. doi:10.1073/pnas.1309648110

Marcondes AM, Karoopongse E, Lesnikova M, Margineantu D, Welte T, Dinarello CA, et al. Alpha-1-antitrypsin (AAT)-modified donor cells suppress GVHD but enhance the GVL effect: a role for mitochondrial bioenergetics. *Blood* (2014) 124(18):2881–91. doi:10.1182/blood-2014-04- 570440

Karlstetter M, Scholz R, Rutar M, Wong WT, Provis JM, Langmann T. Retinal microglia: just bystander or target for therapy? *Prog Retin Eye Res* (2015) 45:30–57. doi:10.1016/j.preteyeres.2014.11.004

Hu X, Leak RK, Shi Y, Suenaga J, Gao Y, Zheng P, et al. Microglial and mac- rophage polarization-new prospects for brain repair. *Nat Rev Neurol* (2015) 11(1):56–64. doi:10.1038/nrneurol.2014.207

Rizzolo LJ, Peng S, Luo Y, Xiao W. Integration of tight junctions and claudins with the barrier functions of the retinal pigment epithelium. *Prog Retin Eye Res* (2011) 30(5):296–323. doi:10.1016/j.preteyeres.2011.06.002

Gunthner R, Anders HJ. Interferon-regulatory factors determine macro-phage phenotype polarization. *Mediators Inflamm* (2013) 2013:731023. doi:10.1155/2013/731023

Chistiakov DA, Myasoedova VA, Revin VV, Orekhov AN, Bobryshev YV. The impact of interferon-regulatory factors to macrophage differentiation and polarization into M1 and M2. *Immunobiology* (2017) 223(1):101–11. doi:10.1016/j.imbio.2017.10.005

Satoh T, Takeuchi O, Vandenbon A, Yasuda K, Tanaka Y, Kumagai Y, et al. The Jmjd3-Irf4 axis regulates M2 macrophage polarization and host responses against helminth infection. *Nat Immunol* (2010) 11(10):936–44. doi:10.1038/ ni.1920

13

Noradrenergic Hypothesis Linking Neurodegeneration-Based Cognitive Decline and Astroglia

Giampiero Leanza [1], Rosario Gulino [2] and Robert Zorec [3,4*]*

[1] Department of Drug Sciences, University of Catania, Catania, Italy, [2] Department of Biomedical and Biotechnological Sciences, University of Catania, Catania, Italy, [3] Laboratory of Neuroendocrinology-Molecular Cell Physiology, Faculty of Medicine, Institute of Pathophysiology, University of Ljubljana, Ljubljana, Slovenia, [4] Laboratory of Cell Engineering, Celica Biomedical, Ljubljana, Slovenia

***Correspondence:**
Giampiero Leanza
gleanza@units.it
Robert Zorec
robert.zorec@mf.uni-lj.si

In the past, manipulation of the cholinergic system was seen as the most likely therapeutic for neurodegeneration-based cognitive decline in Alzheimer's disease (AD) (Whitehouse et al., 1982). However, targeting the noradrenergic system also seems a promising strategy, since more recent studies revealed that in post-mortem tissue from patients with AD and other neurodegenerative disorders there is a robust correlation between cognitive decline and loss of neurons from the Locus coeruleus (LC), a system with diffuse noradrenaline (NA) innervation in the central nervous system (CNS). Therefore, the hypothesis has been considered that increasing NA signaling in the CNS will prevent, or at least halt the progression of neurodegeneration and cognitive decline. A hallmark of the age- and neurodegeneration-related cognitive decline is reduced neurogenesis. We here discuss noradrenergic dysfunction in AD-related cognitive decline in humans and its potential involvement in AD pathology and disease progression. We also focus on animal models to allow the validation of the noradrenergic hypothesis of AD, including those based upon the immunotoxin-mediated ablation of LC based on saporin, a protein synthesis interfering agent, which offers selective and graded demise of LC neurons, Finally, we address how astrocytes, an abundant and functionally heterogeneous cell type of neuroglia maintaining homeostasis, may participate in the regulation of neurogenesis, a new strategy for preventing LC neuron loss.

Keywords: noradrenaline (norepinephrine), cognitive decline, Alzheimer's disease, neurodegeneration, neurogenesis, astroglia

INTRODUCTION

With its widespread efferent projections, the small brainstem nucleus Locus coeruleus (LC) represents the main source of noradrenergic innervation to the entire CNS, and plays a pivotal regulatory role in a variety of physiological processes, including attention, arousal, sleep/wakefulness, consciousness as well as in specific aspects of learning and memory (Amaral and Sinnamon, 1977; Aston-Jones and Cohen, 2005; Benarroch, 2009; Sara, 2009). Notably, increasing clinical and imaging evidences (Peterson and Li, 2018) indicate that LC degeneration constitutes a crucial early event in the pathogenesis of Alzheimer's (AD) and Parkinson's disease and most of the LC-regulated functions have been shown to be severely affected during its progression (Chan-Palay and Asan, 1989; Rub et al., 2001; Wilson et al., 2013; Arendt et al., 2015).

Taken together, these observations have pointed at the noradrenergic system as a viable therapeutic target for the treatment of diseases characterized by memory loss and cognitive decline. However, a feasible animal model recapitulating noradrenergic neuronal and terminal fiber loss and its histopathological and cognitive sequelae has not yet been achieved, mainly due to the objective difficulty to selectively and efficiently target LC neurons. This review seeks to briefly summarize noradrenergic dysfunction in AD-related memory loss and its potential involvement in AD pathology and progression. Also, recent findings emerging from our own studies addressing selective immunotoxic ablation of LC neurons and its effects upon cognitive performance, tissue pathology and hippocampal neurogenesis will be briefly outlined. Then we focus into impaired neurogenesis in neurodegeneration and point to the possible contribution of astroglia to this process. Invariably, much relevant literature on the aforementioned topics will not be mentioned here, and we apologize for this.

LOCUS COERULEUS DYSFUNCTION IN ALZHEIMER'S DISEASE

The existence of an association between noradrenergic depletion and neurodegeneration including AD has long been known (Ishii, 1966; Forno and Alvord, 1971; Mann et al., 1980, 1982; Tomlinson et al., 1981; Iversen et al., 1983), however only in relatively recent years has noradrenergic neuron loss in LC been widely acknowledged as a prominent feature of neurodegeneration and AD, being present often decades prior to the appearance of clinical symptoms, and having been related to neurofibrillary pathology and the severity of cognitive deficits, when overtly present (Haglund et al., 2006; Grudzien et al., 2007; Braak and Del Tredici, 2011a,b; Wilson et al., 2013; Andres-Benito et al., 2017; Peterson and Li, 2018).

There have also been conflicting reports regarding the NA levels in the brain of AD patients. In fact, while some studies have reported marked decline in regional NA brain tissue content, whose magnitude correlated with the severity of cognitive impairments (Martignoni et al., 1992; Nazarali and Reynolds, 1992; Matthews et al., 2002; Chen et al., 2014), others reported no changes (Sparks et al., 1988; Herregodts et al., 1989; Tohgi et al., 1992) or even increased NA levels (Tohgi et al., 1992; Elrod et al., 1997).

Similar conflicting observations in post-mortem brain specimens from AD patients have been reported for adrenergic receptors (Shimohama et al., 1986; Kalaria et al., 1989; Pascual et al., 1992; Leverenz et al., 2001), known to be key mediators of noradrenergic activity, and recently considered relevant candidates as novel therapeutic targets for AD (Chen et al., 2014). These discrepancies, likely reflecting compensatory responses, or their lack, at more advanced stages of the disease (Szot et al., 2006, 2007), have thus driven the need to conclusively dissect the exact role played by the noradrenergic system in the cognitive sequelae and pathogenesis that characterize AD and neurodegeneration in general. During the last decades, numerous animal studies and reviews (Mather and Harley, 2016; Borodovitsyna et al., 2017; Gannon and Wang, 2018) have provided valuable insights into the factors underlying the disease. It has been shown, for example, that - possibly via their direct connections to the prefrontal cortex and hippocampus - LC neurons have a fundamental role in sustaining behavioral responsiveness upon exposure to relevant, reward-predicting, stimuli (Bouret and Sara, 2004; Hagena et al., 2016), including those related to working memory (Milstein et al., 2007; Coradazzi et al., 2016). Furthermore, LC neuron degeneration offers a major contribution to AD pathogenesis and progression (Braak et al., 2011; Iba et al., 2015). In fact, a prevailing hypothesis for noradrenergic neuron and fiber depletion in AD holds that LC neurons are uniquely susceptible to tau toxicity (Chandler et al., 2014) and are especially vulnerable to oxidative stress, possibly owing to their high bioenergetic needs (Sanchez-Padilla et al., 2014). In such scenario, accumulation of abnormally phosphorylated tau in LC neurons and its spreading to most of the brain due to the extremely diffuse efferent projections, would account for the progression of the disease (Braak and Del Tredici, 2015), and the resulting neuronal degeneration and cognitive impairments.

EXPERIMENTAL ANIMAL PARADIGMS TO STUDY NORADRENERGIC DYSFUNCTION

Animal studies have so far been extremely helpful to dissect the importance of NA in cognition and in the pathological events associated to its loss. Experimental manipulation of the noradrenergic system by e.g., pharmacological blockade (Mair et al., 2005; Khakpour-Taleghani et al., 2009), lesioning with N-(2-chloro-ethyl)-N-ethyl-2- bromobenzylamine (DSP4), reportedly an LC-selective neurotoxin (Lapiz et al., 2001; Sontag et al., 2008) or the knockout of the dopamine-β-hydroxylase (DBH) gene (Thomas and Palmiter, 1997; Marino et al., 2005; Hammerschmidt et al., 2013) have all resulted in impaired performance in several learning and memory tasks, demonstrating the existence of an association between NA loss and disturbances in various aspects of cognition. However, in many cases, the impairments observed in these studies have appeared rather inconsistent, both in efficiency and selectivity, thus highlighting the potential limitations inherent to each method (Sontag et al., 2008; Khakpour-Taleghani et al., 2009; Szot et al., 2010; Gannon et al., 2015). In fact, pharmacological agents lack anatomical and neurochemical resolution, acting on most monoaminergic neurons and cells throughout the central and peripheral nervous system. Likewise, DSP4 does not seem to be specific for noradrenergic neurons and has been shown to produce only a modest noradrenergic neuron loss, at best. Finally, although DBH (–/–) knockout mice have provided the unique opportunity to precisely assess the effects of NA loss *per se*, with respect to the various modulators produced and released by the very same LC neurons, they do not seem to offer the possibility to obtain partial or graded neurotransmitter depletions. In light of these limitations, we have chosen an alternative lesioning approach based on the use of the immunotoxin anti-DBH-saporin (Picklo et al., 1994), able to target noradrenergic neurons in the LC with unprecedented selectivity and efficiency. This immunotoxin results from the conjugation of saporin, a powerful ribosome-inactivating plant lectin extracted from *Saponaria officinalis* (Caryophyllaceae) (Lappi et al., 1985; Barthelemy et al., 1993) to a monoclonal antibody raised against DBH (the enzyme converting dopamine to NA) that, in addition to its main localization in the cytosol, is also expressed at the plasma membrane surface of noradrenergic neurons (Weinshilboum, 1978; Studelska and Brimijoin, 1989). Due to its structure, saporin cannot enter the cell (Contestabile and Stirpe, 1993), but when coupled to a carrier molecule (e.g., an antibody), is able to specifically bind a surface antigen protein (such as DBH, in this case), the toxin gains access to the cytosol and binds to the ribosomal 60S subunit, interfering with protein synthesis, and soon leading to cell death (Wiley and Kline, 2000). In initial anatomical investigations, the immunotoxin, infused into the lateral ventricles of either adults (Wrenn et al., 1996) or developing rats (Coradazzi et al., 2010) has been observed to produce highly specific and dose-dependent depletions of LC neurons, with no effects on other cholinergic, dopaminergic or serotonergic neuronal populations (**Figure 1**). Notably, the possibility to induce a partial or total noradrenergic loss (by varying the injected dose) makes this immunotoxic approach extremely suitable to address compensatory events within the noradrenergic projection system, in addition to providing an excellent tissue environment for the survival and integration of implanted NA-rich progenitors (Coradazzi et al., 2010).

NORADRENALINE AND ADULT HIPPOCAMPAL NEUROGENESIS

The subgranular zone of the hippocampal dentate gyrus is one of the brain regions where generation of neural progenitor cells occurs thoughout life in various species, including humans (Altman and Das, 1965; Cameron et al., 1993; Eriksson et al., 1998; Gould et al., 1998). From here, proliferating newborn cells migrate to the granule cell layer, where they differentiate into neurons and glia and functionally integrate into the local tissue environment (Markakis and Gage, 1999; Carlen et al., 2002; van Praag et al., 2002). Many factors have been observed to affect hippocampal neurogenesis, including environmental or stressful stimuli (Dranovsky and Hen, 2006), various kinds of hormones and drugs (Duman et al., 2001), as well as neurotransmitter activity (Brezun and Daszuta, 1999; Mohapel et al., 2005, 2006; Aztiria et al., 2007; Walker et al., 2008). Notably, the hippocampus is the region where explicit memories are apparently acquired and consolidated (Murchison et al., 2004), and one of the brain areas receiving the densest LC-derived noradrenergic innervation (Swanson and Hartman, 1975). Moreover, hippocampal neurogenesis has been proposed to underlie some of the behavioral effects of antidepressant drugs (Santarelli et al., 2003; Warner-Schmidt and Duman, 2006; Airan et al., 2007), whose actions are mainly exerted by increasing extracellular levels of serotonin and/or noradrenaline (Fuller et al., 1994; Sacchetti et al., 1999). It is therefore not surprising that several studies have begun to investigate the contribution of LC neurons to the regulation of hippocampal neurogenesis, and have generally reported a permissive role for NA upon hippocampal neurogenesis (Kulkarni et al., 2002; Jhaveri et al., 2010, 2014; Masuda et al., 2012). Again, however, none of the lesion or pharmacological manipulations adopted in these investigations proved to be region-or transmitter specific, nor were any of the observed effects analyzed also in terms of impact upon cognitive performance. In our study using selective and discrete immunolesioning of LC neurons, associated with a series of hippocampus-dependent spatial navigation tasks (Coradazzi et al., 2016), we found severe deficits in working memory, which correlated with the magnitude of hippocampal noradrenergic depletion and the lesion-induced reduction in the numbers of proliferating cells within the dentate gyrus. Notably, no changes were detected in reference memory abilities, nor did the lesion affect long-term survival or differentiation of granule cell progenitors (Coradazzi et al., 2016). Thus, the noradrenergic regulation of complex aspects of cognitive function (e.g., those related to working memory) may take place via the proliferation of progenitor cells in the hippocampal dentate gyrus.

NORADRENALINE, ASTROGLIA AND NEUROGENESIS

It appears that noradrenergic receptors are densely present in astroglia (Aoki, 1992), therefore it is likely that NA may affect neuronal circuits via astroglia (Ding et al., 2013; Paukert et al., 2014; Pankratov and Lalo, 2015; Gao et al., 2016; Dong et al., 2017) (**Figure 2**. graphical abstract). Current view holds that neurodegeneration in AD is a consequence of neuron-specific deficits. However, it is more likely that preceding or concomitant changes in neuroglia may also contribute to this process (Heneka et al., 2015; Verkhratsky and Parpura, 2015; De Strooper and Karran, 2016; Rodriguez-Vieitez et al., 2016; Stenovec et al., 2016; Verkhratsky et al., 2016). Astroglia, a type of neuroglia, consisting also of oligodendroglia, microglia and NG2 cells, are functionally and morphologically heterogeneous, and are involved in sustaining brain homeostasis at cellular and whole organ levels, by regulating extracellular levels of ions and neurotransmitters, by controlling vascular and metabolic functions, the integrity of blood-brain barrier (BBB) (Terry, 2000; Giaume et al., 2007; Kano and Hashimoto, 2009; Heneka et al., 2010; Nedergaard et al., 2010; Parpura and Zorec, 2010; Verkhratsky and Nedergaard, 2014, 2018; Zorec et al., 2018). Importantly, astrocytes are essential in orchestrating defense in the CNS as well; pathological states in the CNS lead to reactive astrogliosis, a process that contains and isolates events taking place in the damaged brain regions. Moreover, reactive astrogliosis is also augmenting post-damage regeneration and repair of brain tissue (Parpura et al., 2012; Pekny et al., 2016; Verkhratsky et al., 2016); however, under certain conditions reactive astrogliosis can be neurotoxic (Liddelow et al., 2017). Astrocytes were also termed gliocrine cells (Vardjan and Zorec, 2015), since they secrete gliosignalling molecules into the extracellular space and are then convectively distributed thoroughout the brain by the glymphatic system (at least in

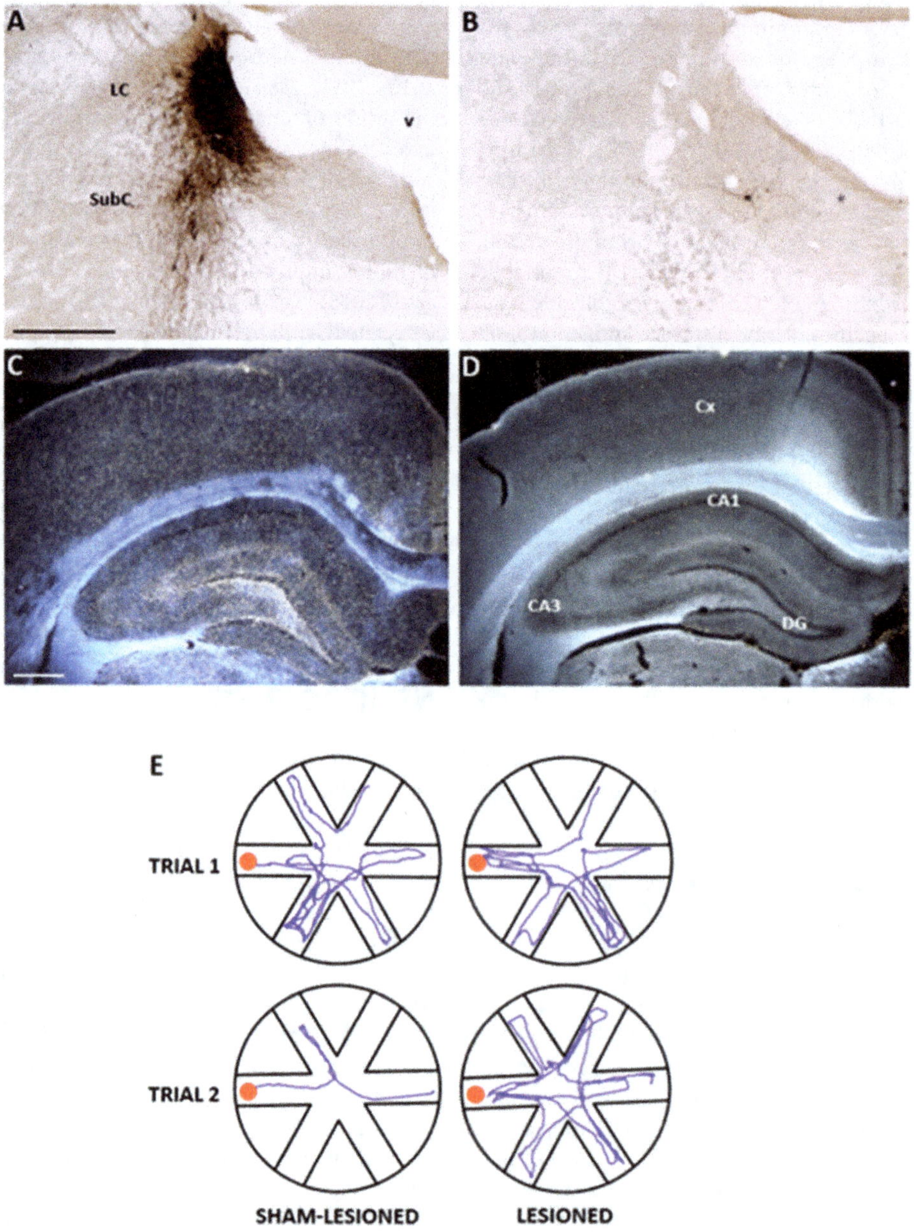

FIGURE 1 | Anatomical and functional effects of the selective noradrenergic lesion in young adult Sprague-Dawley rats. **(A–D)**: Photomicrographs showing the effects of the anti-DBH immunotoxin, injected bilaterally into the LC, on noradrenergic neurons in the LC/SubC **(A,B)** and on the DBH-immunoreactive terminal innervation in the parietal cortex and hippocampus (**C,D**, in dark field). Note in **(B,D)** the nearly complete loss of immunoreactive neurons and fibers induced by the lesion, compared to the normal patterns in the specimens from a sham-lesioned animal **(A,C)**. In **(E)**, the actual swim paths taken by representative sham-lesioned and lesioned animals undergoing the Radial Arm Water Maze (RAWM) task for working memory are illustrated. The sham-lesioned animal rapidly learns the task and significantly improves its performance from the first to the second trial, whereas the lesioned animal does not. LC, locus coeruleus; SubC, subcoeruleus, v fourth ventricle, CA, cornu ammonis of the hippocampus; DG, dentate gyrus. Scale bars in **(A,C)**: 500 μm. Adapted from Coradazzi et al. (2016).

mice), responsible for waste removal (Thrane et al., 2014). Changes in these complex functions of astroglia may lead to a homeostatic failure, leading to disease (Verkhratsky and Parpura, 2015). Hence, primary defect in homeostatic astroglia may lead to a secondary defect in neurons. Interestingly, the role of neuroglial cells in dementia and AD was noted already at least a century ago by A. Alzheimer, who observed glial cells in the proximity of damaged neurons (Alzheimer, 1907, 1910; Strassnig and Ganguli, 2005).

In tissue from post-mortem patients with AD, several morphological observations were made, including astroglial hypertrophy, associated with reactive astrogliosis, the hallmark of which is increased expression of cytoskeletal glial fibrillary acidic protein (GFAP) and S100, a Ca^{2+}-binding protein; these

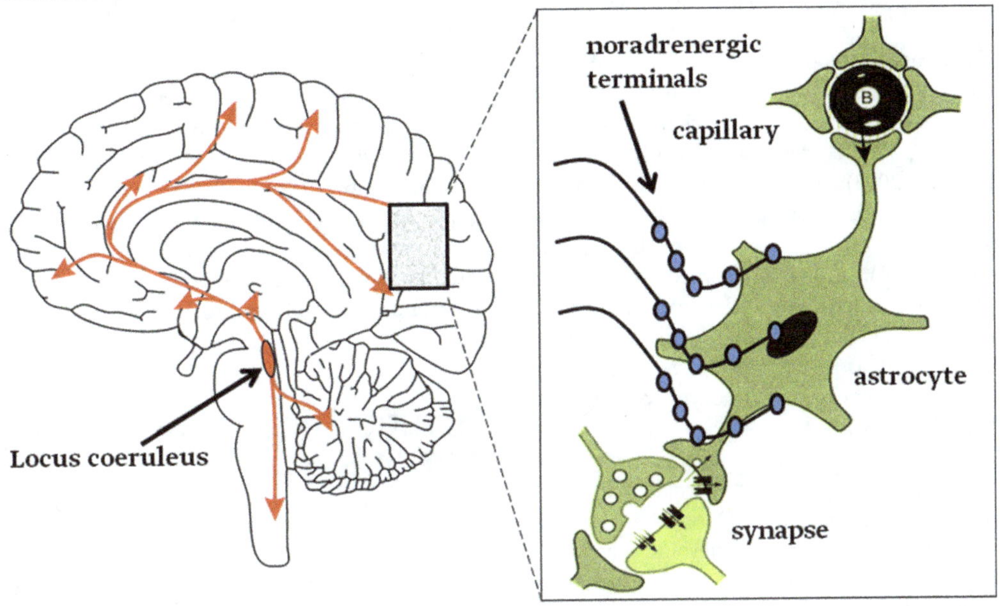

FIGURE 2 | Graphical abstract. **Left**: neurons from the Locus coeruleus (LC) project axons to most, if not all, areas of the brain and into the spinal cord as denoted by the arrows. **Right**: inset magnifies the LC nerve endings with varicosities (swellings on noradrenergic nerve terminals) from which NA is released and mainly activates astroglia (Aoki, 1992; Sherpa et al., 2016). About half of these terminals do not form tight contacts with target cells, resembling synapses (Feinstein et al., 2016).

changes in astrocytes were seen associated with senile plaques (Beach and McGeer, 1988; Griffin et al., 1989; Nagele et al., 2004; Mrak and Griffin, 2005; Verkhratsky et al., 2014). Further, studying patients with AD by imaging methods revealed distinct time- and brain-specific morphological alterations (Rodriguez-Vieitez et al., 2016). Similar changes occurred in an animal mouse model of familial AD as well, including astroglial atrophy in addition to hypertrophy in certain areas of the CNS (Olabarria et al., 2010, 2011; Yeh et al., 2011; Kulijewicz-Nawrot et al., 2012). The appearance of asthenic astrocytes appeared to preced the presence of senile plaques; asthenic astrocytes were at first observed in the entorhinal cortex, an area affected early in AD pathology (Yeh et al., 2011). Ideally, studies of pathological developments associated with neurodegeneration and AD would be best in humans, but these experiments are challenging. Recent attempts to classify protein astrogliopathies, including deposition of amyloid-β, prion protein, tau, α-synuclein, and transactive response DNA-binding protein 43 (TDP-43), demonstrated that these are present in human neurodegenerative diseases (Kovacs et al., 2017). However, animal models seem still very valuable, particularly where many aspects of this pathology can be, to some extent, reproduced, and the time emergence of AD-like characteristic take a much shorter time to develop, providing certain opportunities for studying AD-related neuropathology experimentally. Not only mice, there are several other animal models of AD including nematodes, fruit flies, rabbits, canines, and non-human primates; in each of the models different aspects of AD properties are manifested (reviewed in Woodruff-Pak, 2008). The selective ablation of LC in rats, as presented in previous chapters, offers interesting insights to be explored as a novel animal model of AD and neurodegeneration, not only for functional cognitive impairments (Coradazzi et al., 2016), but also for morphological alterations and aspects of neurogenesis related to astroglia.

Neuron degeneration in the LC during early stages of AD (Heneka et al., 2006) may lead to reduced levels of NA, which is known to be a generic inhibitor of neuroinflammation (De Keyser et al., 2004) and remodeling of the neurovascular unit (del Zoppo, 2009). Therefore, reduced noradrenergic innervation in the CNS may affect the progression of AD through reduced inhibition of neuroinflammation, a tissue remodeling process. This likely operates via a reduced NA contribution of astrocyte adrenergic excitation. Although NA released by the LC neurons acts through α- and β-adrenergic receptors (α/β-ARs), which are expressed in neurons, microglia and astrocytes, but it is the latter cell type that exhibits a high density of ARs, especially the β-ARs (Aoki, 1992) and thus represents a key cell type mediating adrenergic effects on brain tissue. Indeed, activation of α-ARs stimulates Ca^{2+} signaling in astrocytes (Salm and McCarthy, 1989; Kirischuk et al., 1996; Horvat et al., 2016); experiments *in vivo* reported Ca^{2+} waves propagating through astroglial syncytia after stimulating the LC in anesthetized animals (Bekar et al., 2008). In awake animals, electrical stimulation of LC triggered (via activation of $\alpha_?$-ARs) widespread and synchronous astroglial Ca^{2+} signals in practically all astrocytes in the field of study (Ding et al., 2013). This phenomenon may be taken as an event resetting neural networks (Bouret and Sara, 2005). In AD, astrocytic Ca^{2+} signaling is impaired (Lim et al., 2014; Stenovec et al., 2016), which may affect the clearance of Aβ deposits (Mattson, 2004) and subsequent glutamate toxicity (Mattson and Chan, 2003; Mattson, 2004).

Whenever NA is released, not only α-ARs, but also β-ARs are stimulated, although the effects appear in different time-domains (Horvat et al., 2016). These latter receptors stimulate second messenger cAMP, which affects glycogenolysis (Prebil et al., 2011; Kreft et al., 2012), a likely source of glutamate for memory consolidation (Gibbs et al., 2010). Glycogen, an energy reserve, present in astrocytes, but not in neurons (Barros, 2013; Oe et al., 2016), is likely consumed for processes related to morphological plasticity controlled by β-AR/cAMP signaling in astroglia (Vardjan et al., 2014). Not only in memory formation (Zorec et al., 2015), astroglial β-AR/cAMP-dependent morphological changes are central for astrocytic cell oedema attenuation (Vardjan et al., 2016).

Activation of ARs on astrocytes may also affect neurogenesis through neuronal metabolic support by astroglia. Reduced support due to lower NA levels may in AD result in reduced neurogenesis. Astrocytes are the site of aerobic glycolysis, a special metabolic adaptation, present in tissues exhibiting cell division and morphological plasticity. This non-oxidative utilization of glucose, takes place even in the presence of adequate levels of oxygen; it is also known as "the Warburg effect" (Vander Heiden et al., 2009). It is also a characteristic of cancer cells (Salcedo-Sora et al., 2014), hence aerobic glycolysis seems a universal property needed for tissue enlargement and cell shape remodeling.

Astroglial aerobic glycolysis, with the end product L-lactate, is regulated by NA. During exercise, sensory stimulation, alertness and in some pathophysiological states astroglial L-lactate production is up-regulated, requiering the activation of LC neurons (Dienel and Cruz, 2016; Feinstein et al., 2016). LC neurons respond to L-lactate, generated by astroglia when NA activates ARs on astroglial plasma membrane, with elevated electrical activity, an interesting form of communication between the somata of the two cell-types (Tang et al., 2014). It is possible that astroglial L-lactate may also stimulate LC axons away from the somata; as it was recently demonstarted that transcranial direct current stimulation induces NA-dependent elevation of second messenger Ca^{2+} in astrocytes (Monai et al., 2016). Therefore, degenerated LC neurons in AD fail to adequately stimulate astroglia, a source of necessary support for the metabolic needs of neuronal networks, contributing to attenuated neurogenesis. Hypometabolic manifestation is observed clinically in patients with AD (Rodriguez-Vieitez et al., 2016).

Cholesterol is an important building block of membranes and astroglial aerobic glycolysis is likely linked to cholesterol homeostasis in the brain, the most cholesterol-rich organ in the body (Lütjohann et al., 1996). There is little exchange of cholesterol molecules by circulating lipoproteins, between the brain and systemic circulation, since lipoproteins are unable to pass the blood–brain barrier (BBB). Interestingly, the brain exhibits its own cholesterol synthesis, which occurs mainly in glia (Mauch et al., 2001)). Brain cholesterol is not only synthesized *de novo* in the brain, it is also modified/degraded by enzymatic conversion to 24(S)-hydroxycholesterol (Lütjohann et al., 1996), a form that can readily cross the BBB, a major route of cholesterol exit from the brain.

Cholesterol metabolism is linked to the synthesis of the neurosteroids including allopregnanolone, a potent stimulator of neural progenitor cell survival reversing the progress of disease in the 3xTgAD mouse model (Singh et al., 2012). Thus, it is possible to speculate that the hypometabolic state in AD, due to impairment of LC-mediated adrenergic stimulation of astroglia, results in a reduced provision of metabolic intermediates and in attenuated neurosteroid-mediated maintenance of neurogenesis, an avenue that will have to be confirmed experimentally in the future.

FUTURE PROSPECTS AND THERAPY OPTIONS

Based on the above observations of the relationship between central noradrenergic depletion and a series of AD-related changes, including cognitive disturbances, and tissue pathology, it is not surprising that much interest has recently mounted concerning the possibility to restore extracellular NA levels in the brain as a prerequisite to ameliorate cognitive performance as well as to promote cell protection and neurogenesis. Several potential therapeutic venues have been explored so far: one holds that voluntary physical exercise may improve memory by enhancing NA release from LC neurons (Segal et al., 2012), an effect possibly mediated by β-ARs (Van Hoomissen et al., 2004; Ebrahimi et al., 2010). Another study has addressed the possible restorative effects of atomoxetine (a NA reuptake inhibitor) in AD patients undergoing anticholinesterase therapy, reporting no clear-cut cognitive improvements (Mohs et al., 2009), possibly due to a poor effectiveness of the NA uptake inhibitor on LC which is already severely depleted of its neurons as a result of the disease (Braun et al., 2014; Braun and Feinstein, 2017). A similar lack of cognitive improvements has been reported in aged healthy patients following treatment with guanfacine, an α_2-AR agonist (Van Dyck, 2014; http://ClinicalTrials.gov) that, if anything, should reduce, rather than increase NA levels (Starke, 2001). Indeed, α_2-AR antagonists appear to prevent age-related spatial working memory impairments in a transgenic AD mouse model (Scullion et al., 2011). In spite of these inconsistencies, the possibility to ameliorate cognitive performance and cell protection by enhancing noradrenergic neurotransmission remains an exciting prospect: in a recent investigation (Pintus et al., 2018) we have implanted embryonic noradrenergic progenitors bilaterally into the hippocampus of rats whose LC neurons had been almost completely and selectively depleted. We reasoned that if the loss of noradrenergic innervation to the hippocampus was necessary to induce measurable cognitive impairments, its restoration promoted by the implanted embryonic LC neurons should be sufficient to ameliorate/reverse them. As expected (and consistent with our previous findings Coradazzi et al., 2016), the LC lesion induced severe deficits in working memory which were seen fully reversed (up to normal) in transplanted animals and significantly reinstated by a second lesion ablating the implanted neuroblasts. Interestingly, the transplant-promoted noradrenergic reinnervation also normalized the nuclear expression of the transactive response DNA-binding protein 43 (TDP-43) in various hippocampal subregions, whose cytoplasmic (i.e., pathological) occurrence appeared dramatically increased as a result of the lesions. These findings therefore provide support to the view that cognitive and histopathological changes observed in AD patients may require concurrent loss of ascending regulatory noradrenergic inputs from LC and that NA replenishment may be an effective intervention to slow down and/or reverse cognitive decline and tissue pathology, including neurogenesis. Moreover, the NA replenishment may operate via astroglial adrenergic mechanism that include metabolic and neurotrophic support of neural networks.

CONCLUDING REMARKS

Noradrenergic neurons in the LC and their widespread efferent connections play a central role in normal cognition and their disruption is increasingly believed to be critically associated with severe memory loss and neurodegeneration in general (Wilson et al., 2013; Feinstein et al., 2016; Bharani et al., 2017; Satoh and Iijima, 2017). Notably, the presence of functional and densely present adrenergic receptors on astrocytes (Aoki, 1992), whose activation triggers metabolic and energetic responses supporting synaptic functioning and plasticity (Zorec et al., 2017), strongly suggests that many, if not all, the NA-mediated processes may take place also via an action on astrocyes. Thus, addressing astrocytes as viable targets for neurological diseases related to noradrenergic dysfunction is an interesting issue with fundamental therapeutic implications, warranting further research.

AUTHOR CONTRIBUTIONS

GL, RG, and RZ conceived the concept of paper, wrote and edited the manuscript.

REFERENCES

Airan, R. D., Meltzer, L. A., Roy, M., Gong, Y., Chen, H., and Deisseroth, K. (2007). High-speed imaging reveals neurophysiological links to behavior in an animal model of depression. *Science* 317, 819–823. doi: 10.1126/science.1144400

Altman, J., and Das, G. D. (1965). Autoradiographic and histological evidence of postnatal hippocampal neurogenesis in rats. *J. Comp. Neurol.* 124, 319–335. doi: 10.1002/cne.901240303

Alzheimer, A. (1907). Über eine eigenartige Erkrankung der Hirnrinde. Versammlung Südwestdeutscher Irrenärzte in Tübingen am 3. November 1906. *Allgemeine Zeitschrift für Psychiatrie und Psychisch-Gerichtliche Medizin.* 64, 146–148.

Alzheimer, A. (1910). "Beiträge zur Kenntnis der pathologischen Neuroglia und ihrer Beziehungen zu den Abbauvorgängen im Nervengewebe," in *Histologische und histopathologische Arbeiten über die Grosshirnrinde mit besonderer Berücksichtigung der pathologischen Anatomie der Geisteskrankheiten,* ed F. Nissl (Jena: Verlag von G. Fischer), 401–562.

Amaral, D. G., and Sinnamon, H. M. (1977). The locus coeruleus: neurobiology of a central noradrenergic nucleus. *Prog. Neurobiol.* 9, 147–196. doi: 10.1016/0301-0082(77)90016-8

Andrés-Benito P., Fernández-Dueñas V., Carmona, M., Escobar, L. A., Torrejón-Escribano, B., Aso, E., et al. (2017). Locus coeruleus at asymptomatic early and middle Braak stages of neurofibrillary tangle pathology. *Neuropathol. Appl. Neurobiol.* 43, 373–392. doi: 10.1111/nan.12386

Aoki, C. (1992). Beta-adrenergic receptors: astrocytic localization in the adult visual cortex and their relation to catecholamine axon terminals as revealed by electron microscopic immunocytochemistry. *J. Neurosci.* 12, 781–792. doi: 10.1523/JNEUROSCI.12-03-00781.1992

Arendt, T., Bruckner, M. K., Morawski, M., Jager, C., and Gertz, H. J. (2015). Early neurone loss in Alzheimer's disease: cortical or subcortical? *Acta Neuropathol. Commun* 3:10. doi: 10.1186/s40478-015-0187-1

Aston-Jones, G., and Cohen, J. D. (2005). An integrative theory of locus coeruleus-norepinephrine function: adaptive gain and optimal performance. *Annu. Rev. Neurosci.* 28, 403–450. doi: 10.1146/annurev.neuro.28.061604.135709

Aztiria, E., Capodieci, G., Arancio, L., and Leanza, G. (2007). Extensive training in a maze task reduces neurogenesis in the adult rat dentate gyrus probably as a result of stress. *Neurosci. Lett.* 416, 133–137. doi: 10.1016/j.neulet.2007.01.069

Barros, L. F. (2013). Metabolic signaling by lactate in the brain. *Trends Neurosci.* 36, 396–404. doi: 10.1016/j.tins.2013.04.002

Barthelemy, I., Martineau, D., Ong, M., Matsunami, R., Ling, N., Benatti, L., et al. (1993). The expression of saporin, a ribosome-inactivating protein from the plant Saponaria officinalis, in *Escherichia coli. J. Biol. Chem.* 268, 6541–6458.

Beach, T. G., and McGeer, E. G. (1988). Lamina-specific arrangement of astrocytic gliosis and senile plaques in Alzheimer's disease visual cortex. *Brain Res.* 463, 357–361. doi: 10.1016/0006-8993(88)90410-6

Bekar, L. K., He, W., and Nedergaard, M. (2008). Locus coeruleus alpha-adrenergic-mediated activation of cortical astrocytes in vivo. *Cereb. Cortex* 18, 2789–2795. doi: 10.1093/cercor/bhn040

Benarroch, E. E. (2009). The locus ceruleus norepinephrine system: functional organization and potential clinical significance. *Neurology* 73, 1699–1704. doi: 10.1212/WNL.0b013e3181c2937c

Bharani, K. L., Derex, R., Granholm, A. C., and Ledreux, A. (2017). A noradrenergic lesion aggravates the effects of systemic inflammation on the hippocampus of aged rats. *PLoS ONE* 12:e0189821. doi: 10.1371/journal.pone.0189821

Borodovitsyna, O., Flamini, M., and Chandler, D. (2017). Noradrenergic modulation of cognition in health and disease. *Neural Plast.* 2017:6031478. doi: 10.1155/2017/6031478

Bouret, S., and Sara, S. J. (2004). Reward expectation, orientation of attention and locus coeruleus-medial frontal cortex interplay during learning. *Eur. J. Neurosci.* 20, 791–802. doi: 10.1111/j.1460-9568.2004.03526.x

Bouret, S., and Sara, S. J. (2005). Network reset: a simplified overarching theory of locus coeruleus noradrenaline function. *Trends Neurosci.* 28, 574–582. doi: 10.1016/j.tins.2005.09.002

Braak, H., and Del Tredici, K. (2011a). The pathological process underlying Alzheimer's disease in individuals under thirty. *Acta Neuropathol.* 121, 171–181. doi: 10.1007/s00401-010-0789-4

Braak, H., and Del Tredici, K. (2011b). Alzheimer's pathogenesis: is there neuron-to-neuron propagation? *Acta Neuropathol.* 121, 589–595. doi: 10.1007/s00401-011-0825-z

Braak, H., and Del Tredici, K. (2015). The preclinical phase of the pathological process underlying sporadic Alzheimer's disease. *Brain* 138(Pt 10), 2814–2833. doi: 10.1093/brain/awv236

Braak, H., Thal, D. R., Ghebremedhin, E., and Del Tredici, K. (2011). Stages of the pathologic process in Alzheimer disease: age categories from 1 to 100 years. *J. Neuropathol. Exp. Neurol.* 70, 960–969. doi: 10.1097/NEN.0b013e318232a379

Braun, D., and Feinstein, D. L. (2017). The locus coeruleus neuroprotective drug vindeburnol normalizes behavior in the 5xFAD transgenic mouse model of Alzheimer's disease. *Brain Res.* doi: 10.1016/j.brainres.2017.12.028. [Epub ahead of print].

Braun, D., Madrigal, J. L., and Feinstein, D. L. (2014). Noradrenergic regulation of glial activation: molecular mechanisms and therapeutic implications. *Curr. Neuropharmacol.* 12, 342–352. doi: 10.2174/1570159X12666140828220938

Brezun, J. M., and Daszuta, A. (1999). Depletion in serotonin decreases neurogenesis in the dentate gyrus and the subventricular zone of adult rats. *Neuroscience* 89, 999–1002. doi: 10.1016/S0306-4522(98)00693-9

Cameron, H. A., Woolley, C. S., McEwen, B. S., and Gould, E. (1993). Differentiation of newly born neurons and glia in the dentate gyrus of the adult rat. *Neuroscience* 56, 337–344. doi: 10.1016/0306-4522(93)90335-D

Carlen, M., Cassidy, R. M., Brismar, H., Smith, G. A., Enquist, L. W., and Frisen, J. (2002). Functional integration of adult-born neurons. *Curr. Biol* .12, 606–608. doi: 10.1016/S0960-9822(02)00771-6

Chandler, D. J., Gao, W. J., and Waterhouse, B. D. (2014). Heterogeneous organization of the locus coeruleus projections to prefrontal and motor cortices. *Proc. Natl. Acad. Sci. U.S.A.* 111, 6816–6821. doi: 10.1073/pnas.1320827111

Chan-Palay, V., and Asan, E. (1989). Alterations in catecholamine neurons of the locus coeruleus in senile dementia of the Alzheimer type and in Parkinson's disease with and without dementia and depression. *J. Comp. Neurol.* 287, 373–392. doi: 10.1002/cne.902870308

Chen, Y., Peng, Y., Che, P., Gannon, M., Liu, Y., Li, L., et al. (2014). alpha(2A) adrenergic receptor promotes amyloidogenesis through disrupting APP-SorLA interaction. *Proc. Natl. Acad. Sci. U.S.A.* 111, 17296–172301. doi: 10.1073/pnas.1409513111

Contestabile, A., and Stirpe, F. (1993). Ribosome-inactivating proteins from plants as agents for suicide transport and immunolesioning in the nervous system. *Eur. J. Neurosci.* 5, 1292–1301. doi: 10.1111/j.1460-9568.1993.tb00915.x

Coradazzi, M., Gulino, R., Fieramosca, F., Falzacappa, L. V., Riggi, M., and Leanza, G. (2016). Selective noradrenaline depletion impairs working memory and hippocampal neurogenesis. *Neurobiol. Aging* 48, 93–102. doi: 10.1016/j.neurobiolaging.2016.08.012

Coradazzi, M., Gulino, R., Garozzo, S., and Leanza, G. (2010). Selective lesion of the developing central noradrenergic system: short- and long-term effects and reinnervation by noradrenergic-rich tissue grafts. *J. Neurochem.* 114, 761–771. doi: 10.1111/j.1471-4159.2010.06800.x

De Keyser, J., Zeinstra, E., Mostert, J., and Wilczak, N. (2004). Beta 2-adrenoceptor involvement in inflammatory demyelination and axonal degeneration in multiple sclerosis. *Trends Pharmacol. Sci.* 25, 67–71. doi: 10.1016/j.tips.2003.12.002

del Zoppo, G. J. (2009). Inflammation and the neurovascular unit in the setting of focal cerebral ischemia. *Neuroscience* 158, 972–982. doi: 10.1016/j.neuroscience.2008.08.028

De Strooper, B., and Karran, E. (2016). The cellular phase of Alzheimer's disease. *Cell* 164, 603–615. doi: 10.1016/j.cell.2015.12.056

Dienel, G. A., and Cruz, N. F. (2016). Aerobic glycolysis during brain activation: adrenergic regulation and influence of norepinephrine on astrocytic metabolism. *J. Neurochem.* 38, 14–52 doi: 10.1111/jnc.13630

Ding, F., O'Donnell, J., Thrane, A. S., Zeppenfeld, D., Kang, H., Xie, L., et al. (2013). α1-Adrenergic receptors mediate coordinated Ca^{2+} signaling of cortical astrocytes in awake, behaving mice. *Cell Calcium* 54, 387–394. doi: 10.1016/j.ceca.2013.09.001

Dong, J. H., Wang, Y. J., Cui, M., Wang, X. J., Zheng, W. S., Ma, M. L., et al. (2017). Adaptive activation of a stress response pathway improves learning and memory through Gs and beta-arrestin-1-regulated lactate metabolism. *Biol. Psychiatry* 81, 654–670. doi: 10.1016/j.biopsych.2016.09.025

Dranovsky, A., and Hen, R. (2006). Hippocampal neurogenesis: regulation by stress and antidepressants. *Biol. Psychiatry* 59, 1136–1143. doi: 10.1016/j.biopsych.2006.03.082

Duman, R. S., Malberg, J., and Nakagawa, S. (2001). Regulation of adult neurogenesis by psychotropic drugs and stress. *J. Pharmacol. Exp. Ther.* 299, 401–407.

Ebrahimi, S., Rashidy-Pour, A., Vafaei, A. A., and Akhavan, M. M. (2010). Central beta-adrenergic receptors play an important role in the enhancing effect of voluntary exercise on learning and memory in rat. *Behav. Brain Res.* 208, 189–193. doi: 10.1016/j.bbr.2009.11.032

Elrod, R., Peskind, E. R., DiGiacomo, L., Brodkin, K. I., Veith, R. C., and Raskind, M. A. (1997). Effects of Alzheimer's disease severity on cerebrospinal fluid norepinephrine concentration. *Am. J. Psychiatry* 154, 25–30. doi: 10.1176/ajp.154.1.25

Eriksson, P. S., Perfilieva, E., Bjork-Eriksson, T., Alborn, A. M., Nordborg, C., Peterson, D. A., et al. (1998). Neurogenesis in the adult human hippocampus. *Nat. Med.* 4, 1313–1317. doi: 10.1038/3305

Feinstein, D. L., Kalinin, S., and Braun, D. (2016). Causes, consequences, and cures for neuroinflammation mediated via the locus coeruleus: noradrenergic signaling system. *J. Neurochem.* 139(Suppl. 2), 154–178. doi: 10.1111/jnc.13447

Forno, L. S., and Alvord, E. C. Jr. (1971). The pathology of Parkinsonism. Part, I. Some new observations and correlations. *Contemp. Neurol. Ser.* 8, 119–130.

Fuller, R. W., Hemrick-Luecke, S. K., and Snoddy, H. D. (1994). Effects of duloxetine, an antidepressant drug candidate, on concentrations of monoamines and their metabolites in rats and mice. *J. Pharmacol. Exp. Ther.* 269, 132–136.

Gannon, M., Che, P., Chen, Y., Jiao, K., Roberson, E. D., and Wang, Q. (2015). Noradrenergic dysfunction in Alzheimer's disease. *Front. Neurosci.* 9:220. doi: 10.3389/fnins.2015.00220

Gannon, M., and Wang, Q. (2018). Complex noradrenergic dysfunction in Alzheimer's disease: Low norepinephrine input is not always to blame. *Brain Res.* doi: 10.1016/j.brainres.2018.01.001. [Epub ahead of print].

Gao, V., Suzuki, A., Magistretti, P. J., Lengacher, S., Pollonini, G., Steinman, M. Q., et al. (2016). Astrocytic beta2-adrenergic receptors mediate hippocampal long-term memory consolidation. *Proc. Natl. Acad. Sci. U.S.A.* 113, 8526–8531. doi: 10.1073/pnas.1605063113

Giaume, C., Kirchhoff, F., Matute, C., Reichenbach, A., and Verkhratsky, A. (2007). Glia: the fulcrum of brain diseases. *Cell Death Differ.* 14, 1324–1335. doi: 10.1038/sj.cdd.4402144

Gibbs, M. E., Hutchinson, D. S., and Summers, R. J. (2010). Noradrenaline release in the locus coeruleus modulates memory formation and consolidation; roles for α- and β-adrenergic receptors. *Neuroscience* 170, 1209–1222. doi: 10.1016/j.neuroscience.2010.07.052

Gould, E., Tanapat, P., McEwen, B. S., Flügge G., and Fuchs, E. (1998). Proliferation of granule cell precursors in the dentate gyrus of adult monkeys is diminished by stress. *Proc. Natl. Acad. Sci. U.S.A.* 95, 3168–3171. doi: 10.1073/pnas.95.6.3168

Griffin, W. S., Stanley, L. C., Ling, C., White, L., MacLeod, V., Perrot, L. J., et al. (1989). Brain interleukin 1 and S-100 immunoreactivity are elevated in Down syndrome and Alzheimer disease. *Proc. Natl. Acad. Sci. U.S.A.* 86, 7611–7615. doi: 10.1073/pnas.86.19.7611

Grudzien, A., Shaw, P., Weintraub, S., Bigio, E., Mash, D. C., and Mesulam, M. M. (2007). Locus coeruleus neurofibrillary degeneration in aging, mild cognitive impairment and early Alzheimer's disease. *Neurobiol. Aging* 28, 327–335. doi: 10.1016/j.neurobiolaging.2006.02.007

Hagena, H., Hansen, N., and Manahan-Vaughan, D. (2016). beta-adrenergic control of hippocampal function: subserving the choreography of synaptic information storage and memory. *Cereb. Cortex* 26, 1349–1364. doi: 10.1093/cercor/bhv330

Haglund, M., Sjöbeck M, M., and Englund, E. (2006). Locus ceruleus degeneration is ubiquitous in Alzheimer's disease: possible implications for diagnosis and treatment. *Neuropathology* 26, 528–532. doi: 10.1111/j.1440-1789.2006.00725.x

Hammerschmidt, T., Kummer, M. P., Terwel, D., Martinez, A., Gorji, A., Pape, H. C., et al. (2013). Selective loss of noradrenaline exacerbates early cognitive dysfunction and synaptic deficits in APP/PS1 mice. *Biol. Psychiatry* 73, 454–463. doi: 10.1016/j.biopsych.2012.06.013

Heneka, M. T., Carson, M. J., El Khoury, J., Landreth, G. E., Brosseron, F., Feinstein, D. L., et al. (2015). Neuroinflammation in Alzheimer's disease. *Lancet Neurol.* 14, 388–405. doi: 10.1016/S1474-4422(15)70016-5

Heneka, M. T., Ramanathan, M., Jacobs, A. H., Dumitrescu-Ozimek, L., Bilkei-Gorzo, A., Debeir, T., et al. (2006). Locus ceruleus degeneration promotes Alzheimer pathogenesis in amyloid precursor protein 23 transgenic mice. *J. Neurosci.* 26, 1343–1354. doi: 10.1523/JNEUROSCI.4236-05.2006

Heneka, M. T., Rodríguez JJ., and Verkhratsky, A. (2010). Neuroglia in neurodegeneration. *Brain Res. Rev.* 63, 189–211. doi: 10.1016/j.brainresrev.2009.11.004

Herregodts, P., Bruyland, M., De Keyser, J., Solheid, C., Michotte, Y., and Ebinger, G. (1989). Monoaminergic neurotransmitters in Alzheimer's disease. An HPLC study comparing presenile familial and sporadic senile cases. *J. Neurol. Sci.* 92, 101–116. doi: 10.1016/0022-510X(89)90179-2

Horvat, A., Zorec, R., and Vardjan, N. (2016). Adrenergic stimulation of single rat astrocytes results in distinct temporal changes in intracellular Ca^{2+} and cAMP-dependent PKA responses. *Cell Calcium.* 59, 156–163. doi: 10.1016/j.ceca.2016.01.002

Iba, M., McBride, J. D., Guo, J. L., Zhang, B., Trojanowski, J. Q., and Lee, V. M. (2015). Tau pathology spread in PS19 tau transgenic mice following locus coeruleus (LC) injections of synthetic tau fibrils is determined by the LC's afferent and efferent connections. *Acta Neuropathol.* 130, 349–362. doi: 10.1007/s00401-015-1458-4

Ishii, T. (1966). Distribution of Alzheimer's neurofibrillary changes in the brain stem and hypothalamus of senile dementia. *Acta Neuropathol.* 6, 181–187. doi: 10.1007/BF00686763

Iversen, L. L., Rossor, M. N., Reynolds, G. P., Hills, R., Roth, M., Mountjoy, C. Q., et al. (1983). Loss of pigmented dopamine-beta-hydroxylase positive cells from locus coeruleus in senile dementia of Alzheimer's type. *Neurosci. Lett.* 39, 95–100. doi: 10.1016/0304-3940(83)90171-4

Jhaveri, D. J., Mackay, E. W., Hamlin, A. S., Marathe, S. V., Nandam, L. S., Vaidya, V. A., et al. (2010). Norepinephrine directly activates adult hippocampal precursors via beta3-adrenergic receptors. *J. Neurosci.* 30, 2795–2806. doi: 10.1523/JNEUROSCI.3780-09.2010

Jhaveri, D. J., Nanavaty, I., Prosper, B. W., Marathe, S., Husain, B. F., Kernie, S. G., et al. (2014). Opposing effects of alpha2- and beta-adrenergic receptor stimulation on quiescent neural precursor cell activity and adult hippocampal neurogenesis. *PLoS ONE* 9:e98736. doi: 10.1371/journal.pone.0098736

Kalaria, R. N., Andorn, A. C., Tabaton, M., Whitehouse, P. J., Harik, S. I., and Unnerstall, J. R. (1989). Adrenergic receptors in aging and Alzheimer's disease: increased beta 2-receptors in prefrontal cortex and hippocampus. *J. Neurochem.* 53, 1772–17781. doi: 10.1111/j.1471-4159.1989.tb09242.x

Kano, M., and Hashimoto, K. (2009). Synapse elimination in the central nervous system. *Curr. Opin. Neurobiol.* 19, 154–161. doi: 10.1016/j.conb.2009.05.002

Khakpour-Taleghani, B., Lashgari, R., Motamedi, F., and Naghdi, N. (2009). Effect of reversible inactivation of locus ceruleus on spatial reference and working memory. *Neuroscience* 158, 1284–1291. doi: 10.1016/j.neuroscience.2008.11.001

Kirischuk, S., Tuschick, S., Verkhratsky, A., and Kettenmann, H. (1996). Calcium signalling in mouse Bergmann glial cells mediated by alpha1-adrenoreceptors and H1 histamine receptors. *Eur. J. Neurosci.* 8, 1198–1208. doi: 10.1111/j.1460-9568.1996.tb01288.x

Kovacs, G. G., Lee, V. M., and Trojanowski, J. Q. (2017). Protein astrogliopathies in human neurodegenerative diseases and aging. *Brain Pathol.* 27, 675–690. doi: 10.1111/bpa.12536

Kreft, M., Bak, L. K., Waagepetersen, H. S., and Schousboe, A. (2012). Aspects of astrocyte energy metabolism, amino acid neurotransmitter homoeostasis and metabolic compartmentation. *ASN Neuro.* 4:e00086. doi: 10.1042/AN20120007

Kulijewicz-Nawrot, M., Verkhratsky, A., Chvátal, A., Syková, E., and Rodríguez, J. J. (2012). Astrocytic cytoskeletal atrophy in the medial prefrontal cortex of a triple transgenic mouse model of Alzheimer's disease. *J. Anat.* 221, 252–262. doi: 10.1111/j.1469-7580.2012.01536.x

Kulkarni, V. A., Jha, S., and Vaidya, V. A. (2002). Depletion of norepinephrine decreases the proliferation, but does not influence the survival and differentiation, of granule cell progenitors in the adult rat hippocampus. *Eur. J. Neurosci.* 16, 2008–2012. doi: 10.1046/j.1460-9568.2002.02268.x

Lapiz, M. D., Mateo, Y., Durkin, S., Parker, T., and Marsden, C. A. (2001). Effects of central noradrenaline depletion by the selective neurotoxin DSP-4 on the behaviour of the isolated rat in the elevated plus maze and water maze. *Psychopharmacology* 155, 251–259. doi: 10.1007/s00213010 0702

Lappi, D. A., Esch, F. S., Barbieri, L., Stirpe, F., and Soria, M. (1985). Characterization of a Saponaria officinalis seed ribosome-inactivating protein: immunoreactivity and sequence homologies. *Biochem. Biophys. Res. Commun.* 129, 934–942. doi: 10.1016/0006-291X(85)91981-3

Leverenz, J. B., Miller, M. A., Dobie, D. J., Peskind, E. R., and Raskind, M. A. (2001). Increased alpha 2-adrenergic receptor binding in locus coeruleus projection areas in dementia with Lewy bodies. *Neurobiol. Aging* 22, 555–561. doi: 10.1016/S0197-4580(01)00221-4

Liddelow, S. A., Guttenplan, K. A., Clarke, L. E., Bennett, F. C., Bohlen, C. J., Schirmer, L., et al. (2017). Neurotoxic reactive astrocytes are induced by activated microglia. *Nature* 541, 481–487. doi: 10.1038/nature21029

Lim, D., Ronco, V., Grolla, A. A., Verkhratsky, A., and Genazzani, A. A. (2014). Glial calcium signalling in Alzheimer's disease. *Rev. Physiol. Biochem. Pharmacol.* 167, 45–65. doi: 10.1007/112_2014_19

Lütjohann, D., Breuer, O., Ahlborg, G., Nennesmo, I., Sidén, A., Diczfalusy, U., et al. (1996). Cholesterol homeostasis in human brain: evidence for an age-dependent flux of 24S-hydroxycholesterol from the brain into the circulation. *Proc. Natl. Acad. Sci. U.S.A.* 93, 9799–9804. doi: 10.1073/pnas.93.18.9799

Mair, R. D., Zhang, Y., Bailey, K. R., Toupin, M. M., and Mair, R. G. (2005). Effects of clonidine in the locus coeruleus on prefrontal- and hippocampal-dependent measures of attention and memory in the rat. *Psychopharmacology* 181, 280–288. doi: 10.1007/s00213-005-2263-x

Mann, D. M., Lincoln, J., Yates, P. O., Stamp, J. E., and Toper, S. (1980). Changes in the monoamine containing neurones of the human CNS in senile dementia. *Br. J. Psychiatry* 136, 533–541. doi: 10.1192/bjp.136.6.533

Mann, D. M., Yates, P. O., and Hawkes, J. (1982). The noradrenergic system in Alzheimer and multi-infarct dementias. *J. Neurol. Neurosurg. Psychiatry* 45, 113–119. doi: 10.1136/jnnp.45.2.113

Marino, M. D., Bourdelat-Parks, B. N., Cameron Liles, L., and Weinshenker, D. (2005). Genetic reduction of noradrenergic function alters social memory and reduces aggression in mice. *Behav. Brain Res.* 161, 197–203. doi: 10.1016/j.bbr.2005.02.005

Markakis, E. A., and Gage, F. H. (1999). Adult-generated neurons in the dentate gyrus send axonal projections to field CA3 and are surrounded by synaptic vesicles. *J. Comp. Neurol.* 406, 449–460. doi: 10.1002/(SICI)1096-9861(19990419)406:4<449::AID-CNE3>3.0.CO;2-I

Martignoni, E., Blandini, F., Petraglia, F., Pacchetti, C., Bono, G., and Nappi, G. (1992). Cerebrospinal fluid norepinephrine, 3-methoxy-4-hydroxyphenylglycol and neuropeptide Y levels in Parkinson's disease, multiple system atrophy and dementia of the Alzheimer type. *J. Neural Transm. Park. Dis. Dement. Sect.* 4, 191–205. doi: 10.1007/BF02260903

Masuda, T., Nakagawa, S., Boku, S., Nishikawa, H., Takamura, N., Kato, A., et al. (2012). Noradrenaline increases neural precursor cells derived from adult rat dentate gyrus through beta2 receptor. *Prog. Neuropsychopharmacol. Biol. Psychiatry* 36, 44–51. doi: 10.1016/j.pnpbp.2011.08.019

Mather, M., and Harley, C. W. (2016). The locus coeruleus: essential for maintaining cognitive function and the aging brain. *Trends Cogn. Sci.* 20, 214–226. doi: 10.1016/j.tics.2016.01.001

Matthews, K. L., Chen, C. P., Esiri, M. M., Keene, J., Minger, S. L., and Francis, P. T. (2002). Noradrenergic changes, aggressive behavior, and cognition in patients with dementia. *Biol. Psychiatry* 51, 407–416. doi: 10.1016/S0006-3223(01)01235-5

Mattson, M. P. (2004). Pathways towards and away from Alzheimer's disease. *Nature* 430, 631–639. doi: 10.1038/nature02621

Mattson, M. P., and Chan, S. L. (2003). Neuronal and glial calcium signaling in Alzheimer's disease. *Cell Calcium* 34, 385–397. doi: 10.1016/S0143-4160(03)00128-3

Mauch, D. H., Nägler, K., Schumacher, S., Göritz, C., Müller, E. C., Otto, A., et al. (2001). CNS synaptogenesis promoted by glia-derived cholesterol. *Science* 294, 1354–1357. doi: 10.1126/science.294.5545.1354

Milstein, J. A., Lehmann, O., Theobald, D. E., Dalley, J. W., and Robbins, T. W. (2007). Selective depletion of cortical noradrenaline by anti-dopamine beta-hydroxylase-saporin impairs attentional function and enhances the effects of guanfacine in the rat. *Psychopharmacology* 190, 51–63. doi: 10.1007/s00213-006-0594-x

Mohapel, P., Leanza, G., Kokaia, M., and Lindvall, O. (2005). Forebrain acetylcholine regulates adult hippocampal neurogenesis and learning. *Neurobiol Aging* 26, 939–946. doi: 10.1016/j.neurobiolaging.2004.07.015

Mohapel, P., Mundt-Petersen, K., Brundin, P., and Frielingsdorf, H. (2006). Working memory training decreases hippocampal neurogenesis. *Neuroscience* 142, 609–613. doi: 10.1016/j.neuroscience.2006.07.033

Mohs, R. C., Shiovitz, T. M., Tariot, P. N., Porsteinsson, A. P., Baker, K. D., and Feldman, P. D. (2009). Atomoxetine augmentation of cholinesterase inhibitor therapy in patients with Alzheimer disease: 6-month, randomized, double-blind, placebo-controlled, parallel-trial study. *Am. J. Geriatr. Psychiatry* 17, 752–759. doi: 10.1097/JGP.0b013e3181aad585

Monai, H., Ohkura, M., Tanaka, M., Oe, Y., Konno, A., Hirai, H., et al. (2016). Calcium imaging reveals glial involvement in transcranial direct current stimulation-induced plasticity in mouse brain. *Nat. Commun.* 7:11100. doi: 10.1038/ncomms11100

Mrak, R. E., and Griffin, W. S. (2005). Glia and their cytokines in progression of neurodegeneration. *Neurobiol. Aging* 26, 349–354. doi: 10.1016/j.neurobiolaging.2004.05.010

Murchison, C. F., Zhang, X. Y., Zhang, W. P., Ouyang, M., Lee, A., and Thomas, S. A. (2004). A distinct role for norepinephrine in memory retrieval. *Cell* 117, 131–143. doi: 10.1016/S0092-8674(04)00259-4

Nagele, R. G., Wegiel, J., Venkataraman, V., Imaki, H., and Wang, K. C. (2004). Contribution of glial cells to the development of amyloid plaques in Alzheimer's disease. *Neurobiol. Aging* 25, 663–674. doi: 10.1016/j.neurobiolaging.2004.01.007

Nazarali, A. J., and Reynolds, G. P. (1992). Monoamine neurotransmitters and their metabolites in brain regions in Alzheimer's disease: a postmortem study. *Cell Mol. Neurobiol.* 12, 581–587. doi: 10.1007/BF00711237

Nedergaard, M., Rodriguez, J. J., and Verkhratsky, A. (2010). Glial calcium and diseases of the nervous system. *Cell Calcium* 47, 140–149. doi: 10.1016/j.ceca.2009.11.010

Oe, Y., Baba, O., Ashida, H., Nakamura, K. C., and Hirase, H. (2016). Glycogen distribution in the microwave-fixed mouse brain reveals heterogeneous astrocytic patterns. *Glia*. doi: 10.1002/glia.23020

Olabarria, M., Noristani, H. N., Verkhratsky, A., and Rodríguez, J. J. (2010). Concomitant astroglial atrophy and astrogliosis in a triple transgenic animal model of Alzheimer's disease. *Glia* 58, 831–838. doi: 10.1002/glia.20967

Olabarria, M., Noristani, H. N., Verkhratsky, A., and Rodríguez, J. J. (2011). Age-dependent decrease in glutamine synthetase expression in the hippocampal astroglia of the triple transgenic Alzheimer's disease mouse model: mechanism for deficient glutamatergic transmission? *Mol. Neurodegener.* 6:55. doi: 10.1186/1750-1326-6-55

Pankratov, Y., and Lalo, U. (2015). Role for astroglial alpha1-adrenoreceptors in gliotransmission and control of synaptic plasticity in the neocortex. *Front. Cell Neurosci.* 9:230. doi: 10.3389/fncel.2015.00230

Parpura, V., Heneka, M. T., Montana, V., Oliet, S. H., Schousboe, A., Haydon, P. G., et al. (2012). Glial cells in (patho)physiology. *J. Neurochem.* 121, 4–27. doi: 10.1111/j.1471-4159.2012.07664.x

Parpura, V., and Zorec, R. (2010). Gliotransmission: exocytotic release from astrocytes. *Brain Res. Rev.* 63, 83–92. doi: 10.1016/j.brainresrev.2009.11.008

Pascual, J., Grijalba, B., Garcia-Sevilla, J. A., Zarranz, J. J., and Pazos, A. (1992). Loss of high-affinity alpha 2-adrenoceptors in Alzheimer's disease: an autoradiographic study in frontal cortex and hippocampus. *Neurosci. Lett.* 142, 36–40. doi: 10.1016/0304-3940(92)90614-D

Paukert, M., Agarwal, A., Cha, J., Doze, V. A., Kang, J. U., and Bergles, D. E. (2014). Norepinephrine controls astroglial responsiveness to local circuit activity. *Neuron* 82, 1263–1270. doi: 10.1016/j.neuron.2014.04.038

Pekny, M., Pekna, M., Messing, A., Steinhauser, C., Lee, J. M., Parpura, V., et al. (2016). Astrocytes: a central element in neurological diseases. *Acta Neuropathol.* 131, 323–345. doi: 10.1007/s00401-015-1513-1

Peterson, A. C., and Li, C. R. (2018). Noradrenergic dysfunction in Alzheimer's and Parkinson's Diseases-an overview of imaging studies. *Front. Aging Neurosci.* 10:127. doi: 10.3389/fnagi.2018.00127

Picklo, M. J., Wiley, R. G., Lappi, D. A., and Robertson, D. (1994). Noradrenergic lesioning with an anti-dopamine beta-hydroxylase immunotoxin. *Brain Res.* 666, 195–200. doi: 10.1016/0006-8993(94)90772-2

Pintus, R., Riggi, M., Cannarozzo, C., Valeri, A., de Leo, G., Romano, M., et al. (2018). Essential role of hippocampal noradrenaline in the regulation of spatial working memory and TDP-43 tissue pathology. *J. Comp. Neurol.* 526, 1131–1147. doi: 10.1002/cne.24397

Prebil, M., Vardjan, N., Jensen, J., Zorec, R., and Kreft, M. (2011). Dynamic monitoring of cytosolic glucose in single astrocytes. *Glia* 59, 903–913. doi: 10.1002/glia.21161

Rodriguez-Vieitez, E., Saint-Aubert, L., Carter, S. F., Almkvist, O., Farid, K., Schöll, M., et al. (2016). Diverging longitudinal changes in astrocytosis and amyloid PET in autosomal dominant Alzheimer's disease. Brain 139(Pt 3):922–936. doi: 10.1093/brain/awv404

Rub, U., Del Tredici, K., Schultz, C., Thal, D. R., Braak, E., and Braak, H. (2001). The autonomic higher order processing nuclei of the lower brain stem are among the early targets of the Alzheimer's disease-related cytoskeletal pathology. *Acta Neuropathol.* 101, 555–564.

Sacchetti, G., Bernini, M., Bianchetti, A., Parini, S., Invernizzi, R. W., and Samanin, R. (1999). Studies on the acute and chronic effects of reboxetine on extracellular noradrenaline and other monoamines in the rat brain. *Br. J. Pharmacol.* 128, 1332–1338.

Salcedo-Sora, J. E., Caamano-Gutierrez, E., Ward, S. A., and Biagini, G. A. (2014). The proliferating cell hypothesis: a metabolic framework for Plasmodium growth and development. *Trends Parasitol.* 30, 170–175. doi: 10.1016/j.pt.2014.02.001

Salm, A. K., and McCarthy, K. D. (1989). Expression of beta-adrenergic receptors by astrocytes isolated from adult rat cortex. *Glia* 2, 346–352.

Sanchez-Padilla, J., Guzman, J. N., Ilijic, E., Kondapalli, J., Galtieri, D. J., Yang, B., et al. (2014). Mitochondrial oxidant stress in locus coeruleus is regulated by activity and nitric oxide synthase. *Nat. Neurosci.* 17, 832–840. doi: 10.1038/nn.3717

Santarelli, L., Saxe, M., Gross, C., Surget, A., Battaglia, F., Dulawa, S., et al. (2003). Requirement of hippocampal neurogenesis for the behavioral effects of antidepressants. *Science* 301, 805–809. doi: 10.1126/science.1083328

Sara, S. J. (2009). The locus coeruleus and noradrenergic modulation of cognition. *Nat. Rev. Neurosci.* 10, 211–223. doi: 10.1038/nrn2573

Satoh, A., and Iijima, K. M. (2017). Roles of tau pathology in the locus coeruleus (LC) in age-associated pathophysiology and Alzheimer's disease pathogenesis: Potential strategies to protect the LC against aging. *Brain Res.* doi: 10.1016/j.brainres.2017.12.027. [Epub ahead of print].

Scullion, G. A., Kendall, D. A., Marsden, C. A., Sunter, D., and Pardon, M. C. (2011). Chronic treatment with the alpha2-adrenoceptor antagonist fluparoxan prevents age-related deficits in spatial working memory in APPxPS1 transgenic mice without altering beta-amyloid plaque load or astrocytosis. *Neuropharmacology* 60, 223–234. doi: 10.1016/j.neuropharm.2010.09.002

Segal, S. K., Stark, S. M., Kattan, D., Stark, C. E., and Yassa, M. A. (2012). Norepinephrine-mediated emotional arousal facilitates subsequent pattern separation. *Neurobiol Learn. Mem.* 97, 465–469. doi: 10.1016/j.nlm.2012.03.010

Sherpa, A. D., Xiao, F., Joseph, N., Aoki, C., and Hrabetova, S. (2016). Activation of beta-adrenergic receptors in rat visual cortex expands astrocytic processes and reduces extracellular space volume. *Synapse* 70, 307–316. doi: 10.1002/syn.21908

Shimohama, S., Taniguchi, T., Fujiwara, M., and Kameyama, M. (1986). Biochemical characterization of alpha-adrenergic receptors in human brain and changes in Alzheimer-type dementia. *J. Neurochem.* 47, 1295–1301.

Singh, C., Liu, L., Wang, J. M., Irwin, R. W., Yao, J., Chen, S., et al. (2012). Allopregnanolone restores hippocampal-dependent learning and memory and neural progenitor survival in aging 3xTgAD and nonTg mice. *Neurobiol. Aging* 33, 1493–1506. doi: 10.1016/j.neurobiolaging.2011.06.008

Sontag, T. A., Hauser, J., Kaunzinger, I., Gerlach, M., Tucha, O., and Lange, K. W. (2008). Effects of the noradrenergic neurotoxin DSP4 on spatial memory in the rat. *J. Neural. Transm.* 115, 299–303. doi: 10.1007/s00702-007-0830-5

Sparks, D. L., DeKosky, S. T., and Markesbery, W. R. (1988). Alzheimer's disease. Aminergic-cholinergic alterations in hypothalamus. *Arch. Neurol.* 45, 994–999.

Starke, K. (2001). Presynaptic autoreceptors in the third decade: focus on alpha2-adrenoceptors. *J. Neurochem* 78, 685–693.

Stenovec, M., Trkov, S., Lasič, E., Terzieva, S., Kreft, M., Rodríguez Arellano, J. J., et al. (2016). Expression of familial Alzheimer disease presenilin 1 gene attenuates vesicle traffic and reduces peptide secretion in cultured astrocytes devoid of pathologic tissue environment. *Glia* 64, 317–329. doi: 10.1002/glia.22931

Strassnig, M., and Ganguli, M. (2005). About a peculiar disease of the cerebral cortex: Alzheimer's original case revisited. *Psychiatry* 2, 30–33.

Studelska, D. R., and Brimijoin, S. (1989). Partial isolation of two classes of dopamine beta-hydroxylase-containing particles undergoing rapid axonal transport in rat sciatic nerve. *J. Neurochem* .53, 622–631.

Swanson, L. W., and Hartman, B. K. (1975). The central adrenergic system. An immunofluorescence study of the location of cell bodies and their efferent connections in the rat utilizing dopamine-beta-hydroxylase as a marker. *J. Comp. Neurol.* 163, 467–505.

Szot, P., Miguelez, C., White, S. S., Franklin, A., Sikkema, C., Wilkinson, C. W., et al. (2010). A comprehensive analysis of the effect of DSP4 on the locus coeruleus noradrenergic system in the rat. *Neuroscience* 166, 279–291. doi: 10.1016/j.neuroscience.2009

Szot, P., White, S. S., Greenup, J. L., Leverenz, J. B., Peskind, E. R., and Raskind, M. A. (2006). Compensatory changes in the noradrenergic nervous system in the locus ceruleus and hippocampus of postmortem subjects with Alzheimer's disease and dementia with Lewy bodies. *J. Neurosci.* 26, 467–478. doi: 10.1523/JNEUROSCI.4265-05.2006

Szot, P., White, S. S., Greenup, J. L., Leverenz, J. B., Peskind, E. R., and Raskind, M. A. (2007). Changes in adrenoreceptors in the prefrontal cortex of subjects with dementia: evidence of compensatory changes. *Neuroscience* 146, 471–480. doi: 10.1016/j.neuroscience.2007.01.031

Tang, F., Lane, S., Korsak, A., Paton, J. F., Gourine, A. V., Kasparov, S., et al. (2014). Lactate-mediated glia-neuronal signalling in the mammalian brain. *Nat. Commun.* 5:3284. doi: 10.1038/ncomms4284

Terry, R. D. (2000). Cell death or synaptic loss in Alzheimer disease. *J. Neuropathol. Exp. Neurol* .59, 1118–1189.

Thomas, S. A., and Palmiter, R. D. (1997). Disruption of the dopamine beta-hydroxylase gene in mice suggests roles for norepinephrine in motor function, learning, and memory. *Behav. Neurosci.* 111, 579–589.

Thrane, A. S., Rangroo Thrane, V., and Nedergaard, M. (2014). Drowning stars: reassessing the role of astrocytes in brain edema. *Trends Neurosci.* 37, 620–628. doi: 10.1016/j.tins.2014.08.010

Tohgi, H., Ueno, M., Abe, T., Takahashi, S., and Nozaki, Y. (1992). Concentrations of monoamines and their metabolites in the cerebrospinal fluid from patients with senile dementia of the Alzheimer type and vascular dementia of the Binswanger type. *J. Neural Transm. Par.k Dis. Dement. Sect.* 4, 69–77.

Tomlinson, B. E., Irving, D., and Blessed, G. (1981). Cell loss in the locus coeruleus in senile dementia of Alzheimer type. *J. Neurol Sci.* 49, 419–428.

Van Dyck, C. H. (2014). *Guanfacine Treatment for Prefrontal Cognitive Dysfunction in Elderly Subjects*. Available online at: https://clinicaltrials.gov/ct2/show/NCT00935493

Vander Heiden, M. G., Cantley, L. C., and Thompson, C. B. (2009). Understanding the Warburg effect: the metabolic requirements of cell proliferation. *Science* 324, 1029–1033. doi: 10.1126/science.1160809

Van Hoomissen, J. D., Holmes, P. V., Zellner, A. S., Poudevigne, A., and Dishman, R. K. (2004). Effects of beta-adrenoreceptor blockade during chronic exercise on contextual fear conditioning and mRNA for galanin and brain-derived neurotrophic factor. *Behav. Neurosci.* 118, 1378–1390.

van Praag, H., Schinder, A. F., Christie, B. R., Toni, N., Palmer, T. D., and Gage, F. H. (2002). Functional neurogenesis in the adult hippocampus. *Nature* 415, 1030–1034. doi: 10.1038/4151030a

Vardjan, N., Horvat, A., Anderson, J. E., Yu, D., Croom, D., Zeng, X., et al. (2016). Adrenergic activation attenuates astrocyte swelling induced by hypotonicity and neurotrauma. *Glia* 64, 1034–1049. doi: 10.1002/glia.22981

Vardjan, N., Kreft, M., and Zorec, R. (2014). Dynamics of β-adrenergic/cAMP signaling and morphological changes in cultured astrocytes. *Glia.*

Vardjan, N., and Zorec, R. (2015). Excitable astrocytes: Ca^{2+}- and cAMP-regulated exocytosis. *Neurochem. Res.* 40, 2414–2414. doi: 10.1007/s11064-015-1545-x

Verkhratsky, A., Marutle, A., Rodríguez-Arellano, J. J., and Nordberg, A. (2014). Glial Asthenia and Functional Paralysis: a new perspective on neurodegeneration and Alzheimer's disease. *Neuroscientist* 21, 552–568. doi: 10.1177/1073858414547132

Verkhratsky, A., and Nedergaard, M. (2014). Astroglial cradle in the life of the synapse. *Philos. Trans. R. Soc. Lond. B. Biol. Sci.* 369:20130595. doi: 10.1098/rstb.2013.0595

Verkhratsky, A., and Nedergaard, M. (2018). Physiology of Astroglia. *Physiol. Rev.* 98, 239–389. doi: 10.1152/physrev.00042.2016

Verkhratsky, A., and Parpura, V. (2015). Astrogliopathology in neurological, neurodevelopmental and psychiatric disorders. *Neurobiol. Dis.* 85, 254–261. doi: 10.1016/j.nbd.2015.03.025

Verkhratsky, A., Zorec, R., Rodriguez, J. J., and Parpura, V. (2016). Astroglia dynamics in ageing and Alzheimer's disease. *Curr. Opin. Pharmacol.* 26, 74–79. doi: 10.1016/j.coph.2015.09.011

Walker, T. L., White, A., Black, D. M., Wallace, R. H., Sah, P., and Bartlett, P. F. (2008). Latent stem and progenitor cells in the hippocampus are activated by neural excitation. *J. Neurosci* .28, 5240–5247. doi: 10.1523/JNEUROSCI.0344-08.2008

Warner-Schmidt, J. L., and Duman, R. S. (2006). Hippocampal neurogenesis: opposing effects of stress and antidepressant treatment. *Hippocampus* 16, 239–249. doi: 10.1002/hipo.20156

Weinshilboum, R. M. (1978). Serum dopamine beta-hydroxylase. *Pharmacol. Rev.* 30, 133–166.

Whitehouse, P. J., Price, D. L., Struble, R. G., Clark, A. W., Coyle, J. T., and Delon, M. R. (1982). Alzheimer's disease and senile dementia: loss of neurons in the basal forebrain. *Science* 215, 1237–1239.

Wiley, R. G., and Kline, I. R. (2000). Neuronal lesioning with axonally transported toxins. *J. Neurosci. Methods* 103, 73–82

Wilson, R. S., Nag, S., Boyle, P. A., Hizel, L. P., Yu, L., Buchman, A. S., et al. (2013). Neural reserve, neuronal density in the locus ceruleus, and cognitive decline. *Neurology* 80, 1202–1208. doi: 10.1212/WNL.0b013e3182897103

Woodruff-Pak, D. S. (2008). Animal models of Alzheimer's disease: therapeutic implications. *J. Alzheimers. Dis.* 15, 507–521.

Wrenn, C. C., Picklo, M. J., Lappi, D. A., Robertson, D., and Wiley, R. G. (1996). Central noradrenergic lesioning using anti-DBH-saporin: anatomical findings. *Brain Res.* 740, 175–184.

Yeh, C. Y., Vadhwana, B., Verkhratsky, A., and Rodríguez, J. J. (2011). Early astrocytic atrophy in the entorhinal cortex of a triple transgenic animal model of Alzheimer's disease. *ASN Neuro* 3, 271–279. doi: 10.1042/AN20110025

Zorec, R., Horvat, A., Vardjan, N., and Verkhratsky, A. (2015). Memory Formation Shaped by Astroglia. *Front. Integr. Neurosci.* 9:56. doi: 10.3389/fnint.2015.00056

Zorec, R., Parpura, V., and Verkhratsky, A. (2018). Astroglial vesicular network: evolutionary trends, physiology and pathophysiology. *Acta Physiol.* 222. doi: 10.1111/apha.12915

Zorec, R., Vardjan, N., and Verkhratsky, A. (2017). "Locus Coeruleus Noradrenergic Neurons and Astroglia in Health and Disease," in *Noradrenergic Signaling and Astroglia,* eds N. Vardjan and R. Zorec (Oxford: Academic Press), 1–24.

14

Dynamic Calcium Release from Endoplasmic Reticulum Mediated by Ryanodine Receptor 3 is Crucial for Oligodendroglial Differentiation

Tao Li, Lingyun Wang, Teng Ma, Shouyu Wang, Jianqin Niu, Hongli Li and Lan Xiao**

Chongqing Key Laboratory of Neurobiology, Department of Histology and Embryology, Third Military Medical University, Chongqing, China

***Correspondence:**
Hongli Li
lihongli@tmmu.edu.cn
Lan Xiao
xiaolan35@hotmail.com

Increased intracellular Ca^{2+} in oligodendrocyte progenitor cells (OPCs) is important to initiate their differentiation, but the intracellular Ca^{2+} channel involved in this process remains unclear. As a Ca^{2+}-induced Ca^{2+} release (CICR) channel that mediates endoplasmic reticulum (ER) Ca^{2+} release, the role of ryanodine receptors (RyRs) in oligodendroglial development is unexplored. In the present study, we observed that among the three mammalian isoforms, oligodendroglial lineage cells selectively expressed RyR3. Strong RyR3-positive signal was distributed all over the cytoplasm and processes in OPCs and/or immature OLs (imOLs), whereas it gradually decreased and was located mainly around the perinuclear region in mature oligodendrocytes (OLs). In addition, RyR3-mediated intracellular Ca^{2+} waves following caffeine stimulation were correlated with the expression pattern of RyR3, in which high flat Ca^{2+} fluctuations and oscillatory Ca^{2+} waves were more frequently recorded in OPCs and/or imOLs than in OLs. Through further functional exploration, we demonstrated that pretreatment with the RyR antagonist ryanodine could neutralize the increase in intracellular Ca^{2+} induced by OPC differentiation and reduce the number of mature OLs. Moreover, gene-level knockdown of RyR3 by lentivirus in OPCs resulted in inhibition of OPC differentiation. Taken together, our results provide new insight into the crucial role of RyR3-mediated ER Ca^{2+} release in the regulation of OPC differentiation and/or myelination.

Keywords: oligodendrocyte progenitor cells (OPCs), oligodendrocytes (OLs), ryanodine receptor 3 (RyR3), Ca^{2+} release, differentiation, caffeine

INTRODUCTION

In the CNS, myelinating oligodendrocytes (OLs) originate from oligodendrocyte progenitor cells (OPCs) after passing through a series of distinct developmental stages, i.e., OPCs, immature OLs (imOLs) and mature OLs (Stangel and Hartung, 2002). As impairment of OL differentiation has been considered to be the major cause of remyelination failure, which occurs in numerous

Abbreviations: CICR, Ca^{2+}-induced Ca^{2+} release; CNPase, 2′,3′-cyclic nucleotide-3′-phosphodiesterase; ER, endoplasmic reticulum; imOL, immature oligodendrocyte; MBP, myelin basic protein; OL, oligodendrocyte; OPC, oligodendrocyte progenitor cell; PDGFRα, platelet-derived growth factor receptor α; RyR, ryanodine receptor; VOCC, voltage-operated Ca^{2+} channel.

demyelination diseases such as multiple sclerosis (MS; Wolswijk, 1998; Chang et al., 2002; Kuhlmann et al., 2008), promoting OPC differentiation into mature OLs becomes a promising approach for myelin repair. However, the mechanism regulating oligodendroglial differentiation remains to be elucidated. As a critical functional pattern of non-excitable glia cells, Ca^{2+} signaling is essential for oligodendroglial differentiation and myelination (Kirischuk et al., 1995; Cohen et al., 1996; Yoo et al., 1999; Soliven, 2001; Fulton et al., 2010; Cheli et al., 2015; Friess et al., 2016; Baraban et al., 2018; Krasnow et al., 2018). For instance, inhibition of the voltage-operated Ca^{2+} entry in OPCs repressed their maturation and the myelin forming ability (Cheli et al., 2015). Increasing resting intracellular Ca^{2+} through membrane depolarization could facilitate MBP synthesis in OPCs (Friess et al., 2016). Newest studies further provided *in vivo* data showing that the Ca^{2+} transients in OLs could regulate retraction and elongation of the developing myelin sheath (Baraban et al., 2018; Krasnow et al., 2018). However, the Ca^{2+} channels involved in oligodendroglial differentiation is believed to be important but remains largely unexplored.

It is known that endoplasmic reticulum (ER) is the major intracellular Ca^{2+} pool (Meldolesi and Pozzan, 1998) and that ER Ca^{2+} release is driven mainly by inositol-1,4,5,-trisphosphate receptors (IP3Rs) and ryanodine receptors (RyRs; Koulen and Thrower, 2001). In OPCs, both IP3R2 and ryanodine receptor 3 (RyR3) can mediate highly localized Ca^{2+} release, of the types called "puffs" and "sparks", respectively (Haak et al., 2001), but only IP3R2 is able to initiate Ca^{2+} waves under pharmacological treatments (Haak et al., 2001). However, the functions of those channels during oligodendroglial development remain unclear. Series of studies demonstrate that, compared with IP3Rs, the opening of which requires both Ca^{2+} and IP3 (Moraru et al., 1999; Foskett et al., 2007), RyRs are Ca^{2+}-induced Ca^{2+} release (CICR) channels that can be triggered merely by a low concentration of Ca^{2+} ($\sim$1 μM; Meissner et al., 1986, 1997; Bezprozvanny et al., 1991), and this CICR function has been shown to powerfully amplify small inward Ca^{2+} currents in NG2 glial cells (Haberlandt et al., 2011). Therefore, we propose that RyR3 is likely a critical bridge for the formation of intracellular Ca^{2+} signaling and thus participates in the regulation of oligodendroglial development.

In the present study, we sought to characterize the expression and function of RyRs during OPC differentiation. Our results showed that RyR3 was selectively expressed and widely distributed in the soma and processes of the oligodendroglial lineage cells and that its expression level was downregulated following OPC differentiation. Using confocal Ca^{2+} imaging, we found that the ER Ca^{2+} release after caffeine stimulation was much stronger in OPCs and imOLs than in mature OLs. Moreover, inhibiting the function of RyR3 either pharmacologically or by gene knockdown suppressed the differentiation of OPCs. Our results revealed a critical role for RyR3-mediated Ca^{2+} signaling in oligodendroglial differentiation that may provide new insight into therapeutic approaches for demyelinating diseases.

MATERIALS AND METHODS

OPC Culture

Cortical OPCs were purified as previously described (Niu et al., 2012b). Briefly, the mixed glial cells were isolated from cortex of postnatal day 1–3 neonatal Sprague-Dawley (SD) rats and enriched in OPC growth medium followed by two passages to enrich cell numbers. OPC proliferation medium was DMEM/F12 + 1% N2 supplement + PDGFAA. OPCs were induced to differentiate by replacing the medium with OPC differentiation medium: DMEM/F12 + 1% N2 supplement + 5 mg/mL N-acetyl-L-cysteine (Amresco) + 1% fetal bovine serum + 5 mg/mL insulin.

Reagents used were as follows: Dulbecco's modified Eagle's medium/F12 (DMEM/F12; Hyclone, SH30023), N2 supplement (Invitrogen, 17502048), fetal bovine serum (FBS; Hyclone, SV30087), insulin (Sigma, I6634), N-acetyl-l-cysteine (NAC; AMRESCO, 0LA0011), PDGFAA (Peprotech, 100-13A). The SD rats related procedures were performed in accordance with the guidelines approved by the Laboratory Animal Welfare and Ethics Committee of the Third Military Medical University (Niu et al., 2016).

Confocal Ca^{2+} Imaging Measurements

OPCs were grown and differentiated in glass-bottom dish and loaded with the fluorescent Ca^{2+} sensitive dye Fluo-3AM (5 μM, Invitrogen) for 20 min at 37°C in a modified imaging buffer containing (in mM): NaCl, 135; KCl, 3; $MgCl_2$, 2; Glucose, 8; HEPES, 10 and $CaCl_2$, 2 (pH adjusted to 7.4 with NaOH). The dye-loaded cells were washed twice and maintained for at least 20 min at RT in fresh imaging buffer to allow complete dye de-esterification. Ca^{2+} wave and fluorescence images were real-time recorded for at least 15 min in all experiments, using a confocal laser-scanning microscope (FluoView FV1000, Olympus, Japan) with the UplanFl40× objective (N.A. 0.95). The image acquisition frequency is 100–180 ms/image. Cell morphology was detected using differential interference contrast (DIC) under confocal microscopy. Ca^{2+} concentrations were measured by exciting Fluo-3 AM at 488 nm.

Intracellular Ca^{2+} responses in OPCs, imOLs and mature OLs were recorded with caffeine (20 mM, Sigma, Ca^{2+} free) stimulation in a Ca^{2+} free imaging buffer. For spontaneous Ca^{2+} recordings, OPCs were grown in proliferation medium or differentiation medium for 6 h with or without ryanodine treatment (50 μM, TOCRIS, 1329, 10 min). Fluo-3 loading and cell washing was followed by applying normal proliferating medium (Pro-medium) or differentiating medium (Diff-medium) for the following recording.

Cell Processing and Immunocytochemistry

Cells were grown on coverslips, fixed in cold 4% paraformaldehyde for 20 min, rinsed with 0.01 M PBS, blocked with 1% bovine serum albumin (BSA) and 0.2% Triton-X100 for 30 min and then incubated with primary antibodies diluted in 1% BSA overnight at 4°C and then by fluorophore-conjugated secondary antibodies at room temperature (RT) for 2 h. Cell nuclei were stained with 4′,6-diamidino-2-phenylindole (DAPI, Thermo Fisher) for 10 min.

In this study, the following antibodies were used: mouse polyclonal anti-Olig2 (1:500, Millipore, MABN50), rabbit polyclonal anti-RyR3 (1:200, Millipore, AB9082), and goat anti-myelin basic protein (MBP; 1:500, Santa Cruz, sc13914). The secondary antibodies were as follows: Alexa 568-labeled donkey anti-mouse (1:1000, Invitrogen), Alexa 488-labeled donkey anti-goat (1:1000, Invitrogen), Alexa 568-labeled donkey anti-goat (1:1000, Invitrogen), Alexa 568-labeled donkey anti-rabbit (1:1000, Invitrogen) and Cy5-labeled rabbit anti-mouse (1:500, Jackson ImmunoResearch).

Image Acquisition and Quantification

Fluorescent images were captured using an Axio Imager M2 fluorescence microscope (Zeiss, Oberkochen, Germany) or a confocal laser-scanning microscope (Olympus, IV 1000, Shinjuku, Tokyo) with excitation wavelengths appropriate for Alexa Fluor 488 (488 nm), 596 (568 nm), 647 (628 nm) or DAPI (380 nm). Digital images of the oligodendroglial lineage cells in the supplemental figure were acquired with an Olympus IX51 microscope with an Olympus C-7070 camera (Tokyo). For the statistical analysis, randomly selected images in at least three representative fields were acquired from each sample. Detection and quantification were performed using ImageJ software (National Institutes of Health, NIH).

Western Blot Analysis

The cells were lysed using RIPA lysis buffer (Beyotime, P0013B) with freshly added 1% phenylmethylsulfonyl fluoride (PMSF, Amresco, O754) solution. Protein concentration was determined using Coomassie brilliant Blue G-250. SDS-PAGE and Western blotting were carried out as reported previously (Niu et al., 2012a). Proteins were transferred to polyvinylidene difluoride (PVDF) membranes and visualized by chemiluminescence (ECL Plus, GE Healthcare, Marlborough, MA, USA) after incubation with the antibodies. β-actin was used as the loading control. Quantification of band intensity was performed using ImageJ software. The primary antibodies included the following: rabbit polyclonal anti-RyR3 (1:1000, Millipore), rabbit polyclonal anti-platelet-derived growth factor receptor α (PDGFRα; 1:1000, Santa Cruz, sc-338), mouse anti-2′,3′-cyclic nucleotide-3′-phosphodiesterase (CNPase; 1:1000, Abcam, ab6319), goat anti-MBP (1:1000, Santa Cruz) and mouse anti-β-actin (1:2000, Santa Cruz, sc-47778). The secondary antibodies included the following: goat anti-mouse-HRP (1:2000, Santa Cruz, sc-2094), goat anti-rabbit-HRP (1:2000, Santa Cruz, sc-2313) and rabbit anti-goat-HRP (1:2000, Santa Cruz, sc-2020).

RT-PCR Analysis

Total ribonucleic acid (RNA) was isolated from different stages of OPC cultures (OPC differentiated for 1 day, 2 days, or 4 days using TRIzol (Life Technologies). Real-time polymerase chain reaction (RT-PCR) was performed with the C1000 Touch™ Real-time PCR Detection System (Bio-Rad) and GoTaq® qPCR Master Mix (Promega, Sunnyvale, CA, USA). The amplification procedure and melt curve analysis were performed using three independent replicates for each sample.

The oligonucleotide primers used were as follows:

	Forward	Reverse	Tm
RyR1	GAGGGTGATGAAGATGAGAAC	TCCCGCCCGAAGATGTC	60°C
RyR2	GCTGGCCCTGTTTGTTG	ATCCATGCCCAGTAACTCGCT	61.3°C
RyR3	CTGTGTGGTGGGCTATTACTG	TGCTTTGGCCTCTTCTACTG	58.3°C
MBP	GAGACCCTCACAGCGACAC	ATCCAGAGCGGCTGTCTC	59°C
β-actin	CGTTGTACATCCGTAAAGACC	CATCGCACTCCTGCTTGCT	58°C

Lentivirus Mediated shRNA Interference

The shRNA lentivirus was purchased from Obio Technology, Shanghai. Targeted sequence in rat RyR3 gene is: AGATGCTAATTGCATCTC. The primary lentivirus solution was diluted in several gradient concentrations (1:10, 1:100, 1:300, 1:500, 1:1000) to check the best work concentration (normal oligodendroglial viability and good interference efficiency). The lentivirus was diluted 1:300 in the differentiation medium with a primary concentration of $8.67*10^8$ transducing units (TU)/ml for the interference group (shRyR3) and $6.71*10^8$ TU/ml for the control group (shCTL). Lentivirus was removed after transfection for 16 h. OPC were further differentiated for another 32 h and fixed.

Antisense Oligonucleotides

Phosphodiester ODNs protected by terminal phosphorothioate double substitution (capped ODNs) against possible exonuclease-mediated degradation were purchased from Tib-Molbiol (Sigma). The sequences are as follows: anti-RyR3: 5′-A*G*ATGCTAATTGCATC*T*C-3′ (*indicates the phosphorothioate residues) and an 18-mer fully degenerated ODN (dODN), 5′-N*N*NNNNNNNNNNNNNN*N*N-3′ (where N is G, C, A, or T), which was used as a control ODN. ODNs were transported using an artificial cationic lipid (DOTAP; Sigma) to enhance both uptake and stability. Antisense ODNs (aODNs) or dODNs were pre-incubated at 37°C for 30 min. Differentiated OLs were collected on day 3.

Statistics Analysis

All experiments were repeated at least three times. Data are shown as the means ± SEM. Statistical analyses of three groups were performed using one-way analysis of variance (ANOVA) followed by Tukey's *post hoc* test. Comparisons between two experimental groups were made using Student's *t*-test (GraphPad Prism 6). A probability of $p < 0.05$ was considered significant.

RESULTS

Oligodendrocytes Selectively Express RyR3, Which Is Downregulated During Differentiation

Similar to our previous studies (Niu et al., 2012b), different stages of oligodendroglial cells in our culture system were identified based on morphological features using DIC and immunostaining for stage-specific markers. Normally, OPCs have small, round cell bodies with a bipolar morphology (Supplementary Figure S1). After 2 days in the differentiation medium, imOLs with 3–5 primary processes and a sparse arborization predominate this stage (Supplementary Figure S1). After 4 days of differentiation, mainly mature OLs characterized by multipolar processes and a rich arborization were observed (Supplementary Figure S1). These features indicate the accuracy of our culture model for studying the developmental schedule of the OL lineage.

To determine the RyR subtypes expressed in OLs, we first analyzed the mRNA levels in OL lineage cells. We found that only RyR3, the known "brain type", was detected in OL lineage cells. In parallel experiments, RyR3 mRNA was also found in astrocytes, RyR2 mRNA was found in astrocytes and rat brain tissue, and RyR1 mRNA was only found in brain tissue (**Figure 1B**). Interestingly, the RyR3 mRNA level gradually decreased during OPC differentiation, and this downregulation tendency was further confirmed by western blot analysis (**Figures 1C,D**).

Next, we clarified the RyR3 distribution pattern in OL lineage cells by immunostaining. It has been shown that RyR3 is the subtype of ryanodine receptors expressed in OPCs. Specifically, RyR3 is located throughout the cell body and processes of OPCs, except in the region of the nuclear membrane (Haak et al., 2001). Our present results also revealed the enrichment of RyR3 in the processes and non-nuclear area of OPCs and imOLs, while in mature OLs, RyR3 was mainly located in the cell body and primary processes (**Figure 1A**).

Taken together, these results reveal a spatiotemporal regulated expression pattern of RyR3 in OL lineage cells, indicating its potential in regulating OL development.

ER Ca^{2+} Release Following Caffeine Stimulation Is Stage Specific During OPC Differentiation

Although OPCs are reported to have spontaneous oscillatory-like Ca^{2+} activity with peak and plateau transients, while mature OLs show "flat" Ca^{2+} signaling (Niu et al., 2016), the ER Ca^{2+} channels in OPCs (RyR3 and IP_3R2) are relatively quiescent in comparison to those in neurons (Haak et al., 2001). To better study the RyR channel function, we took advantage of the classical RyR agonist caffeine (20 mM, 0 Ca^{2+}; Zucchi and Ronca-Testoni, 1997) and real-time recorded the intracellular Ca^{2+} concentration with confocal microscopy. The Ca^{2+} release peaks are significantly higher in OPC and imOLs than in mature OLs (**Figure 2D**). More importantly, three typical Ca^{2+} responses were recorded in OL lineage cells

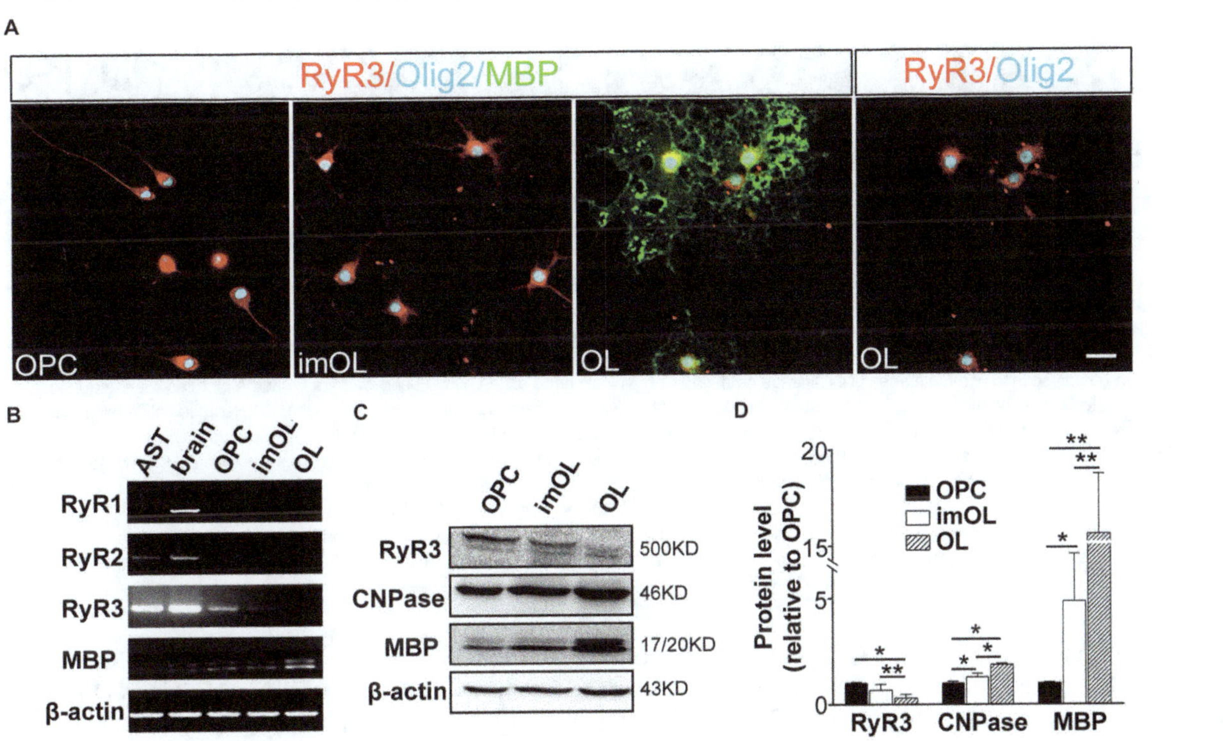

FIGURE 1 | Oligodendroglial cells selectively express ryanodine receptor 3 (RyR3), which is downregulated during differentiation. **(A)** Immunofluorescence staining of RyR3, Olig2 (oligodendroglial lineage marker) and myelin basic protein (MBP; mature OL marker) in three stages of cultured OL lineage cells. RyR3 was strongly distributed all over the cytoplasm and processes in oligodendrocyte progenitor cells (OPCs) and immature oligodendrocytes (imOLs), while in mature OLs, RyR3 signal was mainly located in the cell body and showed lower expression in the primary processes. Scale bar 20 μm. **(B)** mRNA levels of RyRs in OL lineage cells, astrocytes (AST) and rat brain tissue. **(C,D)** Western blot analysis showed increased levels of myelin proteins (CNPase, MBP) and decreased levels of RyR3 protein during OPC development (*$p < 0.05$, **$p < 0.01$).

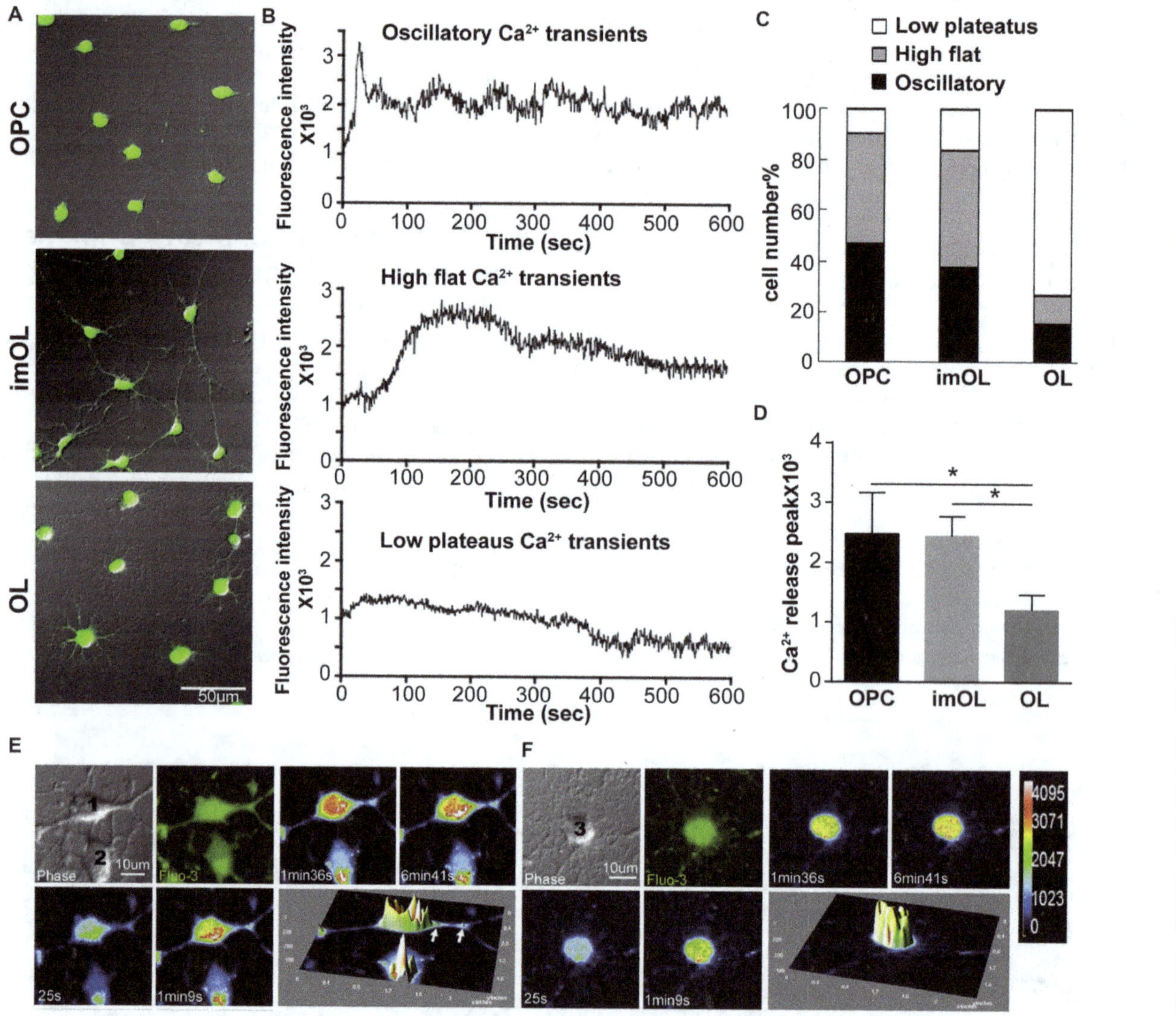

FIGURE 2 | Endoplasmic reticulum (ER) Ca^{2+} release following caffeine stimulation is stage specific during OPC differentiation. **(A)** Differential interference contrast (DIC) with Fluo-3 fluorescence of three stages of OL lineage cells, showing typical shape of OPC, imOL and OL. Note that the pictures were acquired at the end point of Ca^{2+} imaging to ensure the morphology of the oligodendroglial cells, thus Fluo-3 fluorescence was extremely strong. **(B)** Representative examples of Ca^{2+} response following caffeine stimulation. Note that the start point in the x-axis is not the beginning point of recording, baseline recording is omitted from the curve. **(C)** OPCs and imOLs, which tend to respond with oscillatory Ca^{2+} transients and high flat Ca^{2+} transients, reacted stronger than OLs, which tend to respond with low plateaus Ca^{2+} transients. **(D)** OPC and imOL had significantly higher Ca^{2+} release peaks after caffeine stimulation compared with mature OLs ($^*p < 0.05$). Time-lapse Ca^{2+} imaging induced by caffeine (20 mM, 0 Ca^{2+}) in an OPC (1) **(E)** and a mature OL **(F)**. Four successive scans were selected from a series of images obtained within 10 min. Transient fluorescence fluctuations, representing local Ca^{2+} release events, were indicated by the pseudocolored peaks in the OPC cell body and processes (arrows), but they occurred only in the soma in mature OLs. The pseudocolor bar in **(F)** showed the fluorescence intensity of different colors.

after caffeine stimulation, showing stage-specific characteristics (**Figures 2A,B**). The high flat Ca^{2+} response following caffeine application was mainly found in OPCs (53.7%, n = 65) and imOLs (54.8%, n = 72). Likely, the oscillatory Ca^{2+} transients were present at a higher ratio in OPCs (35.5%, n = 65) and imOLs (27.9%, n = 72) than in mature OLs (10.5%, n = 76). The low plateaus Ca^{2+} response was the dominant reaction of mature OLs (76.3%, n = 76) but was barely found in OPCs and imOLs (**Figure 2C**). Notably, the millimolar concentration of caffeine was previously demonstrated to be an inhibitor of the IP3R channel in glioblastoma (1–10 mM; Kang et al., 2010), cerebellar microsomes (10–50 mM; Brown et al., 1992), B lymphocytes (1–25 mM) and other cells (Sei et al., 2001). In our present study, while the corresponding Ca^{2+} influx was abolished by using a Ca^{2+}-free extracellular solution, the caffeine (20 mM)-induced Ca^{2+} transient is likely due to release of Ca^{2+} from ER mediated mainly by RyR3 channel. We assume that this stage-specific response is probably correlated to the downregulated expression of RyR3 during OL development.

Given that RyR3 is not homogeneously distributed in the cell body and cell processes (Haak et al., 2001), especially in mature OLs, as we showed in **Figure 1A**, we wonder whether there is a regional difference in the Ca^{2+} response. By transforming the time-lapse Ca^{2+} imaging into pseudocolor changes and 3D surface plots with Image J software, we found that OPCs respond rapidly and strongly in both the soma and processes (**Figure 2E**), whereas only somal regions showed a slow and weak Ca^{2+} response in mature OLs (**Figure 2F**). Thus, the Ca^{2+} response pattern is highly correlated with the expression level and distribution of RyR3 in OL lineage cells.

At this point, we clarified the spatiotemporal regulated expression pattern of RyR3 and its expression-correlated functional pattern after caffeine stimulation. Our results indicate that RyR3-mediated Ca^{2+} signaling actively participates in the developmental regulation of OL lineage cells. The role of RyR3 under physiological conditions requires further exploration.

Inhibition of RyR3 Function by Ryanodine Suppresses OPC Differentiation

As Ca^{2+} signaling is essential for OPC differentiation and myelination, especially the initiation of OPC differentiation (Cheli et al., 2015; Friess et al., 2016), we next investigated whether there is active Ca^{2+} signaling at the initial stage of OPC differentiation. Spontaneous Ca^{2+} signaling was recorded in OPCs cultured in the proliferation medium and in OPCs that had been cultured in differentiation medium for 6 h. Consistent with our previous research (Niu et al., 2016), approximately 38.5% of OPCs presented occasional Ca^{2+} elevation ($n = 18$), which was characterized by a gradual increase in the Ca^{2+} concentration to a peak and a subsequently

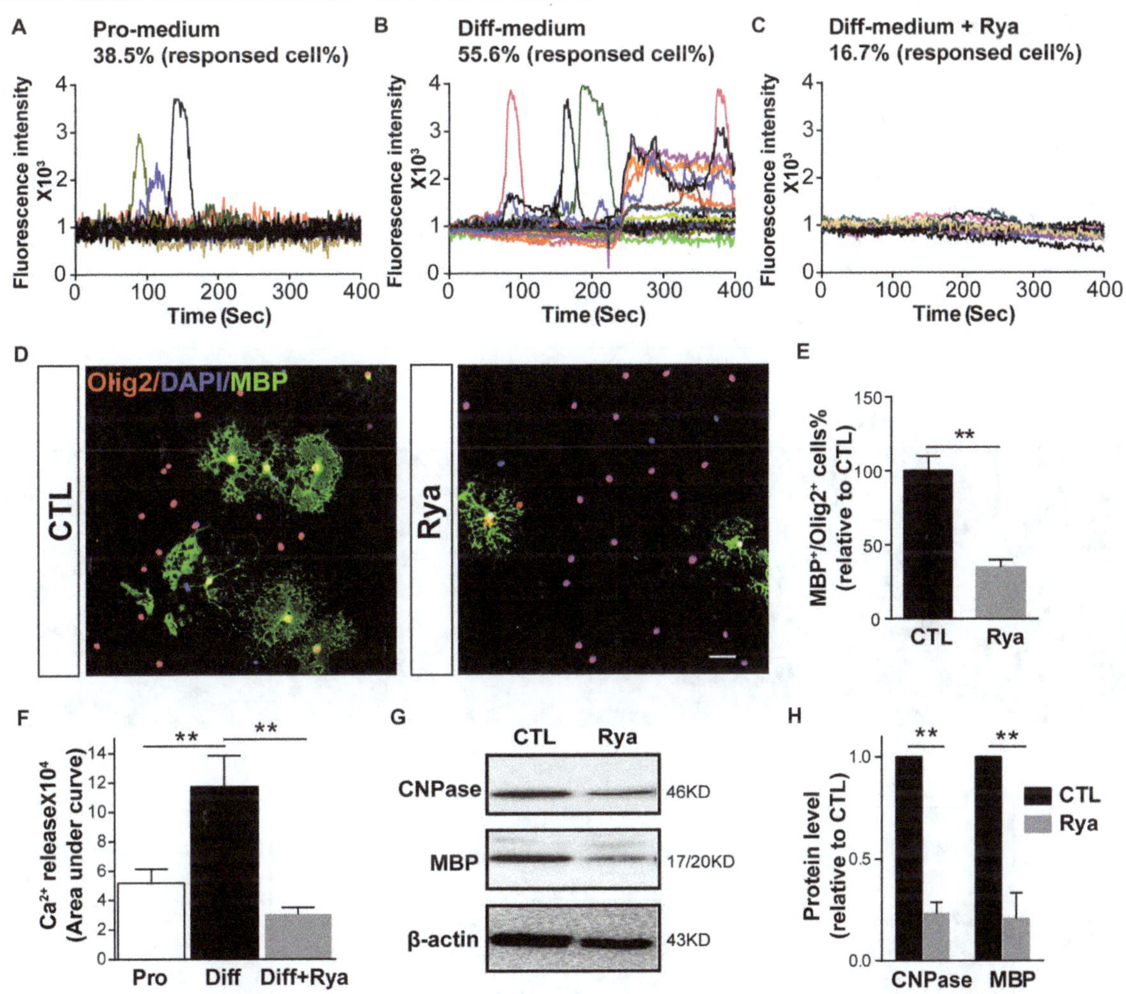

FIGURE 3 | Inhibition of RyR3 by ryanodine suppresses OPC differentiation. **(A)** Ca^{2+} recording of OPCs in proliferation medium (Pro-medium) showing random spontaneous waves. **(B)** OPCs cultured in differentiation medium (Diff-medium) for 6 h showed frequently higher peak waves, which could be inhibited by ryanodine (Rya) **(C)**. **(F)** The total amount of Ca^{2+} release, reflected by the area under curve, was greater in early differentiating OPCs than in proliferating OPCs ($^{**}p < 0.05$), and the release activity in early differentiating OPCs was greatly reduced by ryanodine ($^{**}p < 0.05$). **(D,E)** There were fewer MBP-positive mature OLs after treatment with ryanodine (50 μM) for 72 h. Scale bar 50 μm. **(G,H)** Western blot analysis showed the same inhibitory effect of ryanodine on OPC differentiation ($^{**}p < 0.05$).

decreasing Ca^{2+} signal (**Figure 3A**). Meanwhile, in OPCs that were induced to initiate differentiation, spontaneous Ca^{2+} activity was observed in 55.6% of cells ($n = 20$). Those Ca^{2+} activities appeared at higher frequency; moreover, the elevation of Ca^{2+} signaling was more persistent in certain individual cells (**Figure 3B**). It has been reported that ryanodine, a RyR-specific blocker, locks the RyR channel into a "closed state" at higher concentrations (>50 μM; Meissner, 1986; Lai et al., 1989; McGrew et al., 1989). Thus, we used 50 μM ryanodine as an antagonist to block RyR3. The differentiation-related Ca^{2+} activity was dramatically inhibited ($n = 24$; **Figures 3C,F**). The amount of Ca^{2+} release reflected as the area under curve was markedly greater in early differentiating OPCs than in proliferating OPCs (**Figure 3F**). This result indicates that RyR3-mediated Ca^{2+} release from the ER is a critical process during OPC differentiation. To further verify the function of RyR3 in oligodendroglial differentiation, we treated OPCs with ryanodine (50 μM) and measured the number of mature OLs by immunofluorescence staining after 3 days of differentiation. The number of MBP-positive OLs was significantly decreased in the ryanodine-treated group (**Figures 3D,E**). Importantly, cell viabilities of OPC cultures were not affected by ryanodine treatment as reflected by the nuclear number, which is consistent with previous works (Matyash et al., 2002; Ruiz et al., 2010). Western blot results showed decreased protein levels of MBP and CNPase, which also reflected the inhibition of oligodendroglial maturation (**Figures 3G,H**). Thus, our results demonstrate that RyR3-mediated Ca^{2+} signaling does participate in the differentiation of OPCs.

Knockdown of RyR3 in OPCs Results in Inhibition of OPC Differentiation

To further detect the role of RyR3 in OPC differentiation, we performed lentivirus-mediated gene knockdown of RyR3, which more precisely targets RyR3. OPCs were induced to differentiate after infection with an RNA interference lentivirus (shRyR3) or a control lentivirus (shCTL). After differentiation for 48 h, we observed efficient infection of both lentiviruses, visualized as the GFP expression in the cytoplasm (**Figure 4A**). Immunofluorescence staining for Olig2 (OL lineage marker) and MBP (mature OL marker) showed that the percentage of MBP-positive OLs was significantly decreased and most of the MPB-positive cells showed smaller process areas, and less flat membrane structures after RyR3 knockdown (**Figures 4A,B**), indicating the blockage of OPC differentiation in the absence of RyR3-mediated Ca^{2+} signaling. Cell viability after lentivirus treatment was guaranteed by the unchanged Olig2 positive cell number. As antisense oligonucleotides (aODNs) have the ability to selectively reduce the mRNA level of RyR3 (Galeotti et al., 2008), we also applied aODNs in OPC cultures and confirmed that aODNs reduced the protein level of RyR3. Consistent with the shRNA effect, aODN treatment similarly reduced the number of mature OLs (data not shown) and the levels of MBP (**Figures 4C,D**). Taken together, these observations reveal that gene-level knockdown of RyR3 induces loss of

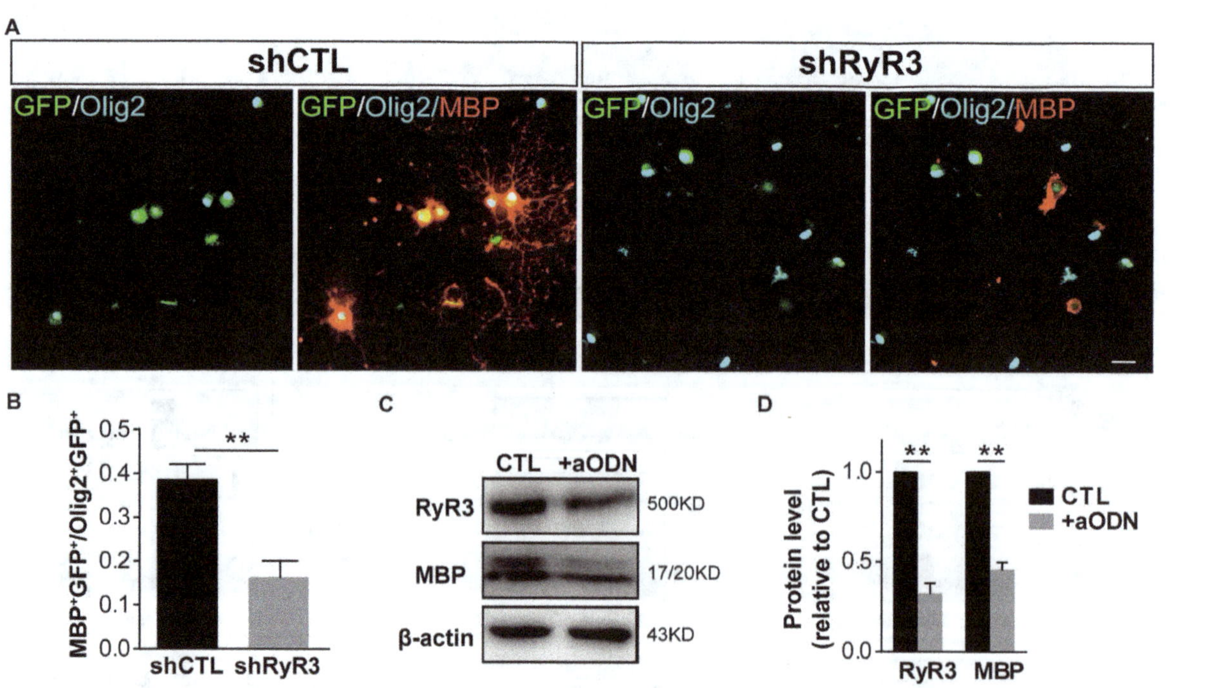

FIGURE 4 | Knockdown of RyR3 in OPCs results in inhibition of OPC differentiation. **(A,B)** Gene knockdown of RyR3 with shRNA interference mediated by lentivirus infection showed fewer MBP-positive mature OLs after 48 h of differentiation. GFP was expressed in lentivirus-infected cells (**$p < 0.05$). Scale bar 20 μm. **(C,D)** Knockdown of RyR3 in OPCs by antisense oligonucleotide (+aODN) resulted in less expression of myelin basic protein (MBP; **$p < 0.05$).

RyR3 channel function in OPCs, finally resulting in inhibition of OPC differentiation, implying that RyR3-mediated Ca^{2+} signaling plays an essential role during OPC differentiation.

DISCUSSION

In non-excitable oligodendroglial cells, how intracellular Ca^{2+} signaling is regulated and how it contributes to oligodendroglial differentiation remains unclear. In our present study, we systematically analyzed the expression and function of the ER Ca^{2+} release channel—RyR3. We found that RyR3 is the only RyR expressed in oligodendroglial cells and dynamically regulated during OPC differentiation. Importantly, inhibition of RyR3 resulted in blockage of OPC differentiation (**Figure 5**). Our results not only demonstrate the essential role of RyR3-mediated Ca^{2+} signaling for oligodendroglial differentiation but also improve our understanding of the intracellular Ca^{2+} channel function in oligodendroglial lineage cells, which has barely been studied before.

RyRs (RyR1–3) are major Ca^{2+} channels responsible for ER Ca^{2+} release, and RyR-mediated transient increase and oscillations of intracellular Ca^{2+} has been considered to be particularly important for cell functions (Zalk et al., 2007; Fulton et al., 2010; Suzuki et al., 2012). In oligodendroglial lineage cells, the RyR functionality was suggested to correlate with their developmental stages both *in vivo* and *in vitro* (Simpson et al., 1998), but the isoform of RyRs involved in this process has not been identified. Among RyRs, RyR3 is abundantly distributed in the CNS, and it functions in the activation and migration of astrocytes (Fill and Copello, 2002; Matyash et al., 2002; Galeotti et al., 2008; Lanner et al., 2010). In oligodendroglial lineage cells, only RyR3 was selectively expressed in cultured rat cortex OPCs (Haak et al., 2001), while another study demonstrated that three isoforms of RyRs (RyR1–3) were expressed in cultured rat optic nerve-derived mature OLs (Ruiz et al., 2010). Here, using rat cortex-derived OPC cultures, we confirmed the selective expression of RyR3 in oligodendroglial lineage cells and further determined the spatiotemporal expression pattern of RyR3 during oligodendroglial differentiation. Additionally, we found that enrichment of RyR3 in OPCs and imOLs corresponded to the stronger Ca^{2+} responses in those cell types than in mature OLs following caffeine stimulation. Interestingly, a previous study in myotubes has revealed that embryonic myotubes, which express RyR3, have considerably more variability in the size and kinetics of their Ca^{2+} sparks than do adult cells, which lose RyR3 expression (Ward et al., 2000). Therefore, expression of RyR3 in OPCs and imOLs likely enabled them to be more hyperactive in terms of Ca^{2+}

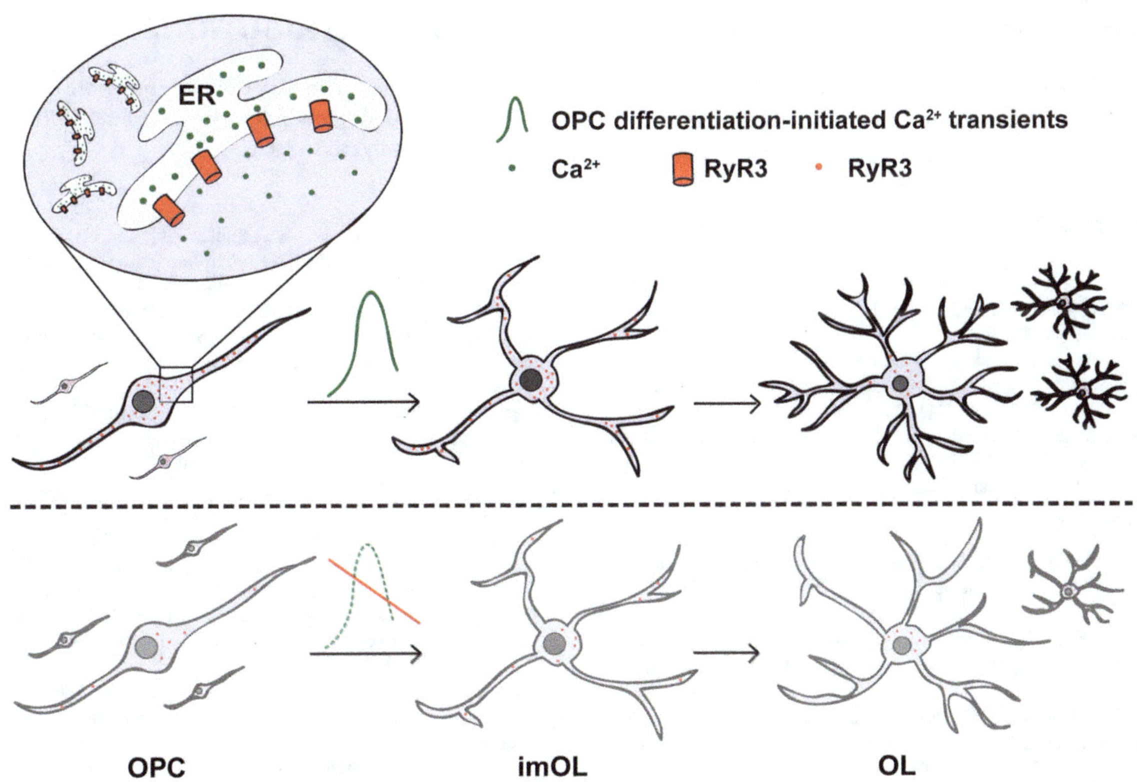

FIGURE 5 | Scheme illustrating the contribution of RyR3-mediated Ca^{2+} signaling during oligodendroglial development. RyR3, a Ca^{2+}-induced Ca^{2+} release (CICR) channel located on the ER membrane during oligodendroglial differentiation, note that it is widely distributed in the cell body and processes of OPC/imOLs but down-regulated and restricted in the cell body of mature OLs. The initiation of OPC differentiation requires hyperactive Ca^{2+} signaling mediated by RyR3. RyR3 knockdown or blockage of RyR3-mediated Ca^{2+} signaling in OPCs results in inhibition of OPC differentiation.

transients and thus program them into the differentiating state.

Previous studies showed that Ca^{2+} responses to depolarization in OPCs were more active than those in mature OLs, and this phenomenon was mainly explained by the downregulation of L-type voltage-operated Ca^{2+} channels (VOCCs) following OPC maturation (Berger et al., 1992; Takeda et al., 1995; Paez et al., 2010; Zhang et al., 2014). Considering that RyR-mediated CICR is a critical amplification point for depolarization-induced intracellular Ca^{2+} elevation (Verkhratsky and Petersen, 2002; Pouvreau et al., 2007; Haberlandt et al., 2011), our results suggested that RyR3-mediated Ca^{2+} release from the ER is also an important contributor to the dynamic intracellular responses during OPC differentiation.

Consistent with a previous study (Simpson et al., 1998), we found that RyR3 was enriched in the entire processes of OPCs and imOLs but was absent in the leading processes of mature OLs, indicating that RyR3-mediated local Ca^{2+} signaling may contribute to enabling the process elongation and/or movements of OPCs (Paez et al., 2007) and to initiating cell differentiation (Friess et al., 2016). In agreement with this, we provide evidence showing that RyR3 knockdown results in failure of oligodendroglial differentiation, indicating that CICR may critically contribute to OPC differentiation.

Even if it has not been studied in this work, we speculate that in physiological *in vivo* condition, the Ca^{2+} signals triggering the opening of RyR3 are likely originated from the nearby neurons. Large amount of evidences have shown that neuronal action potentials could regulate myelin development, the candidate mediators are the neurotransmitters (glutamate, ATP, adenosine...) which could trigger Ca^{2+} influx in oligodendroglial cells and possibly evoke the following CICR through RyR3 (Butt, 2006; Spitzer et al., 2016; Sun et al., 2016; Krasnow et al., 2018). Moreover, axonal activity-induced increases in extracellular K^+ are sufficient to depolarize VOCCs in nearby oligodendroglial cells to produce a significant rise in intracellular Ca^{2+} which also need CICR to form the final Ca^{2+} signal (Cheli et al., 2015). Importantly, RyR3 and IP3R2 are usually co-localized in OPCs, and interactions between them determine the spatial and temporal characteristics of Ca^{2+} signaling (Haak et al., 2001). Although a relative higher concentration of caffeine that was applied in our current study has been shown to stimulate RyR but inhibit the IP3R Ca^{2+} channel in several cell types (Brown et al., 1992; Sei et al., 2001; Kang et al., 2010), we could not exclude the role of IP3R2 in the Ca^{2+} signaling formation in oligodendroglial differentiation. Channel-specific antagonists or a genetic approach may be needed in our future studies.

In summary, we provide direct evidence showing that RyR3-mediated ER Ca^{2+} release is dynamically regulated and plays an essential role in initiating OPC differentiation (**Figure 5**). Our results supplement the current understanding of the function of intracellular Ca^{2+} channels in oligodendroglia and may provide valuable insights into therapeutic strategies for demyelinating diseases.

AUTHOR CONTRIBUTIONS

LX, HL and TL designed experiments and wrote the manuscript. TL, LW, TM and SW conducted the experiments. TL, LW and JN collected and analyzed the data.

REFERENCES

Baraban, M., Koudelka, S., and Lyons, D. A. (2018). Ca^{2+} activity signatures of myelin sheath formation and growth *in vivo*. *Nat. Neurosci.* 21, 19–23. doi: 10.1038/s41593-017-0040-x

Berger, T., Schnitzer, J., Orkand, P. M., and Kettenmann, H. (1992). Sodium and calcium currents in glial cells of the mouse corpus callosum slice. *Eur. J. Neurosci.* 4, 1271–1284. doi: 10.1111/j.1460-9568.1992.tb00153.x

Bezprozvanny, I., Watras, J., and Ehrlich, B. E. (1991). Bell-shaped calcium-response curves of $Ins(1,4,5)P_3$- and calcium-gated channels from endoplasmic reticulum of cerebellum. *Nature* 351, 751–754. doi: 10.1038/351751a0

Brown, G. R., Sayers, L. G., Kirk, C. J., Michell, R. H., and Michelangeli, F. (1992). The opening of the inositol 1,4,5-trisphosphate-sensitive Ca^{2+} channel in rat cerebellum is inhibited by caffeine. *Biochem. J.* 282, 309–312. doi: 10.1042/bj2820309

Butt, A. M. (2006). Neurotransmitter-mediated calcium signalling in oligodendrocyte physiology and pathology. *Glia* 54, 666–675. doi: 10.1002/glia.20424

Chang, A., Tourtellotte, W. W., Rudick, R., and Trapp, B. D. (2002). Premyelinating oligodendrocytes in chronic lesions of multiple sclerosis. *N. Engl. J. Med.* 346, 165–173. doi: 10.1056/nejmoa010994

Cheli, V. T., Santiago González, D. A., Spreuer, V., and Paez, P. M. (2015). Voltage-gated Ca^{2+} entry promotes oligodendrocyte progenitor cell maturation and myelination *in vitro*. *Exp. Neurol.* 265, 69–83. doi: 10.1016/j.expneurol.2014.12.012

Cohen, R. I., Molina-Holgado, E., and Almazan, G. (1996). Carbachol stimulates c-fos expression and proliferation in oligodendrocyte progenitors. *Brain Res. Mol. Brain Res.* 43, 193–201. doi: 10.1016/s0169-328x(96)00176-3

Chang, A., Tourtellotte, W. W., Rudick, R., and Trapp, B. D. (2002). Premyelinating oligodendrocytes in chronic lesions of multiple sclerosis. *N. Engl. J. Med.* 346, 165–173. doi: 10.1056/nejmoa010994

Cheli, V. T., Santiago González, D. A., Spreuer, V., and Paez, P. M. (2015). Voltage-gated Ca^{2+} entry promotes oligodendrocyte progenitor cell maturation and myelination *in vitro*. *Exp. Neurol.* 265, 69–83. doi: 10.1016/j.expneurol.2014.12.012

Cohen, R. I., Molina-Holgado, E., and Almazan, G. (1996). Carbachol stimulates c-fos expression and proliferation in oligodendrocyte progenitors. *Brain Res. Mol. Brain Res.* 43, 193–201. doi: 10.1016/s0169-328x(96)00176-3

Fulton, D., Paez, P. M., Fisher, R., Handley, V., Colwell, C. S., and Campagnoni, A. T. (2010). Regulation of L-type Ca^{++} currents and process morphology in white matter oligodendrocyte precursor cells by golli-myelin proteins. *Glia* 58, 1292–1303. doi: 10.1002/glia.21008

Galeotti, N., Vivoli, E., Bartolini, A., and Ghelardini, C. (2008). A gene-specific cerebral types 1, 2, and 3 RyR protein knockdown induces an antidepressant-like effect in mice. *J. Neurochem.* 106, 2385–2394. doi: 10.1111/j.1471-4159.2008.05581.x

Haak, L. L., Song, L. S., Molinski, T. F., Pessah, I. N., Cheng, H., and Russell, J. T. (2001). Sparks and puffs in oligodendrocyte progenitors: cross talk between ryanodine receptors and inositol trisphosphate receptors. *J. Neurosci.* 21, 3860–3870. doi: 10.1523/JNEUROSCI.21-11-03860.2001

Haberlandt, C., Derouiche, A., Wyczynski, A., Haseleu, J., Pohle, J., Karram, K., et al. (2011). Gray matter NG2 cells display multiple Ca^{2+}-signaling pathways and highly motile processes. *PLoS One* 6:e17575. doi: 10.1371/journal.pone.0017575

Kang, S. S., Han, K. S., Ku, B. M., Lee, Y. K., Hong, J., Shin, H. Y., et al. (2010). Caffeine-mediated inhibition of calcium release channel inositol

1,4,5-trisphosphate receptor subtype 3 blocks glioblastoma invasion and extends survival. *Cancer Res.* 70, 1173–1183. doi: 10.1158/0008-5472.CAN-09-2886

Kirischuk, S., Scherer, J., Möller, T., Verkhratsky, A., and Kettenmann, H. (1995). Subcellular heterogeneity of voltage-gated Ca^{2+} channels in cells of the oligodendrocyte lineage. *Glia* 13, 1–12. doi: 10.1002/glia.440130102

Koulen, P., and Thrower, E. C. (2001). Pharmacological modulation of intracellular Ca^{2+} channels at the single-channel level. *Mol. Neurobiol.* 24, 65–86. doi: 10.1385/mn:24:1-3:065

Krasnow, A. M., Ford, M. C., Valdivia, L. E., Wilson, S. W., and Attwell, D. (2018). Regulation of developing myelin sheath elongation by oligodendrocyte calcium transients *in vivo*. *Nat. Neurosci.* 21, 24–28. doi: 10.1038/s41593-017-0031-y

Kuhlmann, T., Miron, V., Cui, Q., Wegner, C., Antel, J., and Brück, W. (2008). Differentiation block of oligodendroglial progenitor cells as a cause for remyelination failure in chronic multiple sclerosis. *Brain* 131, 1749–1758. doi: 10.1093/brain/awn096

Lai, F. A., Misra, M., Xu, L., Smith, H. A., and Meissner, G. (1989). The ryanodine receptor-Ca^{2+} release channel complex of skeletal muscle sarcoplasmic reticulum. Evidence for a cooperatively coupled, negatively charged homotetramer. *J. Biol. Chem.* 264, 16776–16785.

Lanner, J. T., Georgiou, D. K., Joshi, A. D., and Hamilton, S. L. (2010). Ryanodine receptors: structure, expression, molecular details, and function in calcium release. *Cold Spring Harb. Perspect. Biol.* 2:a003996. doi: 10.1101/cshperspect.a003996

Matyash, M., Matyash, V., Nolte, C., Sorrentino, V., and Kettenmann, H. (2002). Requirement of functional ryanodine receptor type 3 for astrocyte migration. *FASEB J.* 16, 84–86. doi: 10.1096/fj.01-0380fje

McGrew, S. G., Wolleben, C., Siegl, P., Inui, M., and Fleischer, S. (1989). Positive cooperativity of ryanodine binding to the calcium release channel of sarcoplasmic reticulum from heart and skeletal muscle. *Biochemistry* 28, 1686–1691. doi: 10.1021/bi00430a039

Meissner, G. (1986). Ryanodine activation and inhibition of the Ca^{2+} release channel of sarcoplasmic reticulum. *J. Biol. Chem.* 261, 6300–6306.

Meissner, G., Darling, E., and Eveleth, J. (1986). Kinetics of rapid Ca^{2+} release by sarcoplasmic reticulum. Effects of Ca^{2+}, Mg^{2+}, and adenine nucleotides. *Biochemistry* 25, 236–244. doi: 10.1021/bi00349a033

Meissner, G., Rios, E., Tripathy, A., and Pasek, D. A. (1997). Regulation of skeletal muscle Ca^{2+} release channel (ryanodine receptor) by Ca^{2+} and monovalent cations and anions. *J. Biol. Chem.* 272, 1628–1638. doi: 10.1074/jbc.272.3.1628

Meldolesi, J., and Pozzan, T. (1998). The endoplasmic reticulum Ca^{2+} store: a view from the lumen. *Trends Biochem. Sci.* 23, 10–14. doi: 10.1016/s0968-0004(97)01143-2

Moraru, I. I., Kaftan, E. J., Ehrlich, B. E., and Watras, J. (1999). Regulation of type 1 inositol 1,4,5-trisphosphate-gated calcium channels by InsP3 and calcium: simulation of single channel kinetics based on ligand binding and electrophysiological analysis. *J. Gen. Physiol.* 113, 837–849. doi: 10.1085/jgp.113.6.837

Niu, J., Li, T., Yi, C., Huang, N., Koulakoff, A., Weng, C., et al. (2016). Connexin-based channels contribute to metabolic pathways in the oligodendroglial lineage. *J. Cell Sci.* 129, 1902–1914. doi: 10.1242/jcs.178731

Niu, J., Mei, F., Wang, L., Liu, S., Tian, Y., Mo, W., et al. (2012a). Phosphorylated olig1 localizes to the cytosol of oligodendrocytes and promotes membrane expansion and maturation. *Glia* 60, 1427–1436. doi: 10.1002/glia.22364

Niu, J., Wang, L., Liu, S., Li, C., Kong, J., Shen, H. Y., et al. (2012b). An efficient and economical culture approach for the enrichment of purified oligodendrocyte progenitor cells. *J. Neurosci. Methods* 209, 241–249. doi: 10.1016/j.jneumeth.2012.05.032

Paez, P. M., Fulton, D. J., Spreur, V., Handley, V., and Campagnoni, A. T. (2010). Multiple kinase pathways regulate voltage-dependent Ca^{2+} influx and migration in oligodendrocyte precursor cells. *J. Neurosci.* 30, 6422–6433. doi: 10.1523/JNEUROSCI.5086-09.2010

Paez, P. M., Spreuer, V., Handley, V., Feng, J. M., Campagnoni, C., and Campagnoni, A. T. (2007). Increased expression of golli myelin basic proteins enhances calcium influx into oligodendroglial cells. *J. Neurosci.* 27, 12690–12699. doi: 10.1523/JNEUROSCI.2381-07.2007

Pouvreau, S., Royer, L., Yi, J., Brum, G., Meissner, G., Rios, E., et al. (2007). Ca^{2+} sparks operated by membrane depolarization require isoform 3 ryanodine receptor channels in skeletal muscle. *Proc. Natl. Acad. Sci. U S A* 104, 5235–5240. doi: 10.1073/pnas.0706530104

Ruiz, A., Matute, C., and Alberdi, E. (2010). Intracellular Ca^{2+} release through ryanodine receptors contributes to AMPA receptor-mediated mitochondrial dysfunction and ER stress in oligodendrocytes. *Cell Death Dis.* 1:e54. doi: 10.1038/cddis.2010.31

Sei, Y., Gallagher, K. L., and Daly, J. W. (2001). Multiple effects of caffeine on Ca^{2+} release and influx in human B lymphocytes. *Cell Calcium* 29, 149–160. doi: 10.1054/ceca.2000.0175

Simpson, P. B., Holtzclaw, L. A., Langley, D. B., and Russell, J. T. (1998). Characterization of ryanodine receptors in oligodendrocytes, type 2 astrocytes, and O-2A progenitors. *J. Neurosci. Res.* 52, 468–482. doi: 10.1002/(sici)1097-4547(19980515)52:4<468::aid-jnr11>3.0.co;2-#

Soliven, B. (2001). Calcium signalling in cells of oligodendroglial lineage. *Microsc. Res. Tech.* 52, 672–679. doi: 10.1002/jemt.1051

Spitzer, S., Volbracht, K., Lundgaard, I., and Káradóttir, R. T. (2016). Glutamate signalling: a multifaceted modulator of oligodendrocyte lineage cells in health and disease. *Neuropharmacology* 110, 574–585. doi: 10.1016/j.neuropharm.2016.06.014

Stangel, M., and Hartung, H. P. (2002). Remyelinating strategies for the treatment of multiple sclerosis. *Prog. Neurobiol.* 68, 361–376. doi: 10.1016/s0301-0082(02)00105-3

Sun, W., Matthews, E. A., Nicolas, V., Schoch, S., and Dietrich, D. (2016). NG2 glial cells integrate synaptic input in global and dendritic calcium signals. *Elife* 5:e16262. doi: 10.7554/elife.16262

Suzuki, M., Nagai, Y., Wada, K., and Koike, T. (2012). Calcium leak through ryanodine receptor is involved in neuronal death induced by mutant huntingtin. *Biochem. Biophys. Res. Commun.* 429, 18–23. doi: 10.1016/j.bbrc.2012.10.107

Takeda, M., Nelson, D. J., and Soliven, B. (1995). Calcium signaling in cultured rat oligodendrocytes. *Glia* 14, 225–236. doi: 10.1002/glia.440140308

Verkhratsky, A., and Petersen, O. H. (2002). The endoplasmic reticulum as an integrating signalling organelle: from neuronal signalling to neuronal death. *Eur. J. Pharmacol.* 447, 141–154. doi: 10.1016/s0014-2999(02)01838-1

Ward, C. W., Schneider, M. F., Castillo, D., Protasi, F., Wang, Y., Chen, S. R., et al. (2000). Expression of ryanodine receptor RyR3 produces Ca^{2+} sparks in dyspedic myotubes. *J. Physiol.* 525, 91–103. doi: 10.1111/j.1469-7793.2000.t01-2-00091.x

Wolswijk, G. (1998). Chronic stage multiple sclerosis lesions contain a relatively quiescent population of oligodendrocyte precursor cells. *J. Neurosci.* 18, 601–609. doi: 10.1523/JNEUROSCI.18-02-00601.1998

Yoo, A. S., Krieger, C., and Kim, S. U. (1999). Process extension and intracellular Ca^{2+} in cultured murine oligodendrocytes. *Brain Res.* 827, 19–27. doi: 10.1016/s0006-8993(99)01282-2

Zalk, R., Lehnart, S. E., and Marks, A. R. (2007). Modulation of the ryanodine receptor and intracellular calcium. *Annu. Rev. Biochem.* 76, 367–385. doi: 10.1146/annurev.biochem.76.053105.094237

Zhang, Y., Chen, K., Sloan, S. A., Bennett, M. L., Scholze, A. R., O'Keeffe, S., et al. (2014). An RNA-sequencing transcriptome and splicing database of glia, neurons, and vascular cells of the cerebral cortex. *J. Neurosci.* 34, 11929–11947. doi: 10.1523/JNEUROSCI.1860-14.2014

Zucchi, R., and Ronca-Testoni, S. (1997). The sarcoplasmic reticulum Ca^{2+} channel/ryanodine receptor: modulation by endogenous effectors, drugs and disease states. *Pharmacol. Rev.* 49, 1–51.

15

Astrocytic Atrophy Following *Status Epilepticus* Parallels Reduced Ca^{2+} Activity and Impaired Synaptic Plasticity in the Rat Hippocampus

Alex Plata[1†], Albina Lebedeva[1†], Pavel Denisov[1], Olga Nosova[1], Tatiana Y. Postnikova[2,3], Alexey Pimashkin[1], Alexey Brazhe[4], Aleksey V. Zaitsev[2,5], Dmitri A. Rusakov[1,6] and Alexey Semyanov[1,7,8]*

[1] UNN Institute of Neuroscience, N. I. Lobachevsky State University of Nizhny Novgorod, University of Nizhny Novgorod, Nizhny Novgorod, Russia, [2] Laboratory of Molecular Mechanisms of Neural Interactions, Sechenov Institute of Evolutionary Physiology and Biochemistry, Russian Academy of Sciences, St. Petersburg, Russia, [3] Department of Medical Physics, Peter the Great St. Petersburg Polytechnic University, St. Petersburg, Russia, [4] Department of Biophysics, Faculty of Biology, M. V. Lomonosov Moscow State University, Moscow, Russia, [5] Institute of Experimental Medicine, Almazov National Medical Research Centre, St. Petersburg, Russia, [6] UCL Institute of Neurology, University College London, London, United Kingdom, [7] Department of Molecular Neurobiology, Shemyakin-Ovchinnikov Institute of Bioorganic Chemistry, Russian Academy of Sciences, Moscow, Russia, [8] All-Russian Research Institute of Medicinal and Aromatic Plants, Moscow, Russia

***Correspondence:**
Alexey Semyanov
semyanov@ibch.ru

† These authors have contributed equally to this work.

Epilepsy is a group of neurological disorders commonly associated with the neuronal malfunction leading to generation of seizures. Recent reports point to a possible contribution of astrocytes into this pathology. We used the lithium-pilocarpine model of *status epilepticus* (SE) in rats to monitor changes in astrocytes. Experiments were performed in acute hippocampal slices 2–4 weeks after SE induction. Nissl staining revealed significant neurodegeneration in the pyramidal cell layers of hippocampal CA1, CA3 areas, and the hilus, but not in the granular cell layer of the dentate gyrus. A significant increase in the density of astrocytes stained with an astrocyte-specific marker, sulforhodamine 101, was observed in CA1 *stratum (str.) radiatum*. Astrocytes in this area were also whole-cell loaded with a morphological tracer, Alexa Fluor 594, for two-photon excitation imaging. Sholl analyses showed no changes in the size of the astrocytic domain or in the number of primary astrocytic branches, but a significant reduction in the number of distal branches that are resolved with diffraction-limited light microscopy (and are thought to contain Ca^{2+} stores, such as mitochondria and endoplasmic reticulum). The atrophy of astrocytic branches correlated with the reduced size, but not overall frequency of Ca^{2+} events. The volume tissue fraction of nanoscopic (beyond the diffraction limit) astrocytic leaflets showed no difference between control and SE animals. The results of spatial entropy-complexity spectrum analysis were also consistent with changes in ratio of astrocytic branches vs. leaflets. In addition, we observed uncoupling of astrocytes through the gap-junctions, which was suggested as a mechanism for reduced K^+ buffering. However, no significant difference in time-course of synaptically induced K^+ currents in patch-clamped astrocytes argued against possible alterations in K^+ clearance by astrocytes. The magnitude of long-term-potentiation (LTP) was reduced after SE.

Exogenous *D*-serine, a co-agonist of NMDA receptors, has rescued the initial phase of LTP. This suggests that the reduced Ca^{2+}-dependent release of *D*-serine by astrocytes impairs initiation of synaptic plasticity. However, it does not explain the failure of LTP maintenance which may be responsible for cognitive decline associated with epilepsy.

Keywords: astrocyte remodeling, epilepsy, *D*-serine, calcium, plasticity, spatial entropy, spatial complexity, neurodegeneration

INTRODUCTION

Epilepsy is a group of neurological disorders, associated with pathological synchronization of neuronal activity causing seizures. Several cellular mechanisms of this pathology were proposed. The classical view holds that seizures occur because of a shift in the balance between excitation and inhibition in the brain toward excitation (During and Spencer, 1993; DiNuzzo et al., 2014). However, pathological synchronization of excitatory neurons may also result from enhanced inhibition of a relatively small group of inhibitory interneurons, which then synchronously disinhibits numerous excitatory neurons. This is called "rebound" excitation and may make all these cells to fire in synchrony (Paz and Huguenard, 2015). The K^+ hypothesis of epilepsy suggests that extracellular accumulation of this ion in the brain depolarizes neurons and cause epileptiform activity (Green, 1964; Fertziger and Ranck, 1970; Fröhlich et al., 2008). Mutation in astrocytic K^+ channels ($K_{ir}4.1$ type) has been proposed as one of the causes of human epilepsy (Villa and Combi, 2016). Indeed, it has long been reported that periodic elevations of extracellular K^+ concentration could produce long-term changes in neuronal excitability, making the network prone to epileptogenesis (Semyanov and Godukhin, 1997).

An excessive activity of neuronal networks induces excitotoxicity, which leads to neurodegeneration. Hippocampal sclerosis is typically reported in temporal lobe epilepsy (TLE) (Kim, 2001; Blümcke et al., 2002; de Lanerolle and Lee, 2005; Thom, 2014). The loss of neurons is considered as a mechanism for epileptic focus formation and an indication for hippocampus removal in clinic. Neurodegeneration in TLE may also lead to a mild cognitive impairment affecting learning and memory (Hermann and Seidenberg, 2007; Höller and Trinka, 2014). Long-term potentiation (LTP) and depression (LTD) are widely accepted experimental models to explore mechanisms of synaptic memory formation (Bliss and Collingridge, 1993; Ju et al., 2004). Recent studies have demonstrated a significant impairment of long-term synaptic plasticity after SE in different animal models, including a lithium-pilocarpine model of TLE (Zhou et al., 2007; Müller et al., 2013; Cunha et al., 2015; Kryukov et al., 2016; Carpenter-Hyland et al., 2017; Ivanov and Zaitsev, 2017; Postnikova et al., 2017). A reduction or elimination of LTP that can be observed for weeks after SE has frequently been reported (Zhang et al., 2010; Zhou et al., 2011; Suárez et al., 2012). Despite numerous studies, the exact mechanisms of LTP impairment after SE remain unidentified.

The quest for the mechanisms of epileptogenesis have typically been focused on neuronal malfunction, such as altered expression of receptors and channels; extracellular K^+ accumulation synchronizing neuronal populations; a shift in the balance of synaptic excitation and inhibition toward excitation (DiNuzzo et al., 2014). Intriguingly, many of these functions are regulated by astrocytes. Astrocytic uptake of neurotransmitters and extracellular K^+ clearance are key to synaptic function (Cheung et al., 2015; Lebedeva et al., 2018; Verkhratsky and Nedergaard, 2018). Astrocytes control synaptic plasticity in glutamatergic synapses by releasing of *D*-serine, a co-agonist of NMDA receptors (Henneberger et al., 2010, 2012; Papouin et al., 2017). They are involved in uptake and release of GABA, inhibitory neuro- and gliotransmitter (Angulo et al., 2008; Lee et al., 2010; Héja et al., 2012; Kersante et al., 2013; Song et al., 2013). Therefore, astrocytic mechanisms can potentially contribute to epileptogenesis. Indeed, astrocyte control of synaptic NMDA receptors is implicated in the progressive development of TLE (Clasadonte et al., 2013). Astrocytic Ca^{2+}-dependent glutamate release is suggested to trigger synchronous neuronal discharges in rat hippocampal slices following application of a K^+ channel blocker, 4-aminopyridine (4-AP) (Tian et al., 2005). In contrast, some other models of acute epileptiform activity in slices (Mg^{2+}-free solution, picrotoxin, increased extracellular K^+) produce Ca^{2+} oscillations in astrocytes, which are not responsible for paroxysmal activity in neurons (Fellin et al., 2006). Thus, the role of astrocytic Ca^{2+} activity in epileptogenesis remains controversial. Moreover, acute effects have often been obtained in brain slices using convulsants, which may not necessarily reflect changes in astrocytic Ca^{2+} activity after SE.

Another critical aspect of neuron-astrocyte interactions is morphological. Astrocytic processes approach synapses, forming so-called "astrocytic cradle" (Verkhratsky and Nedergaard, 2014). Being highly plastic, perisynaptic astrocytic processes can retract from or extend toward dendritic spines (Bernardinelli et al., 2014a; Heller and Rusakov, 2015). Rearing laboratory animals in complex environment or certain LTP induction protocols appear to increase glial coverage of excitatory synapses (Jones and Greenough, 1996; Lushnikova et al., 2009) whereas some memory consolidation tasks (Ostroff et al., 2014) or experiencing a lactation period (Oliet et al., 2001) appear to decrease it. Suppression of IP_3-dependent Ca^{2+} signaling in astrocytes reduces synaptic coverage (Tanaka et al., 2013). Morphological remodeling of astrocytes following SE has not been systematically studied.

Impairment of K^+ clearance by astrocytes has been proposed as a mechanism for extracellular K^+ accumulation in the epileptic brain (Bedner and Steinhauser, 2013). This may occur due to the redistribution of astrocytic K^+ channels or reduced K^+ buffering due to astrocyte uncoupling through gap-junctions in epileptic tissue (Wallraff et al., 2006; Bedner et al., 2015). Interestingly, K^+ clearance by astrocytes can be linked to astrocytic Ca^{2+} activity, via Ca^{2+}-dependent K^+ channels (Wang et al., 2012). However, this link has not been considered in the context of epileptogenesis.

Although recent reports have convincingly demonstrated the involvement of astrocytes in epileptogenesis, further analysis of cellular and subcellular mechanisms is still needed. Here we report morphological and physiological changes in astrocytes following SE induced by lithium-pilocarpine injection, as well as their possible association with synaptic plasticity changes. Our quantitative morphological assessment employs a novel approach based on the spatial entropy-complexity spectrum analysis.

MATERIALS AND METHODS

Pilocarpine Model of Epilepsy

All procedures were carried out in accordance with University of Nizhny Novgorod regulations. 3–6 weeks old male Sprague-Dawley (Wistar for LTP experiments) rats were injected with lithium chloride (127 mg/kg, Sigma Aldrich) 20–24 h prior to pilocarpine and methylscopolamine (1 mg/kg, Sigma Aldrich) 20 min prior to pilocarpine. Then pilocarpine (Tocris), 10 mg/kg was injected every 30 min (but no more than 60 mg/kg) to induce SE which characterized with generalized seizures lasting for at least 20 min (Supplementary Figure 1). To reduce mortality, phenazepam 1 mg/kg was injected every 10 min for 30–40 min after 20 min of generalized seizures.

Nissl Staining

Brain tissue was prepared according to routine histologic methods (Singh et al., 2008). Briefly, brains were removed immediately after decapitation, immersed in ethanol 96% and embedded in paraffin after dehydration. Paraffin sections (5 μm) were cut in a coronal plane and stained with Nissl's method. Each sixth staining section was chosen for quantitative analysis for each animal. Images of CA1, CA3, hilus, and dentate gyrus were obtained using an x40 magnification. The neurons were counted per 100 μm for cell layer in each area using the plugin "Cell counter" for ImageJ.

Hippocampal Slice Preparation

The slices were prepared 2–4 weeks after SE. The animals were anesthetized with Isoflurane (1-Chloro-2,2,2-trifluoroethyl difluoromethyl ether) and then decapitated. The rest of the procedure was slightly different for whole cell and field potential recordings.

Preparation for Whole-Cell Recording and Imaging

The brains were exposed, and then chilled with ice-cold solution containing (in mM): 50 sucrose; 87 NaCl; 2.5 KCl; 8.48 MgSO4; 1.24 NaH_2PO_4; 26.2 $NaHCO_3$; 0.5 $CaCl_2$; 22 D-Glucose. Hippocampi from both hemispheres were dissected, isolated, and transverse slices (350 μm) were cut using a vibrating microtome (Microm HM650 V; Thermo Fisher Scientific) and left to recover at 34°C for 1 h in a submerged incubation chamber with "storage" solution containing (in mM): 119 NaCl; 2.5 KCl; 1.3 $MgSO_4$; 1 NaH_2PO_4; 26.2 $NaHCO_3$; 1 $CaCl_2$; 1.6 $MgCl_2$; 22 D-Glucose. Then the slices were transferred to the recording chamber and were continuously perfused with a solution containing (in mM): 119 NaCl; 2.5 KCl; 1.3 $MgSO_4$; 1 NaH_2PO_4; 26.2 $NaHCO_3$; 2.5 $CaCl_2$; 11 D-Glucose. All solutions were saturated with carbogen gas mixture containing 95% O_2 and 5% CO_2. Osmolarity was 295 ± 5 mOsm, pH 7.4. All recordings were done at a temperature of 34°C.

Preparation for Field Potential Recording and LTP Induction

The cerebellum and a small section of the frontal cortex were removed. A flat surface for mounting the brain was created by making a cut on the dorsal surface parallel to the horizontal plane. The brain was then mounted onto the stage of the vibratome, and horizontal sections (400 μm thick) were cut in ice-cold artificial cerebrospinal fluid (ACSF). ACSF composed of (in mM): 126 NaCl, 2.5 KCl, 1.25 NaH_2PO_4, 1 $MgSO_4$, 2 $CaCl_2$, 24 $NaHCO_3$, and 10 D-glucose was saturated with carbogen. The prepared slices were immersed in a chamber with ACSF, which was placed in a temperature-controlled water bath (35°C) for 1 h. After the incubation, the slices were transferred to the recording chamber, where they were kept for 15–20 min prior to the electrophysiological study. In this chamber, hippocampal slices were perfused with a constant flow of oxygenated ACSF at a rate of 5 ml/min at room temperature. One to five slices from each rat were used in the experiment.

Sholl Analysis

Sholl analysis was performed on adaptively thresholded maximal projections of Z-stacks, where each XY-plane has been filtered with anisotropic diffusion filtering. All processing steps were performed using image-funcut library [image-funcut, https://github.com/abrazhe/image-funcut] and other custom-written Python scripts, using Scikit-Image [scikit, http://scikit-image.org/] and Sci-Py [scipy, http://www.scipy.org/] libraries (Van Der Walt et al., 2014). The step-by-step procedure is summarized in Supplementary Figure 2. Sholl metric was calculated automatically as a number of intersections of circles with centers at the soma and increasing radii with the thresholded mask obtained as described above.

Shearlet-Based Estimate of Spatial Complexity and Entropy for 2D Patterns

A spatial pattern can be characterized by a pair of statistical properties, namely entropy and statistical complexity (López-Ruiz et al., 1995). An ordered (e.g., periodic) structure with a single spatial scale and preferred feature orientation will have both low entropy and small statistical complexity, as the structure in any part of the system can be reconstructed from a small area. At the other end of the complexity-entropy spectrum, where the state is disordered with no spatial correlations, the entropy of the system will be maximal, while the complexity will again be low (the spatial pattern has the same local statistics). Intermediate cases with high statistical complexity are of more interest, as they represent systems with non-trivial regularities and underlying structure embedded in randomness. We developed an algorithm to map local entropy and complexity values for biologically relevant structures using shearlet transform to induce local probability densities of scale and orientation and Jensen-Shannon divergence to define statistical complexity. Below we describe the two points in more detail.

Entropy and Statistical Complexity

Both entropy and complexity (entropic non-triviality) measures for a 2D pattern were defined statistically for a distribution of spatial features, such as orientation, or scale. Here we denoted such a distribution as

$$P := \{P_i\} \tag{1}$$

for a set of features $i = 1 \ldots N$. Then entropy was defined simply as Shannon information entropy

$$S[P] = -\sum_i P_i \, log_2 \, P_i. \tag{2}$$

Entropy will have its maximum for the equiprobable distribution of all features P_e,

$$S[P_e] = S_{\max} = 2N, \tag{3}$$

where N is a number of possible states or features. This allows to introduce normalized entropy:

$$H_s[P] := S[P]/S[P_e], H_s[P] \in [0, \ldots 1] \tag{4}$$

Following (López-Ruiz et al., 1995) we used the disequilibrium-based complexity measure

$$C[P] := Q[P, P_e]H_s[P] \tag{5}$$

i.e., the one based on the statistical distance between the observed (P) and equiprobable (P_e) distributions. Here, following (Lamberti et al., 2004; Rosso et al., 2007), we employed normalized Jensen-Shannon divergence

$$Q_{JS} = J[P, P_e]/J_{\max} \tag{6}$$

as a measure of distance between two distributions, where Jensen-Shannon divergence is defined as

$$J[P, P_e] = S\left[\frac{P + P_e}{2}\right] - \frac{1}{2}(S[P] + S[P_e]). \tag{7}$$

Clearly, $J[P, P_e] = 0$ if $P = P_e$ and reaches its maximum when only one feature, say m, is present, while all others are absent: $P_i = 1 | i = m$, and $Pi = 0 | i \neq m$.

Shearlet Transform

Shearlet transform provides a convenient probability density function for spatial entropy and complexity estimates, describing local prevalence of structures with some specific scale and orientation. We used fast finite discrete shearlet transform (FFST) described in detail by (Häuser and Steidl, 2013). Here we provide a minimally sufficient description of the FFST and its use in calculation of spatial entropy and complexity.

Discrete shearlet transform was based on convolving the digital 2D image

$$I(x, y) \in R^{(N,N)} \tag{8}$$

with scaled, sheared, and shifted copies of a "mother" shearlet function ψ, thus accounting for different scales and orientations of features contained in the image; one uses the dilation matrix A and shear matrix S to create the sheared, scaled and shifted copies of the mother wavelet ψ_x:

$$\psi_{a,s,t} = a^{-3/4}(Aa^{-1}Ss^{-1}(x - t)). \tag{9}$$

Thus, the scaled and shared copies of ψ pick up dominant anisotropic features at different spatial scales and orientations. In the discrete transform, one uses a fixed number of decomposition scales and shifts as well as scale-dependent number of orientations (more orientations at higher spatial frequencies). Finally, shearlet decomposition of image was given by shearlet coefficients

$$T(I)(j, k, m) = \langle I, \psi_{j,k,m} \rangle \tag{10}$$

where discrete shearlet $\psi_{j,k,m} = \psi_{ajsj,ktm}(x)$ is the shearlet at discrete scale α_j, shear $s_{j,k}$ and shift t_m. Thus, $T(I)$ is a set of K images of the same size as $I(x,y)$, where the value at a specific (x,y) location in the k-th image represents the shearlet coefficient at some specific scale j and shear s.

Following ideas from wavelet entropy (Rosso et al., 2001) and earlier of spectral entropy of (Powell and Percival, 1979), in each location of the studied 2D image $I(x,y)$, we defined $P(x, y) = Pk(x, y)$ as normalized power of the shearlet coefficients at this point:

$$P_k(x, y) = E_k(x, y)/\sum_j E_j(x, y) \tag{11}$$

thus, interpreting a spectrum of local feature scales and orientations as a density function. Here $(K^*_\sigma \cdot)$ denotes convolution with a Gaussian kernel with scale-dependent standard deviation σ_j.

Volume Fraction (VF) of Astrocytic Leaflets

To calculate the VF of the fine process of the astrocyte, we followed a similar method described by Heller and Rusakov (2015) A line of 45 μm length were drawn from the soma on a single Z plane of the stack. Spatial attention was paid to ensure that fluorescence of soma was not saturated. The estimated VF was calculated with the following:

$$GV(i, j) = (F(i, j) - F_0)/(F_{max} - F_0) \tag{12}$$

where $F(I,j)$—the fluorescent in particular pixel of the line, F_{max}—the fluorescence of soma, F_0-the background fluorescence. F_0 was obtained in image area which had no stained astrocytes.

Astrocyte Coupling Analysis

The astrocytes were loaded with 50 μM Alexa Fluor 594 through the patch pipette for 30 min. then Z-stack two-photon images was obtained (emission band-pass filter 565–610 nm, 512 x 512 pixels). The images were then denoised with block matching 4D (BM4D) free scrip for MATLAB (Maggioni et al., 2013; Danielyan et al., 2014). The distance to neighboring astrocytes coupled to the target astrocyte through gap-junctions was calculated in 3D-space using Pythagorean theorem with custom-written MATLAB script. Fluorescence Intensities of all coupled cells were normalized to fluorescence of soma of the patched astrocyte. The relationship between distance fluorescence of coupled astrocyte and distance to this astrocyte was fitted by monoexponential function to obtain coupling length constant (C_λ) (Anders et al., 2014):

$$I(d) = I_0 exp(-d/C_\lambda), \tag{13}$$

were, d—distance, I_0—the normalized fluorescence intensity of the coupled cell.

Electrophysiological Recordings

Whole-Cell Recording

Whole-cell voltage-clamp and current-clamp were performed with Multiclamp 700B amplifier (Molecular Devices). The CA1 *str.radiatum* astrocytes were visualized with BX51WI (Olympus) or Axio Examiner Z1 (Zeiss) microscope equipped with infrared differential interference contrast. Borosilicate patch pipettes (Resistance 3–5 MΩ) were filled with internal solution containing (in mM): 130 KCH_3SO_3, 10 HEPES, 10 Na_2-phosphocreatine, 8 NaCl, 3 l-ascorbic acid, 2 Mg-GTP (pH adjusted to 7.2, osmolarity of 295 ± 3 mOsm). For simultaneous two-photon imaging, 50 μM Alexa Fluor 594 was added to the internal solution.

Bipolar extracellular tungsten electrode (FHC) was placed in *str. radiatum* between CA1 and CA3 areas. Once whole-cell configuration was obtained, the cell was dialyzed for 5 to 10 min before the start of recording. In voltage clamp recordings the astrocytes were held at−80 mV. Voltage steps were applied to obtain current-voltage (I-V) relationship. In current clamp, current steps were applied to corroborate the absence of membrane excitability. Cycles of 1, 4, and 5 electrical stimuli (100 ms, 50 Hz) were applied to Schaffer collaterals. The intensity of stimulation was set to induce synaptic currents in astrocytes of 20 to 50 pA to a single stimulus. Series and input resistances were continuously monitored by a voltage step of−5 mV after each cycle. Signals were sampled at 5 kHz and filtered at 2.5 kHz.

Passive astrocytes were taken at 100–200 μm from the stimulating electrode. They were identified by small soma (5–10 μm), low resting membrane potential (∼−80 mV), low input resistance (<20 MΩ), and linear I-V relationship. Cells with similar characteristics except for higher input resistance (>50 MΩ) were considered NG2 or complex cells and were excluded from the study.

Membrane currents were analyzed using custom-written MATLAB scrips (MathWorks R2016a). Synaptic currents of 1, 4, and 5 stimuli were baseline subtracted and then averaged. I_K (K^+ current) was measured 200 ms after the stimulus. At this time point I_K was not contaminated by the current mediated by field potential and transporter current. From this point the decay was fitted with mono-exponential function and τ_{decay} calculated. To obtain I_K in response to 5th stimulus, the response to 4 stimuli was subtracted from the response to 4 stimuli.

Field Potential Recording

Field excitatory postsynaptic potentials (fEPSPs) were recorded from CA1 *str. radiatum* using glass microelectrodes (0.2–1.0 MΩ) filled with ACSF. Synaptic responses were evoked with extracellular stimulation of the Schaffer collaterals using a bipolar twisted stimulating electrode made of insulated nichrome wire placed in the *str. radiatum* at the CA1–CA2 border. The stimulation was performed with rectangular paired pulses (duration, 0.1 ms; interstimulus interval, 50 ms) every 20 s via an A365 stimulus isolator (WPI). Responses were amplified using a microelectrode AC amplifier model 1800 (A-M Systems) and were digitized and recorded on a personal computer using ADC/DAC NI USB-6211 (National Instruments) and WinWCP v5.2.2 software by John Dempster (University of Strathclyde). Electrophysiological data were analyzed with the Clampfit 10.2 program (Axon Instruments).

The dependence of field response amplitude on stimulation strength was determined by increasing the current intensity from 25 to 300 μA. For each fEPSP, the amplitude and the slope of the rising phase at a level of 20–80% of the peak amplitude were measured. The presynaptic fiber volley (PrV) was quantified by the peak amplitude. The maximum rise slope of the input-output (I/O) relationships (fEPSP amplitude vs. PrV amplitude) was calculated for every slice by fitting with a sigmoidal Gompertz function (Equation 14) using OriginPro 8 (OriginLab Corporation).

$$y = ae^{-e^{(-k(x-x_c))}}, \tag{14}$$

where a is an asymptote of the maximum fEPSP amplitude; e is Euler's Number ($e = 2.71828\ldots$); k and x_c are positive numbers describing the shape of the curve; x_c is the PrV amplitude at which the maximum slope of the curve is observed; ak/e is a maximum slope of the curve.

The stimulus intensity used in the experiment was chosen so that the amplitude of fEPSP would be 40–50% of the amplitude where the population spike appeared for the first time. The strength of stimulation was unvaried during the experiments, usually being 50–150 μA. The paired-pulse ratio (PPR) was measured as a ratio of the second and first fEPSP amplitude.

The LTP induction was started only if a stable amplitude of the baseline fEPSP had been recorded for 20 min. Three trains of high-frequency stimulation (HFS, 100 pulses at 100 Hz, with an inter-train interval of 20 s protocol) was applied to induce

LTP. The fEPSPs were recorded after induction protocol during 60 min. The baseline fEPSP and the potentiated fEPSP (recorded 47–60 min after HFS) were averaged separately to measure LTP in a slice. Plasticity value was calculated as a ratio of the slope of the rising phase in the averaged potentiated and baseline fEPSP.

Ca^{2+} Imaging

Ca^{2+} activity was recorded with a confocal microscope, Zeiss LSM DuoScan 510, in CA1 *str.radiatum* of acute hippocampal slices pre-incubated with Ca^{2+} dye, Oregon Green 488 BAPTA-1 AM (Invitrogen) and an astrocyte specific marker, sulforhodamine 101 (100 nM, Invitrogen). After the preparation, the slices were transferred to a 3 ml incubation chamber with constantly gassed ACSF containing both dyes. Oregon Green 488 BAPTA-1 AM was initially dissolved to 0.795 mM in 0.8% Pluronic F-127 in DMSO. Then 3 μl of the dye was added to the chamber. After incubation for 40–45 min at 37°C in the dark, the slices were transferred to the recording/imaging chamber for time-lapse imaging (one frame/s). Oregon Green 488 BAPTA1 was excited with a 488 nm argon laser and imaged with an emission band-pass filter 500–530 nm; sulforhodamine 101 was excited with a 543 nm HeNe laser and imaged with an emission band-pass filter 650–710 nm. The imaging was performed for 10 min at 34°C in normal ASCF, then 30 dark noise images were recorded.

The raw imaging data were exported to MATLAB. The median of the dark noise was calculated for each pixel and subtracted from the corresponding pixel intensity value of the fluorescence images. Then denoising was done with the BM3D algorithm (Danielyan et al., 2014). The movement artifacts were corrected with the single-step DFT algorithm (Guizar-Sicairos et al., 2008). The whole Ca^{2+} events (x-y-time 3D Ca^{2+} signals) were detected with the adapted algorithm which we described previously (Wu et al., 2014). Briefly, each pixel of the image series was analyzed independently. Firstly, we roughly estimated a baseline fluorescence F_{0temp} applying 60-s 3rd order Savitzky-Golay polynomial filter which smoothed all Ca^{2+} signals on the fluorescent signal F. Then, we estimated a temporary $(\Delta F/F)_{temp} = (F - F_{0temp}) / F_{0temp}$ to find Ca^{2+} transients exceeding a statistical threshold. Then these transients were excluded from the baseline which was further smoothed with 100-s filter. This filter interpolated the intervals left by the excluded transients, and, thus, we obtained the uninterrupted final baseline F_0 which was used to obtain $\Delta F/F = (F - F_0) / F_0$. Ca^{2+} transients exceeding a statistical threshold were detected and binarized. The active neighboring pixels were grouped into x-y 2D Ca^{2+} events, which were reconstructed into x-y-time 3D Ca^{2+} events. For each Ca^{2+} event the maximal projection (S_{max}), the integral and the duration were calculated. To avoid noise detection the events excluded from further analysis if the integral was less than 4 μm^2 s, or the S_{max} was less than 10 μm^2, or the duration less than 2 s.

The probability density of the events sizes and the durations appeared linear in log-log scale, suggesting that the obtained distributions can be described by a power law. Therefore, the power law fit was applied, and the corresponding exponent was calculated for each slice.

Statistical Analysis

All data are presented as mean ± standard error of mean (SEM). Statistical significance was assessed using non-parametric Mann-Whitney test, parametric Student's *t*-test and repeated measures two-way ANOVA as stated in the text. $P < 0.05$ was considered statistically significant.

RESULTS

Rat hippocampal slices were prepared 2–4 weeks after pilocarpine-induced SE when the animals typically started to develop spontaneous seizures. Nissl staining confirmed statistically significant neurodegeneration in pyramidal layers of three hippocampal regions: CA1, CA3, and hilus (CA1: 7.9 ± 0.36 cells per 100 μm in control, $n = 6$; 6.0 ± 0.4 cells per 100 μm after SE, $n = 6$; $p = 0.02$; CA3: 8.9 ± 0.5 cells per 100 μm in control, $n = 6$; 6.0 ± 0.4 cells per 100 μm after SE, $n = 6$; $p = 0.01$, Mann-Whitney test; hilus: 7.5 ± 0.6 cells per 100 μm in control, $n = 6$; 5.3 ± 0.5 cells per 100 μm after SE, $n = 6$; $p = 0.04$, Mann-Whitney test; **Figures 1A–C**). No significant changes in the cell density were observed in granular cell layer of the dentate gyrus (17.2 ± 1.7 cells per 100 μm in control, $n = 6$; 15.5 ± 0.7 cells per 100 μm after SE, $n = 6$; $p = 0.37$, Mann-Whitney test; **Figure 1D**). Consistent with previously reported astrogliosis, the density of CA1 *str. radiatum* astrocytes stained with an astrocyte-specific marker, sulforhodamine 101, was significantly higher after SE (1.2 ± 0.1 cells per 100 μm^2 in control, $n = 15$; 1.5 ± 0.1 cells per 100 μm^2 after SE, $n = 20$; $p = 0.02$, Mann-Whitney test; **Figures 1E,F**) (Mazzuferi et al., 2010; Pekny et al., 2016). However, this astrogliosis would be considered rather minor.

Then we performed Sholl analysis on two-photon images of astrocytes loaded trough patch pipette with 50 μM Alexa Fluor 594 (morphological tracer, see Materials and Methods, Supplementary Figure 2 and **Figure 2A**). There was no significant difference in the number of primary branches (connected to soma), the peak number of the branches and size of the astrocytic domain in control and SE animals (Supplementary Figure 3). However, the number of distal branches was significantly lower after SE (control, $n = 6$; SE, $n = 6$; two-way repeated measures (RM) ANOVA, $F_{(1,5)} = 6.862$, $p = 0.047$, partial $\eta^2 = 0.578$, with a mean difference of 5.61 ± 2.14; **Figure 2B**). This morphological rearrangement can also be assessed using the analysis of spatial complexity-entropy spectrum (**Figure 2C**). Spatial complexity and spatial entropy are both low in highly ordered or anisotropic systems. As the system loses the order, both entropy and complexity start to increase. When the elements of the systems become randomly distributed ("noise") the entropy is the highest, while the complexity decreases. Remodeling of astrocytic processes after SE significantly increased both entropy and complexity (entropy: 0.51 ± 0.02 in control, $n = 11$; 0.56 ± 0.001 after SE $n = 8$; $p = 0.009$, Mann-Whitney test; complexity 0.343 ± 0.003 in control; 0.352 ± 0.003 after SE; $p = 0.02$, Mann-Whitney test; **Figure 2D**). This finding suggests that astrocytic

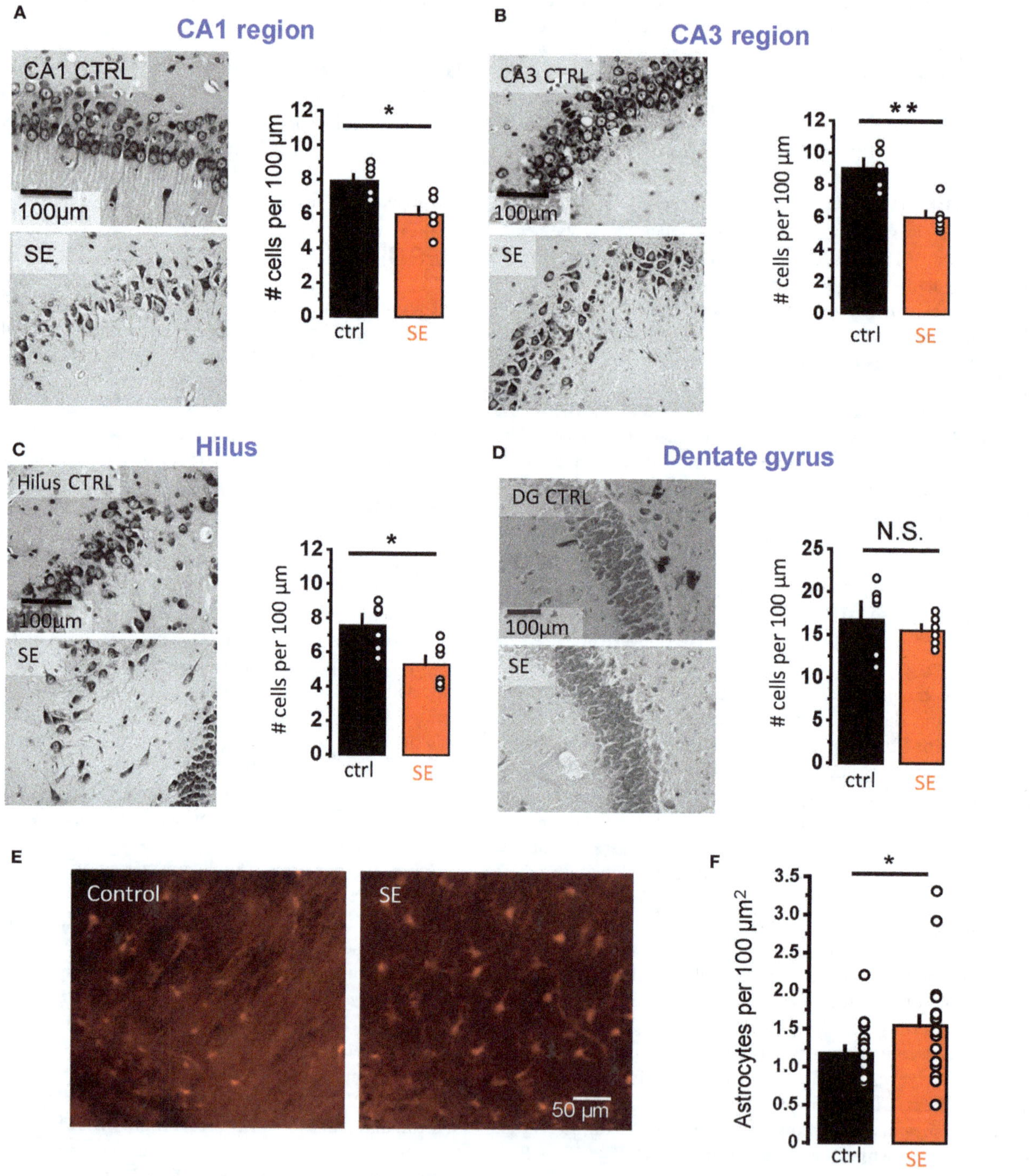

FIGURE 1 | Neurodegeneration and astrogliosis after SE. **(A–D)** Nissl staining showing neurodegeneration in the *str. pyramidale* of CA1 **(A)**, CA3 **(B)**, and hilus **(C)** after SE. No significant neurodegeneration was observed in the granular cell layer of dentate gyrus **(D)**. *Right*, illustrations of stained cells in control (*top*) and SE-rats (*bottom*). *Left*, the summary data from several rats. The cells were counted along the cell layers and normalized to the length of the layers. **(E)** Fluorescent images of astrocytes stained with sulforhodamine 101: in control (*left image*) and SE-rats (*right image*). **(F)** the summary data on several rats. The circles show values in individual rats. The bars with error bars are means ± SEMs. $^{**}p < 0.01$, $^{*}p < 0.05$, N.S. $p > 0.05$ Mann-Whitney test.

processes become less orderly organized after SE. It can be potentially explained by the decrease in the ratio of primary astrocytic branches which are resolved with diffraction-limited light microscopy and thin astrocytic leaflets which appear as a chaotic fluorescent pattern because of their sizes are beyond diffraction-limited light microscopy resolution. To estimate

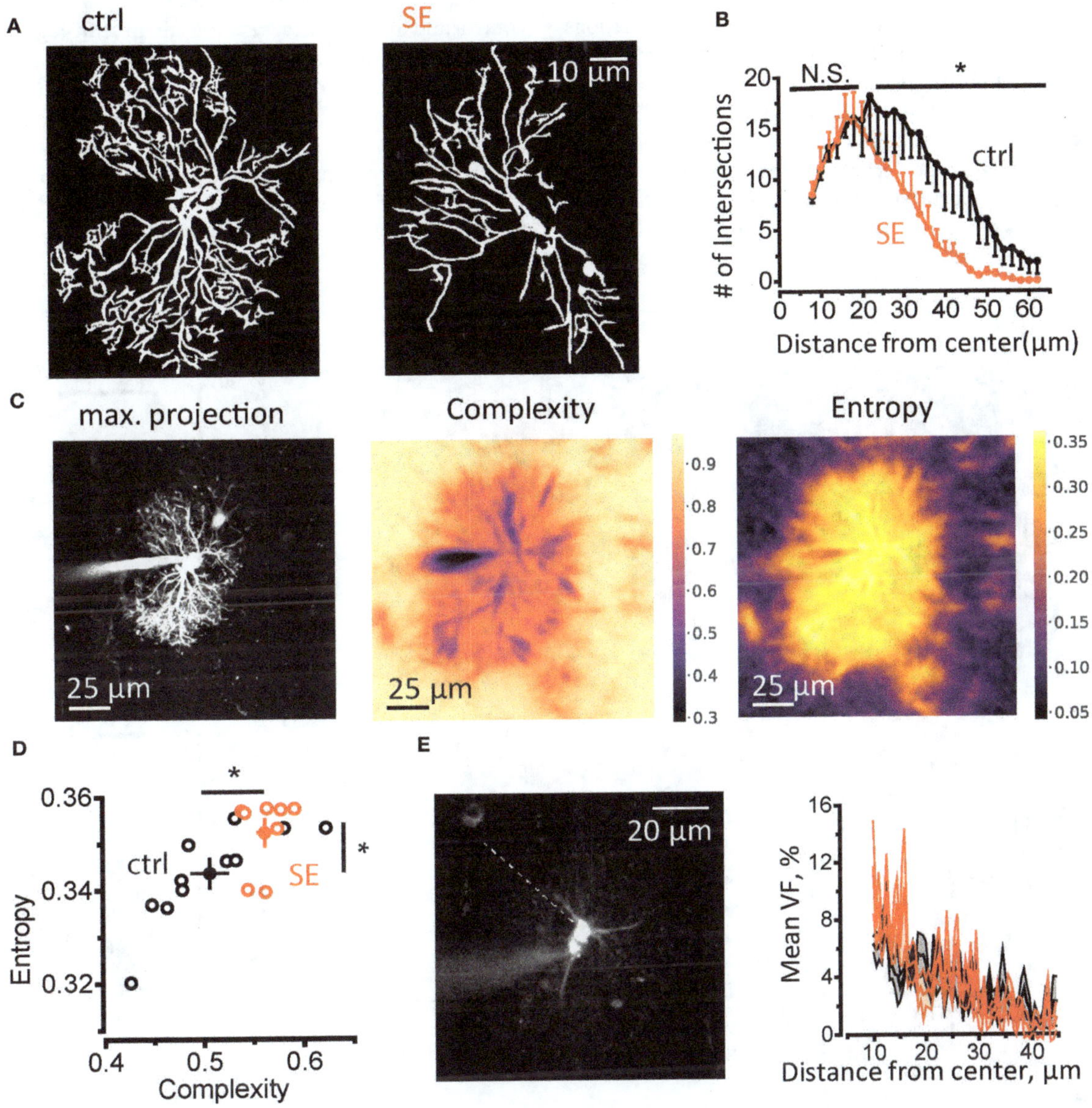

FIGURE 2 | Morphological remodeling of astrocytes after SE. **(A)** Masks of astrocytic branches in control (left) and SE rats (right) which were used for Sholl analysis. The masks were obtained from maximal projections of z-stack of fluorescence images of astrocytes loaded with 50 μM Alexa Fluor 594 through patch pipette. **(B)** The summary data for the number of intersections of circles drawn around center of the astrocyte soma with astrocytic branches. *$p < 0.05$, N.S. $p > 0.05$, two-way ANOVA. **(C)** Spatial entropy-complexity analysis. *Left*, maximal projection of Z-stack of fluorescence images of an astrocyte. *Middle*, The spatial complexity profile of the astrocyte. *Right*, The spatial entropy profile of the astrocyte. **(D)** The summary graph of spatial entropy-complexity pairs of astrocytes on control (black circles) and SE-rats (red circles). Empty circles are individual astrocytes. Filled circles are means ± SEMs. **(E)** Estimation of astrocyte leaflets' VF. *Left*, construction of fluorescence profile across an astrocyte. Dashed line indicates the place there the profile was obtained. It passes through the soma and the area or unresolved processes devoid of identifiable branches. *Right*, the mean ± SEM fluorescent profiles normalized to the fluorescence in soma for control (*black trace*) and SE-rats (*red trace*).

possible changes in leaflets volume fraction, we performed line scan through soma and the area of unresolved astrocytic leaflets avoiding astrocytic branches (**Figure 2E**). This method is based on the assumption that unsaturated fluorescence level in soma corresponds to 100% volume fraction (VF) (Medvedev et al., 2014; Heller and Rusakov, 2015). The ratio between the fluorescence of leaflets area and fluorescence of soma was considered VF of leaflets and did not differ between control and SE rats (mean leaflets VF: 3.4 ± 0.2%, $n = 6$ in control; 3.6 ± 0.2%, $n = 6$ after SE; $p = 0.59$, Mann-Whitney test; **Figure 2E**). The equal VF of leaflets does not, however, rule out their spatial rearrangement after SE which can be only assessed with super-resolution light microscopy or electron microscopy.

Morphological remodeling of astrocytes can be linked to astrocytic uncoupling and disruption of astrocytic syncytium after SE (Wallraff et al., 2006; Bedner et al., 2015). To assess astrocyte coupling, we monitored diffusion of Alexa Fluor 594 from patched astrocyte to its neighbors through gap-junctions (**Figure 3A**). The number of stained astrocytes was significantly lower after SE compared to control (15.3 ± 2.8, $n = 7$, in control; 6.6 ± 2.9, $n = 8$ after SE; $p = 0.03$ Mann-Whitney test; **Figure 3B**). We also estimated the "strength" of gap-junction connections by the decay of fluorescence measured in somas of coupled astrocytes as a function of distance from the astrocyte loaded with Alexa Fluor 594 through patch pipette. The exponential decay of fluorescence was observed both in control and SE rats (linear relationship in semi-logarithmic scale, **Figure 3C**). Although length constant (λ) tended to be smaller after SE, the difference did not reach significance (λ: 33 ± 5 µm, $n = 7$, in control; 26 ± 5 µm, $n = 6$, after SE, $p = 0.07$, Mann-Whitney test, **Figure 3D**).

Decreased coupling of astrocytes through gap-junctions can lead to reduced spatial buffering of K^+ released during synaptic transmission (Wallraff et al., 2006; Shih et al., 2013; Bedner

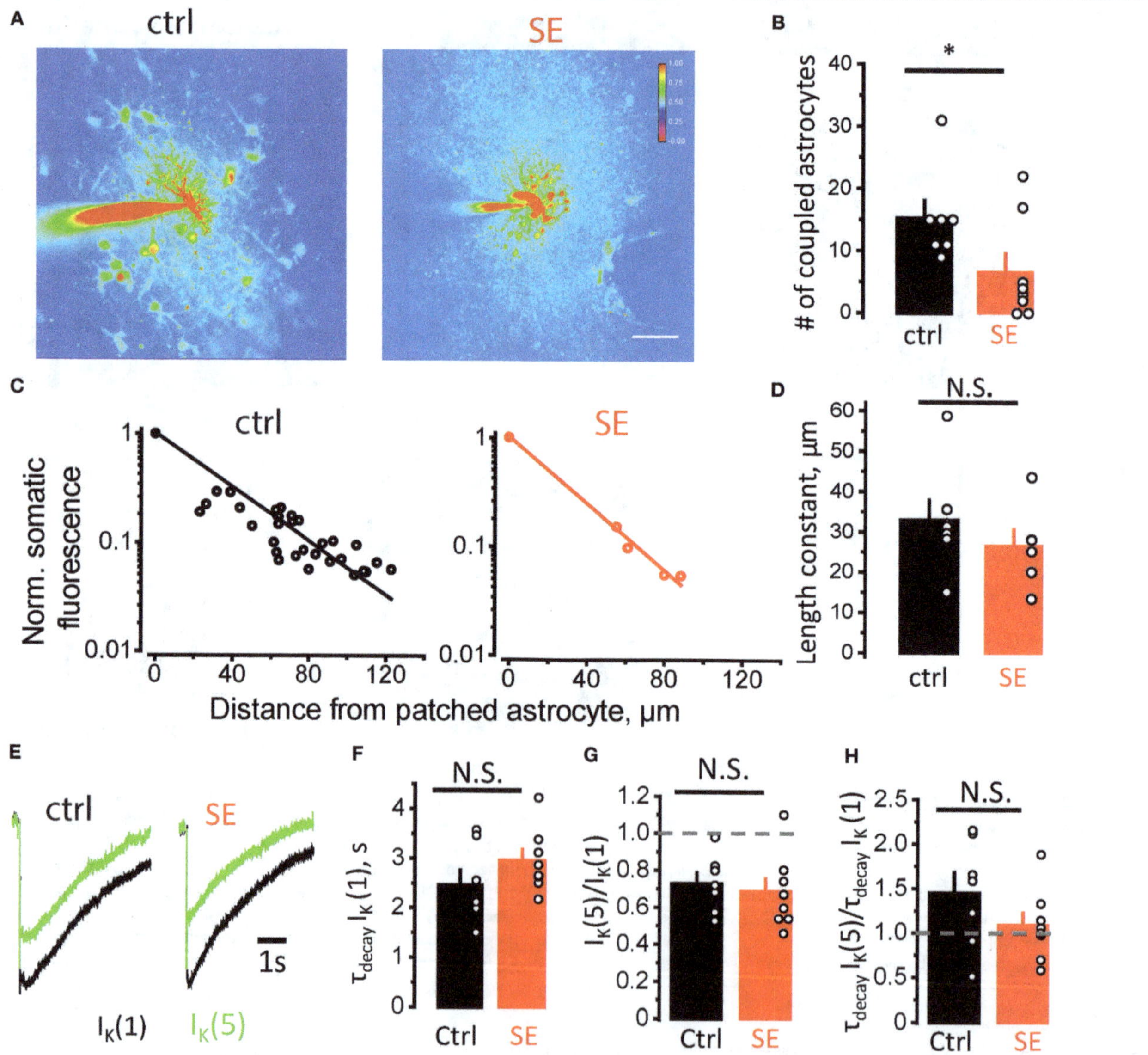

FIGURE 3 | Astrocytic uncoupling through the gap-junctions does not affect K^+ clearance by astrocytes. **(A)** Fluorescence image of an astrocyte stained with 50 µM Alexa Fluor 594 through patch pipette in control (*left*) and SE-rats (*left*). Because the dye goes through gap-junctions, the coupled astrocytes are also stained. The number of stained astrocytes is lower after SE. **(B)** The summary data showing the number of coupled astrocytes in control (*black*) and after SE (*red*). **(C)** Decay of fluorescence in somas of coupled astrocytes with distance from the astrocyte loaded with Alexa Fluor 594. The slope of the linear fit in semi-logarithmic scale determines the length constant. **(D)** The summary data showing the length constant in control (*black*) and after SE (*red*). **(E)** The sample traces of the I_K recorded to the single stimulus [$I_K(1)$, black trace] and to the fifth stimulus [the response to 4 stimuli was subtracted from the response to 5 stimuli, $I_K(5)$, green trace]. **(F)** The summary data of decay time constant of $I_K(1)$ [τ_{decay} $I_K(1)$] in control (*black*) and after SE (*red*). **(G,H)** The summary data of $I_K(5)/I_K(1)$ and τ_{decay} $I_K(5)/\tau_{decay}$ $I_K(1)$ ratios, respectively. The circles show values in individual rats. The bars with error bars are means ± SEMs. *$p < 0.05$, N.S. $p > 0.05$ Mann-Whitney test.

et al., 2015; Cheung et al., 2015; Lebedeva et al., 2018). To address possible changes in K^+ dynamics after SE, we recorded K^+ current (I_K) in CA1 *str.radiatum* astrocytes in response to stimulation of Schaffer collaterals (**Figure 3E**). The astrocytic response consists of three overlapping currents: field potential-induced current, glutamate transporter current, and I_K (Sibille et al., 2014). First two currents are short and typically end within 100 ms, while I_K lasts for several seconds (Afzalov et al., 2013; Lebedeva et al., 2018). Therefore, the amplitude of I_K was measured 200 ms after the stimulus. The I_K was also fitted from this point with mono-exponential function, which was used to calculate the decay time-constant (τ_{decay} I_K). No significant difference was observed in τ_{decay} I_K in response to single stimulus after SE [τ_{decay} $I_K(1)$: 2.5 ± 0.3 s, $n = 7$, in control; 3.0 ± 0.2 s, $n = 8$, after SE; $p = 0.09$, Mann-Whitney test, **Figure 3F**]. This result suggests that K^+ clearance is not affected during single synaptic events after SE, but it may be affected during repeated activity. To address this issue, we stimulated Schaffer collaterals 4 times and 5 times at 50 Hz. Then the response to 4 stimuli was subtracted from the response to 5 stimuli to obtain isolated I_K to 5th stimulus [$I_K(5)$].The ratio of $I_K(5)/I_K(1)$ demonstrated activity-dependent depression of I_K and was not significantly different between control and SE rats [$I_K(5)/I_K(1)$: 0.72 ± 0.06, $n = 7$, in control; 0.68 ± 0.07, $n = 8$, after SE; $p = 0.26$, Mann-Whitney test, **Figure 3G**]. The ratio of τ_{decay} $I_K(5)/\tau_{decay}$ $I_K(1)$ was also not significantly different between control and SE rats [τ_{decay} $I_K(5)/\tau_{decay}$ $I_K(1)$: 1.45 ± 0.23, $n = 7$, in control; 1.09 ± 0.14, $n = 8$, after SE; $p = 0.14$, Mann-Whitney test, **Figure 3H**]. These findings suggest that possible reduction in K^+ spatial buffering in astrocytic syncytium does not affect K^+ clearance during moderate activity of the neuronal network.

Astrocytic atrophy and uncoupling can affect Ca^{2+} signaling in an astrocytic syncytium. To test this hypothesis, we measured Ca^{2+} signals in CA1 *str. radiatum* stained with membrane-permeable Ca^{2+} dye Oregon Green 488 BAPTA-1, AM (**Figure 4A**). Although this dye predominantly stains astrocytes, we could not rule out the contribution of neuronal responses to a fluorescent signal. Fortunately, astrocyte can generate much slower Ca^{2+} signals than neurons (seconds vs. hundreds of milliseconds) (Bazargani and Attwell, 2016). Such long signals are likely to represent only a proportion of overall Ca^{2+} activity in astrocytes and require Ca^{2+} release from endogenous Ca^{2+} stores. Thus, astrocytic Ca^{2+} events can be separated from neuronal based on their duration (Monai et al., 2016). Time-lapse Ca^{2+} imaging was performed with a confocal microscope at the rate of 1 frame per second. Events lasting ≥ 2 s were considered astrocytic (Supplementary Video). Whole (x-y-time) Ca^{2+} events were identified in local astrocytic syncytium as previously described for single astrocytes (**Figure 4B**) (Wu et al., 2014). The frequency of Ca^{2+} events normalized to the imaged area (frequency density) was not significantly different between control and SE rats (frequency density: 5.3 ± 1.8 s^{-1} mm^{-2}, $n = 7$, in control; 3.8 ± 0.8 s^{-1} mm^{-2}, $n = 7$, after SE; $p = 0.5$, Mann-Whitney test, **Figure 4C**). Consistent with the previous report the distributions of event sizes (S_{max}, maximal projection) and of events durations followed a power law (Wu et al., 2014):

$$P(x) \sim x^{-\alpha}, \tag{15}$$

where *P*–probability, *x*–analyzed parameter (S_{max} or duration), α–power law exponent. α defines mean value and standard deviation of the sample. Smaller α suggests that distribution has higher proportion of larger or longer events, and vice versa. To calculate α the sample was log-binned and fitted with power function. The α of S_{max} was significantly larger after SE, suggesting a smaller proportion of large Ca^{2+} events in the distribution [$\alpha(S_{max})$: 2.7 ± 0.3, $n = 7$, in control; 3.5 ± 0.2, $n = 7$, after SE; $p = 0.015$, Mann-Whitney test, **Figure 4D**]. No significant difference in α of durations was observed after SE [α(durations): 2.75 ± 0.23, $n = 7$, in control; 3.02 ± 0.24, $n = 7$, after SE; $p = 0.15$, Mann-Whitney test, **Figure 4E**]. This finding suggests that morphological changes in astrocytes correlate with reduction of large-sized Ca^{2+} events in astrocytic syncytium after SE.

Ca^{2+} activity in astrocytes is responsible for a number of functions including the release of gliotransmitters such as *D*-serine (Papouin et al., 2017). *D*-serine acts as a co-agonist of NMDA receptors and is required for long-term potentiation (LTP) (Henneberger et al., 2010, 2012). Thus, the decrease of astrocytic Ca^{2+} activity can reduce the amount of *D*-serine released and thus impair LTP after SE (Sherwood et al., 2017). To test this hypothesis, we recorded field (f)EPSPs in CA1 *str.radiatum* in response to stimulation of Shaffer collaterals (**Figure 5A**) The relationship between fEPSP amplitude and stimulus strength has significantly decreased after SE [$F_{(11,297)} = 4.39$, $p < 0.001$, two-way ANOVA; **Figures 5B,C**], but no significant difference in relationship between presynaptic fiber volley (PrV) and stimulus strength was observed [$F_{(11,275)} = 0.37$, $p = 0.97$, two-way ANOVA; **Figures 5B,C**]. The maximum rise slope of the input-output (I/O) relationships (fEPSP amplitude vs. PrV amplitude) was lower after SE (control: 3.32 ± 0.32, $n = 13$, SE: 1.61 ± 0.33, $n = 14$, $p < 0.01$, *t*-test; **Figure 5C**). These results suggest that the number of fibers or excitability of Shaffer collaterals did not change (unless the decreased number of fibers is compensated by their higher excitability), but the number of activated synapses decreased after SE. This result is consistent with neurodegeneration observed after SE. 10 μM *D*-serine changed the relationship between fEPSP amplitude vs. stimulus strength both in control [$F_{(11,231)} = 2.23$, $p < 0.05$, two-way ANOVA; **Figures 5B,C**] and after SE [$F_{(11,330)} = 2.13$, $p < 0.05$, two-way ANOVA; **Figures 5B,C**], however, *post hoc* LSD test did not confirm the effect of *D*-serine at any level of stimulation.

Three trains of high-frequency stimulation (HFS, 100 pulses at 100 Hz, with an inter-train interval of 20 s) were applied to induce LTP (Henneberger et al., 2010). The LTP magnitude was significantly lower after SE compared to control animals (1.73 ± 0.12, $n = 14$, in control 50–60 min after induction; 1.29 ± 0.07, $n = 15$, after SE; $p < 0.01$, *t*-test; **Figures 5D–G**). *D*-serine did not significantly affect the magnitude of LTP in control ($p = 0.48$, *t*-test; **Figures 5D–G**), but restored initial phase of LTP (5–15 min) after SE (1.92 ± 0.17, $n = 13$) to the level of control animals (1.99 ± 0.13, $n = 14$, *t*-test = 0.35, $p = 0.73$). LTP enhancement by exogenous *D*-serine in a later phase of LTP (50–60 min) was not however significant (1.37 ± 0.14, $n = 13$, *t*-test = 0.54, $p = 0.60$, **Figure 5G**).

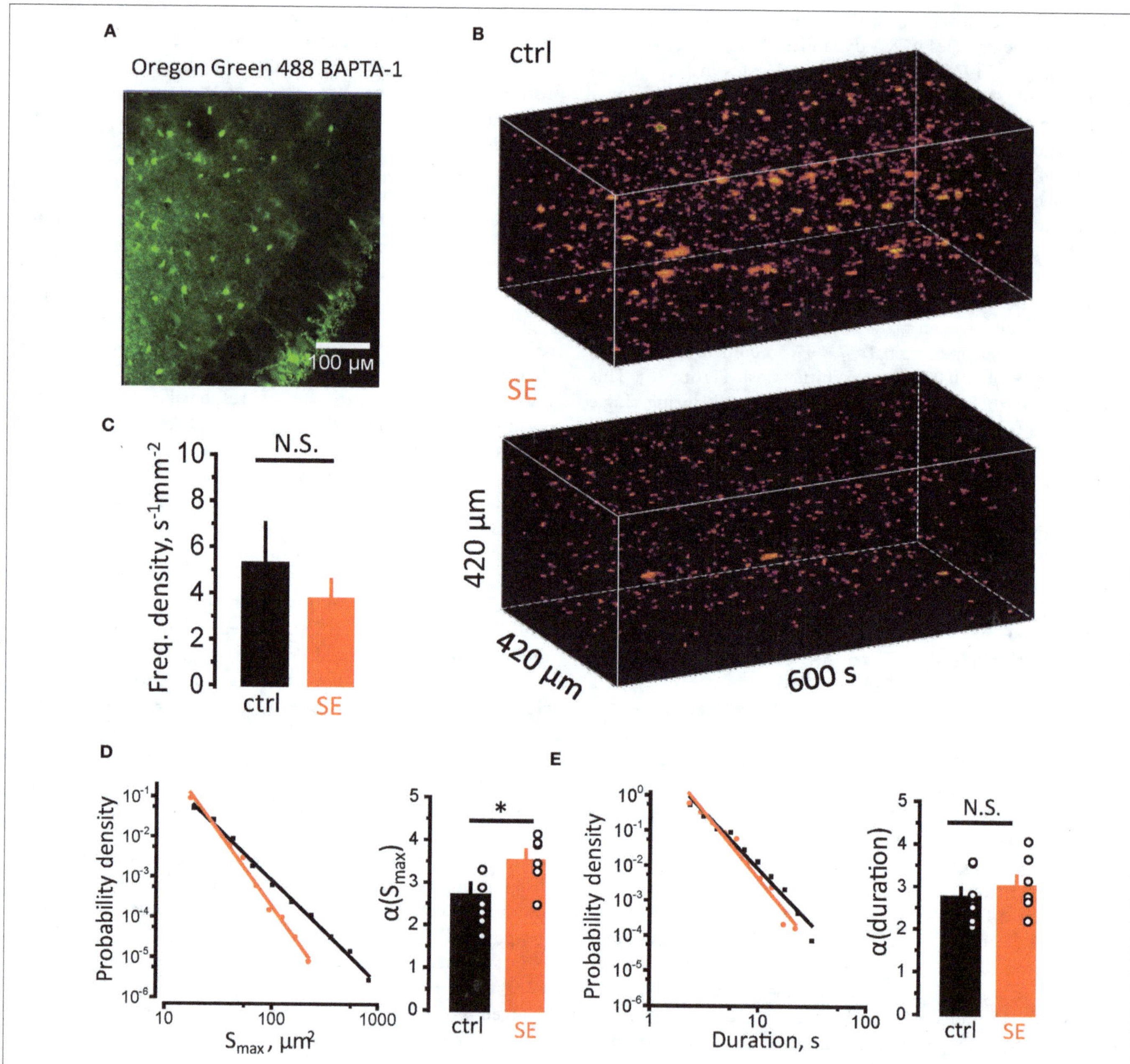

FIGURE 4 | Reduction in sizes of spontaneous Ca^{2+} events in astrocytic syncytium. **(A)** Hippocampal slice stained with Oregon Green 488 BAPTA-1 AM. The image shows that astrocytes in *str.radiatum*, but not neurons in *str.pyramidale* stained. **(B)** 3D reconstruction of Ca^{2+} events in *str.radiatum* astrocytes (x-y-time). *Top*, Time-course of Ca^{2+} events in control slice. *Bottom*, in slice after SE. **(C)** The summary data of the frequency density of astrocytic Ca^{2+} events in control (black) and SE-rats (red). **(D,E)** *Right*, Probability density distributions of S_{max} **(D)** and durations **(E)** of Ca^{2+} events in log–log scale (log-binned data) for all recorded slices. Solid black and red lines are the power function fits for control and SE data, respectively. *Left*, The summary data of the exponents $\alpha(S_{max})$ **(D)** and α(durations) **(E)**. The circles show values in individual rats. The bars with error bars are means ± SEMs. $^{*}p < 0.05$, N.S. $p > 0.05$ Mann-Whitney test.

Our findings suggest that neurodegeneration 2–4 weeks after pilocarpine-induced SE is accompanied by re-modeling of astrocytes and reduction of their coupling through gap-junctions. It does not affect the ability of astrocytes to clean up extracellular K^+ despite the expected decrease in its spatial buffering. Decreased number of astrocytic branches correlates with reduced sizes of spontaneous Ca^{2+} events in astrocytes. This in turn may be responsible for *D*-serine deficiency and impaired LTP after SE.

DISCUSSION

Astrocytic processes often comprise of thicker astrocytic branches and thinner, nanoscale leaflets (Bernardinelli et al.,

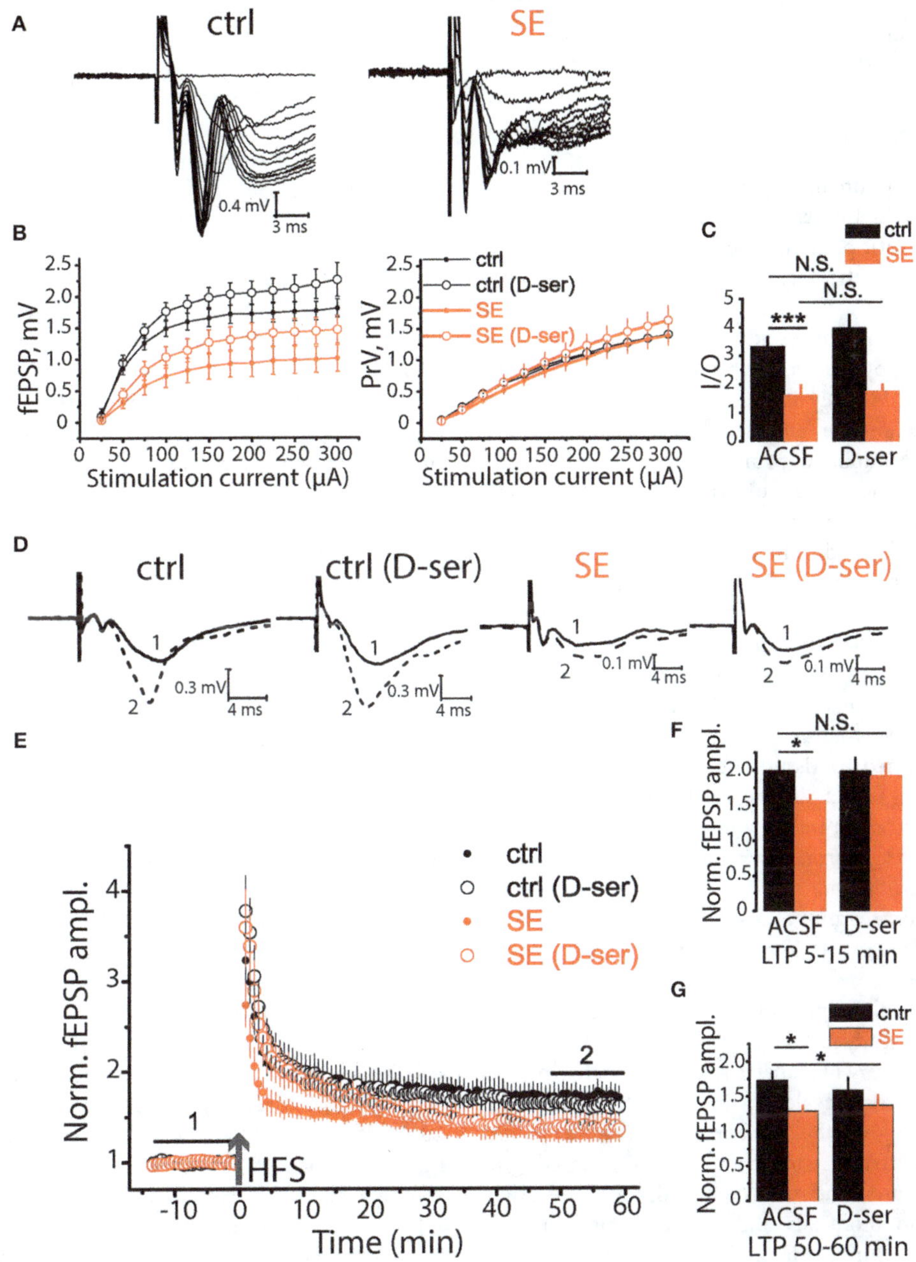

FIGURE 5 | Effects of D-serine on synaptic neurotransmission in the hippocampus of control and SE-rats. **(A)** Representative examples of fEPSPs recorded in the hippocampal CA1 of control (ctrl) and SE-rats (SE). **(B)** Input-output (I/O) curves for the fEPSPs (*left*) and PrV (*right*). **(C)** Bar graphs showing the maximum slope of I/O curves for the fEPSPs. **(D)** Traces are the average of fEPSPs recorded during baseline (1) and 47–60 min after HFS (2). **(E)** LTP induced by HFS in the CA1 region of the hippocampus of control and SE-rats. Note that application of D-serine (D-ser) does not affect the magnitude of LTP in control rats but fully restored initial phase of LTP (5–15 min) in SE-rats. **(F,G)** The summary bar graphs showing significant reductions in LTP after SE as measured by the average normalized fEPSP amplitude and the effect of D-serine on initial **(F)** and later **(G)** phase of LTP. The bars with error bars are means ± SEMs. $^{***}p < 0.01$ $^{*}p < 0.05$, N.S. $p > 0.05$ t-test.

2014a; Khakh and Sofroniew, 2015). Electron microscopy studies have suggested that thicker astrocytic branches contain endoplasmic reticulum and mitochondria serving as endogenous Ca^{2+} stores (Reichenbach et al., 2010; Patrushev et al., 2013; Bernardinelli et al., 2014b). These structures are thought to be responsible for the amplification of Ca^{2+} signals and their propagation within the astrocyte. The leaflets are thin, sheet-like structures connected to branches. They have the minimal volume of cytoplasm and seem to be largely devoid of organelles. The high surface-to-volume ratio of leaflets ensures the high surface of their plasma membrane within the limited volume of brain tissue (Lehre and Rusakov, 2002). This makes leaflets highly efficient for neurotransmitter uptake and K^+ clearance. Perisynaptic leaflets form an "astrocytic cradle" around synapses

(Verkhratsky and Nedergaard, 2014). Like dendritic spines, astrocytic leaflets are highly plastic; morphological plasticity of perisynaptic leaflets regulates synaptic coverage (Witcher et al., 2007; Reichenbach et al., 2010; Bernardinelli et al., 2014a; Heller and Rusakov, 2015). For example, reduced IP_3-dependent Ca^{2+} activity in astrocytes leads to reduced coverage of spines by astrocytic processes and enhanced glutamate spillover (Tanaka et al., 2013). Changes in the astrocytic environment are also reported following synaptic plasticity (Bernardinelli et al., 2014a).

Our morphological analysis revealed a striking difference between SE and control astrocytes. The Sholl analysis was performed on astrocytic processes identified with two-photon laser scanning microscopy (astrocytic branches ~ >0.5 μm in diameter). Although this analysis did not reveal a difference between the samples in the enclosing radius (size of the astrocytic domain) and number of primary astrocytic branches, a significant reduction in the number of distal astrocytic branches was detected. However, astrocytic leaflets are very thin and are beyond the diffraction-limited light microscopy resolution. To assess possible changes in leaflets VF, we plotted a profile of leaflets fluorescence normalized to the fluorescence of soma (Medvedev et al., 2014; Heller and Rusakov, 2015). This method assumes that, with a ~1 μm thick two-photon excitation plane, somatic fluorescence originates from the dye that occupies ~100% of the visualized volume whereas fluorescence of the leaflets area is determined by tissue volume fraction occupied by the cytoplasm of all leaflets in the excitation region. Thus, the ratio between the non-saturated soma fluorescence and the fluorescence of leaflet area is roughly proportional to the VF of leaflets. No significant difference in this ratio was found between astrocytes of SE and control animals. Nevertheless, equal VFs of leaflets do not rule out nanoscopic redistribution of leaflets that affect synaptic coverage.

To get further insights into the spatial organization of astrocytic processes, we advanced a novel, spatial complexity-entropy spectra analysis. This analysis evaluates the nature of order and structure prevalence in a system. Entropy monotonically increases as the system goes from a highly ordered crystalline-like structure to a completely disordered state lacking internal structure. Statistical complexity, on the other hand, changes non-monotonically and reaches a maximum at intermediate levels of entropy, falling near zero for the two extremes of orderliness in a system. Intuitively, the complexity of regular patterns is low due to its predictability and the complexity of a disordered state is low due to the simplicity of its statistical description. Between the two extremes, a given entropy level can correspond to a range of complexity levels. Systems near critical states or in dynamical chaos regimes are characterized by high levels of complexity (Rosso et al., 2007). In such systems, disorder is non-trivial, and the systems contain both structure and randomness. Branching patterns of astrocytic processes are an example of such a non-trivial structure, where large astrocytic branches correspond to locally ordered structures, whereas fine ramifications of leaflets are seen as more complex and chaotic. Thus, entropy-complexity spectra will reflect the ratio between astrocytic branches and leaflets. Consistent with other morphological analysis we observed a shift in entropy-complexity spectra of astrocytes corresponding to a decrease in branches/leaflets ratio.

Astrocytic coupling through gap junctions has been suggested as a mechanism for K^+ "spatial buffering" in the central nervous system (Kofuji and Newman, 2004; Meeks and Mennerick, 2007). K^+ released during action potential propagation and during activation of postsynaptic glutamate receptors is removed by astrocytes through various mechanisms (Walz, 2000; Dallérac et al., 2013; Shih et al., 2013; Pekny et al., 2016; Lebedeva et al., 2018). Once taken by astrocytes, K^+ quickly re-equilibrates within an astrocytic syncytium through gap-junctions which are permeable to this ion. One counterargument to this hypothesis is that intracellular K^+ concentration does not considerably change due to the removal of K^+ from narrow intercellular cleft to the large volume of cytoplasm. To test if astrocytic remodeling and uncoupling through gap-junctions affect K^+ clearance, we recorded I_K in astrocytes in response to synaptic stimulation. No significant difference in the amplitude and the timecourse of I_K was detected between SE and control animals. Thus, in conditions of synchronous synaptic activity, K^+ clearance did not change after SE. It rasises the question whether gap junction suppression could contribute significantly for excessive K^+ accumulation in epileptogenesis.

Astrocytic remodeling may occur as a result of neurodegeneration which is pronounced 2 weeks after SE (Curia et al., 2008). Fewer neurons make fewer synaptic connections which need less astrocytic processes to support them. Reduction in a number of distal astrocytic branches and gap-junction coupling may also decrease Ca^{2+} events spread in single astrocytes and astrocytes syncytium. Spreading of Ca^{2+} events depends on Ca^{2+} release from endogenous stores. When some of the distal astrocytic branches containing Ca^{2+} stores are abolished, Ca^{2+} transients starting in leaflets have less chance to be amplified after SE. Astrocytic gap-junction may play a role in the propagation of Ca^{2+} signals in astrocytic syncytium (Enkvist and Mccarthy, 1992; Höfer et al., 2002; Fujii et al., 2017). When astrocytes become uncoupled, Ca^{2+} events spread may be limited. This, however, does not exclude a possibility of Ca^{2+} activity spread by the release of gliotransmitters (e.g., ATP) (Guthrie et al., 1999; Cotrina et al., 2000). Overall, we found that morphological changes correlated with the reduced spread of Ca^{2+} events in astrocytic syncytium observed after SE. Diminished Ca^{2+} signaling in astrocytes may affect several astrocytic functions such as release of gliotransmitters (Zorec et al., 2012; Araque et al., 2014), Ca^{2+} dependent K^+ clearance (Wang et al., 2012) or outgrow of perisynaptic astrocytic processes (Tanaka et al., 2013; Heller and Rusakov, 2015). This can not only affect signaling in astrocytic syncytium, but also neuronal excitability, activation of extrasynaptic receptors, synaptic transmission, and plasticity (Rusakov et al., 2011; Zorec et al., 2012; Verkhratsky and Nedergaard, 2018).

In the present study, we have demonstrated hippocampal synaptic dysfunctions in pilocarpine-treated rats, as assessed by the reductions in both basal transmission and LTP at Schaffer collateral-CA1 synapses. The decrease in basal transmission is consistent with a reduction in the number of principal neurons in the CA3 hippocampal area of pilocarpine-treated animals observed earlier by us and other investigators in different hippocampal areas (see for review Curia et al., 2014). LTP induction at Schaffer collateral-CA1 synapses is an NMDA receptor-dependent form of synaptic plasticity (Malenka and Bear, 2004) and, therefore, the impairment of LTP most likely indicates the disturbance in NMDA receptor signaling. NMDA receptor signaling might be disturbed in many ways in epilepsy. For example, it might be affected because of the perturbations in the glutamate-glutamine cycle, such as increased extracellular levels of glutamate, loss of astrocytic glutamine synthetase, and changes in glutaminase and glutamate dehydrogenase, which are frequently encountered in patients with epilepsy (Coulter and Eid, 2012; Eid et al., 2016). Another possible reason is the changes in the production of individual subunits of NMDA receptors, which have been shown in several experimental models of epilepsy including the lithium-pilocarpine model (Di Maio et al., 2013; Müller et al., 2013; Peng et al., 2016; Amakhin et al., 2017). Hippocampal astrocytes in brain slices retain the ability to control LTP within or near their individual territories involving Ca^{2+}-dependent *D*-serine release (Henneberger et al., 2010). This implies that astrocytes are at least capable of regulating local *D*-serine supply and might indeed be able to deliver *D*-serine to specific NMDA receptor populations. Therefore, our next set of experiments was designed to evaluate the role of *D*-serine signaling. We found that application of *D*-serine fully restored the initial phase of LTP (5–15 min) after SE.

Morphological and functional integrity of astrocytes is a key to the healthy brain. Astrocytes occupy non-overlapping or slightly overlapping spatial domains (Bushong et al., 2002) which they fill with highly ramified processes (Khakh and Sofroniew, 2015). Such architecture helps to support neurons metabolically, maintain local homeostasis, synaptic plasticity etc. (Verkhratsky and Nedergaard, 2018). Disruption of the astrocytic domain organization, morphological alterations of astrocytes and changes in the number of these cells are characteristic to many of brain disorders (Verkhratsky et al., 2017). It appears that the changes in astrocytes often precede neurodegeneration and clinical symptoms (Rossi and Volterra, 2009). Astrocytic pathologies can be of several types: (1) changes in the number of astrocytes, e.g., astrogliosis or astrodegeneration; (2) astrocytic remodeling, e.g., atrophy; and (3) reactive astrogliosis. A combination of such astrocytic pathologies might signify of a particular disease or of a stage of the disease (Verkhratsky and Parpura, 2016). Astrocyte degeneration and atrophy have been described in the hSOD1 mouse, an experimental model of amyotrophic lateral sclerosis (Rossi et al., 2008). Astrocytic markers are reduced in the Parkinson's disease (Tong et al., 2015). Region- and stage-specific alterations in astrocytic morphology have been reported in the Alzheimer's disease (Olabarria et al., 2010; Rodríguez-Arellano et al., 2016). Loss of astrocytic domain organization has been observed in the epileptic brain (Oberheim et al., 2008). Here we find that the lithium-pilocarpine model of SE is associated with the atrophy of distal astrocytic branches and with astrocyte uncoupling reported earlier (Bedner et al., 2015). This, in turn, diminishes Ca^{2+} activity in these cells and correlates with the *D*-serine dependent impairment of LTP initiation. However, the LTP maintenance phase failure was not related to the insufficient *D*-serine supply. The maintenance phase could be potentially linked to reduced supply of energy substrates from atrophic astrocytes or to a malfunction of any other metabolic or homeostatic support of synapses. However, this hypothesis requires further experimental testing. Our results demonstrate that the astrocyte remodeling after SE does not affect VF of astrocytic leaflets and K^+ clearance. This indicates that astrocytes can effectively maintain synaptic microenvironment at moderate levels of synaptic activity. However, if astrocytes can prevent the K^+ accumulation during synchronized epileptiform activity remains unclear. In agreement with previous reports we observe astrocyte uncoupling through the gap-junctions (Bedner et al., 2015). This may affect spatial buffering of excessive amounts of K^+ that are released during focal seizures (Fröhlich et al., 2008). In addition, optical methods, which we used, assess the synaptic microenvironment very indirectly. More careful studies on the integrity of "astrocytic cradle" around synapses are required with the use of super-resolution light microscopy or electron microscopy.

AUTHOR CONTRIBUTIONS

APl, AL, PD, ON, and TP performed the experiments, analyzed the data, prepared the figures. AP wrote the MATLAB script for Ca^{2+} analysis. AB wrote the scripts for Sholl analysis and entropy-complexity analysis, analyzed the data, prepared the figures. AZ analyzed the data, prepared the figures. DR planned the experiments. AS planned the experiments, supervised the project. AB, AZ, DR, and AS wrote the paper.

ACKNOWLEDGMENTS

The authors are grateful to Dr. Nikolay Gavrilov for help with 3D visualization of astrocytic Ca^{2+} activity and to Prof. Dmitrii Korzhevskii for help with Nissl staining visualization. The project was supported by COMFI grant 17-00-00412 (K) from RFBR for joint research of AS (grant 17-00-00409), AB (grant 17-00-00407), and AZ (grant 17-00-00408), Wellcome Trust PRF and ERC Advanced (DR).

REFERENCES

Afzalov, R., Pryazhnikov, E., Shih, P. Y., Kondratskaya, E., Zobova, S., Leino, S., et al. (2013). Low micromolar Ba^{2+} potentiates glutamate transporter current in hippocampal astrocytes. *Front. Cell. Neurosci.* 7:135. doi: 10.3389/fncel.2013.00135

Amakhin, D. V., Malkin, S. L., Ergina, J. L., Kryukov, K. A., Veniaminova, E. A., Zubareva, O. E., et al. (2017). Alterations in properties of glutamatergic transmission in the temporal cortex and hippocampus following pilocarpine-induced acute seizures in Wistar rats. *Front. Cell. Neurosci.* 11:264. doi: 10.3389/fncel.2017.00264

Anders, S., Minge, D., Griemsmann, S., Herde, M. K., Steinhäuser, C., and Henneberger, C. (2014). Spatial properties of astrocyte gap junction coupling in the rat hippocampus. *Philos. Trans. R. Soc. Lond. B Biol. Sci.* 369:20130600. doi: 10.1098/rstb.2013.0600

Angulo, M. C., Le Meur, K., Kozlov, A. S., Charpak, S., and Audinat, E. (2008). GABA, a forgotten gliotransmitter. *Prog. Neurobiol.* 86, 297–303. doi: 10.1016/j.pneurobio.2008.08.002

Araque, A., Carmignoto, G., Haydon, P. G., Oliet, S. H., Robitaille, R., and Volterra, A. (2014). Gliotransmitters travel in time and space. *Neuron* 81, 728–739. doi: 10.1016/j.neuron.2014.02.007

Bazargani, N., and Attwell, D. (2016). Astrocyte calcium signaling: the third wave. *Nat. Neurosci.* 19, 182–189. doi: 10.1038/nn.4201

Bedner, P., Dupper, A., Hüttmann, K., Müller, J., Herde, M. K., Dublin, P., et al. (2015). Astrocyte uncoupling as a cause of human temporal lobe epilepsy. *Brain* 138, 1208–1222. doi: 10.1093/brain/awv067

Bedner, P., and Steinhäuser, C. (2013). Altered Kir and gap junction channels in temporal lobe epilepsy. *Neurochem. Int.* 63, 682–687. doi: 10.1016/j.neuint.2013.01.011

Bernardinelli, Y., Muller, D., and Nikonenko, I. (2014a). Astrocyte-synapse structural plasticity. *Neural Plast.* 2014:232105. doi: 10.1155/2014/232105

Bernardinelli, Y., Randall, J., Janett, E., Nikonenko, I., König, S., Jones, E. V., et al. (2014b). Activity-dependent structural plasticity of perisynaptic astrocytic domains promotes excitatory synapse stability. *Curr. Biol.* 24, 1679–1688. doi: 10.1016/j.cub.2014.06.025

Bliss, T. V., and Collingridge, G. L. (1993). A synaptic model of memory: long-term potentiation in the hippocampus. *Nature* 361, 31–39. doi: 10.1038/361031a0

Blümcke, I., Thom, M., and Wiestler, O. D. (2002). Ammon's horn sclerosis: a maldevelopmental disorder associated with temporal lobe epilepsy. *Brain Pathol.* 12, 199–211. doi: 10.1111/j.1750-3639.2002.tb00436.x

Bushong, E. A., Martone, M. E., Jones, Y. Z., and Ellisman, M. H. (2002). Protoplasmic astrocytes in CA1 stratum radiatum occupy separate anatomical domains. *J. Neurosci.* 22, 183–192. doi: 10.1523/JNEUROSCI.22-01-00183.2002

Carpenter-Hyland, E., Bichler, E. K., Smith, M., Sloviter, R. S., and Benveniste, M. (2017). Epileptic pilocarpine-treated rats exhibit aberrant hippocampal EPSP-spike potentiation but retain long-term potentiation. *Physiol. Rep.* 5:e13490. doi: 10.14814/phy2.13490

Cheung, G., Sibille, J., Zapata, J., and Rouach, N. (2015). Activity-dependent plasticity of astroglial potassium and glutamate clearance. *Neural Plast.* 2015:109106. doi: 10.1155/2015/109106

Clasadonte, J., Dong, J., Hines, D. J., and Haydon, P. G. (2013). Astrocyte control of synaptic NMDA receptors contributes to the progressive development of temporal lobe epilepsy. *Proc. Natl. Acad. Sci. U.S.A.* 110, 17540–17545. doi: 10.1073/pnas.1311967110

Cotrina, M. L., Lin, J. H., López-García, J. C., Naus, C. C., and Nedergaard, M. (2000). ATP-mediated glia signaling. *J. Neurosci.* 20, 2835–2844. doi: 10.1523/JNEUROSCI.20-08-02835.2000

Coulter, D. A., and Eid, T. (2012). Astrocytic regulation of glutamate homeostasis in epilepsy. *Glia* 60, 1215–1226. doi: 10.1002/glia.22341

Cunha, A. O., De Oliveira, J. A., Almeida, S. S., Garcia-Cairasco, N., and Leão, R. M. (2015). Inhibition of long-term potentiation in the schaffer-CA1 pathway by repetitive high-intensity sound stimulation. *Neuroscience* 310, 114–127. doi: 10.1016/j.neuroscience.2015.09.040

Curia, G., Longo, D., Biagini, G., Jones, R. S., and Avoli, M. (2008). The pilocarpine model of temporal lobe epilepsy. *J. Neurosci. Methods* 172, 143–157. doi: 10.1016/j.jneumeth.2008.04.019

Curia, G., Lucchi, C., Vinet, J., Gualtieri, F., Marinelli, C., Torsello, A., et al. (2014). Pathophysiogenesis of mesial temporal lobe epilepsy: is prevention of damage antiepileptogenic? *Curr. Med. Chem.* 21, 663–688. doi: 10.2174/0929867320666131119152201

Dallérac, G., Chever, O., and Rouach, N. (2013). How do astrocytes shape synaptic transmission? Insights from electrophysiology. *Front. Cell. Neurosci.* 7:159. doi: 10.3389/fncel.2013.00159

Danielyan, A., Wu, Y.-W., Shih, P. Y., Dembitskaya, Y., and Semyanov, A. (2014). Denoising of two-photon fluorescence images with Block-Matching 3D filtering. *Methods* 68, 308–316. doi: 10.1016/j.ymeth.2014.03.010

de Lanerolle, N. C., and Lee, T. S. (2005). New facets of the neuropathology and molecular profile of human temporal lobe epilepsy. *Epilepsy Behav.* 7, 190–203. doi: 10.1016/j.yebeh.2005.06.003

Di Maio, R., Mastroberardino, P. G., Hu, X., Montero, L. M., and Greenamyre, J. T. (2013). Thiol oxidation and altered NR2B/NMDA receptor functions in *in vitro* and *in vivo* pilocarpine models: implications for epileptogenesis. *Neurobiol. Dis.* 49, 87–98. doi: 10.1016/j.nbd.2012.07.013

DiNuzzo, M., Mangia, S., Maraviglia, B., and Giove, F. (2014). Physiological bases of the K+ and the glutamate/GABA hypotheses of epilepsy. *Epilepsy Res.* 108, 995–1012. doi: 10.1016/j.eplepsyres.2014.04.001

During, M. J., and Spencer, D. D. (1993). Extracellular hippocampal glutamate and spontaneous seizure in the conscious human brain. *Lancet* 341, 1607–1610. doi: 10.1016/0140-6736(93)90754-5

Eid, T., Gruenbaum, S. E., Dhaher, R., Lee, T. W., Zhou, Y., and Danbolt, N. C. (2016). The Glutamate-Glutamine Cycle in Epilepsy. *Adv. Neurobiol.* 13, 351–400. doi: 10.1007/978-3-319-45096-4_14

Enkvist, M. O., and Mccarthy, K. D. (1992). Activation of protein kinase C blocks astroglial gap junction communication and inhibits the spread of calcium waves. *J. Neurochem.* 59, 519–526. doi: 10.1111/j.1471-4159.1992.tb09401.x

Fellin, T., Gomez-Gonzalo, M., Gobbo, S., Carmignoto, G., and Haydon, P. G. (2006). Astrocytic glutamate is not necessary for the generation of epileptiform neuronal activity in hippocampal slices. *J. Neurosci.* 26, 9312–9322. doi: 10.1523/JNEUROSCI.2836-06.2006

Fertziger, A. P., and Ranck, J. B. Jr. (1970). Potassium accumulation in interstitial space during epileptiform seizures. *Exp. Neurol.* 26, 571–585. doi: 10.1016/0014-4886(70)90150-0

Fröhlich, F., Bazhenov, M., Iragui-Madoz, V., and Sejnowski, T. J. (2008). Potassium dynamics in the epileptic cortex: new insights on an old topic. *Neuroscientist* 14, 422–433. doi: 10.1177/1073858408317955

Fujii, Y., Maekawa, S., and Morita, M. (2017). Astrocyte calcium waves propagate proximally by gap junction and distally by extracellular diffusion of ATP released from volume-regulated anion channels. *Sci. Rep.* 7:13115. doi: 10.1038/s41598-017-13243-0

Green, J. D. (1964). The hippocampus. *Physiol. Rev.* 44, 561–608. doi: 10.1152/physrev.1964.44.4.561

Guizar-Sicairos, M., Thurman, S. T., and Fienup, J. R. (2008). Efficient subpixel image registration algorithms. *Opt. Lett.* 33, 156–158. doi: 10.1364/OL.33.000156

Guthrie, P. B., Knappenberger, J., Segal, M., Bennett, M. V., Charles, A. C., and Kater, S. B. (1999). ATP released from astrocytes mediates glial calcium waves. *J. Neurosci.* 19, 520–528. doi: 10.1523/JNEUROSCI.19-02-00520.1999

Häuser, S., and Steidl, G. (2013). Convex multiclass segmentation with shearlet regularization. *Int. J. Comput. Math.* 90, 62–81. doi: 10.1080/00207160.2012.688960

Héja, L., Nyitrai, G., Kékesi, O., Dobolyi, A., Szabó, P., Fiáth, R., et al. (2012). Astrocytes convert network excitation to tonic inhibition of neurons. *BMC Biol.* 10:26. doi: 10.1186/1741-7007-10-26

Heller, J. P., and Rusakov, D. A. (2015). Morphological plasticity of astroglia: Understanding synaptic microenvironment. *Glia* 63, 2133–2151. doi: 10.1002/glia.22821

Henneberger, C., Bard, L., and Rusakov, D. A. (2012). D-Serine: a key to synaptic plasticity? *Int. J. Biochem. Cell Biol.* 44, 587–590. doi: 10.1016/j.biocel.2012.01.005

Henneberger, C., Papouin, T., Oliet, S. H., and Rusakov, D. A. (2010). Long-term potentiation depends on release of D-serine from astrocytes. *Nature* 463, 232–236. doi: 10.1038/nature08673

Hermann, B., and Seidenberg, M. (2007). Epilepsy and cognition. *Epilepsy Curr.* 7, 1–6. doi: 10.1111/j.1535-7511.2007.00151.x

Höfer, T., Venance, L., and Giaume, C. (2002). Control and plasticity of intercellular calcium waves in astrocytes: a modeling approach. *J. Neurosci.* 22, 4850–4859. doi: 10.1523/JNEUROSCI.22-12-04850.2002

Höller, Y., and Trinka, E. (2014). What do temporal lobe epilepsy and progressive mild cognitive impairment have in common? *Front. Syst. Neurosci.* 8:58. doi: 10.3389/fnsys.2014.00058

Ivanov, A. D., and Zaitsev, A. V. (2017). NMDAR-independent hippocampal long-term depression impairment after status epilepticus in a lithium-pilocarpine model of temporal lobe epilepsy. *Synapse.* 71:e21982. doi: 10.1002/syn.21982

Jones, T. A., and Greenough, W. T. (1996). Ultrastructural evidence for increased contact between astrocytes and synapses in rats reared in a complex environment. *Neurobiol. Learn. Mem.* 65, 48–56. doi: 10.1006/nlme.1996.0005

Ju, W., Morishita, W., Tsui, J., Gaietta, G., Deerinck, T. J., Adams, S. R., et al. (2004). Activity-dependent regulation of dendritic synthesis and trafficking of AMPA receptors. *Nat. Neurosci.* 7, 244–253. doi: 10.1038/nn1189

Kersante, F., Rowley, S. C. S., Pavlov, I., Gutierrez-Mecinas, M., Semyanov, A., Reul, J. M. H. M., et al. (2013). A functional role for both gamma-aminobutyric acid (GABA) transporter-1 and GABA transporter-3 in the modulation of extracellular GABA and GABAergic tonic conductances in the rat hippocampus. *J. Physiol. Lond.* 591, 2429–2441. doi: 10.1113/jphysiol.2012.246298

Khakh, B. S., and Sofroniew, M. V. (2015). Diversity of astrocyte functions and phenotypes in neural circuits. *Nat. Neurosci.* 18, 942–952. doi: 10.1038/nn.4043

Kim, J. H. (2001). Pathology of epilepsy. *Exp. Mol. Pathol.* 70, 345–367. doi: 10.1006/exmp.2001.2372

Kofuji, P., and Newman, E. A. (2004). Potassium buffering in the central nervous system. *Neuroscience* 129, 1045–1056. doi: 10.1016/j.neuroscience.2004.06.008

Kryukov, K. A., Kim, K. K., Magazanik, L. G., and Zaitsev, A. V. (2016). Status epilepticus alters hippocampal long-term synaptic potentiation in a rat lithium-pilocarpine model. *Neuroreport* 27, 1191–1195. doi: 10.1097/WNR.0000000000000656

Lamberti, P. W., Martin, M. T., Plastino, A., and Rosso, O. A. (2004). Intensive entropic non-triviality measure. *Physica A* 334, 119–131. doi: 10.1016/j.physa.2003.11.005

Lebedeva, A., Plata, A., Nosova, O., Tyurikova, O., and Semyanov, A. (2018). Activity-dependent changes in transporter and potassium currents in hippocampal astrocytes. *Brain Res. Bull.* 136, 37–43. doi: 10.1016/j.brainresbull.2017.08.015

Lee, S., Yoon, B. E., Berglund, K., Oh, S. J., Park, H., Shin, H. S., et al. (2010). Channel-mediated tonic GABA release from glia. *Science* 330, 790–796. doi: 10.1126/science.1184334

Lehre, K. P., and Rusakov, D. A. (2002). Asymmetry of glia near central synapses favors presynaptically directed glutamate escape. *Biophys. J.* 83, 125–134. doi: 10.1016/S0006-3495(02)75154-0

López-Ruiz, R., Mancini, H. L., and Calbet, X. (1995). A statistical measure of complexity. *Phys. Lett. A* 209, 321–326. doi: 10.1016/0375-9601(95)00867-5

Lushnikova, I., Skibo, G., Muller, D., and Nikonenko, I. (2009). Synaptic potentiation induces increased glial coverage of excitatory synapses in CA1 hippocampus. *Hippocampus* 19, 753–762. doi: 10.1002/hipo.20551

Maggioni, M., Katkovnik, V., Egiazarian, K., and Foi, A. (2013). Nonlocal Transform-Domain Filter for Volumetric Data Denoising and Reconstruction. *IEEE Trans. Image Process.* 22, 119–133. doi: 10.1109/TIP.2012.2210725

Malenka, R. C., and Bear, M. F. (2004). LTP and LTD: an embarrassment of riches. *Neuron* 44, 5–21. doi: 10.1016/j.neuron.2004.09.012

Mazzuferi, M., Palma, E., Martinello, K., Maiolino, F., Roseti, C., Fucile, S., et al. (2010). Enhancement of GABA(A)-current run-down in the hippocampus occurs at the first spontaneous seizure in a model of temporal lobe epilepsy. *Proc. Natl. Acad. Sci. U.S.A.* 107, 3180–3185. doi: 10.1073/pnas.0914710107

Medvedev, N., Popov, V., Henneberger, C., Kraev, I., Rusakov, D. A., and Stewart, M. G. (2014). Glia selectively approach synapses on thin dendritic spines. *Philos. Trans. R. Soc. Lond. B Biol. Sci.* 369, 20140047. doi: 10.1098/rstb.2014.0047

Meeks, J. P., and Mennerick, S. (2007). Astrocyte membrane responses and potassium accumulation during neuronal activity. *Hippocampus* 17, 1100–1108. doi: 10.1002/hipo.20344

Monai, H., Ohkura, M., Tanaka, M., Oe, Y., Konno, A., Hirai, H., et al. (2016). Calcium imaging reveals glial involvement in transcranial direct current stimulation-induced plasticity in mouse brain. *Nat. Commun.* 7:11100. doi: 10.1038/ncomms11100

Müller, L., Tokay, T., Porath, K., Kohling, R., and Kirschstein, T. (2013). Enhanced NMDA receptor-dependent LTP in the epileptic CA1 area via upregulation of NR2B. *Neurobiol. Dis.* 54, 183–193. doi: 10.1016/j.nbd.2012.12.011

Oberheim, N. A., Tian, G. F., Han, X., Peng, W., Takano, T., Ransom, B., et al. (2008). Loss of astrocytic domain organization in the epileptic brain. *J. Neurosci.* 28, 3264–3276. doi: 10.1523/JNEUROSCI.4980-07.2008

Olabarria, M., Noristani, H. N., Verkhratsky, A., and Rodríguez, J. J. (2010). Concomitant astroglial atrophy and astrogliosis in a triple transgenic animal model of Alzheimer's disease. *Glia* 58, 831–838. doi: 10.1002/glia.20967

Oliet, S. H., Piet, R., and Poulain, D. A. (2001). Control of glutamate clearance and synaptic efficacy by glial coverage of neurons. *Science* 292, 923–926. doi: 10.1126/science.1059162

Ostroff, L. E., Manzur, M. K., Cain, C. K., and Ledoux, J. E. (2014). Synapses lacking astrocyte appear in the amygdala during consolidation of Pavlovian threat conditioning. *J. Comp. Neurol.* 522, 2152–2163. doi: 10.1002/cne.23523

Papouin, T., Henneberger, C., Rusakov, D. A., and Oliet, S. H. R. (2017). Astroglial versus neuronal D-serine: fact checking. *Trends Neurosci.* 40, 517–520. doi: 10.1016/j.tins.2017.05.007

Patrushev, I., Gavrilov, N., Turlapov, V., and Semyanov, A. (2013). Subcellular location of astrocytic calcium stores favors extrasynaptic neuron-astrocyte communication. *Cell Calcium* 54, 343–349. doi: 10.1016/j.ceca.2013.08.003

Paz, J. T., and Huguenard, J. R. (2015). Microcircuits and their interactions in epilepsy: is the focus out of focus? *Nat. Neurosci.* 18, 351–359. doi: 10.1038/nn.3950

Pekny, M., Pekna, M., Messing, A., Steinhäuser, C., Lee, J. M., Parpura, V., et al. (2016). Astrocytes: a central element in neurological diseases. *Acta Neuropathol.* 131, 323–345. doi: 10.1007/s00401-015-1513-1

Peng, W. F., Ding, J., Li, X., Fan, F., Zhang, Q. Q., and Wang, X. (2016). N-methyl-D-aspartate receptor NR2B subunit involved in depression-like behaviours in lithium chloride-pilocarpine chronic rat epilepsy model. *Epilepsy Res.* 119, 77–85. doi: 10.1016/j.eplepsyres.2015.09.013

Postnikova, T. Y., Zubareva, O. E., Kovalenko, A. A., Kim, K. K., Magazanik, L. G., and Zaitsev, A. V. (2017). Status epilepticus impairs synaptic plasticity in rat hippocampus and is followed by changes in expression of NMDA receptors. *Biochemistry* 82, 282–290. doi: 10.1134/S0006297917030063

Powell, G. E., and Percival, I. C. (1979). A spectral entropy method for distinguishing regular and irregular motion of Hamiltonian systems. *J. Phys. A Math. Gen.* 12, 2053. doi: 10.1088/0305-4470/12/11/017

Reichenbach, A., Derouiche, A., and Kirchhoff, F. (2010). Morphology and dynamics of perisynaptic glia. *Brain Res. Rev.* 63, 11–25. doi: 10.1016/j.brainresrev.2010.02.003

Rodríguez-Arellano, J. J., Parpura, V., Zorec, R., and Verkhratsky, A. (2016). Astrocytes in physiological aging and Alzheimer's disease. *Neuroscience* 323, 170–182. doi: 10.1016/j.neuroscience.2015.01.007

Rossi, D., Brambilla, L., Valori, C. F., Roncoroni, C., Crugnola, A., Yokota, T., et al. (2008). Focal degeneration of astrocytes in amyotrophic lateral sclerosis. *Cell Death Differ.* 15, 1691–1700. doi: 10.1038/cdd.2008.99

Rossi, D., and Volterra, A. (2009). Astrocytic dysfunction: insights on the role in neurodegeneration. *Brain Res. Bull.* 80, 224–232. doi: 10.1016/j.brainresbull.2009.07.012

Rosso, O. A., Blanco, S., Yordanova, J., Kolev, V., Figliola, A., Schürmann, M., et al. (2001). Wavelet entropy: a new tool for analysis of short duration brain electrical signals. *J. Neurosci. Methods* 105, 65–75. doi: 10.1016/S0165-0270(00)00356-3

Rosso, O. A., Larrondo, H. A., Martin, M. T., Plastino, A., and Fuentes, M. A. (2007). Distinguishing noise from chaos. *Phys. Rev. Lett.* 99:154102. doi: 10.1103/PhysRevLett.99.154102

Rusakov, D. A., Zheng, K., and Henneberger, C. (2011). Astrocytes as regulators of synaptic function: a quest for the Ca^{2+} master key. *Neuroscientist* 17, 513–523. doi: 10.1177/1073858410387304

Semyanov, A., and Godukhin, O. (1997). Kindling-like state in rat hippocampal CA1 slices induced by the repeated short-term extracellular K+ increases: The role of L-type Ca^{2+}-channels. *Neurosci. Lett.* 223, 177–180. doi: 10.1016/S0304-3940(97)13436-X

Sherwood, M. W., Arizono, M., Hisatsune, C., Bannai, H., Ebisui, E., Sherwood, J. L., et al. (2017). Astrocytic IP3 Rs: contribution to Ca^{2+} signalling and hippocampal LTP. *Glia* 65, 502–513. doi: 10.1002/glia.23107

Shih, P. Y., Savtchenko, L. P., Kamasawa, N., Dembitskaya, Y., Mchugh, T. J., Rusakov, D. A., et al. (2013). Retrograde synaptic signaling mediated by K+ efflux through postsynaptic NMDA receptors. *Cell Rep.* 5, 941–951. doi: 10.1016/j.celrep.2013.10.026

Sibille, J., Pannasch, U., and Rouach, N. (2014). Astroglial potassium clearance contributes to short-term plasticity of synaptically evoked currents at the tripartite synapse. *J. Physiol.* 592, 87–102. doi: 10.1113/jphysiol.2013.261735

Singh, N. A., Otto, J. F., Dahle, E. J., Pappas, C., Leslie, J. D., Vilaythong, A., et al. (2008). Mouse models of human KCNQ2 and KCNQ3 mutations for benign familial neonatal convulsions show seizures and neuronal plasticity without synaptic reorganization. *J. Physiol.* 586, 3405–3423. doi: 10.1113/jphysiol.2008.154971

Song, I., Volynski, K., Brenner, T., Ushkaryov, Y., Walker, M., and Semyanov, A. (2013). Different transporter systems regulate extracellular GABA from vesicular and non-vesicular sources. *Front. Cell. Neurosci.* 7:23. doi: 10.3389/fncel.2013.00023

Suárez, L. M., Cid, E., Gal, B., Inostroza, M., Brotons-Mas, J. R., Gómez-Domínguez, D., et al. (2012). Systemic injection of kainic acid differently affects LTP magnitude depending on its epileptogenic efficiency. *PLoS ONE* 7:e48128. doi: 10.1371/journal.pone.0048128

Tanaka, M., Shih, P.-Y., Gomi, H., Yoshida, T., Nakai, J., Ando, R., et al. (2013). Astrocytic Ca^{2+} signals are required for the functional integrity of tripartite synapses. *Mol. Brain* 6:6. doi: 10.1186/1756-6606-6-6

Thom, M. (2014). Review: Hippocampal sclerosis in epilepsy: a neuropathology review. *Neuropathol. Appl. Neurobiol.* 40, 520–543. doi: 10.1111/nan.12150

Tian, G. F., Azmi, H., Takano, T., Xu, Q., Peng, W., Lin, J., et al. (2005). An astrocytic basis of epilepsy. *Nat. Med.* 11, 973–981. doi: 10.1038/nm1277

Tong, J., Ang, L. C., Williams, B., Furukawa, Y., Fitzmaurice, P., Guttman, M., et al. (2015). Low levels of astroglial markers in Parkinson's disease: relationship to alpha-synuclein accumulation. *Neurobiol. Dis.* 82, 243–253. doi: 10.1016/j.nbd.2015.06.010

Van Der Walt, S., Schönberger, J. L., Nunez-Iglesias, J., Boulogne, F., Warner, J. D., Yager, N., et al. (2014). scikit-image: image processing in Python. *PeerJ* 2:e453. doi: 10.7717/peerj.453

Verkhratsky, A., and Nedergaard, M. (2014). Astroglial cradle in the life of the synapse. *Philos. Trans. R. Soc. Lond. B Biol. Sci.* 369:20130595. doi: 10.1098/rstb.2013.0595

Verkhratsky, A., and Nedergaard, M. (2018). Physiology of astroglia. *Physiol. Rev.* 98, 239–389. doi: 10.1152/physrev.00042.2016

Verkhratsky, A., and Parpura, V. (2016). Astrogliopathology in neurological, neurodevelopmental and psychiatric disorders. *Neurobiol. Dis.* 85, 254–261. doi: 10.1016/j.nbd.2015.03.025

Verkhratsky, A., Zorec, R., Rodriguez, J. J., and Parpura, V. (2017). Neuroglia: functional paralysis and reactivity in Alzheimer's disease and other neurodegenerative pathologies. *Adv. Neurobiol.* 15, 427–449. doi: 10.1007/978-3-319-57193-5_17

Villa, C., and Combi, R. (2016). Potassium channels and human epileptic phenotypes: an updated overview. *Front. Cell. Neurosci.* 10:81. doi: 10.3389/fncel.2016.00081

Wallraff, A., Köhling, R., Heinemann, U., Theis, M., Willecke, K., and Steinhäuser, C. (2006). The impact of astrocytic gap junctional coupling on potassium buffering in the hippocampus. *J. Neurosci.* 26, 5438–5447. doi: 10.1523/JNEUROSCI.0037-06.2006

Walz, W. (2000). Role of astrocytes in the clearance of excess extracellular potassium. *Neurochem. Int.* 36, 291–300. doi: 10.1016/S0197-0186(99)00137-0

Wang, F., Smith, N. A., Xu, Q., Fujita, T., Baba, A., Matsuda, T., et al. (2012). Astrocytes modulate neural network activity by Ca(2)(+)-dependent uptake of extracellular K(+). *Sci. Signal.* 5, ra26. doi: 10.1126/scisignal.2002334

Witcher, M. R., Kirov, S. A., and Harris, K. M. (2007). Plasticity of perisynaptic astroglia during synaptogenesis in the mature rat hippocampus. *Glia* 55, 13–23. doi: 10.1002/glia.20415

Wu, Y. W., Tang, X., Arizono, M., Bannai, H., Shih, P. Y., Dembitskaya, Y., et al. (2014). Spatiotemporal calcium dynamics in single astrocytes and its modulation by neuronal activity. *Cell Calcium* 55, 119–129. doi: 10.1016/j.ceca.2013.12.006

Zhang, Y., Cai, G. E., Yang, Q., Lu, Q. C., Li, S. T., and Ju, G. (2010). Time-dependent changes in learning ability and induction of long-term potentiation in the lithium-pilocarpine-induced epileptic mouse model. *Epilepsy Behav.* 17, 448–454. doi: 10.1016/j.yebeh.2010.02.008

Zhou, C., Lippman, J. J., Sun, H., and Jensen, F. E. (2011). Hypoxia-induced neonatal seizures diminish silent synapses and long-term potentiation in hippocampal CA1 neurons. *J. Neurosci.* 31, 18211–18222. doi: 10.1523/JNEUROSCI.4838-11.2011

Zhou, J. L., Shatskikh, T. N., Liu, X., and Holmes, G. L. (2007). Impaired single cell firing and long-term potentiation parallels memory impairment following recurrent seizures. *Eur. J. Neurosci.* 25, 3667–3677. doi: 10.1111/j.1460-9568.2007.05598.x

Zorec, R., Araque, A., Carmignoto, G., Haydon, P. G., Verkhratsky, A., and Parpura, V. (2012). Astroglial excitability and gliotransmission: an appraisal of Ca^{2+} as a signalling route. *ASN Neuro* 4:art:e00080. doi: 10.1042/AN20110061

16

Deconvolution of Human Brain Cell Type Transcriptomes Unraveled Microglia-Specific Potential Biomarkers

R. Ayana[1], Shailja Singh[1,2] and Soumya Pati[1]**

[1] *Department of Life Sciences, School of Natural Sciences, Shiv Nadar University, Lucknow, India,*
[2] *Special Centre for Molecular Medicine, Jawaharlal Nehru University, New Delhi, India*

Correspondence:
Shailja Singh
shailja.jnu@gmail.com;
Soumya Pati
soumya.pati@snu.edu.in

Microglial cells form a context-dependent network of brain immunoeffector cells. Despite their indispensable roles, unresolved questions exist around biomarker discovery relevant to their cellular localization, self-renewing potential, and brain developmental dynamics. To resolve the existent gap in the annotation of candidate biomarkers, we conducted a meta-analysis of brain cells using available high-throughput data sets for deciphering microglia-specific expression profiles. We have identified 3,290 significant genes specific to microglia and further selected the top 20 dysregulated genes on the basis of *p*-value and $\log_2$FC. To this list, we added 7 known microglia-specific markers making the candidate list comprising 27 genes for further downstream analyses. Next, we established a connectome of these potential markers with their putative protein partners, which demonstrated strong associations of upregulated genes like Dedicator of cytokinesis 2 (*DOCK2*) with early/mature microglial markers such as Sphingosine kinase 1 (*SPHK1*), *CD68*, and *CD45*. To elucidate their respective brain anatomical location, we deconvoluted the BrainSpan Atlas expression data. This analysis showed high expression of the majority of candidate genes in microglia-dense regions (Amygdala, Hippocampus, Striatum) in the postnatal brain. Furthermore, to decipher their localized expression across brain ages, we constructed a developmental dynamics map (DDM) comprising extensive gene expression profiles throughout prenatal to postnatal stages, which resulted in the discovery of novel microglia-specific gene signatures. One of the interesting readout from DDM is that all the microglia-dense regions exhibit dynamic regulation of few genes at 37 post conception week (pcw), the transition period between pre- and postnatal stages. To validate these findings and correlate them as potential biomarkers, we analyzed the expression of corresponding proteins in hESC-derived human microglia precursors. The cultured microglial precursors showed expression of Pentraxin 3 (*PTX3*) and *SPHK1* as well as several known markers like *CD68*, Allograft inflammatory factor 1 (*AIF1/IBA1*). In summary, this study has furnished critical insights into microglia dynamics across human brain ages and cataloged potential transcriptomic fingerprints that can be further exploited for designing novel neurotherapeutics.

Keywords: *SPHK1*, age, microglia stem cell-like progenitors, *PTX3*, transcriptomic analysis, amygdala, hippocampus, striatum

INTRODUCTION

The complex human brain is composed of neurons embedded in a framework of glial cells (astrocytes, oligodendrocytes, and microglia) and blood vessels. Microglial cells, the third and highly motile type of glial cell populating the brain were first identified and simultaneously reported by F. Robertson and F. Nissl, who first gave them the name "Staebchenzellen" based on their rod-like nuclei shape (1, 2). Later, these cells were renamed as "microglia" by Pio del Rio-Hortega, who further characterized and delimited them from other glial types (3). It is noteworthy that the primary role of microglia is to dynamically monitor CNS-invading threats. Studies have proven that microglia play critical functions like maintenance of neuronal development by continuous patrolling of active neurons, scavenging dead neurons and dendritic spine pruning under normal physiological conditions (4–6). Abnormal microglial activities can either under-prune or over-trim the dendritic spines leading to progressive neurodevelopmental defects (5). Besides this, these brain scavengers can play multidimensional roles in maintaining brain homeostasis. One of the newly discovered roles suggests that microglia regulates the number of healthy neural precursor cells (NPCs) in adult brain (7, 8). First, microglia control the number of NPCs by selectively colonizing near proliferative NPC zones and phagocytizing the neural precursors *in utero*, and second, the microglial activity gets regulated in the brain by NPC-derived chemokines, such as *VEGF*, as shown by increased microglial density surrounding NPC pool (9, 10). Notably, this NPC-microglia crosstalk is most active during neuroinflammation, as microglia is one of the primary innate immunoeffector cells (10, 11). They usually accumulate around degenerating neurons, leading to the obvious misconception of causing detrimental effects. Albeit monumental efforts made toward understanding microglia dynamics in the developing brain, knowledge regarding their cellular localization, self-renewal/expansion property, and other stage-specific roles is still in its infancy.

Origin and continual existence of microglia throughout human brain span are being currently investigated. Interestingly, microglial ontogeny describes three lines of opinions. This includes its emergence either during the embryonic day (E7.0–E10.5) from yolk sac (YS) (12–15) or their infiltration during diseased conditions from bone-marrow (16). There is also a third view which suggests that microglia is of neuroectodermal origin, derived either from glioblasts or from the germinal matrix (17). In a recent study, Tay et al. established an *in vivo* fate mapping system and studied proliferation of microglia in the healthy and diseased brain which demonstrated a context-dependent growth pattern of microglia under normal physiological conditions and selective clonal expansion during disease states (18).

The burning questions of the hour deal with life-long inhabitation and maintenance of microglia in the human brain. Answers to these will shed light underlying disease onset and manifestations in several neurological diseases including Alzheimer's disease and Parkinson's disease (AD and PD). One of the challenges in microglial biology is to track their expression and activity throughout the brain development (fetal to adult and aged brain stages). Till date, hardly any study exists that has shed light on *in silico* or *in vivo* tracking of the microglial population (19). Specifically, microglial age and variance over brain regions can be the guidance cue to understanding onset and progression of neurodegenerative disorders (20–23). To track microglia-specific biomarker expression, their localization, and generation of a regulatory dynamic map, this study presents an integrated meta-analysis approach wherein, we analyzed the differentially expressed microglia-specific gene subsets using available data repositories. This led to a unique catalog of genes (27) which are dynamically/consistently expressed in the microglial population during different stages of brain development. For the first time, we have employed a novel strategy for *in silico* monitoring of microglial gene signatures to decipher their developmental dynamics map (DDM). Interestingly, protein–protein interaction network analysis revealed six different protein clusters, which exposed primary interacting partners like Tissue factor (coagulation) protein family, Retinoid X receptor gamma signaling linked *MGST* protein family, ATP-dependent protein family, and hippocampal BMP signaling molecules like *BMP4*. Furthermore, region specificity of the above candidates was evaluated using Allen Brain Atlas's (ABA) BrainSpan which revealed localized expression of several genes in microglia-centric regions. Finally, to validate the emergence of early/mature microglial signatures *in vitro*, we induced hESC-derived microglial differentiation using an in-house protocol. In conclusion, the study has validated the expression of stage-specific markers as well as presented an updated catalog of potential age-specific microglial biomarkers, which can be used as future prognostic biomarkers. The overall strategy adopted for *in silico* analyses has been compiled (**Figure 1**).

MATERIALS AND METHODS

Literature Survey

We utilized repositories including NCBI Pubmed and HighWirePress to understand the prior literature linked to microglia biology. Other high-throughput data repositories like NCBI's Gene Expression Omnibus (GEO) and Sequence Read Archive (SRA) were scanned for biological sequence data specific to transcriptomics of human microglia.

High-Throughput Data Analysis

First, using available RNA-sequencing data set linked to microglia transcriptomics (24), we delineated the brain cell-type specific expression in accordance with different brain cell types namely, microglia, astrocytes, endothelium, oligodendrocytes, and neurons. Specifically, we evaluated the differential expression between Rest of the Brain Cells comprising astrocytes, endothelial cells, oligodendrocytes and neurons (ROBC), and microglial cell type. Log_2FoldChange (FC) values threshold of ± 1.5 was considered with a p-value significance of <0.05. This formed our *primary* data set comprising microglia cell type-specific genes. Our *secondary* data set was utilized to gauge the correlation of differential expression between brain cell types and differential expression of induced pluripotent stem cells (iPSCs) versus derived microglia. This data series deposited in GEO database contains human microglia-specific microarray Affy HuGene-2_0-st data (GSE78115) (25). Specifically, we considered the 12 samples (3 replicates each) derived from 2 different cell lines NCRM-5 and iNC-01 of iPSC origin. Using OLIGO package of R

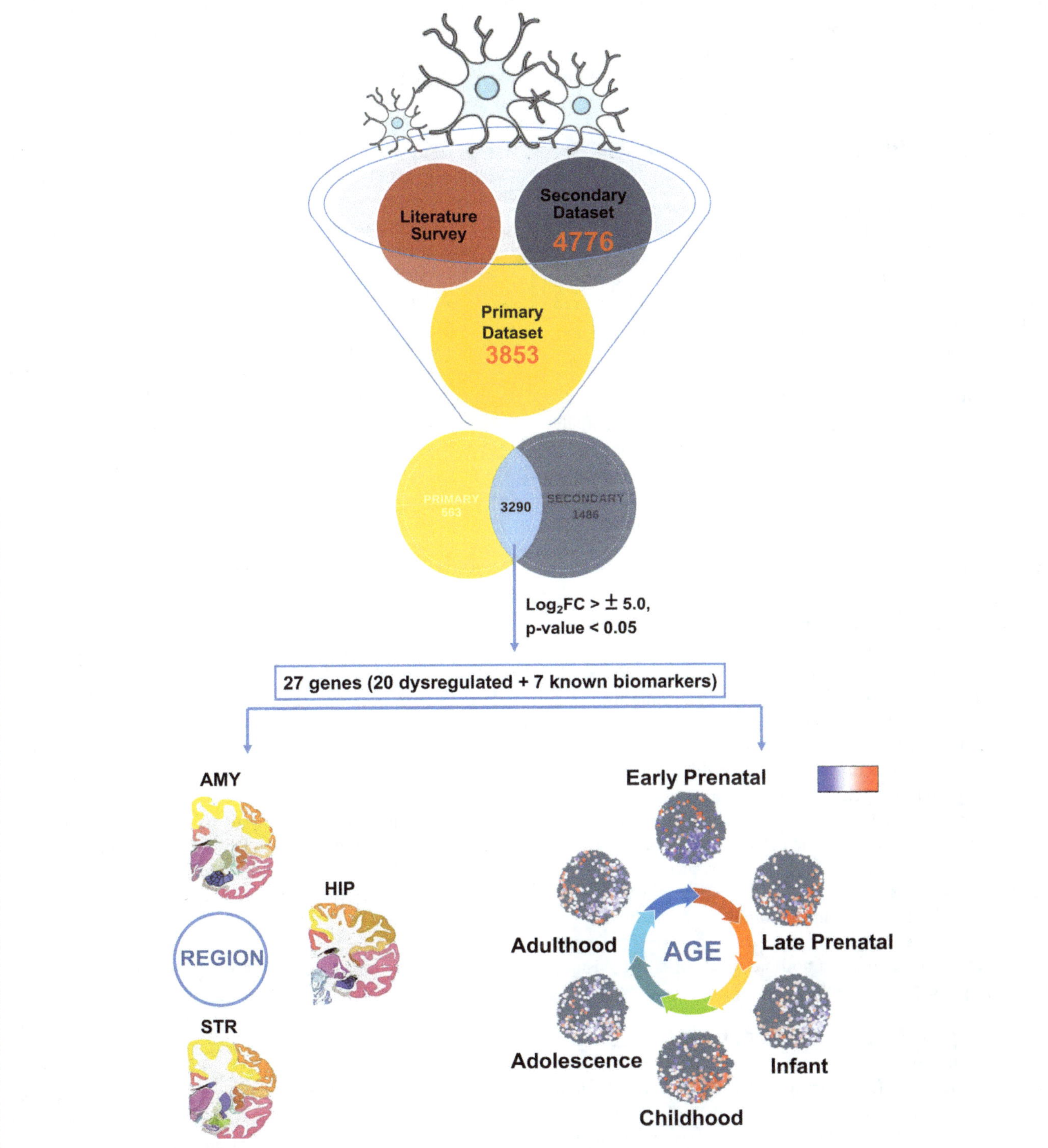

FIGURE 1 | Primary strategy (*in silico* approach) employed to track microglial gene expression dynamics within the developing human brain.

Bioconductor, we performed data analysis (including data normalization) to assess the differential expression pattern specific to human microglia with respect to iPSCs. The OLIGO package is freely available through Bioconductor website (26). Differential expression cutoff of ± 1.5 $\log_2$FC and p-value <0.05 was considered significant in all cases. The differentially expressed genes from both primary and secondary human microglia sample sets were combined to ensure minimal data loss. This integration yielded a common pool of microglial specific genes (3,290), which was further used for downstream analyses. Pearson correlation coefficient was calculated for the integrated gene list (3,290) using R. We further analyzed the expression patterning between the *primary* and *secondary* data sets using the top 20 (out of 3,290) up and downregulated genes, which were shortlisted on the basis of p-value and log2FC values below 0.01 and above ±5.0, respectively. To this list, we also added a list of seven known biomarker genes specific to microglia for understanding their development expression dynamics. Taken together, the total list used for brain age and region-specific dynamics consisted of 27 genes.

Protein Classification and Pathway Analysis

We utilized Protein ANalysis THrough Evolutionary Relationships classification system and mapped the total differentially expressed list of 3,290 genes to various protein classes for elucidating their functional significance (27). For pathway analysis, we inputted the same gene list to Consensus PathDB-human to identify their key associative pathways (28). The Bonferroni corrected p-value cutoff <0.01 was applied for both analyses.

Gene Set Enrichment Analysis

In order to ascertain significant gene sets in our differentially expressed gene list, we implemented Gene Set Enrichment analysis using the Broad Institute GSEA v2.07 software,[1] the molecular signatures database,[2] and the C5: GO gene sets database, comprised of 1,454 gene sets named by genes and respective GO terms.[3]

Protein–Protein Interaction Network Analysis

The differentially expressed gene list (3290) was mapped to STRING for whole proteome network analysis. Furthermore, the shortlisted microglia-specific gene list (27) was also analyzed for protein partner associations using STRING (29). Primarily, using all the top upregulated and downregulated protein clusters, a single protein–protein network was constructed. A stringent cutoff of 0.7 for edge/interaction score was considered for building the primary protein–protein interaction network without clustering and no more than 20 primary interactor proteins. This combined confidence score is calculated on the basis of various parameters including phylogenetic co-occurrence, gene fusion, homology, co-expression, experimentally determined interaction, and neighborhood on any chromosome. The graph was primarily constructed according to interaction/edge score while specifying the source and target nodes. All networks were visualized using Cytoscape 3.4.0 (30).

Brain Region Expression Mapping

The Allen Brain Atlas contains an extensive catalog of human brain expression data (Microarray and RNA-Seq) with region specificity (31, 32). We collected the raw data of the developmental transcriptome from Allen Brain Atlas and segregated the data into different ages from prenatal to postnatal (adult) brain (31). Here, we mapped the top gene set of 27 genes onto all the prenatal and postnatal brain regions, with a special emphasis on human microglia-enriched regions like Hippocampus, Striatum, and Amygdala. Other than these genes, we also show the expression of known microglial marker genes like *TMEM119*, *P2RY12*, Phospholipase D4 (*PLD4*), *AIF1*, *PTPRC1*, Sphingosine kinase 1 (*SPHK1*), *PTX3*, and G protein-coupled receptor 34 (*GPR34*).

Age-Wise Dynamics

To assess the microglia dynamics throughout the age (pre- and postnatal) of a human being, we utilized the human developmental transcriptome data (BrainSpan) deposited in the Allen Brain database (33, 34). Replicate samples for each developmental stage (prenatal or postnatal) were considered for analysis. The age groups ranged from prenatal stages, including 14th postconception week (pcw), 22 pcw to 40 years of age. Specifically, for each microglia-enriched region of the brain (hippocampus, striatum, and amygdala), we constructed individual graphs for age-wise DDM of potential and known microglia-specific markers which included shortlisted gene set of 27, containing 20 (p-value < 0.01, $\log_2$FC > ±5.0) and 7 known markers. Following this, correlation analysis was performed using the same gene list as input to find correlations among the human ages spanning prenatal stages like 12 pcw, 37 pcw to postnatal stages, such as 1, 15–19, and 40 years of age. Pearson correlation matrices or correlograms were constructed and visualized using the Corrplot package of R Bioconductor suite. All graphs have been made using RStudio 1.0.153 (35).

Culturing of Human Embryonic Stem Cells (H9)

Human embryonic stem cells (h9-hESC, NIH approved) were obtained from human embryonic stem cell core, Baylor College of Medicine. H9 clones were propagated based on feeder-free culture system, using human ESC-qualified Matrix, BD Matrigel (BD Biosciences, San Jose, CA, USA), and grown in mTeSR-1 complete media supplements (Stemcell Technologies, Vancouver, BC, Canada) at 37°C and 5% CO_2. Clones were maintained and expanded according to manufacturer's protocol, in accordance with NIH guidelines. Complete media was changed every day and cells were sub-cultured using Dispase (Stemcell Technologies, Vancouver, BC, Canada) every 5–6 days. During each stage of subculture, uniform H9 colonies were collected without any visible differentiation.

Embryonic Body (EB) Formation

To induce the formation of EBs or neuroectodermal spheres, the H9 colonies with 70% confluency were used. To achieve neural induction, H9 colonies were incubated in neural induction me-dium [NIM; DMEM/F12: Neurobasal (1:1), 2% B27, 1% N2 (Invitrogen)] for almost 24 h and dissociated into small clumps by manual scraping. Small clumps were further grown as suspension culture with NIM in non-coated bacterial Petri dishes for a period of 4 days as exactly described by Cho et al. (36). 50% of the media were replaced in each of the EB-containing Petri dishes at a timely interval of 48 h.

Microglial Differentiation and Generation of Microglia Precursors From hESC

To induce microglial differentiation in culture from H9 derived EBs and obtain human microglial precursors *in vitro*, we used an in-house standardized protocol (unpublished), modified based on an earlier established protocol (37). In brief, the whole process

[1]http://www.broadinstitute.org/gsea (Accessed: July 17, 2017).
[2]http://www.broadinstitute.org/gsea/msigdb (Accessed: July 17, 2017).
[3]www.geneontology.org (Accessed: July 17, 2017).

goes through a long-term culture (approximately 8 weeks) covering four major steps, (i) formation of hESC (H9) derived EBs, (ii) neuroectodermal lineage commitment, (iii) microglial differentiation, and (iv) formation of induced microglia precursors (iMPs), as suggested in the previous article by Beutner et al. (37). We have used stage-specific conditioned media supplements as suggested by Beutner et al., including differentiation media (DIFF), Insulin/Transferrin/Selenite/Fibronectin media ($ITSF_n$) and N2 Media. However, we have introduced several modifications during the stage-specific progression of microglial differentiation. These include the following: (1) Usage of NIM media till 5 Days *in vitro* (DIV) for maintaining EBs and then switching them to $ITSF_n$ on sixth day, (2) addition of human stem cell factor (hSCF, Peprotech, USA) and basic fibroblast growth factor (Peprotech) in culture from DIV 7 to DIV 14 with concentrations of 50 and 25 ng/ml, respectively, (3) treatment of cultures from DIV 14 to DIV 21 with hSCF, hIL3, and human granulocyte macrophage stimulating factor (hGM-CSF, Peprotech) with concentrations of 50 ng/ml for both hSCF and hIL3, and 25 ng/ml for hGM-CSF, and finally (4) addition of only hGM-CSF from DIV 21 onward till DIV 40 and furthermore.

Immunocytochemistry/ Immunofluorescence Analysis

Cells from different stages of microglia induction and differentiation were fixed for performing immunocytochemistry and immunofluorescence-based analyses using primary antibodies and fluorophore-conjugated secondary antibodies. Detailed information related to fixation, antibodies, dilution, and image analysis are listed within Data sheet 1 (Supplementary Table 1) in Supplementary Material.

Statistical Analyses

Differential expression cutoff of ± 1.5 $\log_2$FC and p-value < 0.05 was considered significant unless specified. Bonferroni corrected p-value cutoff $= 0.01$ was applied in protein classification and pathway analyses. At least triplicate donors were considered for each developmental stage (prenatal or postnatal) for age/region analyses. In case of the *in vitro* experiments, we have maintained biological replicate of $n = 3$.

RESULTS

Cell-Type Specific Gene Set Enrichment Analysis of Human Microglia

The complete GEO and SRA high-throughput databases were scanned to identify microglia-specific data sets. Out of this, two data sets were shortlisted for further in-depth *in silico* analyses. Initially, we analyzed the *primary* data set, i.e., the total gene set from the cell-type specific data comprised of significantly expressed genes in ROBC versus microglia (**Figure 2A**). A total of 3,853 genes were differentially expressed, with a range comprising 6.0 $\log_2$FC as the maximum expression to -9.4 $\log_2$FC as the minimum expression. The samples of various cellular origins were tested for homogeneity *via* 2D-Principal Component Analysis which resulted in noticeable differences in microglia sample cluster from other cell types in Principal components, PC 1 and PC 2 (**Figure 2B**). We specifically considered the *primary data set* to be the cell-type data to ensure elimination of non-microglia related noise in the human brain. Furthermore, we identified known biomarkers including *SPHK1*, *PLD4*, *TMEM119*, and *CD68*, expressed in the range of 2–6 FC and used them as a positive control for *in silico* tracking of microglia.

Sphingosine kinase 1 is a known biomarker for quiescent neural stem cell (qNSC) pool and the key enzyme responsible for phosphorylation of sphingosine to sphingosine-1-phosphate (*S1P*) (38). It has been also seen to play an important role in the regulation of proinflammatory cytokines in activated microglia (39). Purinergic receptor 12 (*P2RY12*) is known to suppress microglial process motility and delay in closure of blood–brain barrier (BBB) (40). *P2RY12*-mediated microglial cell activation contributes to the rapid closure of BBB small leaks, by aggregation of microglial cells at the injury site. Expression of Phospholipase D4 (*PLD4*) and its localization changes are correlated with the activation state of microglia (41, 42). This transmembrane glycoprotein localized in the endoplasmic reticulum and Golgi apparatus is primarily seen to play an important role in microglia dynamics during early postnatal brain development. Transmembrane protein 119 (*TMEM119*) is an established, reliable microglial marker that discriminates resident microglia from blood-derived macrophages in the human brain (43). A glia-derived Pentraxin 3 (*PTX3*) expressed in the microglial secretome, has been seen to modulate phagocytic functions of microglia having crucial implications in the regulation of microglial activity in brain diseases (44).

The *secondary* data set was a redundant list of total 4,776 genes which was seen to be differentially expressed among 2 human microarray data sets. This data set contains six induced microglia samples derived from human iPSCs (25). We integrated the differentially expressed microglia-specific data enriched from both *primary* and *secondary* data sets containing 3,853 and 4,776 genes, respectively. This meta-analysis yielded a common pool of microglial specific genes (3,290) which was further used for downstream analyses (**Figure 1**).

Pathway Mapping and Gene Set Enrichment Analysis

The readout of gene candidates (3,290) was used to target enriched pathways. Among the candidate pathways, GPCR, Glycerophospholipid, NOTCH1, RAP1, glycosaminoglycan metabolism, VEGFR signaling, etc. were significant (**Figure 2C**). This analysis also demonstrated enrichment of 278 out of 3,290 genes within the immune system pathway, suggesting their potential roles in microglia functioning or immune surveillance (**Figure 2C**).

The list of significantly expressed genes (3,290) was used for gene/protein type-based clustering. This analysis revealed 25 crucial protein classes that majorly comprise of 10.7% enzyme regulators, followed by 9.9% nucleic acid binding proteins, and enzymes like transferases, hydrolases, transcription factors, and signaling molecules (Figure S1 in Supplementary Material). The other segments included cell junction, immunity/defense and extracellular matrix proteins containing 28, 40 and 46 proteins, respectively. These classes acted as guidance cues for

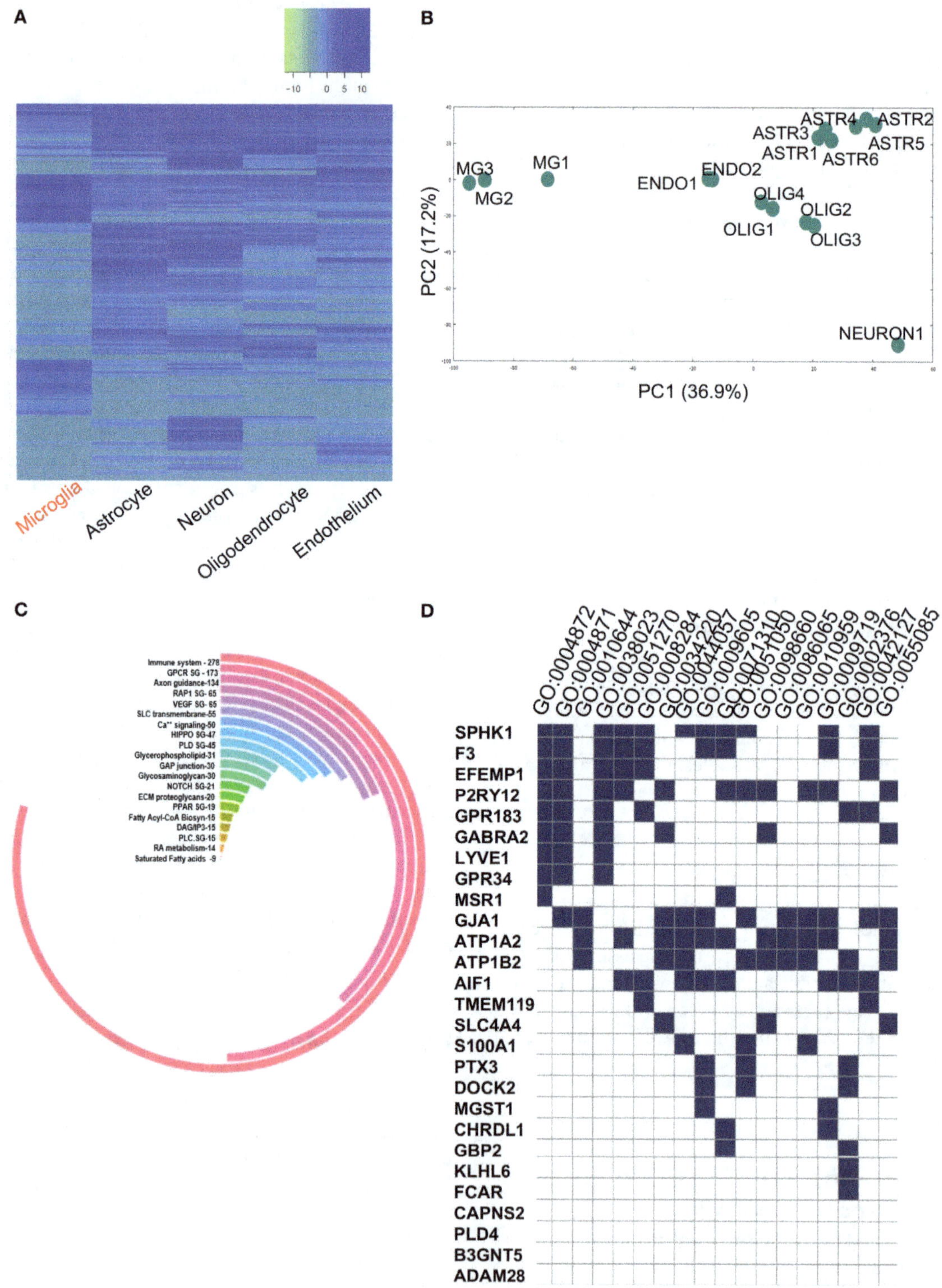

FIGURE 2 | **(A)** Heatmap representing significantly expressed upregulated and downregulated gene profiles of various brain cell types. **(B)** Principal component analysis plot showing homogeneity within each cell type and diversity among the five brain cell types namely, microglia, endothelial cells, oligodendrocytes, neurons, and astrocytes. **(C)** Circular plot showing pathway analysis of total dysregulated gene list (3,290) between ROBC and microglia. **(D)** GSE Analysis showing enriched GO BP/MF terms ($p < 0.05$) in 27 shortlisted genes overlapping between primary and secondary datasets.

the protein–protein based network analysis as well as pathway mapping and exposed many critical clusters of gene regulatory elements. We also performed GSEA on critically altered gene set of 20 dysregulated genes and 7 known microglia-specific markers to find critical GO terms like Immune System process, Receptor activity, Regulation of cell proliferation which further substantiated the data (**Figure 2D**; Supplementary Table 2 in Supplementary Material). The aforementioned shortlisted gene set (27) was used for further analyses.

Correlation Analysis of Microglial Biomarkers Among the Data Sets

To establish correlations among the *primary* and *secondary* data sets, we calculated the Pearson correlation coefficient for the common gene pool between the two data sets (3,290), which resulted in a value of −0.22 (**Figures 3A,B**). First, the primary and secondary datasets were analyzed for the similarity in expression of shortlisted gene set comprising 27 (20 novel + 7 known) genes (**Figure 3C**). Within the upregulated gene set, all these biomarker genes excluding *GPR34* demonstrated similarity in expression when compared to the *secondary* data set [iPSC derived microglia (iMG) data] (**Figure 3B**). To highlight, top upregulated genes, including Dedicator of cytokinesis 2 (*DOCK2*), Guanylate binding protein 2 (*GBP2*), A disintegrin and metalloproteinase 28 (*ADAM28*), G Protein-coupled receptor 183 (*GPR183*), Calpain small subunit 2 (*CAPNS2*), and Fc-α receptor mediator (FCAR), were found to be consistently expressed in primary and secondary data set.

Further literature mining of this top altered gene set (27) substantiated their mechanistic roles relevant to neuroinflammation and neurotoxicity. Among these, *DOCK2*, an atypical Rho guanine nucleotide exchange factor (GEF) for *RAC* and/or *CDC42* GTPases are known to regulate phagocytic activity, cytokine release, and paracrine neurotoxicity in microglia (45). In one of the microglia-mediated neuroinflammatory cascades, the *DOCK2-RAC1* pathway was seen to be elicited by S1P through an integrin-dependent association (46). Precisely, S1P which is the phosphorylated end-product of *SPHK1* is known to regulate immune cell trafficking upon binding to G-coupled receptors (*GPRs*) (47). Interestingly, we found fivefold expression of a novel marker *GPR183* in microglia whose role is yet to be elucidated. *DOCK2* is directly regulated by the Prostaglandin E2 receptor (*EP2*) and has also been linked with AD pathophysiology (45).

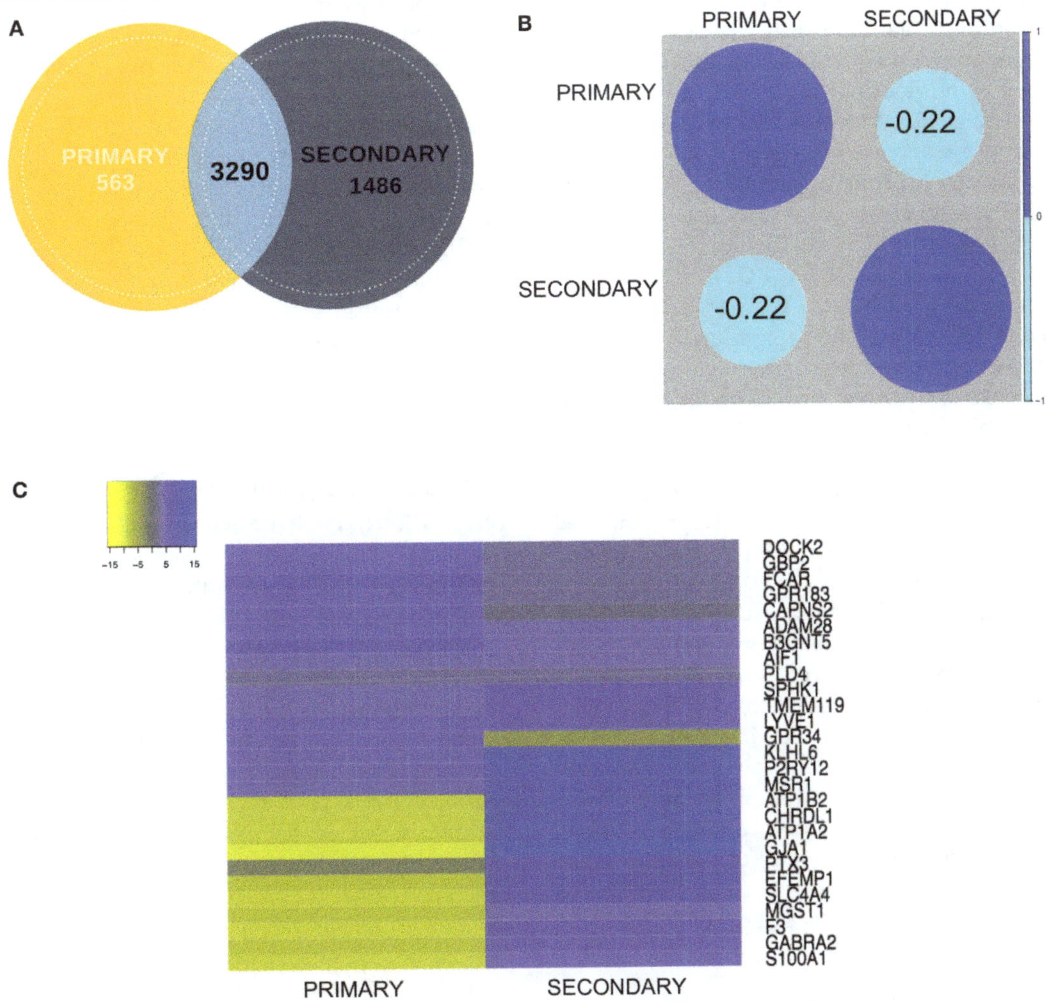

FIGURE 3 | (A) Venn diagram showing contribution of each dataset (primary and secondary) to the overlapping gene pool. **(B)** Correlation analysis between the primary and secondary datasets showed mildly negative correlation. **(C)** Representative expression heatmap between the final shortlisted 27 genes has been elaborated in the expression map for in-depth correlation analysis.

Concurrent evidence involving Prostaglandin signaling in microglia also suggested that balanced induction of Peroxisome proliferator-activated receptor gamma (*PPARG*) by *LPS/IL13* or activation of *EP2* can lead to neuroprotection or death of microglia *via* calpain-dependent pathway (48). Our data also validated the elevated expression of *PPARG* as well as its downstream target calpain associated genes like *CAPNS2* in the microglial pool versus the other cell types.

Another immunoeffector molecule, *FCAR* (*CD89*) was seen to be highly expressed among the top 20 altered genes in the microglial population, which also corroborated by a recent report by Galatro et al. (49). Previous reports suggest that human microglial cells expressing truncated IRF3 show decreased expression of *GBP2*, thus proposing a possible link of this protein to *TLR3/4*-dependent neuroinflammation. Our study has shown another important marker of the microglial cell type known to be involved in CNS invasion, i.e., *ADAM28*, a member of the metalloproteinase family (50). Another enriched marker gene *N*-acetylglucosaminyltransferase *B3GNT5* was seen to be a prominent player in the expression of lactoseries sulfoglucuronylglycolipids within glycosphingolipid pathway during the prenatal brain develop-ment (51). High expression of Kelch-like family member 6 (*KLHL6*) was also detected in the microglial pool, which is supported by previous AD mice model-based studies linked it to AD pathology (52). We also found an interesting factor known as macrophage scavenger receptor 1 (*MSR1*) involved in the phagocytic machinery of microglia (53).

Contrary to this, in case of top 10 downregulated genes, the expression values for the *secondary* data set were found to be non-correlative to the primary data set (ROBC vs. microglia) (**Figure 3C**). Microsomal Glutathione S-Transferase 1 (*MGST1*), a regulator of Retinoid X receptor gamma signaling during remyelination showed non-correlative expression between the two data sets (54). Genes like S100 Calcium Binding Protein A1 (*S100A1*) (55) and Gamma-aminobutyric acid receptor subunit alpha-2 (*GABRA2*) (24, 56) which are known to be highly expressed in astrocytes were also seen to be enriched in the secondary data set and repressed in the primary data set, thus justifying the need to enrich a pure microglia cell type marker-enriched population for *in silico* and *in vivo* tracking. We also detected a novel gene EGF Containing Fibulin Like Extracellular Matrix Protein 1 (*EFEMP1*), a fibulin glycoprotein family member which was found to be downregulated in ROBC versus microglia. This was earlier shown to be upregulated in malignant gliomas and age-related macular degeneration (57). Another interesting observation suggested expression of several synaptosomal and synaptic receptor genes like *SLC4A4* and *GABRA2* significantly downregulated in the primary data set in coherence with the study by Ji et al. (58). This analysis sheds light on probable roles of the top 20 genes in other brain cell types as well as exposes a filtered population of markers linked to microglia function in the developing human brain.

Protein–Protein Interaction Network Analysis

To develop an in-depth understanding of the microglia-specific markers and their primary neighborhood protein partners, we constructed a primary protein-protein interaction network. Out of the input list of 3,290 genes, 994 coding proteins showed interactions with a confidence score above 0.70 (**Figure 4**). Network statistical analysis showed the Network centralization score to be 0.07 and clustering coefficient to be 0.423 with a total number of hub nodes to be 43.

Subset PPI networks were constructed based on the top 20 differentially expressed gene list in combination with the known biomarkers. This network exposed primary neighborhood interacting partners of *DOCK2*, *F3*, *GJA1*, *CHRDL1*, *MGST1*, and *ATP1A2* (**Figure 4**). Among these, one of the highly upregulated genes, *DOCK2* showed multiple interacting partnerships with known biomarkers like *SPHK1*, *PTPRC*, and *CD68*, which is substantiated by numerous connections (edges) and their respective confidence scores (**Figure 4**). *DOCK2* is also seen to be connected to *SPHK1* through *LYN*, and to *PTPRC*, and *CD68* through *FYN* (**Figure 4**). In a second cluster, we found *GJA1* to be interacting with *CDH2* and *TJP1*. The *GJA1* mutation has been seen to cause Oculodentodigital dysplasia, an autosomal dominant disorder, with high penetrance in intra- and interfamilial phenotypes (59). In another cluster, F3 or thrombosis factor 3 was prominently associated with its own family members (**Figure 4**). F3 gene encodes coagulation factor 3 which is a membrane-bound glycoprotein is part of the Tissue factor family. Although no specific links were found between *F3* and brain development, it has been shown to be regulated during brain neoplasms (60) or in the innate immune cells (61, 62). Based on GSEA, correlation, and PPI analysis, we conclude that *DOCK2* acts as a molecular hub for channelized signaling in microglia-based neuroprotection, neuroinflammation, and neurological diseases, including AD and PD, Autism spectrum disorders, and schizophrenia (63). This newly discovered microglia-specific regulator is known to be associated with disease pathology of AD patients independent of COX signaling (45).

Trends in Region-Specificity for Potential, Novel Microglia Signatures

Cellular localization of microglia is still debatable in human brain development. Notably, the prenatal human brain displays smaller brain size with a higher number of sub-regions (26), whereas the postnatal brain contains 18 developed sub-regions. Microglia population in the healthy brain is known to be residing primarily in hippocampus, amygdala, substantia nigra, and striatum. However, developmental tracking of human microglial activity in these regions is still elusive. To elucidate the region-specific dynamic regulation of microglial markers, we analyzed the distribution of the identified *in silico* candidates. Our data revealed that in the prenatal and postnatal brain regions, the expression profiles of selected dysregulated gene set (20 + 7) was dynamically regulated as seen by their variable expression across sub-regions. Understanding this variability in their expression pattern would be crucial for building the microglial dynamic map in human brain development.

In the prenatal brain, the expression levels of known biomarker genes like *PLD4*, *SPHK1* were high in regions of importance like caudal ganglionic eminence, medial ganglionic eminence, lateral

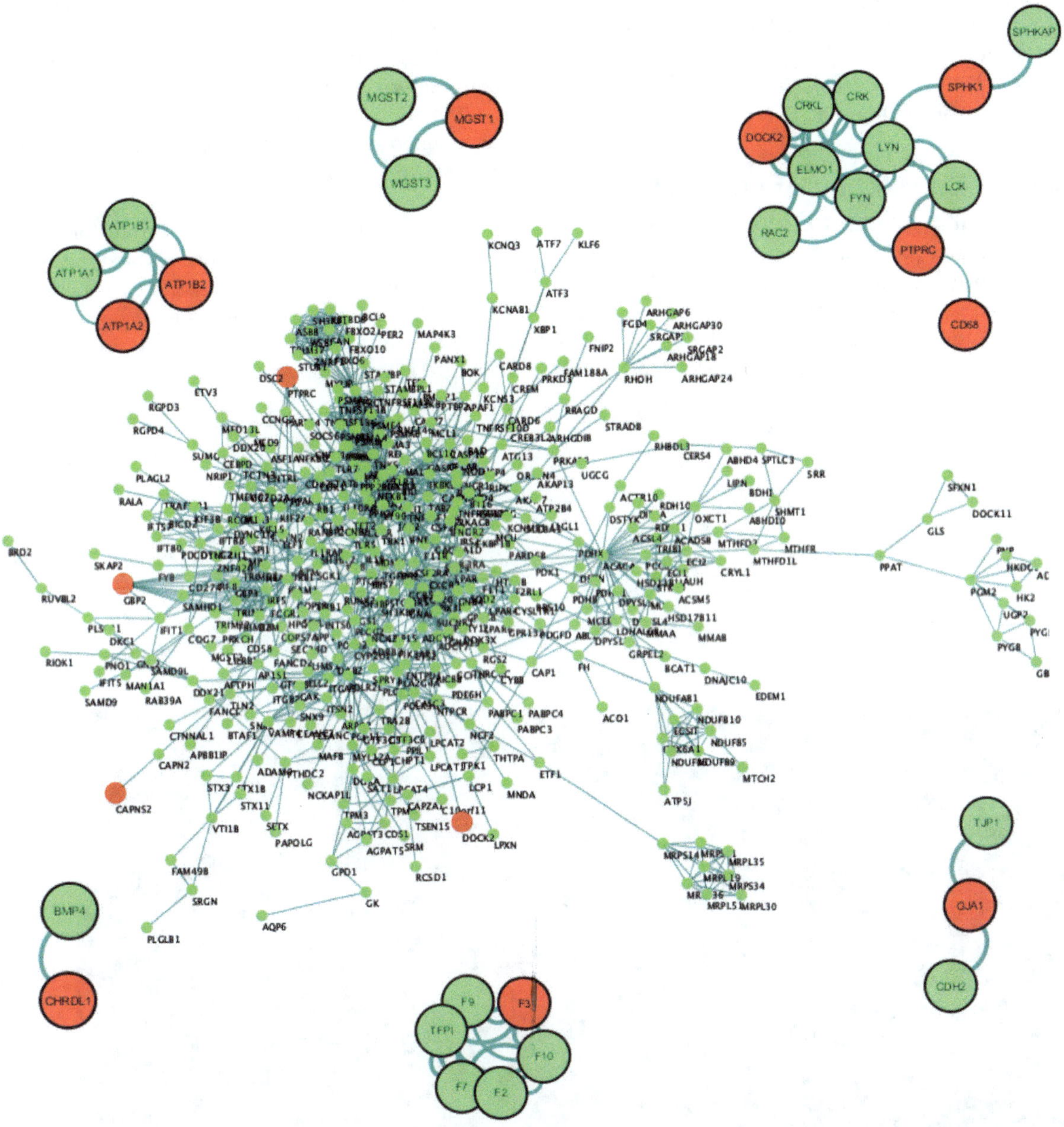

FIGURE 4 | PPI network construction of dysregulated proteome coded by 3,290 genes. Only 994 out of 3,290 genes passed the threshold interaction score of 0.70. The second representative protein–protein interaction network of the top differentially expressed gene set (27 genes) showed 11 out of 27 proteins coded by differential expressed genes to be actively interacting with each other or with varied protein partners. Red: differentially expressed gene in the primary data set, green: interacting partner, edge indicates the source to target node strength. Combined interaction score > 0.70, No. of first level interactors = 20.

ganglionic eminence and lower in regions like posteroinferior parietal cortex, inferolateral temporal cortex, medial frontal cortex (MFC) (**Figure 5A**). However, *GPCR* associated genes like *GPR183* and *GPR34*, and EFEMP1 were seen to be highly expressed only in the primary visual cortex (V1C) region in the prenatal brain (**Figure 5A**). To highlight, *EFEMP1* mutations are associated with Doyne honeycomb retinal dystrophy, an autosomal dominant disorder in which there are drusen (lipid) deposits in the macula eventually leading to vision loss (64, 65). In the postnatal brain, genes such as *GBP2*, astrocyte marker *S100A1* and *MGST1* are highly expressed in the cerebellar cortex region, while genes like *GPR183*, *EFEMP1*, and *GPR34* are only expressed in MFC (**Figure 5B**).

In summary, this analysis revealed consistent high expression of 14 out of 27 microglial markers in regions like hippocampus (HIP), amygdala (AMY), and striatum (STR) during postnatal development (**Figure 5B**). However, all of these 14 genes are low in expression throughout prenatal development (**Figure 5A**). Interestingly, amygdaloid complex (AMY) of the prenatal brain showed very low expression of all 27 genes, whereas the same region showed pronounced expression of numerous genes, like *P2RY12*, *DOCK2*, *MSR1*, *B3GNT5*, *AIF1*, etc. during postnatal development (**Figure 5B**). Overall, this analysis has provided region-based developmental tracking of microglial expression dynamics in the human brain.

Age-Specific Correlation of Microglial Signatures During Brain Development

Ontogeny, renewal, and maintenance of microglial population have long been debated. In this direction, we decided to correlate region-specific expression of microglial markers to developmental stages of the human brain. The ABA developmental transcriptome data yielded a humungous gene pool constituting age-wise microglial marker expression specifically in AMY, HIP and STR regions. We, further, focused on the gene regulation status of top 10 upregulated and downregulated gene sets across brain ages encompassing prenatal (8–37 pcw) and postnatal (1–40 years). Furthermore, we built correlation matrices between several important ages, including 13 and 37 pcw, marking the prenatal ages, and 1, 8, 15, 19, and 40 years indicating infant to the adult brain.

The correlation matrices with a correlation coefficient scale (−1, 1) for all three brain regions showed positive correlation at varying degrees. The representative correlogram for prenatal stages of *HIP* showed moderately positive correlation ($R = 0.65$) between 13 and 37 pcw and aged brain stage of 40 years ($R = 0.70$) (Figure S2 in Supplementary Material). Interestingly, a very strong

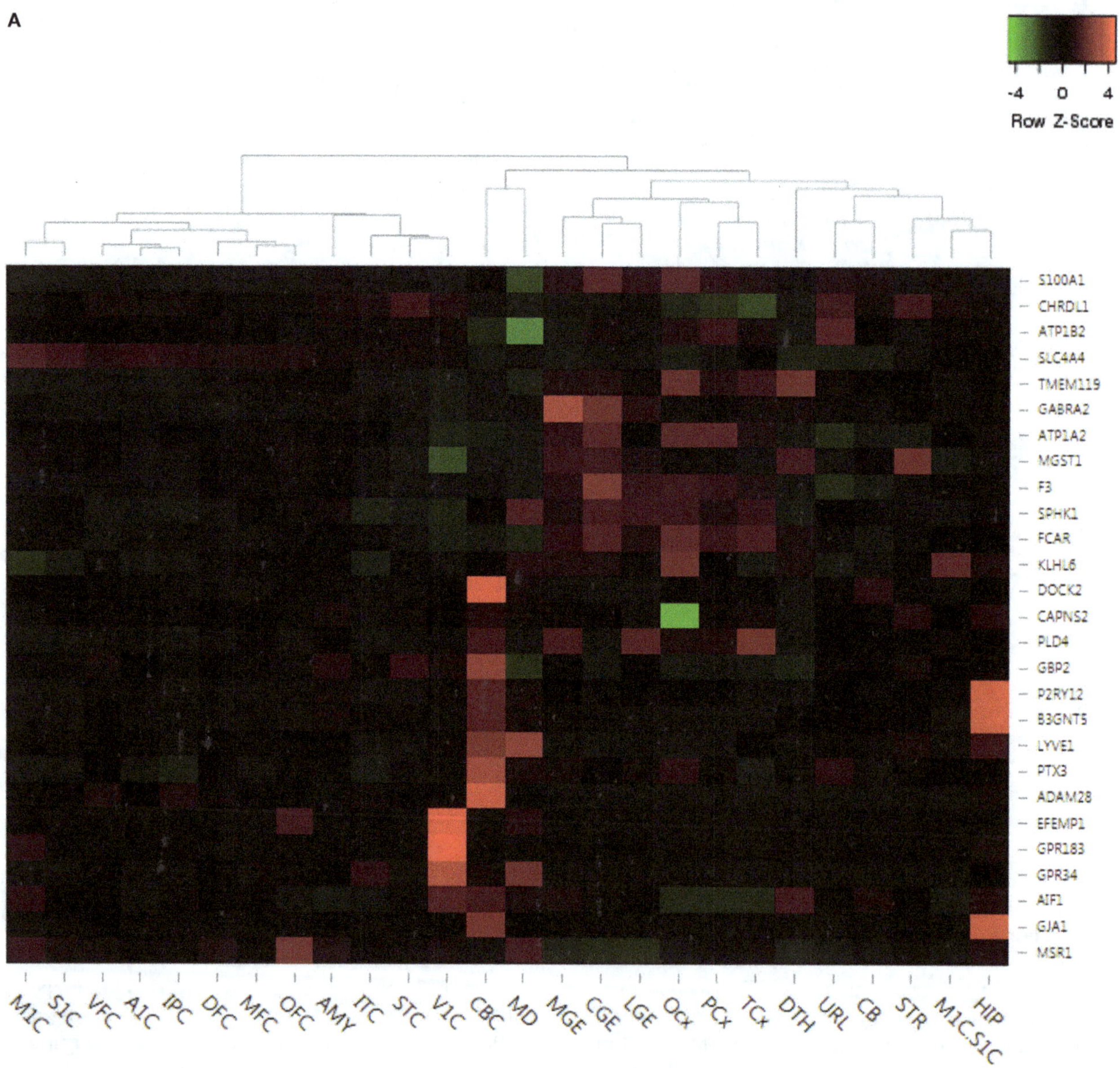

FIGURE 5 | Continued

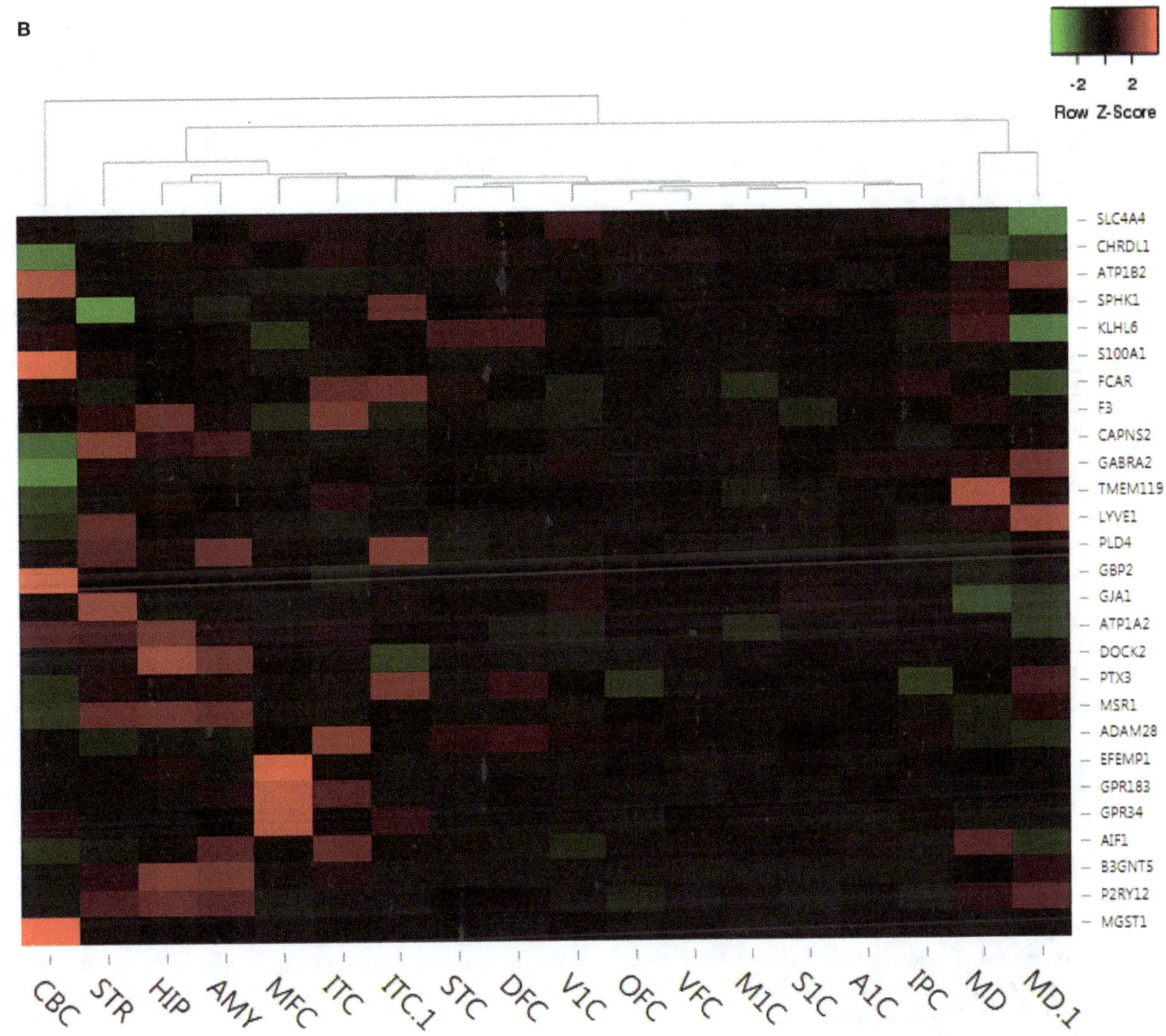

FIGURE 5 | Delineation of top dysregulated gene expression profiles across crucial brain anatomic regions associated to neurodevelopment. **(A)** Prenatal region-wise expression of the top upregulated and downregulated genes. **(B)** Postnatal region-wise expression of the top upregulated and downregulated genes. Abbreviations: DFC, dorsolateral prefrontal cortex; VFC, ventrolateral prefrontal cortex; MFC, anterior (rostral) cingulate (medial frontal cortex); OFC, orbital frontal cortex; M1C, primary motor cortex (M1); S1C, primary somatosensory cortex (S1); IPC, posteroinferior (ventral) parietal cortex; A1C, primary auditory cortex (core); STC, posterior (caudal) superior temporal cortex (Tac); ITC, inferolateral temporal cortex (Tev); V1C, primary visual cortex (striate cortex, V1); HIP, hippocampus (hippocampal formation); AMY, amygdaloid complex; STR, striatum; MD, mediodorsal nucleus of thalamus; CBC, cerebellar cortex.

correlation was also noticed between the early adult ages of 8 years and later ages like 18 and 40 years ($R = 0.90$) in the *HIP* region. Respective correlogram for *STR* displayed increasing correlation (0.77) between 13 and 37 pcw and all adult ages, and notably, highest correlation score was observed between 19 and 40 years of age ($R = 0.97$) (Figure S2 in Supplementary Material). In case of *AMY*, we noticed very strong correlation between the postnatal ages of 1, 5, 15, and 40 years ($R = 0.88$). However, a moderate correlation was observed among 13 and 37 pcw ($R = 0.67$), and 13 pcw and 40 years stages in case of *AMY* region (Figure S2 in Supplementary Material).

Upregulated Genes in Age-Wise Distribution

In the *HIP*, known biomarker genes like *AIF1* was found to be consistently high ($\log_2$FPKM ~ 5.0), while the expression levels of *SPHK1* and *GBP2* remained low ($\log_2$FPKM < 0). In *STR*, *GBP2* was highly expressed, while *AIF1* and *SPHK1* were low in expression ($\log_2$FPKM < 0). In *AMY*, *AIF1* and *SPHK1* showed consistently high expression ($\log_2$FPKM ~ 3.5, 3.0), while GBP2 remained low in expression throughout prenatal to postnatal brain developmental stages ($\log_2$FPKM < 0). In case of the top upregulated genes (10), variation in gene expression was strongly evident (**Figure 6**).

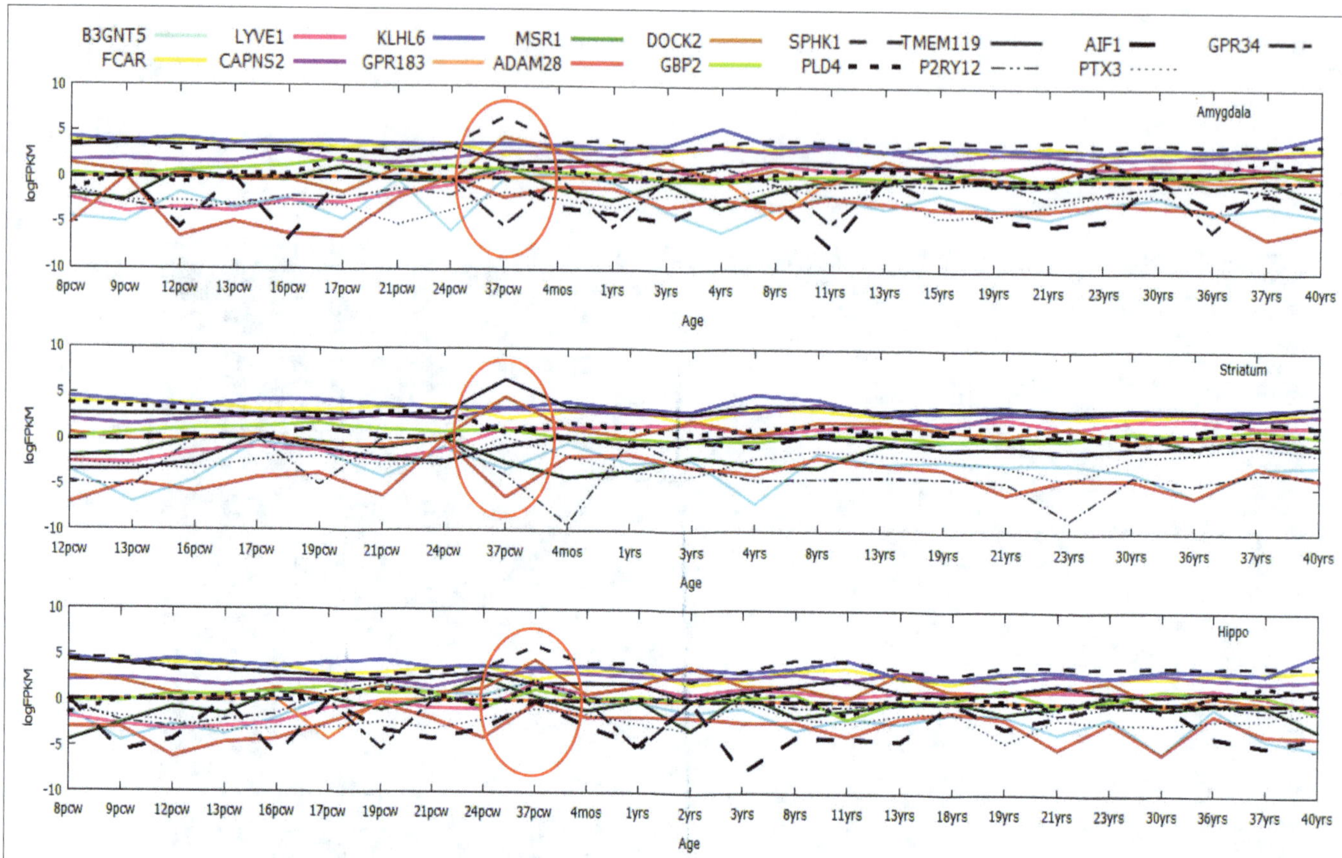

FIGURE 6 | Age-wise expression dynamics of highly upregulated genes across microglia-enriched regions of the human brain. Gene expression profiling of top 10 upregulated genes and known expressed biomarkers (17) in the developing/degenerating brain including prenatal stages like 9 pcw fetus and postnatal brain stages like 3 years to a grown adult 40-year-old brain.

Among the top regulated genes, Lymphatic Vessel Endothelial Hyaluronan Receptor 1 (*LYVE1*), a type I integral membrane glycoprotein was found to be consistently repressed till the prenatal stages, followed by gradual increment in expression during postnatal stages (after 37 pcw) with minor inconsistencies in all the three brain regions (**Figure 6**). A supporting study from mice also showed similarity in the variation of *LYVE1* expression profiles in hippocampal and cerebellar regions (19). Consistent high expression of *FCAR* and *KLHL6* throughout pre- and postnatal development was distinct in all three aforementioned regions. The most inconsistently expressed genes across all the three regions included *DOCK2* and *GPR183* exhibiting a rapid decrease in expression at various ages like 12 pcw, 17 pcw, 24 pcw, 8 years, 21 years, 30 years, and 37 years suggestive of their dynamic roles in brain aging.

Downregulated Genes in Age-Wise Distribution

In case of the top downregulated gene set (10), a uniform pattern of expression was observed. In the *HIP*, all the genes were found to be consistently high in expression (log_2FPKM ~ 4.5) throughout all the age groups excluding two inconsistently expressed genes, *GJA1* and *EFEMP1* (**Figure 7**). We also found that *EFEMP1* showed unaltered expression till the age of 19 years, followed by a sharp dip in expression at the ages of 21 and 40 years. On the other hand, *GJA1* showed prominent inconsistency with a gradual decline toward 13 pcw, increase at 17 pcw, sharp dip at the age of 4 months and finally, remained unaltered in the rest of the postnatal-adult ages. In *AMY*, all the genes showed consistently high expression (log_2FPKM ~ 4.0), which then gradually decreased toward the postnatal stages (log_2FPKM ~ 3.0) excluding F3 gene. Surprisingly, expression of *F3* was seen to be significantly low (log_2FPKM < -2) during prenatal stages, with inconsistency in expression during postnatal ages (**Figure 7**). In *STR*, all genes except *EFEMP1*, *GJA1*, and *GABRA2* were seen to be continually expressed at a high level throughout all age groups. Especially, *GABRA2* showed sharp dips at 17 pcw, 37 pcw, and 3 years whereas, *GJA1* remained repressed (log_2FPKM < 0) with gradual increment in expression during the postnatal stages (log_2FPKM ~ 1.0) (**Figure 7**). Early pieces of evidences from *GJA1*/Connexin 43 null and conditional knock-out mouse brains revealed its strong regulatory role in overall brain development (66). The age-wise expression map (DDM) represented a set of gene fingerprints specific to microglia which consisted of seven genes, namely, *EFEMP1*, *GJA1*, *PTX3*, *KLHL6*, *SPHK1*, *FCAR*, and *CAPNS2* (**Figure 8A**). This gene set showed consistently altered expression (both up and downregulated) in all three

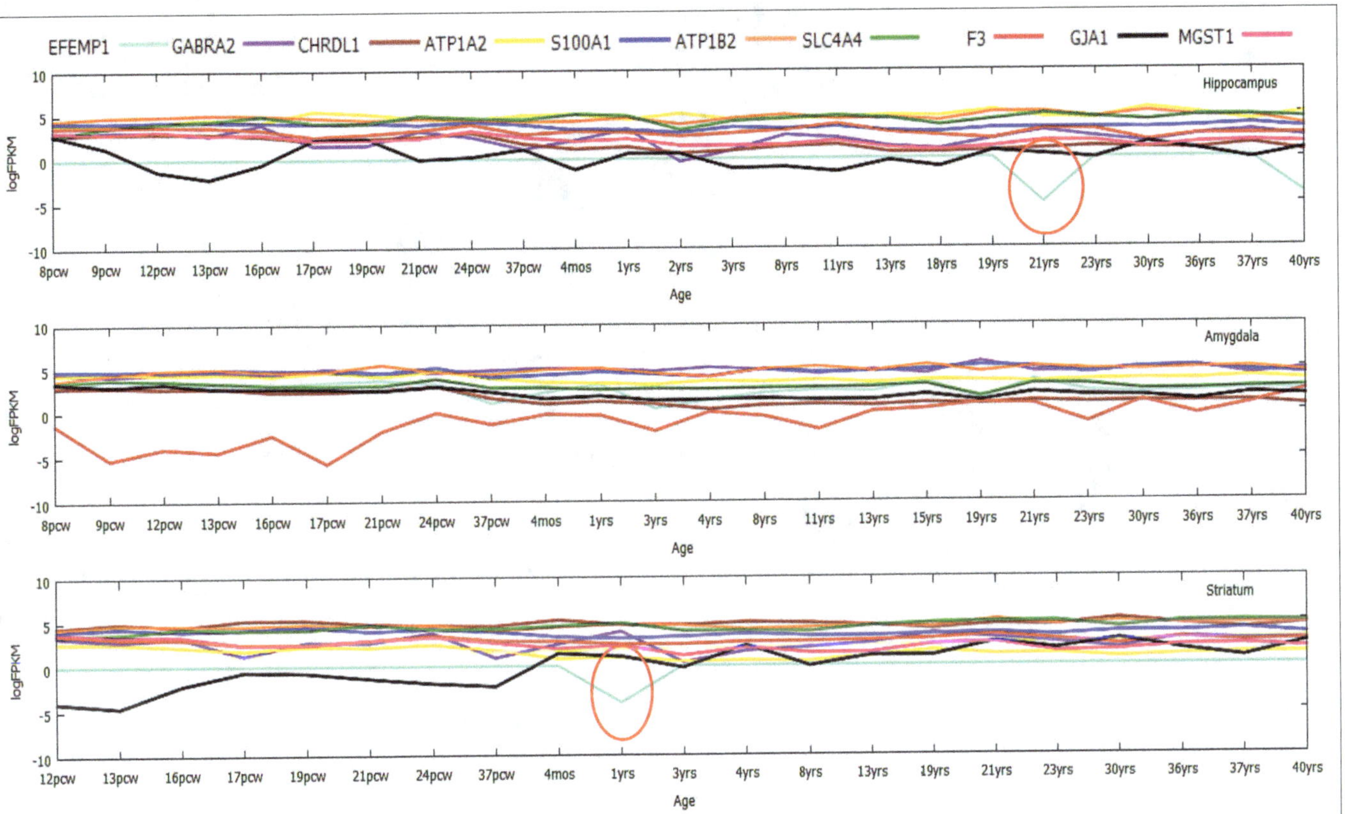

FIGURE 7 | Gene expression profiling of top 10 downregulated genes during prenatal to postnatal brain developmental stages displaying numerous inconsistently or dynamically expressed genes like *DOCK2*.

microglia-dense regions, across the human brain development. To highlight, DDM also revealed three most dynamically regulated genes namely, *DOCK2*, *LYVE1*, and *TMEM119* demonstrating an irregular pattern of expression across the brain ages (**Figure 8B**). These gene signatures can be further explored *in vitro* to elucidate their specific roles in microgliogenesis.

In Vitro Validation of Microglia-Specific Biomarkers in the Cellular Model of Microglia

To validate the emergence of microglia-specific biomarkers during brain development, we have established a cellular model of human microglia using an *in-house* standardized method. To generate human microglial precursors *in vitro*, hESCs were induced for microglial commitment using a long-term culture platform, which could be maintained for a period of 4 to 6 weeks. To authenticate the generation of microglial precursors *in vitro*, we have evaluated the stage-specific lineage commitment that comprised of (i) formation of EBs, (ii) neural induction, and (iii) generation of early and late microglial precursors using specific biomarkers. Details of the culture stages have been explained in the scheme (**Figure 9**). Interestingly, the *in silico* age dynamics study has projected *SPHK1* as a key marker to be consistently upregulated at 37 pcw, specifically in *AMY* and *HIP* (**Figure 6**). It is already evident that *SPHK1* plays crucial roles in microglia-dependent neuroinflammatory cascades. Additionally, we also found *PTX3*, which is another marker of microglia involved in regulation of phagocytic activity to be consistently downregulated across DDM of the human brain, specifically in *AMY* and *HIP* (**Figure 6**). Based on these findings, we further tried to elucidate expression of *SPHK1* and *PTX3* in long-term culture of human microglia.

Our results have demonstrated the appearance of 3D cellular aggregates on fifth DIV with differential expression of *CD24* and *CD184* in both flow cytometric and confocal imaging analysis (**Figure 10**). These cellular aggregates (EBs) were further induced to form early microglial precursors as evident by expression of *CD68* and *AIF1/IBA1* on 21DIV as shown in the flow cytometry micrographs (**Figure 10A**). These precursors later gave rise to microglial structures on 34DIV with distinct expression of two mature microglial markers *SPHK1* and *PTX3*, as evident in immunofluorescence-based studies (**Figure 10B**). This study has explored the optimal microenvironment for establishing cellular model mimicking human microglia and the associated important biomarkers.

DISCUSSION

Albeit microglia constitute approximately 10% of the total cell population in the CNS (67), there are major lacunae in the knowledge regarding their expression across human brain ages

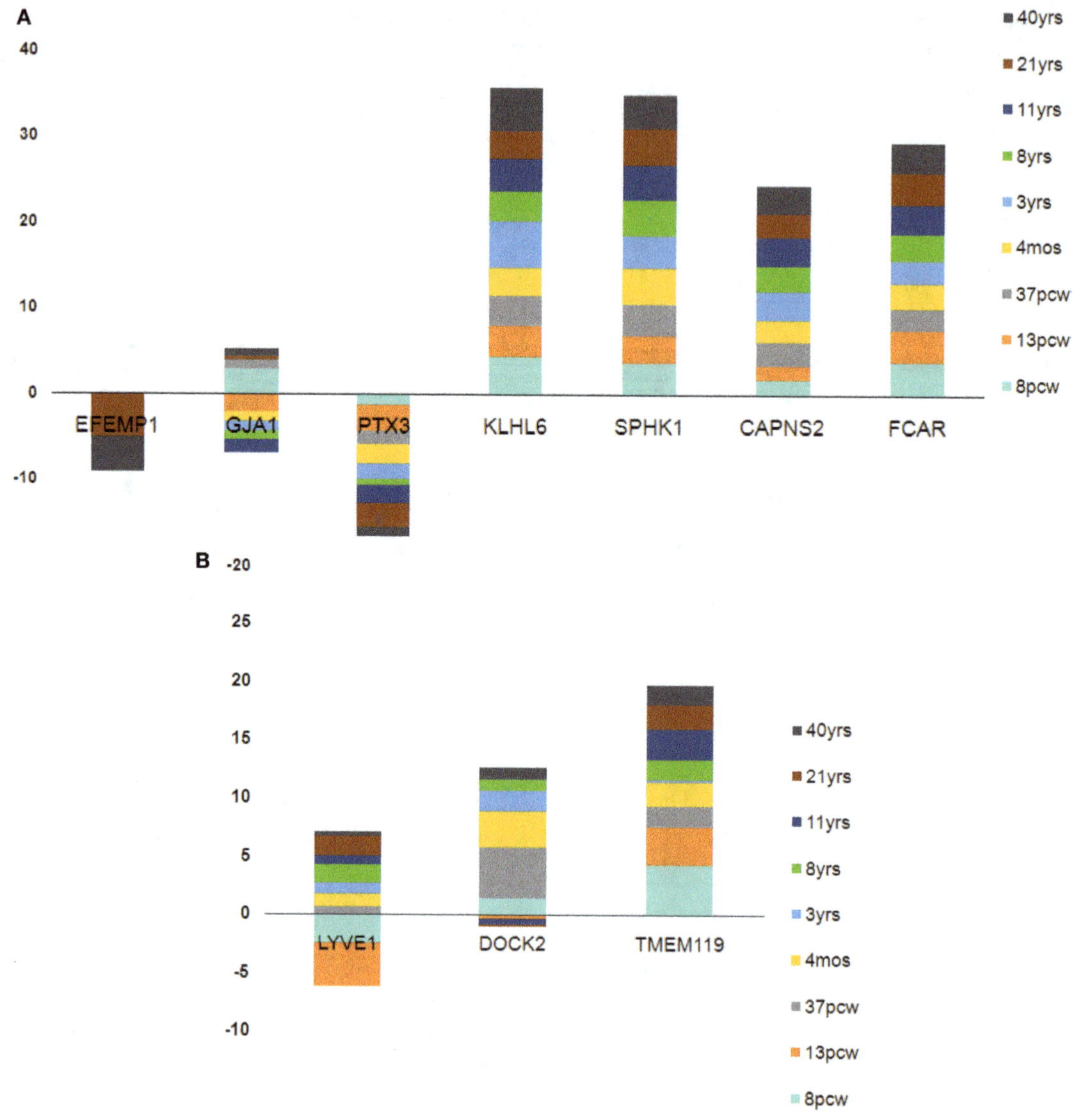

FIGURE 8 | Age-wise dynamics yielded crucial microglia-specific signatures. **(A)** Seven genes found to be consistently high in expression from pre- to prenatal brain developmental stages. **(B)** Three genes found to be dynamically expressed with sharp dips at the transition stage of 37 postconception week (pcw) except in case of *DOCK2* which was inconsistently expressed throughout pre- and postnatal stages. Each stack in the stacked bar plot represents a single brain age whose height denotes the amount of expression.

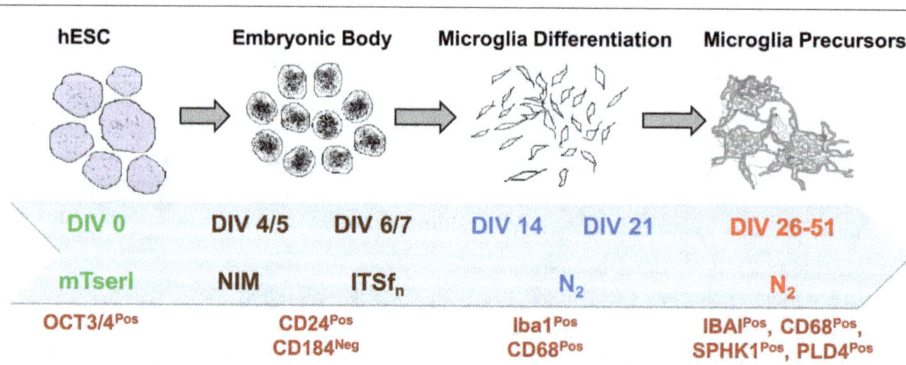

FIGURE 9 | Brief scheme demonstrating the protocol of directed differentiation of hESCs into human microglia which is characterized by several stage-specific markers.

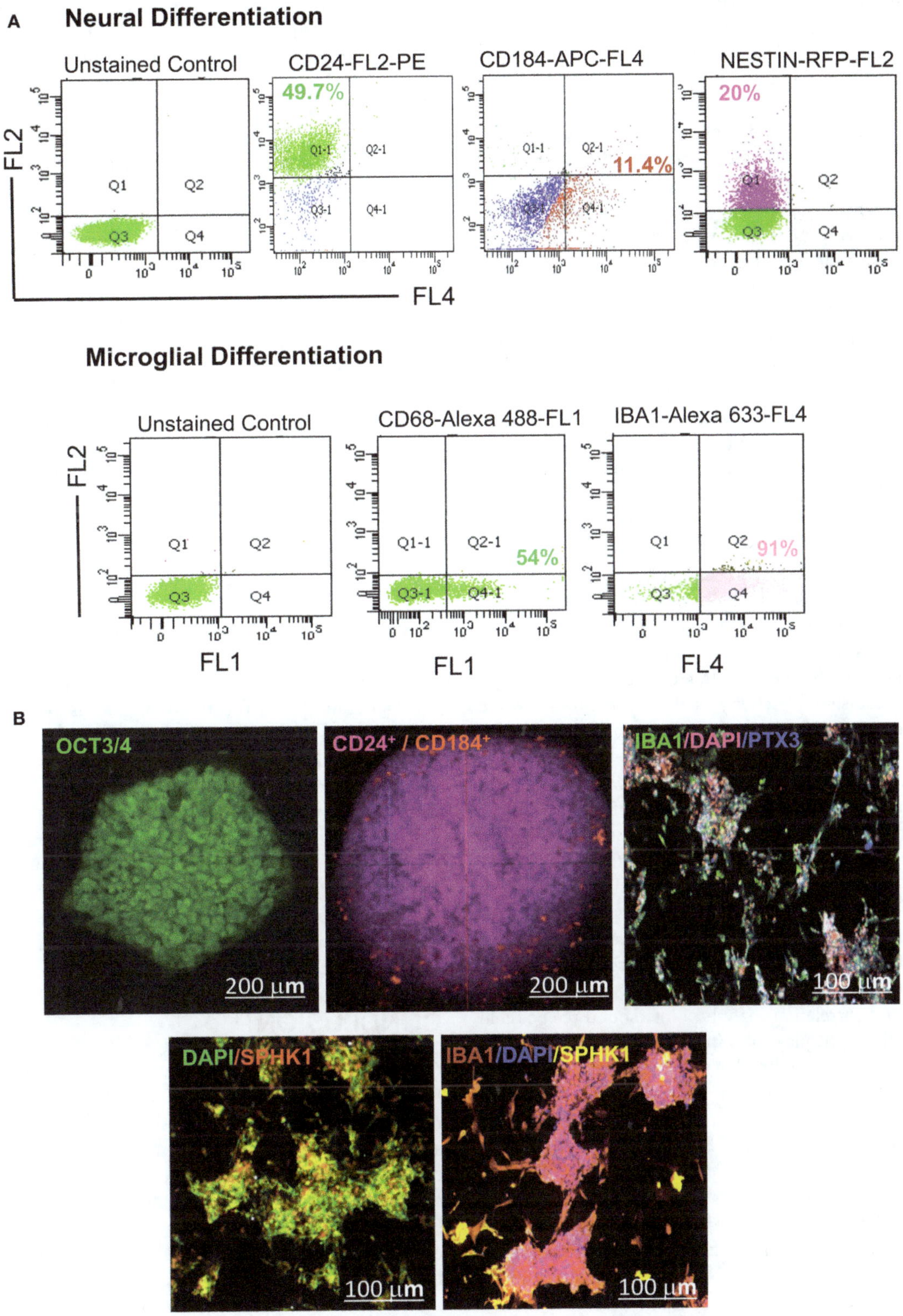

FIGURE 10 | Identification of hESC-derived microglia precursors *in vitro* using different stage-specific markers. **(A)** Flow cytometry based analysis demonstrated characterization of microglia induction and differentiation of microglial precursors. The analysis indicated differentiation of hESCs into embryonic bodies (EB, 5 DIV), as shown by expression of ~49.7% *CD24*^{Pos} and 11.4% *CD184*^{Pos} population. This is followed by neural induction as shown by 20% of *NESTIN*^{Pos} cells. Neurally induced cells further showed the generation of a distinct population having 54% of *CD68*^{Pos} cells and 91% *IBA1*^{Pos} cells. **(B)** Confocal imaging of hESC-derived EBs showed expression of *CD24* and *CD184 in vitro* at 5 DIV. The emergence of mature microglial precursors could be detected on 34DIV, as shown by *IBA1*, *SPHK1*, and *PTX3* expression. DAPI = nuclear marker, DIV = days *in vitro*. The scale bar represents size in micrometers.

and regions (68). Many critical questions involving their ontogeny, localization, and self-renewal still need to be addressed. Discovery of novel microglial biomarkers and elucidation of their specific roles in brain developmental processes will shed light on the onset of neurological disease and progression.

Studies involving human microglia-derived transcriptomics and proteomics are still in its infancy, and given the low scope of availability of healthy/diseased human brain tissue, it is imperative to perform high-throughput data analysis to be able to identify and narrow down crucial microglia signatures (33). In this direction, using available tissue based transcriptomics (69) as well as induced microglia microarray data generated from pluripotent stem cells (25), we analyzed the differentially regulated gene sets pertinent to microglia biology and homeostasis, which resulted in a novel catalog of microglial biomarkers. Further using GSEA on the ascertained list, we have identified different clusters of gene sets specific to human microglia, which revealed several crucial clusters such as metalloproteinases, GPCRs, proteoglycans, immunoeffectors, gap junction family genes, and synaptosomal receptors, etc. These prominent gene clusters explained multidimensional roles and biological activity of microglia. This GSEA map acted as a primer for further downstream analyses. Correlation analysis further explained the differential expression behavior of many genes between the human brain cell types and induced microglia data sets, *viz.*, *primary* and *secondary* data sets, respectively. Another interesting inference indicates that there might be different gene subsets required for maintaining microglial homeostasis and activity in the human brain, while enforced microglial differentiation *in vitro* requires an exclusive set of gene regulators. To understand the protein–protein interactions among the shortlisted candidates, we constructed a primary protein neighborhood network and performed pathway analysis. The study highlights some novel pathways associated with microgliasuch as Glycosaminoglycan metabolism, RAP1, Phospholipase C/D, *PPAR*, Fatty acid CoA signaling pathways. Upon assessment of the DDM across brain age groups, we have identified the top 20 altered genes as molecular signatures to developmentally track microglia in fetal, infant, child, adolescent, and adult brains.

Both GSEA and correlation analysis pointed out three possible hypotheses which might play roles in microglia-NPC crosstalk. The first hypothesis suggests that enhanced expression of *PTGS1/2* is linked to *EP2* receptors in microglia and might be the key to crosstalk with the qNSCs that express specifically *PGE2*. The second hypothesis discusses the interaction between microglia and qNSCs through plausible crosstalk between *DOCK2* and S1P, as evidence suggests qNSCs specifically express S1P as biomarker and microglia express SPHK1, *DOCK2*, and *S1PR1/5* (S1P receptor 1/5). The third hypothesis suggests that *GPR183* acts a conditioned rheostat for NSC quiescence and activity in *NOTCH1* dependent pathway. Since differential expression analysis showed high expression of *GPR183* in microglia, and *NOTCH1* regulates self-renewal of hippocampal NSCs, there can be possible molecular crosstalk (70, 71).

Studies from past few decades have indicated both specific and context-dependent localization and activity of microglia during neuroinflammatory circumstances (18, 39). However, hardly any study exists which could track their activity throughout brain development. Another major roadblock that is faced is the unavailability of human brain cells. To address such issues, we established a cellular model of the human microglial cell population and deciphered the expression of existing biomarkers, especially *CD68*, *AIF1 (IBA1)*, *SPHK1*, and *PTX3*. Earlier findings have indicated that SPHK1 is found to be abundantly expressed in several rat brain cell types, including neurons in the hippocampus, cerebellar granule cells, and astrocytes and also in the amoeboid microglial cells of the corpus callosum of the postnatal brain (72). The *in silico* data strongly corroborated with this existing evidence, which suggested that *SPHK1* expression is significantly upregulated in microglia-dense regions, namely hippocampus and amygdala (**Figure 6**). The *in vitro* long-term culture study also supported these findings and suggested that human microglia precursors from post 26DIV showed consistent expression of *SPHK1* in late microglial precursors (**Figure 10B**). The results have also represented another unique marker, *PTX3* in the in cultured human microglial precursor cells, which is known to regulate the phagocytic activity of microglia (44). This observation supported the DDM data which has shown *PTX3* to be consistently downregulated across different ages in HIP and AMY of healthy brains. In summary, our *in vitro* study has validated the emergence of the microglial signatures in hESC-derived human microglial precursors and also represented a robust human microglial model for future investigations involving microglial biology and activity.

Microglia has been associated with onset, aggravation, and progression of neurodegenerative diseases like Alzheimer's, Dementia, and Glioma (73–76). This is primarily due to dysregulated microglial activity during development as well as disease leading to severe neurocognitive impairments and neuronal dysfunction. Overall, this study has unraveled fundamental information underlying the labyrinthine molecular circuitry in microglia functioning, physiology, homeostasis, and genetics across brain developmental stages and regions.

CONCLUSION

Our study has introduced a novel integrative biology approach for tracking human microglia signatures across 25 developmental stages spanning fetal, infant, adolescence, and adult, and delineated their localized expression in microglia-dense regions. The *in vitro* data have also validated the expression of some of the biomarkers in human microglial precursors. Precisely, tracking of microglia transcriptomic dynamics based on brain anatomy and age could establish an essential DDM comprising highly altered gene clusters. This map can be further utilized to develop set of putative prognostic markers in accordance with the age of the human brain. One of the interesting readouts of the DDM was the dynamic expression associated with 37 pcw, i.e., a transition period between prenatal and postnatal brain. To highlight, the study has led to the discovery of microglia age-specific gene signatures, including *EFEMP1*,

GJA1, *PTX3*, *KLHL6*, *SPHK1*, *FCAR*, and *CAPNS2*, which can be further explored for understanding their roles in microglial biology in the human brain.

FUTURE PERSPECTIVES

In essence, this study has established a repertoire of microglia-specific signatures involved in the developing human brain. Characterization of mechanistic roles of these gene fingerprints in experimental models would further bridge the gap in understanding the molecular mechanisms involved in microglia ontogeny, renewal, and maintenance. Age-wise expression profiling of these novel signatures (DDM) in both pre- and postnatal brain stages can act as guidance cues for designing future prognostic and diagnostic biomarkers against neurodegenerative diseases. Finally, the data associated with age and region-specificity can pave the way to personalized medicine interventions for debilitating neurological diseases.

AUTHOR CONTRIBUTIONS

RA: performed analyses, computations, and graphics. RA and SP: designed the methodological approach and wrote the manuscript. SP and SS: read, critically analyzed, and approved the final manuscript.

ACKNOWLEDGMENTS

RA is grateful to Shiv Nadar University for the Ph.D. fellowship. The Center for Informatics, Shiv Nadar University is also duly acknowledged. SS would like to acknowledge Innovative Young Biotechnologist Award (IYBA) from Department of Biotechnology. SP would like to sincerely thank Shiv Nadar Foundation for the ongoing work. The authors sincerely thank Dr. Mirjana Maletic-Savatic for providing support for conducting hESC-related experiments and acknowledge the Human Embryonic Stem Cell Core, Baylor College of Medicine for giving the h9 clones.

SUPPLEMENTARY MATERIAL

FIGURE S1 | Donut plot showing various protein classes associated with the differentially expressed gene set or 3,290 genes.

FIGURE S2 | Correlation matrix/correlogram showing varying correlation ranging from very weak to very strong (0.0–0.99) among significant brain ages within three microglia-enriched regions, namely AMY, HIP, and STR. The color bar depicts the range of correlation coefficient (−1, 1) and in each correlogram, the colored slice in the pie increases in size and intensity in accordance with the correlation coefficient value.

REFERENCES

Nissl F. Ueber einige Beziehungen zwishcen Nerven zellerkrankungen und gliosen Erscheinnungen bei verschiedenen Psychosen. *Arch Psychiatry* (1899) 32:1–21.

Robertson F. A microscopic demonstration of the normal and pathological histology of mesoglia cells. *Br J Psychiatry* (1900) 46(195):724.

Tremblay ME, Lecours C, Samson L, Sanchez-Zafra V, Sierra A. From the Cajal alumni Achucarro and Rio-Hortega to the rediscovery of never-resting microglia. *Front Neuroanat* (2015) 9:45. doi:10.3389/fnana.2015.00045

Chen Z, Trapp BD. Microglia and neuroprotection. *J Neurochem* (2016) 136 (Suppl 1):10–7. doi:10.1111/jnc.13062

Gomez-Nicola D, Perry VH. Microglial dynamics and role in the healthy and diseased brain: a paradigm of functional plasticity. *Neuroscientist* (2015) 21(2):169–84. doi:10.1177/1073858414530512

Paolicelli RC, Bolasco G, Pagani F, Maggi L, Scianni M, Panzanelli P, et al. Synaptic pruning by microglia is necessary for normal brain development. *Science* (2011) 333(6048):1456–8. doi:10.1126/science.1202529

Mosher KI, Andres RH, Fukuhara T, Bieri G, Hasegawa-Moriyama M, He Y, et al. Neural progenitor cells regulate microglia functions and activity. *Nat Neurosci* (2012) 15(11):1485–7. doi:10.1038/nn.3233

Cunningham CL, Martínez-Cerdeño V, Noctor SC. Microglia regulate the number of neural precursor cells in the developing cerebral cortex. *J Neurosci* (2013) 33(10):4216–33. doi:10.1523/JNEUROSCI.3441-12.2013

Eyo UB, Wu LJ. Bidirectional microglia-neuron communication in the heal- thy brain. *Neural Plast* (2013) 2013:456857. doi:10.1155/2013/456857

Suzumura A. Neuron-microglia interaction in neuroinflammation. *Curr Pro- tein Pept Sci* (2013) 14(1):16–20. doi:10.2174/1389203711314010004

Andreasson KI, Bachstetter AD, Colonna M, Ginhoux F, Holmes C, Lamb B, et al. Targeting innate immunity for neurodegenerative disorders of the central nervous system. *J Neurochem* (2016) 138(5):653–93. doi:10.1111/jnc.13667

Alliot F, Godin I, Pessac B. Microglia derive from progenitors, originating from the yolk sac, and which proliferate in the brain. *Brain Res Dev Brain Res* (1999) 117(2):145–52. doi:10.1016/S0165-3806(99)00113-3

Alliot F, Lecain E, Grima B, Pessac B. Microglial progenitors with a high proliferative potential in the embryonic and adult mouse brain. *Proc Natl Acad Sci U S A* (1991) 88(4):1541–5. doi:10.1073/pnas.88.4.1541

Tay TL, Savage JC, Hui CW, Bisht K, Tremblay ME. Microglia across the life- span: from origin to function in brain development, plasticity and cognition. *J Physiol* (2017) 595(6):1929–45. doi:10.1113/JP272134

Schulz C, Gomez Perdiguero E, Chorro L, Szabo-Rogers H, Cagnard N, Kierdorf K, et al. A lineage of myeloid cells independent of Myb and hematopoi- etic stem cells. *Science* (2012) 336(6077):86–90. doi:10.1126/science.1219179

Prinz M, Mildner A. Microglia in the CNS: immigrants from another world. *Glia* (2011) 59(2):177–87. doi:10.1002/glia.21104

Rubenstein J, Rakic P. *Patterning and Cell Type Specification in the Developing CNS and PNS: Comprehensive Developmental Neuroscience*. 1st ed. Amsterdam: Academic Press (2013).

Tay TL, Mai D, Dautzenberg J, Fernández-Klett F, Lin G, Datta M, et al. A new fate mapping system reveals context-dependent random or clonal expansion of microglia. *Nat Neurosci* (2017) 20(6):793–803. doi:10.1038/nn.4547

Grabert K, Michoel T, Karavolos MH, Clohisey S, Baillie JK, Stevens MP, et al. Microglial brain region-dependent diversity and selective regional sensitivi- ties to aging. *Nat Neurosci* (2016) 19(3):504–16. doi:10.1038/nn.4222

Hart AD, Wyttenbach A, Perry VH, Teeling JL. Age related changes in micro- glial phenotype vary between CNS regions: grey versus white matter differ- ences. *Brain Behav Immun* (2012) 26(5):754–65. doi:10.1016/j.bbi.2011.11.006

Harry GJ. Microglia during development and aging. *Pharmacol Ther* (2013) 139(3):313–26. doi:10.1016/j.pharmthera.2013.04.013

Koellhoffer EC, McCullough LD, Ritzel RM. Old maids: aging and its impact on microglia function. *Int J Mol Sci* (2017) 18(4):769. doi:10.3390/ijms18040769

Norden DM, Godbout JP. Review: microglia of the aged brain: primed to be activated and resistant to regulation. *Neuropathol Appl Neurobiol* (2013) 39(1):19–34. doi:10.1111/j.1365-2990.2012.01306.x

Zhang Y, Sloan SA, Clarke LE, Caneda C, Plaza CA, Blumenthal PD, et al. Purification and characterization of progenitor and mature human astrocytes reveals transcriptional and functional differences with mouse. *Neuron* (2016) 89(1):37–53. doi:10.1016/j.neuron.2015.11.013

Pandya H, Shen MJ, Ichikawa DM, Sedlock AB, Choi Y, Johnson KR, et al. Differentiation of human and murine induced pluripotent stem cells to microglia-like cells. *Nat Neurosci* (2017) 20(5):753–9. doi:10.1038/nn.4534

Carvalho BS, Irizarry RA. A framework for oligonucleotide microarray preprocessing. *Bioinformatics* (2010) 26(19):2363–7. doi:10.1093/bioinformatics/btq431

Mi H, Huang X, Muruganujan A, Tang H, Mills C, Kang D, et al. PANTHER version 11: expanded annotation data from gene ontology and reactome pathways, and data analysis tool enhancements. *Nucleic Acids Res* (2017) 45 (D1):D183–9. doi:10.1093/nar/gkw1138

Kamburov A, Stelzl U, Lehrach H, Herwig R. The ConsensusPathDB inte- raction database: 2013 update. *Nucleic Acids Res* (2013) 41(D1):D793–800. doi:10.1093/nar/gks1055

Szklarczyk D, Morris JH, Cook H, Kuhn M, Wyder S, Simonovic M, et al. The STRING database in 2017: quality-controlled protein–protein association networks, made broadly accessible. *Nucleic Acids Res* (2017) 45(D1):D362–8. doi:10.1093/nar/gkw937

Cline MS, Smoot M, Cerami E, Kuchinsky A, Landys N, Workman C, et al. Integration of biological networks and gene expression data using Cytoscape. *Nat Protoc* (2007) 2(10):2366–82. doi:10.1038/nprot.2007.324

Sunkin SM, Ng L, Lau C, Dolbeare T, Gilbert TL, Thompson CL, et al. Allen Brain Atlas: an integrated spatio-temporal portal for exploring the central nervous system. *Nucleic Acids Res* (2013) 41(D1):D996–1008. doi:10.1093/ nar/gks1042

Shen EH, Overly CC, Jones AR. The Allen Human Brain Atlas: comprehen- sive gene expression mapping of the human brain. *Trends Neurosci* (2012) 35(12):711–4. doi:10.1016/j.tins.2012.09.005

Mahfouz A, Huisman SMH, Lelieveldt BPF, Reinders MJT. Brain transcrip- tome atlases: a computational perspective. *Brain Struct Funct* (2017) 222(4): 1557–80. doi:10.1007/s00429-016-1338-2

Miller JA, Ding SL, Sunkin SM, Smith KA, Ng L, Szafer A, et al. Transcriptional landscape of the prenatal human brain. *Nature* (2014) 508(7495):199–206. doi:10.1038/nature13185

Team R. *RStudio: Integrated Development for R.* Boston, MA: RStudio, Inc (2015). Available from: http://www.rstudio.com (Accessed: May 12, 2017).

Cho EG, Zaremba JD, McKercher SR, Talantova M, Tu S, Masliah E, et al. MEF2C enhances dopaminergic neuron differentiation of human embry- onic stem cells in a parkinsonian rat model. *PLoS One* (2011) 6(8):e24027. doi:10.1371/journal.pone.0024027

Beutner C, Roy K, Linnartz B, Napoli I, Neumann H. Generation of micro- glial cells from mouse embryonic stem cells. *Nat Protoc* (2010) 5(9):1481–94. doi:10.1038/nprot.2010.90

Bryan L, Kordula T, Spiegel S, Milstien S. Regulation and functions of sphin- gosine kinases in the brain. *Biochim Biophys Acta* (2008) 1781(9):459–66. doi:10.1016/j.bbalip.2008.04.008

Lv M, Zhang D, Dai D, Zhang W, Zhang L. Sphingosine kinase 1/sphingosine-1-phosphate regulates the expression of interleukin-17A in activated micro- glia in cerebral ischemia/reperfusion. *Inflamm Res* (2016) 65(7):551–62. doi:10.1007/s00011-016-0939-9

Lou N, Takano T, Pei Y, Xavier AL, Goldman SA, Nedergaard M. Purinergic receptor P2RY12-dependent microglial closure of the injured blood–brain barrier. *Proc Natl Acad Sci U S A* (2016) 113(4):1074–9. doi:10.1073/ pnas.1520398113

Yoshikawa F, Banno Y, Otani Y, Yamaguchi Y, Nagakura-Takagi Y, Morita N, et al. Phospholipase D family member 4, a transmembrane glycoprotein with no phospholipase D activity, expression in spleen and early postnatal microglia. *PLoS One* (2010) 5(11):e13932. doi:10.1371/journal.pone.0013932

Otani Y, Yamaguchi Y, Sato Y, Furuichi T, Ikenaka K, Kitani H, et al. PLD4 is involved in phagocytosis of microglia: expression and localization changes of PLD4 are correlated with activation state of microglia. *PLoS One* (2011) 6(11):e27544. doi:10.1371/journal.pone.0027544

Satoh J, Kino Y, Asahina N, Takitani M, Miyoshi J, Ishida T, et al. TMEM119 marks a subset of microglia in the human brain. *Neuropathology* (2016) 36(1):39–49. doi:10.1111/neup.12235

Jeon H, Lee S, Lee WH, Suk K. Analysis of glial secretome: the long pentraxin PTX3 modulates phagocytic activity of microglia. *J Neuroimmunol* (2010) 229(1–2):63–72. doi:10.1016/j.jneuroim.2010.07.001

Cimino PJ, Sokal I, Leverenz J, Fukui Y, Montine TJ. DOCK2 is a microg- lial specific regulator of central nervous system innate immunity found in normal and Alzheimer's disease brain. *Am J Pathol* (2009) 175(4):1622–30. doi:10.2353/ajpath.2009.090443

Bayless KJ, Davis GE. Sphingosine-1-phosphate markedly induces matrix metalloproteinase and integrin-dependent human endothelial cell invasion and lumen formation in three-dimensional collagen and fibrin matrices. *Biochem Biophys Res Commun* (2003) 312(4):903–13. doi:10.1016/j.bbrc.2003.11.017

Chi H. Sphingosine 1-phosphate and immune regulation: trafficking and beyond. *Trends Pharmacol Sci* (2011) 32(1):16–24. doi:10.1016/j.tips.2010. 11.002

Liu SH, Yang CN, Pan HC, Sung YJ, Liao KK, Chen WB, et al. IL-13 down- regulates PPAR-gamma/heme oxygenase-1 via ER stress-stimulated calpain activation: aggravation of activated microglia death. *Cell Mol Life Sci* (2010) 67(9):1465–76. doi:10.1007/s00018-009-0255-4

Galatro TF, Holtman IR, Lerario AM, Vainchtein ID, Brouwer N, Sola PR, et al. Transcriptomic analysis of purified human cortical microglia reveals age-associated changes. *Nat Neurosci* (2017) 20(8):1162–71. doi:10.1038/nn.4597

Nuttall RK, Silva C, Hader W, Bar-Or A, Patel KD, Edwards DR, et al. Metalloproteinases are enriched in microglia compared with leukocytes and they regulate cytokine levels in activated microglia. *Glia* (2007) 55(5):516–26. doi:10.1002/glia.20478

Biellmann F, Hulsmeier AJ, Zhou D, Cinelli P, Hennet T. The Lc3-synthase gene B3gnt5 is essential to pre-implantation development of the murine embryo. *BMC Dev Biol* (2008) 8:109. doi:10.1186/1471-213X-8-109

Wirz KT, Bossers K, Stargardt A, Kamphuis W, Swaab DF, Hol EM, et al. Cortical beta amyloid protein triggers an immune response, but no synaptic changes in the APPswe/PS1dE9 Alzheimer's disease mouse model. *Neuro- biol Aging* (2013) 34(5):1328–42. doi:10.1016/j.neurobiolaging.2012.11.008

Kocur M, Schneider R, Pulm A-K, Bauer J, Kropp S, Gliem M, et al. IFNβ secreted by microglia mediates clearance of myelin debris in CNS autoimmu- nity. *Acta Neuropathol Commun* (2015) 3(1):20. doi:10.1186/s40478-015-0192-4

Huang JK, Jarjour AA, Oumesmar BN, Kerninon C, Williams A, Krezel W, et al. Retinoid X receptor gamma signaling accelerates CNS remyelination. *Nat Neurosci* (2011) 14(1):45–53. doi:10.1038/nn.2702

Ackermann GE, Marenholz I, Wolfer DP, Chan WY, Schäfer B, Erne P, et al. S100A1-deficient male mice exhibit increased exploratory activity and reduced anxiety-related responses. *Biochim Biophys Acta* (2006) 1763(11): 1307–19. doi:10.1016/j.bbamcr.2006.08.048

Gonzalez-Nunez V. Role of gabra2, GABAA receptor alpha-2 subunit, in CNS development. *Biochem Biophys Rep* (2015) 3:190–201. doi:10.1016/j.bbrep.2015.08.003

Hu Y, Pioli PD, Siegel E, Zhang Q, Nelson J, Chaturbedi A, et al. EFEMP1 suppresses malignant glioma growth and exerts its action within the tumor extracellular compartment. *Mol Cancer* (2011) 10(1):123. doi:10.1186/ 1476-4598-10-123

Ji P, Schachtschneider KM, Schook LB, Walker FR, Johnson RW. Peri- pheral viral infection induced microglial sensome genes and enhanced microglial cell activity in the hippocampus of neonatal piglets. *Brain Behav Immun* (2016) 54:243–51. doi:10.1016/j.bbi.2016.02.010

Paznekas WA, Boyadjiev SA, Shapiro RE, Daniels O, Wollnik B, Keegan CE, et al. Connexin 43 (GJA1) mutations cause the pleiotropic phenotype of oculodentodigital dysplasia. *Am J Hum Genet* (2003) 72(2):408–18. doi:10.1086/346090

D'Asti E, Fang Y, Rak J. Brain neoplasms and coagulation—lessons from heterogeneity. *Rambam Maimonides Med J* (2014) 5(4):e0030. doi:10.5041/RMMJ.10164

Fagerberg L, Hallstrom BM, Oksvold P, Kampf C, Djureinovic D, Odeberg J, et al. Analysis of the human tissue-specific expression by genome-wide integration of transcriptomics and antibody-based proteomics. *Mol Cell Proteomics* (2014) 13(2):397–406. doi:10.1074/mcp.M113.035600

Han X, Zha H, Yang F, Guo B, Zhu B. Tumor-derived tissue factor aberrantly activates complement and facilitates lung tumor progression via recruitment of myeloid-derived suppressor cells. *Int J Mol Sci* (2017) 18(1):22. doi:10.3390/ijms18010022

Shi L. Dock protein family in brain development and neurological disease. *Commun Integr Biol* (2013) 6(6):e26839. doi:10.4161/cib.26839

Stone EM, Lotery AJ, Munier FL, Heon E, Piguet B, Guymer RH, et al. A single EFEMP1 mutation associated with both malattia leventinese and Doyne honeycomb retinal dystrophy. *Nat Genet* (1999) 22(2):199–202. doi:10.1038/9722

Fu L, Garland D, Yang Z, Shukla D, Rajendran A, Pearson E, et al. The R345W mutation in EFEMP1 is pathogenic and causes AMD-like deposits in mice. *Hum Mol Genet* (2007) 16(20):2411–22. doi:10.1093/hmg/ddm198

Iacobas S, Iacobas DA, Spray DC, Scemes E. The connexin43-dependent transcriptome during brain development: importance of genetic background. *Brain Res* (2012) 1487:131–9. doi:10.1016/j.brainres.2012.05.062

Aguzzi A, Barres BA, Bennett ML. Microglia: scapegoat, saboteur, or some- thing else? *Science* (2013) 339(6116):156–61. doi:10.1126/science.1227901

Hickman SE, Kingery ND, Ohsumi TK, Borowsky ML, Wang LC, Means TK, et al. The microglial sensome revealed by direct RNA sequencing. *Nat Neurosci* (2013) 16(12):1896–905. doi:10.1038/nn.3554

Hawrylycz MJ, Lein ES, Guillozet-Bongaarts AL, Shen EH, Ng L, Miller JA, et al. An anatomically comprehensive atlas of the adult human brain trans- criptome. *Nature* (2012) 489(7416):391–9. doi:10.1038/nature11405

Zhang P, He Q, Chen D, Liu W, Wang L, Zhang C, et al. G protein-coupled receptor 183 facilitates endothelial-to-hematopoietic transition via Notch1 inhibition. *Cell Res* (2015) 25(10):1093–107. doi:10.1038/cr.2015.109

Ables JL, Decarolis NA, Johnson MA, Rivera PD, Gao Z, Cooper DC, et al. Notch1 is required for maintenance of the reservoir of adult hip- pocampal stem cells. *J Neurosci* (2010) 30(31):10484–92. doi:10.1523/ JNEUROSCI.4721-09.2010

Lin H, Baby N, Lu J, Kaur C, Zhang C, Xu J, et al. Expression of sphin- gosine kinase 1 in amoeboid microglial cells in the corpus callosum of post- natal rats. *J Neuroinflammation* (2011) 8:13. doi:10.1186/1742-2094-8-13

Walker DG, Lue L-F. Immune phenotypes of microglia in human neuro-degenerative disease: challenges to detecting microglial polarization in human brains. *Alzheimers Res Ther* (2015) 7(1):56. doi:10.1186/s13195-015-0139-9

Sanchez-Guajardo V, Barnum Christopher J, Tansey Malú G, Romero- Ramos M. Neuroimmunological processes in Parkinson's disease and their relation to α-synuclein: microglia as the referee between neuronal processes and peripheral immunity. *ASN Neuro* (2013) 5(2):e00112. doi:10.1042/ AN20120066

Madeira MH, Boia R, Santos PF, Ambrósio AF, Santiago AR. Contribution of microglia-mediated neuroinflammation to retinal degenerative diseases. *Mediators Inflamm* (2015) 2015:673090. doi:10.1155/2015/673090

Kumar A, Stoica BA, Loane DJ, Yang M, Abulwerdi G, Khan N, et al. Microglial-derived microparticles mediate neuroinflammation after trau- matic brain injury. *J Neuroinflammation* (2017) 14:47. doi:10.1186/s12974- 017-0819-4

17

SDF1-CXCR4 Signaling Maintains Central Post-Stroke Pain through Mediation of Glial-Neuronal Interactions

Fei Yang[1†], Wen-Jun Luo[1†], Wei Sun[1,2†], Yan Wang[1,2], Jiang-Lin Wang[1], Fan Yang[1,2], Chun-Li Li[1,2], Na Wei[1,2], Xiao-Liang Wang[1,2], Su-Min Guan[3] and Jun Chen[1,2,4]**

[1]Institute for Biomedical Sciences of Pain, Tangdu Hospital, The Fourth Military Medical University, Xi'an, China, [2]Key Laboratory of Brain Stress and Behavior, People's Liberation Army (PLA), Xi'an, China, [3]School of Stomatology, The Fourth Military Medical University, Xi'an, China, [4]Beijing Institute for Brain Disorders, Beijing, China

***Correspondence:**
Su-Min Guan
jchsmg@fmmu.edu.cn
Jun Chen
junchen@fmmu.edu.cn

†These authors have contributed equally to this work.

Central post-stroke pain (CPSP) is an intractable central neuropathic pain that has been poorly studied mechanistically. Here we showed that stromal cell-derived factor 1 (SDF1 or CXCL12), a member of the CXC chemokine family, and its receptor CXCR4 played a key role in the development and maintenance of thalamic hemorrhagic CPSP through hypoxia inducible factor 1α (HIF-1α) mediated microglial-astrocytic-neuronal interactions. First, both intra-thalamic collagenase (ITC) and SDF1 injections could induce CPSP that was blockable and reversible by intra-thalamic administration of both AMD3100 (a selective CXCR4 antagonist) and inhibitors of microglial or astrocytic activation. Second, long-term increased-expression of SDF1 and CXCR4 that was accompanied by activations of both microglia and astrocytes following ITC could be blocked by both AMD-3100 and YC-1, a selective inhibitor of HIF-1α. AMD-3100 could also inhibit release of proinflammatory mediators (TNFα, IL1β and IL-6). Increased-expression of HIF-1α, SDF1, CXCR4, Iba1 and GFAP proteins could be induced by both ITC and intra-thalamic $CoCl_2$, an inducer of HIF-1α that was blockable by both HIF-1α inhibition and CXCR4 antagonism. Finally, inhibition of HIF-1α was only effective in prevention, but not in treatment of ITC-induced CPSP. Taken together, the present study demonstrated that in the initial process of thalamic hemorrhagic state HIF-1α up-regulated SDF1-CXCR4 signaling, while in the late process SDF1-CXCR4 signaling-mediated positive feedback plays more important role in glial-glial and glial-neuronal interactions and might be a novel promising molecular target for treatment of CPSP in clinic.

Keywords: central post-stroke pain, intra-thalamic injection, mechanical pain hypersensitivity, SDF1-CXCR4 signaling, hypoxia inducible factor 1α, glial-neuronal crosstalk

INTRODUCTION

Stroke is the leading cause of adult disability worldwide (Feigin et al., 2014). In both developed countries and high income developing countries such as China, acute treatment of stroke has been improved substantially, resulting in decrease in mortality and increase in the proportion of survivors with disability (Feigin et al., 2014; Yang et al., 2014; Ferro et al., 2016; Yang Y. et al., 2016). Central post-stroke pain (CPSP) is one of the most troublesome sequelae of both ischemic and hemorrhagic stroke caused by primary lesions affecting the central somatosensory system (Yang et al., 2014; Hosomi et al., 2015; Yang Y. et al., 2016). About 11%–32% of thalamic stroke survivors experience persistent spontaneous pain, hyperalgesia, allodynia and dysesthesia and about 50%–60% patients with CPSP are resistant to the existing analgesic therapies (Chung et al., 1996; Nasreddine and Saver, 1997). So far, the underlying mechanisms of CPSP remain largely unknown due to lack of valid animal studies (De Vloo et al., 2017). In a critical review on the animal models for CPSP, the experimental thalamic hemorrhage model induced by intra-thalamic collagenase (ITC) injection (Wasserman and Koeberle, 2009; Yang et al., 2014; Kuan et al., 2015; Yang Y. et al., 2016) was believed to be the only one with relatively high reproducibility (De Vloo et al., 2017).

As introduced initially in rats with ITC-induced hemorrhagic CPSP (Wasserman and Koeberle, 2009), activation of microglial cells and astrocytes (glial barrier) surrounding the edge of the hematoma was identified on the first day and reached peak on the seventh day after ITC. Post-treatment with systemic minocycline, a microglial inhibitor, has been shown to attenuate ITC-induced mechanical allodynia and thermal hyperalgesia, indicating that the activated microglial cells contribute to the maintenance of CPSP (Hanada et al., 2014). However, what kind of chemical substances (signaling pathways) are involved in mediation of glial-glial and glial-neuronal interactions (crosstalk) remains unknown and requires to be addressed.

Stromal cell-derived factor 1 (SDF1, also named as CXCL12), a member of the CXC chemokine family, is constitutively expressed in various kinds of cells in the peripheral and central nervous system (Reaux-Le Goazigo et al., 2012; Réaux-Le Goazigo et al., 2013). Under pathological states, activated glial cells are the main source of SDF1. In LPS-induced systemic inflammation model and bone cancer pain model, SDF1 was predominantly released by activated astrocytes in the spinal cord (Shen et al., 2014; Hang et al., 2016; Yang L. et al., 2016). Microglia was also a main source of SDF1 production in the spinal cord of rats with spared nerve injury or ischemia-reperfusion injury (Bai et al., 2016; Li et al., 2016). SDF1 exerts multiple biological functions by binding to CXCR4 (Reaux-Le Goazigo et al., 2012). Recently, several studies reported that systemic treatment with AMD3100, a highly selective CXCR4 antagonist, significantly improved the functional outcome following experimental stroke through attenuating the inflammatory response and microglial activation (Huang et al., 2013; Ruscher et al., 2013; Walter et al., 2015). Importantly, the pro-algesic effect of SDF1-CXCR4 signaling has also been verified in several pathological pain models (Oh et al., 2001; Bhangoo S. et al., 2007; Bhangoo S. K. et al., 2007; Bhangoo et al., 2009; Knerlich-Lukoschus et al., 2011; Wilson et al., 2011; Menichella et al., 2014). In a rat model of bone cancer pain or ischemia-reperfusion induced inflammatory pain, both SDF1 neutralizing antibody and AMD3100 could suppress the development of pain hypersensitivity through inhibiting astrocytic and microglial activation (Shen et al., 2014; Hu et al., 2015; Li et al., 2016). In a neuropathic pain model induced by partial sciatic nerve ligation, it has been demonstrated that SDF1-CXCR4 signaling is involved in the induction of neuropathic pain through mediating microglia-astrocyte crosstalk (Luo et al., 2016). Our recent studies have also demonstrated that SDF1-CXCR4 signaling plays critical roles in mediating the satellite glial cell-neuronal crosstalk in the dorsal root ganglion (DRG) that contributes to the hyperexcitability of primary nociceptors and transition from acute to chronic processes of inflammatory pain (Dubovy et al., 2010; Yang et al., 2015, 2017). Hypoxia inducible factor 1α (HIF-1α), a transcription factor which is closely related to the expression of SDF1 and CXCR4 under hypoxia condition, has been observed to be significantly up-regulated following ITC-induced hemorrhagic stroke (Jiang et al., 2002; Liu et al., 2010; Youn et al., 2011). Taking all these into account, we hypothesized that ITC-induced thalamic hemorrhage could induce up-regulation of SDF1-CXCR4 signaling via HIF-1α which would then maintain CPSP by neuroinflammatory microenvironment caused by microglial-astrocytic-neuronal interactions through positive feedback regulation.

MATERIALS AND METHODS

Animals

Male Sprague-Dawley rats (280–320 g) were provided by the Laboratory Animal Center of the Fourth Military Medical University (FMMU). This study was carried out in accordance with the recommendations of the Animal Care and Use Committee of FMMU and performed in accordance with the updated Guide for the Care and Use of Laboratory Animals (8th edition, the National Academies Press, 2011). The number of rats used and their suffering were minimized. During the whole experiment, the rats were randomized.

Surgery

Surgery was performed according to the methods described previously (Yang et al., 2014; Yang Y. et al., 2016). Rats were anesthetized with chloral hydrate (0.3 g/kg, i.p.) and placed in a stereotaxic apparatus (Narishige Scientific Instrument Lab, Japan). Collagenase type IV (0.025 U/0.25 μl, Sigma-Aldrich China, Shanghai) or saline (0.25 μl) was microinjected into the ventral posterior lateral nucleus (VPL) of the right thalamus (bregma −3.48 mm antero-posterior; 3.6 mm lateral to the midline, and 6.2 mm ventral to the brain surface) according to the stereotaxic coordinates (Paxinos and Watson, 2005). After each injection, the syringe remained for 5 min to prevent spread of the agent to the brain surface. Then the needle was slowly withdrawn, the skin closed using 4.0 sutures, and all rats were allowed to recover in individual cages for at least 3 days. Naïve rats were fed under the same conditions in a parallel manner.

Mechanical Pain Sensitivity Testing

The mechanical pain sensitivity testing was performed by experimenters blinded to the experimental treatments (Yang et al., 2014; Yang Y. et al., 2016). The rats were placed on a metal mesh floor and habituated for 1 h before testing. Ascending graded von Frey filaments were applied from underneath the metal mesh floor to the plantar area of the appropriate hindpaw.

Each von Frey filament was applied 10 times (once every several seconds). The bending force value of the von Frey filament that caused an appropriate 50% occurrence of paw withdrawal was expressed as the paw withdrawal mechanical threshold (PWMT, g). Baseline PWMT of all experimental rats were measured prior to the surgery and the effects of different drugs on PWMT were evaluated at the time point indicated in corresponding figures.

Intra-Thalamic Drug Injections

Intra-thalamic drug injections were performed at 10 days after collagenase injection when CPSP was stably established or at 30 min before collagenase injection prior to CPSP induction. The rats were injected with minocycline hydrochloride (Sigma, 10 μg), fluorocitrate (Sigma, 1 nM), AMD3100 (Abcam, 7.5 or 15 μg) or YC-1 (Abcam, 0.2 mM) in a total volume of 1 μl over a period of 10 min. The doses used in the present study were based on or slightly modified from previous studies (Zhao et al., 2007; Trecki and Unterwald, 2009; Huang et al., 2014; Luo et al., 2014; Choi et al., 2015). Minocycline hydrochloride selectively inhibits microglial activation and proliferation, with no direct effect on neurons or astrocytes. Fluorocitrate, an inhibitor of astrocyte metabolism, inhibits the tricarboxylic acid cycle enzyme aconitase in astrocytes. AMD3100, which is a highly selective CXCR4 antagonist, was used to evaluate the roles of CXCR4 in the development of CPSP. YC-1, an established HIF-1α inhibitor, was applied to verify the regulatory effects of HIF-1α in the development of CPSP. A separate group of naïve rats underwent intra-thalamic injection of recombinant active SDF1 protein (Abcam, 50 ng) or $CoCl_2$ (Sinopharm Chemical Reagent Co. Ltd, 100 mM), a well-known inducer of HIF-1α, following minocycline, fluorocitrate or AMD3100 injections to explore the cellular mechanisms underlying the pro-algesic effect of SDF1-CXCR4 signaling. The intra-thalamic injection methods were the same as above-mentioned and the injection locations were assessed histologically. Fluorocitrate was dissolved in 0.3% 2 M HCl in phosphate-buffered saline (PBS) and YC-1 was dissolved in 1% dimethyl sulfoxide (DMSO), whereas all other drugs were dissolved in sterile saline.

Immunohistochemistry

As reported previously (Yang et al., 2015), the rats were anesthetized with chloral hydrate, then perfused with physiological saline, followed by 4% paraformaldehyde in 0.1 M PB solution. The brain was removed and postfixed overnight at 4°C and then immersed in 30% sucrose in 0.1 M PB. Brain tissue were cut into transverse sections (30 μm thick) on CM1900 freezing microtome (Leica, Germany). After blocking with 10% goat serum in PBS, the sections were incubated overnight at 4°C with the following primary antibodies: mouse anti-GFAP (1:500 Millipore), rabbit anti-Iba-1 (1:500, WAKO), mouse anti-NeuN (1:200, Millipore), Goat anti-CXCR4 (1:200, Abcam) and rabbit anti-c-Fos (1:500, Abcam). On the following day, Cy3- or FITC-conjugated secondary antibodies were incubated for 2–3 h at room temperature. For double immunostaining, sections were incubated with a mixture of primary antibodies overnight at 4°C, followed by a mixture of secondary antibodies. The images were examined under a laser scan confocal fluorescent microscope (Olympus FV1000, Japan).

Western Blotting

As reported previously, the thalamus tissues around the hemorrhagic lesion sites were collected and homogenized in RIPA lysis buffer containing protease inhibitors (Applygen Technologies Inc., China; Yang et al., 2015, 2017; Yang Y. et al., 2016). Protein concentrations of the lysate were determined using a BCA protein assay kit (Thermo Scientific, Rockford, IL, USA). Equal amounts of protein were separated by 10% separation gels and then transferred to polyvinylidene difluoride membranes (Bio-Rad, Hercules, CA, USA). After blocking with 5% nonfat milk (Bio-Rad, Hercules, CA, USA) for 1 h at room temperature, the membranes were incubated with primary antibodies at 4°C overnight. The primary antibodies included mouse anti-HIF-1α (1:500, Abcam), rabbit anti-SDF1 (1:200, Abcam) and the antibodies mentioned in immunohistochemistry section. On the following day, the membranes were washed three times in PBST and then incubated with an HRP-conjugated secondary antibody (Bio-Rad) for 2 h at room temperature. The membranes were visualized with enhanced chemiluminescence solution (Alpha Innotech Corp) and the signals were captured with FluorChem FC2 (Alpha Innotech Corp). The density of specific bands was measured with a computer-assisted imaging analysis system and normalized to β-tubulin or β-actin intensity.

ELISA Assay

The thalamus tissues around the hemorrhagic lesion sites were collected and homogenized in RIPA lysis buffer containing protease and phosphatase inhibitors (Applygen Technologies Inc.). After quantitative measurement of the total protein concentrations with a BCA protein assay kit (Thermo Scientific), the homogenized thalamus tissue were assayed for TNF-α, IL-1β and IL-6 (Westang, Shanghai, China) by ELISA. Each protein of interest was expressed as picograms per milligram of total proteins.

Statistical Analysis

Data were analyzed using GraphPad Prism version 5 (GraphPad, San Diego, CA, USA) and all data were expressed as means ± SEM. Differences in changes of values of each group were tested using *t*-tests and one-way or two-way repeated ANOVA, followed by individual *post hoc* comparisons (Bonferroni or Tukey test). Linear relationships were assessed using Pearson's correlation test. A level of $P < 0.05$ was accepted as significant.

RESULTS

Long-Term Activation of Microglia and Astrocytes in Peri-Thalamic Lesion Sites Caused by ITC

Similar to our previous reports (Yang et al., 2014), unilateral ITC injection confined to the VPL thalamic nucleus (**Figure 1A**)

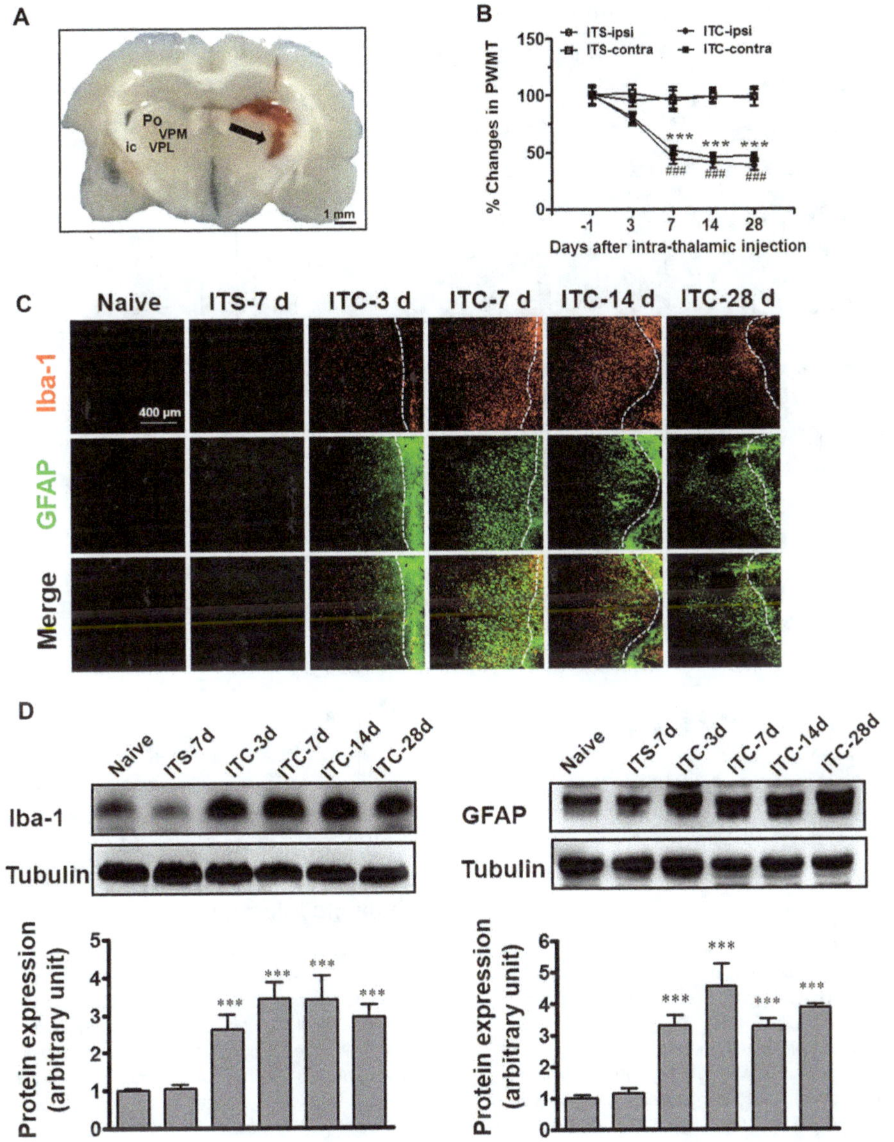

FIGURE 1 | Thalamic hemorrhagic rats exhibit bilateral mechanical pain hypersensitivity and microglial and astrocytic activation in peri-thalamic lesion sites. **(A)** Photomicrograph of brain slice showing the hemorrhagic lesion site in the thalamus following ITC. Scale bar, 1 mm; ic, internal capsule; Po, posterior thalamic nuclear group; VPL, ventral posterolateral nucleus of the thalamus; VPM, ventral posteromedial nucleus of the thalamus. **(B)** Development of bilateral mechanical pain hypersensitivity induced by ITC. Saline injection served as control. ITS, intra-thalamic saline; ITC, intra-thalamic collagenase injection; contra, contralateral; ipsi, ipsilateral; PWMT, paw-withdrawal mechanical threshold; $^{***}P < 0.001$ ITC-ipsi vs. ITS-ipsi; $^{\#\#\#}P < 0.001$ ITC-contra vs. ITS-contra; $n = 10$ rats/group. **(C)** Representative immunofluorescent photomicrographs showing the time course expression of Iba-1 (red) and GFAP (green), markers of microglia and astrocytes respectively, in the peri-thalamic lesion sites. The hemorrhagic lesion core is on the right side of the white line in each image. Scale bar, 400 μm. **(D)** Iba-1 and GFAP expression as examined using Western blot assay. Representative bands are shown on the top, and data summary is shown on the bottom. $^{***}P < 0.001$ vs. ITS-7 d group; $n = 4$/group.

resulted in bilateral reductions in PWMT, which were identified on day 7 post-injection and remained unchanged until day 28 post-injection, suggesting a chronic, persistent bilateral mechanical allodynia in this model (**Figure 1B**). To examine the involvement of microglia and astrocytes, the expression of Iba-1 and GFAP in the peri-thalamic lesion (hematoma) sites were quantified by immunohistochemistry and Western blot on 3, 7, 14 and 28 days after ITC. Compared with intra-thalamic saline (ITS) injection group, the ITC group showed marked increases in Iba-1 and GFAP expression at each time point examined. The expression level of Iba-1 and GFAP was substantially increased in peri-thalamic lesion sites on day 3 after ITC, reached peak on day 7 and remained unchanged until day 28 (**Figures 1C,D**). However, the expression level of Iba-1 and GFAP in the contralateral thalamus remained at basal levels (**Figure 1**).

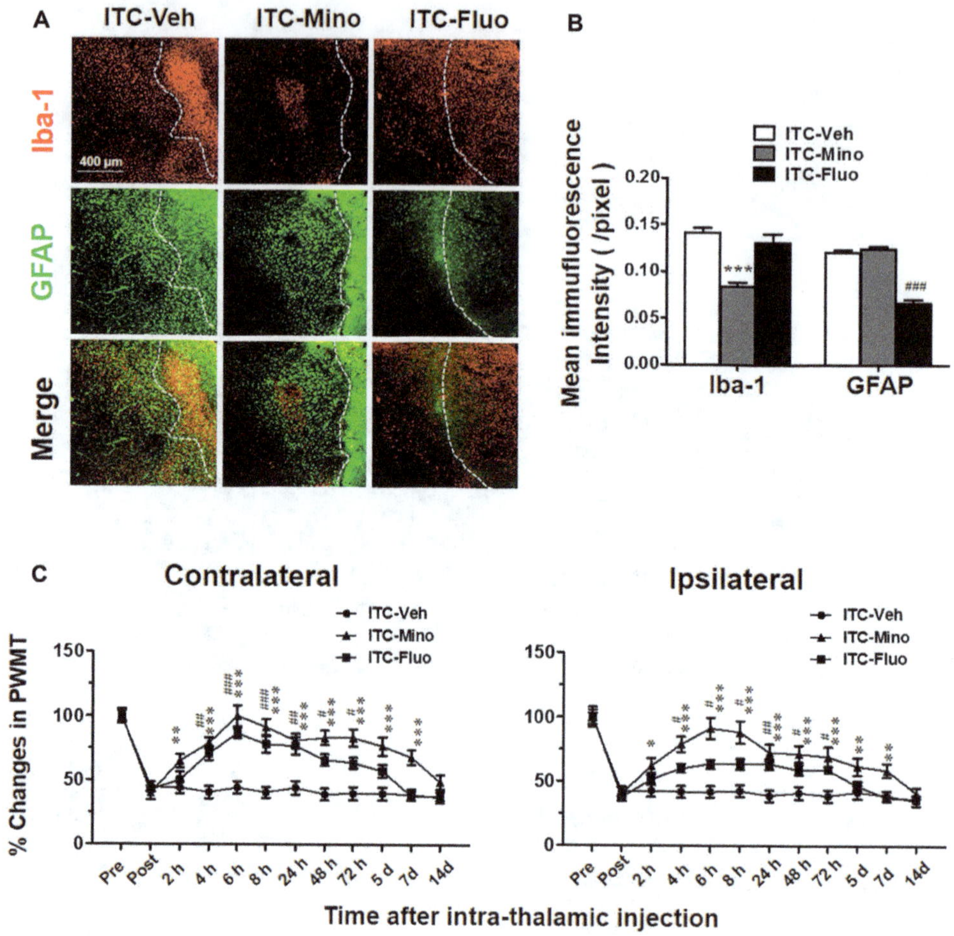

FIGURE 2 | Effects of intra-thalamic injection of minocycline and fluorocitrate on Iba-1 and GFAP expression and the established central post-stroke pain (CPSP). **(A)** Representative immunofluorescent photomicrographs showing expression of Iba-1 and GFAP in peri-thalamic lesion sites following intra-thalamic minocycline (10 μg, 1 μl) and fluorocitrate (1 nM, 1μl) injection. The hemorrhagic lesion core is on the right side of the white line in each image. Scale bar, 400 μm. **(B)** Quantification of the mean immunofluorescent intensity of Iba-1 and GFAP following intra-thalamic minocycline and fluorocitrate injection. $^{***}P < 0.001$, ITC-Mino vs. ITC-Veh; $^{\#\#\#}P < 0.001$, ITC-Fluo vs. ITC-Veh; $n = 4$/group. **(C)** Temporal pattern of effects of intra-thalamic minocycline and fluorocitrate injection at 10 days post thalamic hemorrhage on the established bilateral mechanical pain hypersensitivity. PWMT, paw-withdrawal mechanical threshold; $^{*}P < 0.05$, $^{**}P < 0.01$, $^{***}P < 0.001$, ITC-Mino vs. ITC-Veh; $^{\#}P < 0.05$, $^{\#\#}P < 0.01$, $^{\#\#\#}P < 0.001$, ITC-Fluo vs. ITC-Veh; $n = 8$/group.

Intra-Thalamic Administration of Minocycline or Fluorocitrate Reversed ITC-Induced CPSP via Suppressing the Activation of Microglial Cells and Astrocytes

After CPSP was well established by 10 days after ITC, intra-thalamic injection of minocycline selectively blocked the upregulation of Iba-1 but without any effect on GFAP expression in the peri-thalamic lesion sites, while intra-thalamic injection of fluorocitrate significantly diminished ITC-induced activation of astrocytes labeled by GFAP but without any effect on Iba-1 expression (**Figures 2A,B**). To investigate the roles of microglial and astrocytic activation in the CPSP, we explored the time-related effects of minocycline and fluorocitrate on ITC-induced bilateral mechanical pain hypersensitivity. Our results showed that single intra-thalamic injection with minocycline alleviated the established bilateral mechanical pain hypersensitivity. The anti-allodynic effect of minocycline reached peak at 6 h after injection and maintained at a significant level for at least 7 days. The established bilateral mechanical pain hypersensitivity was also remarkably reduced by fluorocitrate, which lasted from 4 h to 3 days after injection (**Figure 2C**). All these results suggest that both microglia and astrocytes play important roles in maintaining the thalamic hemorrhage-induced CPSP.

CPSP Induced by Intra-Thalamic Injection of SDF1 in Naïve Rats Was Blocked by CXCR4 Antagonism

As SDF1-CXCR4 signaling was closely related to microglia and astrocyte activation, we next tried to determine whether

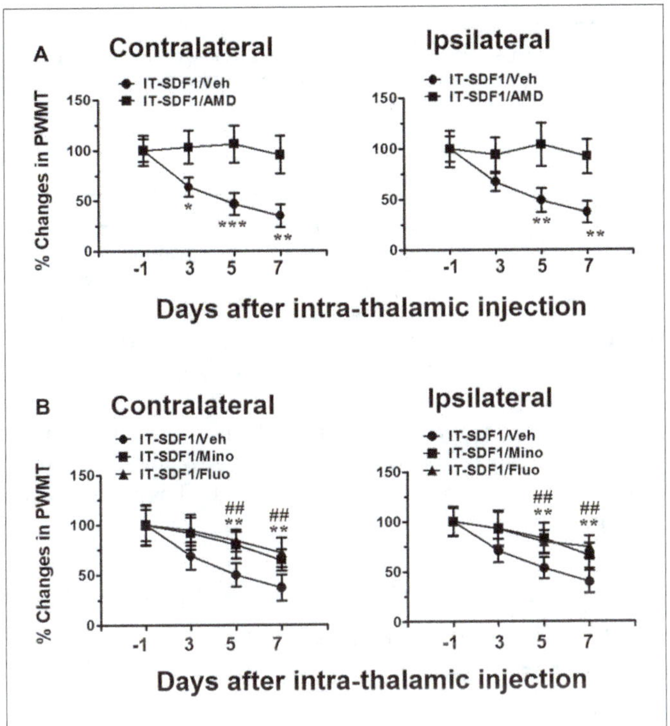

FIGURE 3 | Cellular mechanisms underlying intra-thalamic stromal cell-derived factor 1 (SDF1) injection-induced mechanical pain hypersensitivity in naïve rats. **(A)** Pre-injection of AMD3100 (15 μg/μl) prevents intra-thalamic SDF1 injection-induced bilateral mechanical pain hypersensitivity. $^{*}P < 0.05$, $^{**}P < 0.01$, $^{***}P < 0.001$ IT-SDF1/AMD vs. IT-SDF1/Veh; *n* = 7/group. **(B)** Pre-injection of minocycline and fluorocitrate prevent intra-thalamic SDF1 protein injection-induced bilateral mechanical pain hypersensitivity. $^{**}P < 0.01$ for IT-SDF1/Mino. IT-SDF1/Mino vs. IT-SDF1/Veh; $^{\#\#}P < 0.01$ IT-SDF1/Fluo. IT-SDF1/Fluo vs. IT-SDF1/Veh, PWMT, paw-withdrawal mechanical threshold; *n* = 7/group.

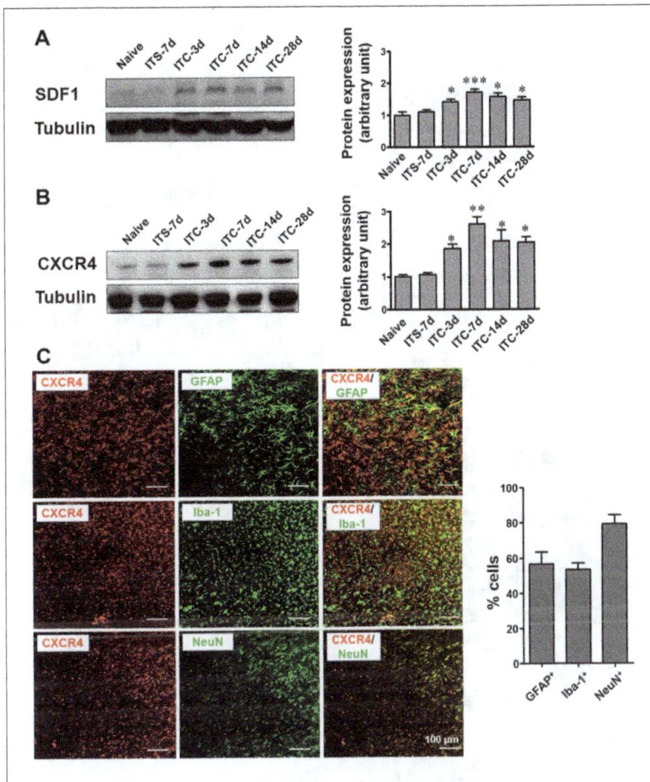

FIGURE 4 | Temporal and spatial changes in SDF1 and CXCR4 expression in peri-thalamic lesion sites under CPSP condition. **(A)** Western blot assay showing the time course expression of SDF1 following thalamic hemorrhage. Representative bands are shown on the left, and data summary is shown on the right. $^{*}P < 0.05$, $^{***}P < 0.001$ vs. ITS group. **(B)** Western blot assay showing the time course expression of CXCR4 following thalamic hemorrhage. Representative bands are shown on the left, and data summary is shown on the right. $^{*}P < 0.05$, $^{**}P < 0.01$ vs. ITS group; *n* = 4/group. **(C)** Left panel showing the immunofluorescence micrographs of the double-staining of CXCR4 with GFAP, Iba-1 and NeuN. Right panel quantifying the percentages of GFAP-positive, Iba-1-positive and NeuN-positive cells expressing CXCR4. Scale bar, 100 μm.

SDF1-CXCR4 signaling was involved in the induction of pain in naïve rats through regulating glial cell activation. As shown in **Figure 3A**, intra-thalamic SDF1 injection elicited bilateral mechanical pain hypersentivity from day 3 to day 7 which could be completely blocked by pre-administration of AMD3100 into the SDF1 injection sites. Moreover, pre-administration of minocycline or fluorocitrate in the SDF1 injection sites also significantly suppressed SDF1-induced bilateral mechanical pain hypersentivity from day 5 to day 7 (**Figure 3B**).

Temporal and Spatial Changes in SDF1 and CXCR4 Expressions in the Thalamus Following ITC

Under the CPSP condition induced by ITC, SDF1 expression was significantly increased in the peri-thalamic lesion sites at 3, 7, 14 and 28 days post-ITC relative to the saline control (**Figure 4A**). The level of CXCR4 expression in peri-thalamic lesion sites under CPSP condition was also examined by Western blot analysis. As shown in **Figure 4B**, increased CXCR4 protein expression was also seen at 3, 7, 14 and 28 days post-ITC relative to the saline control.

To define the cellular distribution of CXCR4, double staining of CXCR4 with different cell markers was performed. CXCR4 was not only co-localized with neuronal marker NeuN but also with astrocytic marker GFAP and microglial marker Iba-1 (**Figure 4C**).

Antagonizing CXCR4 by Intra-Thalamic Administration of AMD3100 Suppressed Both Glial Cell Activation and Neuronal Hyperactivity

To determine whether CXCR4 contributes to microglial and astrocytic activations under CPSP condition, intra-thalamic VPL injection of AMD3100 (15 μg) was administered 30 min prior to ITC. As shown in **Figure 5A**, CXCR4 antagonism prevented increases in both Iba-1 and GFAP expression in the peri-thalamic lesion sites on day 3 and 7 post-ITC. Moreover, the increased release of pro-inflammatory mediators including TNF-α, IL-6 and IL-1β in the peri-thalamic lesion sites was also prevented by intra-VPL AMD3100 administration on day 7 and 14 post-ITC (**Figure 5B**). Furthermore, antagonizing CXCR4 by intra-thalamic AMD3100 on day 10 post-ITC also suppressed

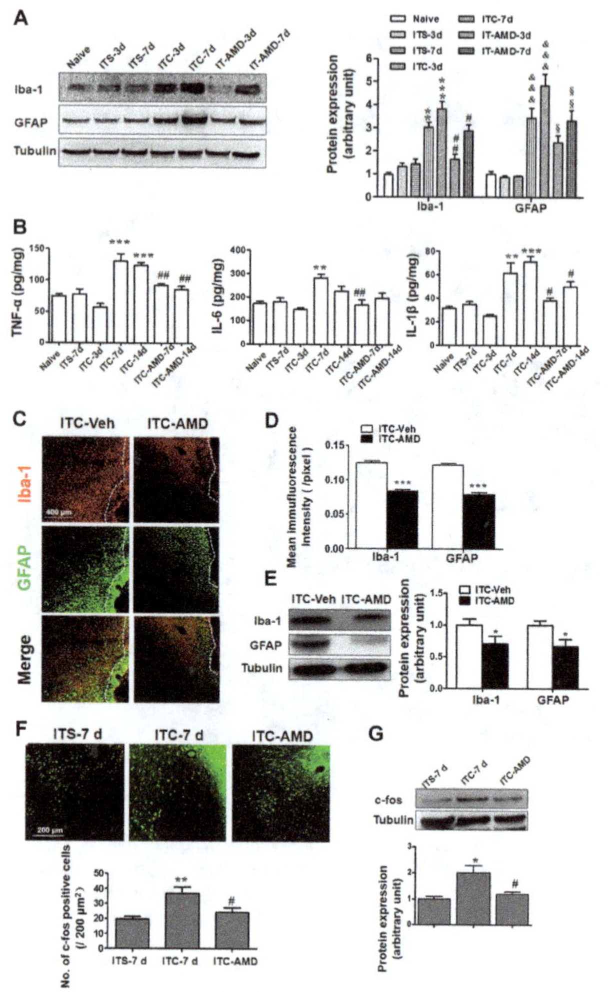

FIGURE 5 | Effects of intra-thalamic injection of AMD3100 on thalamic hemorrhage-induced glial cells activation and neuronal hyperactivity. **(A)** Western blot assay showing the effect of AMD3100 (15 μg/μl), injected 30 min before ITC, on the expression level of Iba-1 and GFAP in peri-thalamic lesion sites. $^{**}P < 0.01$, $^{***}P < 0.001$ vs. ITS at corresponding time points; $^{\&\&\&}P < 0.001$ vs. ITS at corresponding time points; $^{\#}P < 0.05$, $^{\#\#}P < 0.01$ vs. ITC at corresponding time points; $^{\S}P < 0.05$, $^{\S\S}P < 0.01$ vs. ITC at corresponding time points. **(B)** ELISA assay showing the effect of AMD3100 (15 μg/μl) pre-injection on the expression level of TNF-α, IL-6 and IL-1β. $^{**}P < 0.01$, $^{***}P < 0.001$ vs. ITS-7 d; $^{\#}P < 0.05$, $^{\#\#}P < 0.01$ vs. ITC at corresponding time points. **(C)** Representative immunofluorescent photomicrographs showing expression of Iba-1 and GFAP in peri-thalamic lesion sites following intra-thalamic AMD3100 (15 μg/μl) injection at 10 days after ITC. The hemorrhagic lesion core is on the right side of the white line in each image. Scale bar, 400 μm. **(D)** Quantification of the mean immunofluorescent intensity of Iba-1 and GFAP following intra-thalamic AMD3100 (15 μg/μl) post-injection. $^{***}P < 0.001$ ITC-AMD vs. ITC-Veh; *n* = 4/group. **(E)** Western blot assay showing the effect of AMD3100 (15 μg/μl) post-injection on the expression level of Iba-1 and GFAP in peri-thalamic lesion sites. Representative bands are shown on the left, and data summary is shown on the right. $^{*}P < 0.05$, vs. ITC-Veh. **(F)** Representative immunofluorescent photomicrographs showing co-expression of c-fos (red) and NenN (green) in peri-thalamic lesion sites following intra-thalamic AMD3100 (15 μg/μl) post-injection. $^{**}P < 0.01$, vs. ITS-7 d; $^{\#}P < 0.05$ vs. ITC-7 d;. Scale bar, 200 μm. **(G)** Western blot assay showing the effect of AMD3100 (15 μg/μl) post-injection on the expression *(Continued)*

FIGURE 5 | Continued
level of c-Fos in peri-thalamic lesion sites. Representative bands are shown on the top, and data summary is shown on the bottom. $^{*}P < 0.05$, vs. ITS-7 d; $^{\#}P < 0.05$ vs. ITC-7 d; *n* = 3/group.

the increase in Iba-1 and GFAP expressions, implicating that SDF1-CXCR4 signaling is directly involved in the ITC-induced activations of both microglia and astrocytes (**Figures 5C–E**).

The role of CXCR4 in the neuronal activity was also examined by measuring c-Fos expression surrounding the glial barrier. As shown in **Figure 5F**, the c-Fos immunoreactivity was barely observed in rats receiving ITS, while significant increase in c-Fos-labeled neurons was observed in the peri-thalamic lesion sites following ITC. Intra-thalamic microinjection of AMD3100 reduced the number of c-Fos-labeled neurons (**Figure 5F**) and decreased the ITC-induced c-Fos expression as measured by Western blot (**Figure 5G**).

Antagonizing CXCR4 by Intra-Thalamic Administration of AMD3100 Could Completely Reverse the Well-Established CPSP

To determine whether CXCR4 contributes to the induction of thalamic hemorrhage-induced CPSP, AMD3100 (15 μg) was microinjected into the VPL 30 min prior to ITC. Our behavioral assay verified that pre-injection of AMD3100 partially inhibited the ITC-induced bilateral mechanical pain hypersensitivity for 21 days relative to ITS control (**Figure 6A**), implying that CXCR4 plays a partial role in the initiation of CPSP.

However, the maintaining role of CXCR4 in the thalamic hemorrhage-induced CPSP was more significant. As shown in **Figure 6B**, intra-thalamic injection of AMD3100 with two doses (7.5 μg and 15 μg) on day 10 after ITC almost completely reversed the well-established CPSP. The analgesic effect of AMD3100 was dose-related and lasted for 48 h to 72 h, implicating the therapeutic value of AMD3100 in treatment of CPSP (**Figures 6B,C**).

Inhibition of HIF-1α Blocked the Initiation but Not the Maintenance of ITC-Induced CPSP

In the present study, using Western blot analysis (**Figure 7A**), we examined HIF-1α expression in the peri-thalamic lesion sites at 3, 7, 14 and 28 days after ITC. It was revealed that ITC resulted in a short-term increase in HIF-1α expression, reaching peak during day 3 and 7, returning to the control level since day 14 post-ITC (**Figure 7A**).

We next explored whether HIF-1α is involved in the development of ITC-induced CPSP. It was interesting to find that intra-thalamic microinjection of YC-1, a potent inhibitor of HIF-1α production, blocked the occurrence of ITC-induced CPSP when being administered 30 min prior to ITC (**Figure 7B**), however it had no reversal effect on the well-established CPSP when being administered 10 days after ITC (**Figure 7C**).

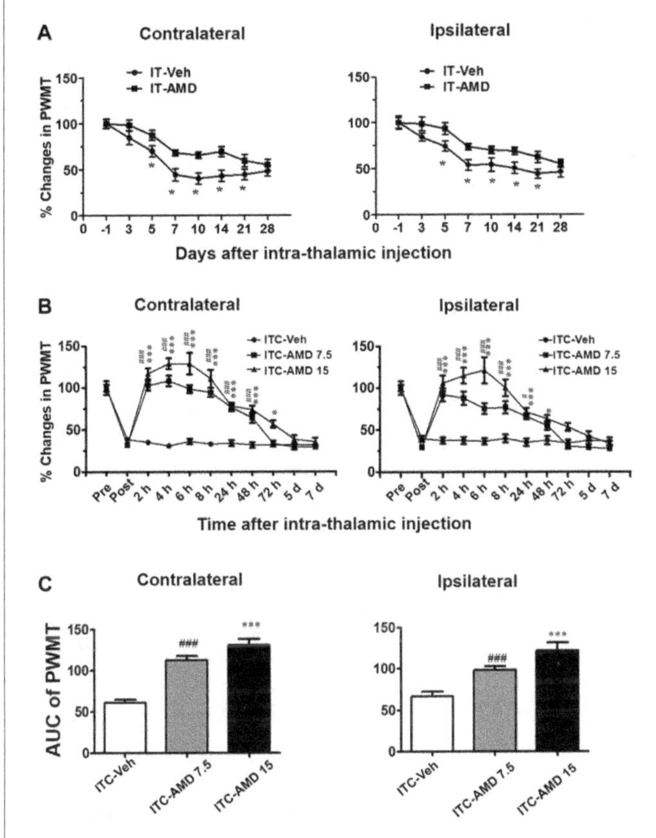

FIGURE 6 | Effects of intra-thalamic AMD3100 injection on the induction and maintenance of ITC-induced CPSP. **(A)** Temporal pattern of effects of intra-thalamic AMD3100 (15 μg/μl) injection at 30 min before ITC on the induction of bilateral mechanical pain hypersensitivity. $*P < 0.05$, IT-AMD vs. IT-Veh; $n = 10$/group. **(B)** Temporal pattern of effects of intra-thalamic AMD3100 (7.5 and 15 μg/μl) injection at 10 days post thalamic hemorrhage on the established bilateral mechanical pain hypersensitivity. $*P < 0.05$, $***P < 0.001$, ITC-AMD 15 vs. ITC-Veh; $^{\#}P < 0.05$, $^{\#\#\#}P < 0.001$, ITC-AMD 7.5 vs. ITC-Veh; $n = 10$/group. **(C)** Quantification of area under the analgesic effects of AMD3100 (7.5 and 15 μg/μl). $***P < 0.001$ vs. ITC-Veh; $^{\#\#\#}P < 0.001$ vs. ITC-Veh; PWMT, paw-withdrawal mechanical threshold.

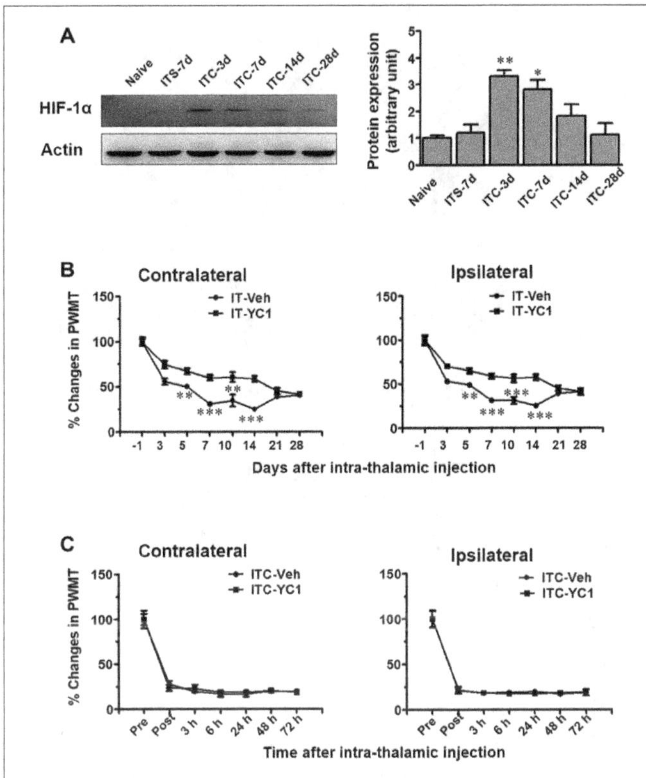

FIGURE 7 | Temporal pattern of hypoxia inducible factor 1α (HIF-1α) expression in peri-thalamic lesion sites following thalamic hemorrhage and effects of intra-thalamic YC-1 injection on the induction and maintenance of CPSP. **(A)** Western blot assay showing the time course expression of HIF-1α following thalamic hemorrhage. Representative bands are shown on the left, and data summary is shown on the right. $*P < 0.05$, $**P < 0.01$ vs. ITS-7 d; $n = 4$/group. **(B)** Effect of intra-thalamic pre-injection of YC-1 (0.2 mM, 1 μl) on the induction of bilateral mechanical pain induced by thalamic hemorrhage. $**P < 0.01$, $***P < 0.001$ IT-YC1 vs. IT-Veh; $n = 7$/group. **(C)** Effect of intra-thalamic YC-1 injection at 10 days post thalamic hemorrhage on the established bilateral mechanical pain. $n = 7$/group. PWMT, paw-withdrawal mechanical threshold.

Inhibition of HIF-1α Prior to ITC Blocked the Up-Regulation of SDF1-CXCR4 Signaling and Activations of Microglia and Astrocytes

To verify whether HIF-1α is directly involved in regulating the expression of SDF1-CXCR4 signaling under CPSP condition, intra-thalamic injection of YC-1 was administered 30 min prior to ITC. Western blot results showed that, similar to CXCR4 antagonism by AMD3100, local inhibition of HIF-1α expression by YC-1 resulted in significant reduction in the expression of both SDF1/CXCR4 (**Figures 8A,B**) and Iba-1/GFAP (**Figures 8A,C**) on day 3 and day 7 post-ITC.

Intra-thalamic injection of $CoCl_2$, a well-known inducer of HIF-1α, was shown to result in elevation of HIF-1α, SDF1 and CXCR4 expressions examined on day 3 and 7 post-injection (**Figures 9A,B**). Pre-blockade of HIF-1α production by YC-1 through intra-thalamic injection prevented the enhancement of HIF-1α, SDF1 and CXCR4 expression, whereas, pre-blockade of CXCR4 with AMD3100 failed to change the enhanced expression of HIF-1α although SDF1 and CXCR4 up-regulation was prevented (**Figures 9A,B**). Similar to the results of ITC, intra-thalamic injection of $CoCl_2$ induced overexpression of Iba-1 and GFAP (**Figures 9A,C**), suggesting activation of both microglia cells and astrocytes by $CoCl_2$. The $CoCl_2$-induced overexpression of SDF1 and CXCR4 and activations of microglia and astrocytes were also blocked by pre-injection of AMD3100 or YC-1, respectively (**Figures 9A–C**).

Similar to the results of ITC, intra-thalamic injection of $CoCl_2$ induced bilateral mechanical pain hypersensitivity that could be blocked by pre-treatment of AMD3100 through intra-thalamic microinjection (**Figure 9D**).

DISCUSSION

The present study demonstrated that ITC-induced CPSP hypersensitivity was initiated by microglial-astrocytic-neuronal

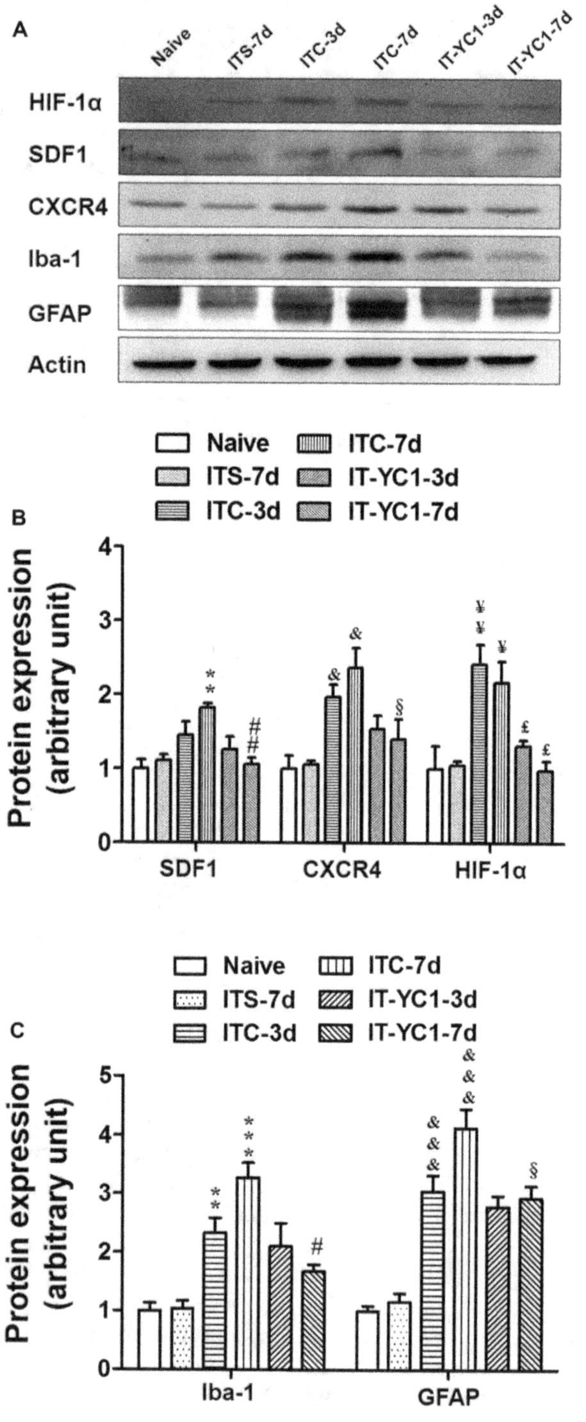

FIGURE 8 | Effects of intra-thalamic pre-injection of YC-1 on the increased SDF1 and CXCR4 expression and glial cells activation induced by ITC. **(A)** Representative bands showing the effect of intra-thalamic pre-injection of YC-1 (0.2 mM, 1 μl) on thalamic hemorrhage-induced HIF-1α, SDF1 and CXCR4 up-regulation as well as on microglial marker Iba-1 and astrocytes marker GFAP up-regulation. **(B)** Quantification of Western blot assay showing that intra-thalamic pre-injection of YC-1 prevented thalamic hemorrhage-induced HIF-1α, SDF1 and CXCR4 up-regulation at 3 and 7 days. $^{\&}P < 0.05$, $^{¥}P < 0.05$, $^{¥¥}P < 0.01$, $^{**}P < 0.01$ vs. ITS-7 d group; *(Continued)*

FIGURE 8 | Continued
$^{£}P < 0.05$, $^{\#\#}P < 0.01$, $^{\S}P < 0.05$ vs. ITC at corresponding time points; $n = 5$/group. **(C)** Quantification of Western blot assay showing that intra-thalamic pre-injection of YC-1 prevented thalamic hemorrhage-induced Iba-1 and GFAP up-regulation at 3 and 7 days. $^{**}P < 0.01$, $^{***}P < 0.001$, $^{\&\&\&}P < 0.001$ vs. ITS; $^{\#}P < 0.05$, $^{\S}P < 0.05$ vs. ITC at corresponding time points; $n = 5$/group.

interactions in the peri-lesion sites through HIF-1α-dependent up-regulation of SDF1-CXCR4 signaling in microglia, astrocytes and neurons at the early stage, while it was maintained by persistent release of pro-inflammatory mediators (TNFα, IL-1β and IL-6) through SDF1-CXCR4 signaling-mediated glial-glial and glial-neuronal positive feedback regulation at the late stage. The direct involvement of SDF1-CXCR4 signaling in the induction and maintenance of CPSP was confirmed using exogenous SDF1, which was sufficient to induce mechanical pain hypersensitivity after intra-thalamic microinjection. Meanwhile, the involvement of HIF-1α in the induction, but not the maintenance, of CPSP was also confirmed by intra-thalamic YC-1 produced HIF-1α suppression and intra-thalamic $CoCl_2$ produced HIF-1α over-expression. The proposed underlying mechanisms of CPSP that is mediated by microglial-astrocytic-neuronal interactions through HIF-1α-SDF1-CXCR4 signaling mediation is shown in **Figure 10**.

HIF-1α Induces SDF1-CXCR4 Signaling Over-Expression and Development of CPSP

HIF-1α is a key transcription factor for the adaptive responses to hypoxia (Singh et al., 2012). In the present study, we found that the protein level of HIF-1α was significantly increased at the early phase, but not in the late phase, after hemorrhagic stroke induced by ITC. As supporting evidence, Jiang et al. (2002) reported that HIF-1α was significantly up-regulated in the basal ganglia following experimental intracerebral hemorrhage in rats receiving injections of autologous blood, thrombin, or lysed red blood cells. Here, we further found that inhibition of HIF-1α could prevent the development of CPSP but without analgesic effect during the maintenance of CPSP, implicating that up-regulation of the downstream gene expression of HIF-1α is initiated by HIF-1α but maintained by other mechanisms. By utilizing YC-1, the most specific inhibitor available for HIF-1α (Zhang et al., 2014), it was demonstrated that the increased expression of SDF1 and CXCR4 in the peri-thalamic lesion sites was almost completely suppressed. In fact, it has been demonstrated that both SDF1 and CXCR4 promoters contain hypoxia response elements (HREs) which have a HIF-1 DNA binding site (Hitchon et al., 2002; Schioppa et al., 2003; Speth et al., 2014) and blockade of HIF-1α activation can abrogate both hypoxia-induced CXCR4 up-regulation and SDF1-dependent cell migration in microglia *in vitro* (Wang et al., 2008). In the current study, the thalamic hemorrhage-induced microglial and astrocytic activations was prevented by blockade of both HIF-1α and SDF1-CXCR4 signaling, suggesting that HIF-1α plays a critical role in the up-regulation of SDF1-CXCR4 signaling that

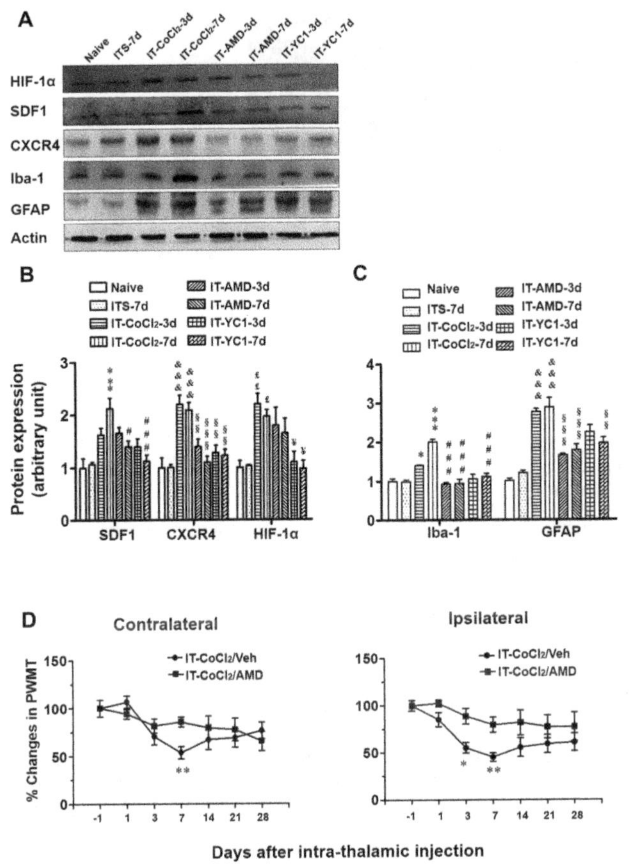

FIGURE 9 | Effects of intra-thalamic injection of $CoCl_2$ on HIF-1α, SDF1 and CXCR4, as well as on microglial marker Iba-1 and astrocytes marker GFAP expression in naïve rats. **(A)** Representative bands showing the effect of intra-thalamic injection of $CoCl_2$ (100 mM, 1 μl) on HIF-1α, SDF1, CXCR4, as well as on Iba-1 and GFAP expression in the thalamic injection sites. **(B)** Quantification of Western blot assay showing that intra-thalamic $CoCl_2$ injection increased HIF-1α, SDF1 and CXCR4 expression, and which were prevented by pre-injection of AMD3100 and YC-1. £$P < 0.05$, ££$P < 0.01$, &&&$P < 0.001$, ***$P < 0.001$ vs. ITS-7 d; §§$P < 0.01$, §§§$P < 0.001$, #$P < 0.05$, ###$P < 0.001$, ¥$P < 0.05$ vs. IT-$CoCl_2$ at corresponding time points; $n = 5$/group. **(C)** Quantification of Western blot assay showing that intra-thalamic $CoCl_2$ injection increased Iba-1 and GFAP expression, and which were prevented by pre-injection of AMD3100 and YC-1. *$P < 0.05$, ***$P < 0.001$, &&&$P < 0.001$ vs. ITS group; §§$P < 0.01$, §§§$P < 0.001$, ###$P < 0.001$, vs. IT-$CoCl_2$ at corresponding time points; $n = 5$/group. **(D)** Intra-thalamic injection of $CoCl_2$ induced bilateral mechanical pain hypersensitivity in naïve rats which were prevented by pre-injection of AMD3100. *$P < 0.05$, **$P < 0.01$, IT-$CoCl_2$/AMD vs. IT-$CoCl_2$/Veh; PWMT, paw-withdrawal mechanical threshold; $n = 7$/group.

subsequently mediates the glial cell activations at the late stage of thalamic hemorrhage condition. To further verify the role of HIF-1α, we injected $CoCl_2$, a well-known inducer of HIF-1α, into the VPL of normal rats. As expected, the expression of SDF1 and CXCR4 protein was up-regulated by intra-VPL $CoCl_2$ microinjection and which was prevented by pre-inhibition of HIF-1α. Likewise, microglia and astrocytes was also activated by intra-VPL $CoCl_2$ microinjection and the activation could be blocked by pre-treatment with CXCR4 inhibitor, indicating that HIF-1α is essential for the induction of SDF1-CXCR4 signaling-mediated glial cell activations. This result can be supported by a previous report showing that both astrocytes and microglia could be activated by microinjection of $CoCl_2$ into the frontoparietal cortex of rats (Caltana et al., 2009). Most interestingly, our present study provided a new line of experimental evidence demonstrating that intra-thalamic $CoCl_2$ microinjection in normal rats is sufficient to induce transient bilateral mechanical pain hypersensitivity through HIF-1α activation. HIF-1α has also been shown to be involved in the development of peripheral neuropathic pain, demonstrated by HIF-1α gene knockout in primary nociceptive neurons (Kanngiesser et al., 2014). Moreover, inhibiting accumulation of HIF-1α in the sciatic nerve also relieved both spared nerve injury-evoked and chronic constriction injury-induced neuropathic pain (Hsieh et al., 2012). However, none of these two previous studies clearly identified how HIF-1α worked to produce pain. Thus the link between up-regulation of SDF1-CXCR4 signaling by HIF-1α activation and the development of CPSP in the current study would shed new light on the treatment of CPSP in clinic.

SDF1-CXCR4 Signaling Contributes to Maintenance of CPSP through Mediating Microglial-Astrocytic-Neuronal Interactions

To determine the roles of microglia and astrocytes in the maintenance of CPSP, we evaluated the analgesic effect of minocycline, a microglial inhibitor, and fluorocitrate, an astrocytic inhibitor, by intra-thalamic microinjection of these agents, respectively. As expected, both minocycline and fluorocitrate reversed the well-established mechanical pain hypersensitivity induced by ITC-induced hemorrhage, indicating involvement of activated microglial and astrocytic cells in the maintenance of CPSP. This result can be supported by a previous study showing attenuation of thalamic hemorrhage-induced mechanical allodynia following post-treatment with systemic minocycline (Hanada et al., 2014). Also, another line of evidence demonstrated that intra-thalamic microinjection of minocycline was sufficient to reverse sciatic nerve injury-induced thermal hyperalgesia for 30 min (LeBlanc et al., 2011). However, in an animal model of the spinal cord injury-induced mechanical pain hypersensitivity, it was shown that intra-thalamic minocycline microinjection produced anti-hyperalgesic effect for 8 h (Zhao et al., 2007). Herein, our results showed the duration of the analgesic effect of minocycline on CPSP was much longer and lasted at least for 7 days. The discrepancy in duration of analgesia for minocycline might be duo to various reasons such as differences in the time course of microglial cell activation in different pain models, drug delivery route and other unknown mechanisms involved. Unlike the sciatic nerve and spinal cord injury, thalamic hemorrhage was the primary injury in CPSP model and glial activation in the thalamus would be the critical underlying mechanism for CPSP which was the reason for us to adopt intra-thalamic injection in the present study. It has been already shown that the activated glial cells of both microglia and astrocytes can produce and release various types of chemokines and cytokines that not only trigger neuronal hyperexcitability but

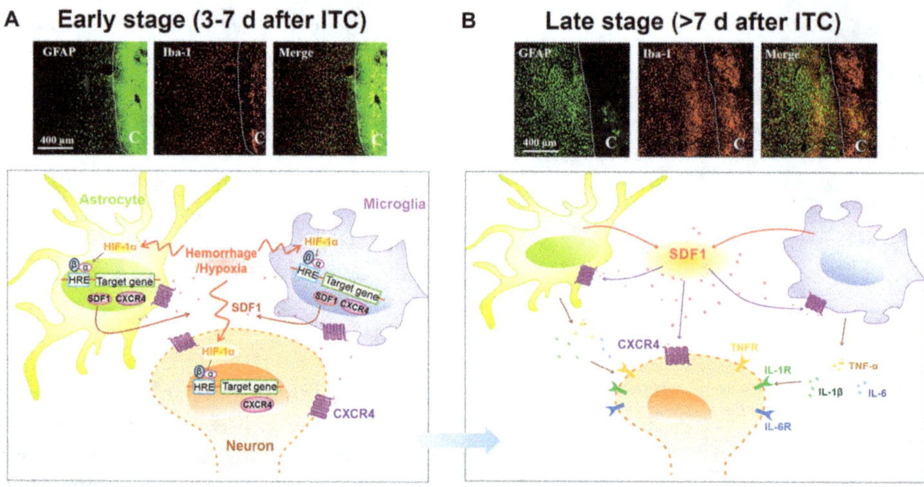

FIGURE 10 | Schematic illustration of microglial-astrocytic-neuronal interactions mediated by HIF-1α-SDF1-CXCR4 signaling following thalamic hemorrhage. **(A)** The upper panel shows representative immunofluorescent photomicrographs of microglial marker Iba-1 (red) and astrocytic marker GFAP (green) expression in thalamic sections at the early stage after hemorrhage (<7 days). The hemorrhagic lesion core is on the right side of the white line in each immunofluorescent image. At this phase, hypoxia microenvironment was established in peri-thalamic lesion site which induced accumulation of HIF-1α in the cytoplasm. As a transcription factor, HIF-1α translocates into the nucleus where it dimerizes with HIF-1β subunit and forms the active HIF-1 complex which binds to the hypoxia response element (HRE) in the DNA sequence and then induces CXCR4 expression in the neurons and glial cells as well as SDF1 expression in glial cells. The synthetic CXCR4 would integrate into cell membrane forming a functional receptor for neuron and glial cells while synthetic SDF1 would be secreted into extracellular space as a ligand (lower panel). Scale bar, 400 μm. **(B)** The upper panel shows representative immunofluorescent photomicrographs of microglial marker Iba-1(red) and astrocytic marker GFAP (green) expression in thalamic sections at the late stage after hemorrhage (>7 days). The hemorrhagic lesion core is on the right side of the white line in each immunofluorescent image. At this phase, abundant SDF1 from astrocytes and microglia activates CXCR4 on glial cells, resulting in the activation of glial cells by a positive feedback activation pattern and subsequent release of proinflammatory cytokines, including TNF-α, IL-1β and IL-6, which could subsequently increase neuronal excitation and glial activation in peri-thalamic lesion site. For the survival neurons in the peri-thalamic lesion site, the released SDF1 from glial cells could bind to CXCR4 and directly evoke neuronal hyperactivity which combined with glial activation contribute to the maintenance of CPSP (lower panel). Scale bar, 400 μm.

also facilitate inter-glial cell activations through positive feedback regulation (Guo et al., 2007; Sheridan and Murphy, 2013; Macht, 2016). Here we further demonstrated that chemokine SDF1 and its cognate receptor CXCR4 were both up-regulated in the peri-thalamic lesion sites after ITC and CXCR4 was localized not only in neurons but also in astrocytes and microglia. In line with previous studies, the temporal pattern of SDF1 up-regulation was similar to the up-regulation of GFAP and Iba1, suggesting that activated microglia and astrocytes are the main SDF1-producing cells in the peri-thalamic lesion sites (Shen et al., 2014; Bai et al., 2016; Yang L. et al., 2016). Based on the cellular localization of CXCR4 in both neurons and glial cells and the abundant production of SDF1 from activated astrocytes and microglia, we propose that through positive feedback regulation, SDF1-CXCR4 signaling would play critical roles in mediating microglial-astrocytic-neuronal interactions, because selectively blocking CXCR4 with intra-thalamic AMD3100 microinjection could: (1) remarkably suppress thalamic hemorrhage-induced microglia and astrocyte activation; (2) significantly inhibit the release of pro-inflammatory mediators such as TNF-α, IL-1β and IL-6 from activated glial cells; and (3) reverse thalamic hemorrhage-induced increase in neuronal activities labeled with c-Fos in the peri-thalamic lesion sites. The persistent release of pro-inflammatory mediators could cause neuronal hyperexcitability and sensitization through binding to their respective receptors (Zhang et al., 2011; Clark et al., 2013; Chirila et al., 2014). As a new line of evidence, we have previously demonstrated that blocking CXCR4 with AMD3100 directly inhibited the hyperexcitability of primary nociceptor neurons under peripheral inflammatory pain state (Yang et al., 2015). Thus, it is conceivable that SDF1-CXCR4 signaling could mediate microglial-astrocytic-neuronal interactions (crosstalk) through two possible mechanisms: (1) glial-neuronal interactions mediated by direct binding of astrocyte/microglia-derived SDF1 with neuronal CXCR4; and (2) positive feedback glial-glial interactions through direct binding of astrocyte- or microglia-derived SDF1 with glial CXCR4 (**Figure 10**). The glial-glial interactions may serve as the source of pro-inflammatory mediators that in turn cause neuronal hyperexcitability and sensitization.

Given the potential role of SDF1-CXCR4 signaling in the glial-glial interactions and glial-neuronal crosstalk, we here demonstrated that pre-treatment with CXCR4 inhibitor AMD3100 could prevent the development of CPSP accompanying with the reduction of neuroinflammation which was essential for CPSP. To further confirm whether SDF1-CXCR4 signaling participates in the CPSP process, we injected exogenous SDF1 into VPL of naïve rats. As expected, single intra-thalamic SDF1 injection was sufficient to induce bilateral mechanical pain hypersensitivity through CXCR4 agonism. In addition, the SDF1-induced pain hypersensitivity was prevented by inhibiting microglial and astrocytic activation. It was interesting to note that both the analgesic effect of intra-thalamic AMD3100 and hyperalgesic effect of intra-thalamic SDF1 remained for a relative longer duration after administration although AMD3100 and SDF1 themselves have a short-term half-life (Hendrix et al., 2004; Kirkpatrick et al., 2010). Why? One possibility is that SDF1-CXCR4 signaling serves as a priming signal to mediate a positive feedback between glial-glial interactions under CPSP state. Once being activated, the downstream inflammatory cascades (TNF-α, IL-1β and IL-6) would be recruited to promote the local persistent neuroinflammation in the lesioned thalamus (**Figure 10**). This was supported by a previous study showing that intra-thalamic injection of CCL21 induced significant microglia activation and pain hypersensitivity which lasted longer than its own half-life and intra-thalamic injection of minocycline could also suppress CCL21-induced changes (Zhao et al., 2007). Moreover, here we also showed that post-treatment with AMD3100 could reverse the established CPSP in a dose-related manner. Taken all the existing data into account, we conclude that SDF1-CXCR4 signaling-mediated glial-glial/glial-neuronal communications are required for the development and maintenance of CPSP. Since it has been demonstrated that treatment with P2X7 receptor antagonist and IL-1β antibody could also reduce thalamic hemorrhage-induced microglia aggregation, neuronal excitability and allodynia phenomena in CPSP rats (Kuan et al., 2015), a common upstream regulation by HIF-1α-SDF1-CXCR4 signaling might be shared.

Because the analgesic effect of intra-thalamic injection of lidocaine on the thalamic hemorrhage-induced tactile hypersensitivity is short-lasting (Gritsch et al., 2016), the use of AMD3100 (also named as plerixafor) to treat CPSP would be more promising. Its safety and efficacy was approved by the U.S. Food and Drug Administration in 2008 for the mobilization of hematopoietic stem cells (Hummel et al., 2014). Moreover, several other CXCR4 antagonists are currently in clinical trials for cancer, HIV and WHIM syndrome treatment, highly supporting its potential use for the treatment of CPSP in clinic (Debnath et al., 2013). Since intraperitoneal delivery of AMD3100 has been demonstrated to be effective in suppression of pain (Dubovy et al., 2010; Xie et al., 2016), SDF1-CXCR4 signaling pathway as a new therapeutic target for the treatment of CPSP might be promising.

AUTHOR CONTRIBUTIONS

JC and FeiY conceived and designed the experiments. FanY, W-JL, WS, YW, C-LL and NW performed the experiments, FeiY and JC analyzed and interpreted the data. J-LW, FanY and X-LW supplied SPF animals. JC, S-MG and FeiY wrote the article. All authors listed, have made substantial, direct and intellectual contribution to the work, and approved it for publication.

ACKNOWLEDGMENTS

This work was supported by National Basic Research Development Program of China (2013CB835100), the National Natural Science Foundation of China (81171049, 81571072, 31300919, 31400948) and the National Key Technology R&D Program (2013BAI04B04).

REFERENCES

Bai, L., Wang, X., Li, Z., Kong, C., Zhao, Y., Qian, J. L., et al. (2016). Upregulation of chemokine CXCL12 in the dorsal root ganglia and spinal cord contributes to the development and maintenance of neuropathic pain following spared nerve injury in rats. *Neurosci. Bull.* 32, 27–40. doi: 10.1007/s12264-015-0007-4

Bhangoo, S. K., Ren, D., Miller, R. J., Chan, D. M., Ripsch, M. S., Weiss, C., et al. (2007). CXCR4 chemokine receptor signaling mediates pain hypersensitivity in association with antiretroviral toxic neuropathy. *Brain Behav. Immun.* 21, 581–591. doi: 10.1016/j.bbi.2006.12.003

Bhangoo, S., Ren, D., Miller, R. J., Henry, K. J., Lineswala, J., Hamdouchi, C., et al. (2007). Delayed functional expression of neuronal chemokine receptors following focal nerve demyelination in the rat: a mechanism for the development of chronic sensitization of peripheral nociceptors. *Mol. Pain* 3:38. doi: 10.1186/1744-8069-3-38

Bhangoo, S. K., Ripsch, M. S., Buchanan, D. J., Miller, R. J., and White, F. A. (2009). Increased chemokine signaling in a model of HIV1-associated peripheral neuropathy. *Mol. Pain* 5:48. doi: 10.1186/1744-8069-5-48

Caltana, L., Merelli, A., Lazarowski, A., and Brusco, A. (2009). Neuronal and glial alterations due to focal cortical hypoxia induced by direct cobalt chloride (CoCl2) brain injection. *Neurotox. Res.* 15, 348–358. doi: 10.1007/s12640-009-9038-9

Chirila, A. M., Brown, T. E., Bishop, R. A., Bellono, N. W., Pucci, F. G., and Kauer, J. A. (2014). Long-term potentiation of glycinergic synapses triggered by interleukin 1beta. *Proc. Natl. Acad. Sci. U S A* 111, 8263–8268. doi: 10.1073/pnas.1401013111

Choi, H. S., Roh, D. H., Yoon, S. Y., Moon, J. Y., Choi, S. R., Kwon, S. G., et al. (2015). Microglial interleukin-1beta in the ipsilateral dorsal horn inhibits the development of mirror-image contralateral mechanical allodynia through astrocyte activation in a rat model of inflammatory pain. *Pain* 156, 1046–1059. doi: 10.1097/j.pain.0000000000000148

Chung, C. S., Caplan, L. R., Han, W., Pessin, M. S., Lee, K. H., and Kim, J. M. (1996). Thalamic haemorrhage. *Brain* 119, 1873–1886. doi: 10.1093/brain/119.6.1873

Clark, A. K., Old, E. A., and Malcangio, M. (2013). Neuropathic pain and cytokines: current perspectives. *J. Pain Res.* 6, 803–814. doi: 10.2147/JPR.S53660

Debnath, B., Xu, S., Grande, F., Garofalo, A., and Neamati, N. (2013). Small molecule inhibitors of CXCR4. *Theranostics* 3, 47–75. doi: 10.7150/thno.5376

De Vloo, P., Morlion, B., van Loon, J., and Nuttin, B. (2017). Animal models for central poststroke pain: a critical comprehensive review. *Pain* 158, 17–29. doi: 10.1097/j.pain.0000000000000722

Dubovy, P., Klusakova, I., Svizenska, I., and Brazda, V. (2010). Spatio-temporal changes of SDF1 and its CXCR4 receptor in the dorsal root ganglia following unilateral sciatic nerve injury as a model of neuropathic pain. *Histochem. Cell Biol.* 133, 323–337. doi: 10.1007/s00418-010-0675-0

Feigin, V. L., Forouzanfar, M. H., Krishnamurthi, R., Mensah, G. A., Connor, M., Bennett, D. A., et al. (2014). Global and regional burden of stroke during 1990–2010: findings from the Global Burden of Disease Study 2010. *Lancet* 383, 245–254. doi: 10.1016/s0140-6736(13)61953-4

Ferro, J. M., Caeiro, L., and Figueira, M. L. (2016). Neuropsychiatric sequelae of stroke. *Nat. Rev. Neurol.* 12, 269–280. doi: 10.1038/nrneurol.2016.46

Gritsch, S., Bali, K. K., Kuner, R., and Vardeh, D. (2016). Functional characterization of a mouse model for central post-stroke pain. *Mol. Pain* 12:1744806916629049. doi: 10.1177/1744806916629049

Guo, W., Wang, H., Watanabe, M., Shimizu, K., Zou, S., LaGraize, S. C., et al. (2007). Glial-cytokine-neuronal interactions underlying the mechanisms of persistent pain. *J. Neurosci.* 27, 6006–6018. doi: 10.1523/jneurosci.0176-07.2007

Hanada, T., Kurihara, T., Tokudome, M., Tokimura, H., Arita, K., and Miyata, A. (2014). Development and pharmacological verification of a new mouse model of central post-stroke pain. *Neurosci. Res.* 78, 72–80. doi: 10.1016/j.neures.2013.09.005

Hang, L.-H., Li, S.-N., Luo, H., Shu, W.-W., Mao, Z.-M., Chen, Y.-F., et al. (2016). Connexin 43 mediates CXCL12 production from spinal dorsal horn to maintain bone cancer pain in rats. *Neurochem. Res.* 41, 1200–1208. doi: 10.1007/s11064-015-1815-7

Hendrix, C. W., Collier, A. C., Lederman, M. M., Schols, D., Pollard, R. B., Brown, S., et al. (2004). Safety, pharmacokinetics, and antiviral activity of AMD3100, a selective CXCR4 receptor inhibitor, in HIV-1 infection. *J. Acquir. Immune Defic. Syndr.* 37, 1253–1262. doi: 10.1097/01.qai.0000137371.80695.ef

Hitchon, C., Wong, K., Ma, G., Reed, J., Lyttle, D., and El-Gabalawy, H. (2002). Hypoxia-induced production of stromal cell-derived factor 1 (CXCL12) and vascular endothelial growth factor by synovial fibroblasts. *Arthritis Rheum* 46, 2587–2597. doi: 10.1002/art.10520

Hosomi, K., Seymour, B., and Saitoh, Y. (2015). Modulating the pain network—neurostimulation for central poststroke pain. *Nat. Rev. Neurol.* 11, 290–299. doi: 10.1038/nrneurol.2015.58

Hsieh, Y. L., Chou, L. W., Chang, P. L., Yang, C. C., Kao, M. J., and Hong, C. Z. (2012). Low-level laser therapy alleviates neuropathic pain and promotes function recovery in rats with chronic constriction injury: possible involvements in hypoxia-inducible factor 1alpha (HIF-1alpha). *J. Comp. Neurol.* 520, 2903–2916. doi: 10.1002/cne.23072

Hu, X. M., Liu, Y. N., Zhang, H. L., Cao, S. B., Zhang, T., Chen, L. P., et al. (2015). CXCL12/CXCR4 chemokine signaling in spinal glia induces pain hypersensitivity through MAPKs-mediated neuroinflammation in bone cancer rats. *J. Neurochem.* 132, 452–463. doi: 10.1111/jnc.12985

Huang, X., Li, Y., Li, J., Feng, Y., and Xu, X. (2014). Tanshinone IIA dampens the cell proliferation induced by ischemic insult in rat astrocytes via blocking the activation of HIF-1alpha/SDF-1 signaling. *Life Sci.* 112, 59–67. doi: 10.1016/j.lfs.2014.07.020

Huang, J., Li, Y., Tang, Y., Tang, G., Yang, G. Y., and Wang, Y. (2013). CXCR4 antagonist AMD3100 protects blood-brain barrier integrity and reduces inflammatory response after focal ischemia in mice. *Stroke* 44, 190–197. doi: 10.1161/STROKEAHA.112.670299

Hummel, S., Van Aken, H., and Zarbock, A. (2014). Inhibitors of CXC chemokine receptor type 4: putative therapeutic approaches in inflammatory diseases. *Curr. Opin. Hematol.* 21, 29–36. doi: 10.1097/MOH.0000000000000002

Jiang, Y., Wu, J., Keep, R. F., Hua, Y., Hoff, J. T., and Xi, G. (2002). Hypoxia-inducible factor-1α accumulation in the brain after experimental intracerebral hemorrhage. *J. Cereb. Blood Flow Metab.* 22, 689–696. doi: 10.1097/00004647-200206000-00007

Kanngiesser, M., Mair, N., Lim, H. Y., Zschiebsch, K., Blees, J., Häussler, A., et al. (2014). Hypoxia-inducible factor 1 regulates heat and cold pain sensitivity and persistence. *Antioxid. Redox Signal.* 20, 2555–2571. doi: 10.1089/ars.2013.5494

Kirkpatrick, B., Nguyen, L., Kondrikova, G., Herberg, S., and Hill, W. D. (2010). Stability of human stromal-derived factor-1α (CXCL12α) after blood sampling. *Ann. Clin. Lab. Sci.* 40, 257–260.

Knerlich-Lukoschus, F., von der Ropp-Brenner, B., Lucius, R., Mehdorn, H. M., and Held-Feindt, J. (2011). Spatiotemporal CCR1, CCL3(MIP-1α), CXCR4, CXCL12(SDF-1α) expression patterns in a rat spinal cord injury model of posttraumatic neuropathic pain. *J. Neurosurg. Spine* 14, 583–597. doi: 10.3171/2010.12.SPINE10480

Kuan, Y. H., Shih, H. C., Tang, S. C., Jeng, J. S., and Shyu, B. C. (2015). Targeting P(2)X(7) receptor for the treatment of central post-stroke pain in a rodent model. *Neurobiol. Dis.* 78, 134–145. doi: 10.1016/j.nbd.2015.02.028

LeBlanc, B. W., Zerah, M. L., Kadasi, L. M., Chai, N., and Saab, C. Y. (2011). Minocycline injection in the ventral posterolateral thalamus reverses microglial reactivity and thermal hyperalgesia secondary to sciatic neuropathy. *Neurosci. Lett.* 498, 138–142. doi: 10.1016/j.neulet.2011.04.077

Li, X. Q., Zhang, Z. L., Tan, W. F., Sun, X. J., and Ma, H. (2016). Down-regulation of CXCL12/CXCR4 expression alleviates ischemia-reperfusion-induced inflammatory pain via inhibiting glial TLR4 activation in the spinal cord. *PLoS One* 11:e0163807. doi: 10.1371/journal.pone.0163807

Liu, H., Xue, W., Ge, G., Luo, X., Li, Y., Xiang, H., et al. (2010). Hypoxic preconditioning advances CXCR4 and CXCR7 expression by activating HIF-1alpha in MSCs. *Biochem. Biophys. Res. Commun.* 401, 509–515. doi: 10.1016/j.bbrc.2010.09.076

Luo, X., Tai, W. L., Sun, L., Pan, Z., Xia, Z., Chung, S. K., et al. (2016). Crosstalk between astrocytic CXCL12 and microglial CXCR4 contributes to the development of neuropathic pain. *Mol. Pain* 12:1744806916636385. doi: 10.1177/1744806916636385

Luo, X., Tai, W. L., Sun, L., Qiu, Q., Xia, Z., Chung, S. K., et al. (2014). Central administration of C-X-C chemokine receptor type 4 antagonist alleviates the development and maintenance of peripheral neuropathic pain in mice. *PLoS One* 9:e104860. doi: 10.1371/journal.pone.0104860

Macht, V. A. (2016). Neuro-immune interactions across development: a look at glutamate in the prefrontal cortex. *Neurosci. Biobehav. Rev.* 71, 267–280. doi: 10.1016/j.neubiorev.2016.08.039

Menichella, D. M., Abdelhak, B., Ren, D., Shum, A., Frietag, C., and Miller, R. J. (2014). CXCR4 chemokine receptor signaling mediates pain in diabetic neuropathy. *Mol. Pain* 10:42. doi: 10.1186/1744-8069-10-42

Nasreddine, Z. S., and Saver, J. L. (1997). Pain after thalamic stroke: right diencephalic predominance and clinical features in 180 patients. *Neurology* 48, 1196–1199. doi: 10.1212/WNL.48.5.1196

Oh, S. B., Tran, P. B., Gillard, S. E., Hurley, R. W., Hammond, D. L., and Miller, R. J. (2001). Chemokines and glycoprotein120 produce pain hypersensitivity by directly exciting primary nociceptive neurons. *J. Neurosci.* 21, 5027–5035.

Paxinos, G., and Watson, C. (2005). *The Rat Brain in Stereotaxic Coordinates*, 5th Edn., Amsterdam: Elsevier Academic Press.

Reaux-Le Goazigo, A., Rivat, C., Kitabgi, P., Pohl, M., and Melik Parsadaniantz, S. (2012). Cellular and subcellular localization of CXCL12 and CXCR4 in rat nociceptive structures: physiological relevance. *Eur. J. Neurosci.* 36, 2619–2631. doi: 10.1111/j.1460-9568.2012.08179.x

Réaux-Le Goazigo, A., Van Steenwinckel, J., Rostène, W., and Mélik Parsadaniantz, S. (2013). Current status of chemokines in the adult CNS. *Prog. Neurobiol.* 104, 67–92. doi: 10.1016/j.pneurobio.2013.02.001

Ruscher, K., Kuric, E., Liu, Y., Walter, H. L., Issazadeh-Navikas, S., Englund, E., et al. (2013). Inhibition of CXCL12 signaling attenuates the postischemic immune response and improves functional recovery after stroke. *J. Cereb. Blood Flow Metab.* 33, 1225–1234. doi: 10.1038/jcbfm.2013.71

Schioppa, T., Uranchimeg, B., Saccani, A., Biswas, S. K., Doni, A., Rapisarda, A., et al. (2003). Regulation of the chemokine receptor CXCR4 by hypoxia. *J. Exp. Med.* 198, 1391–1402. doi: 10.1084/jem.20030267

Shen, W., Hu, X. M., Liu, Y. N., Han, Y., Chen, L. P., Wang, C. C., et al. (2014). CXCL12 in astrocytes contributes to bone cancer pain through CXCR4-mediated neuronal sensitization and glial activation in rat spinal cord. *J. Neuroinflammation* 11:75. doi: 10.1186/1742-2094-11-75

Sheridan, G. K., and Murphy, K. J. (2013). Neuron-glia crosstalk in health and disease: fractalkine and CX3CR1 take centre stage. *Open Biol.* 3:130181. doi: 10.1098/rsob.130181

Singh, N., Sharma, G., and Mishra, V. (2012). Hypoxia inducible factor-1: its potential role in cerebral ischemia. *Cell. Mol. Neurobiol.* 32, 491–507. doi: 10.1007/s10571-012-9803-9

Speth, J. M., Hoggatt, J., Singh, P., and Pelus, L. M. (2014). Pharmacologic increase in HIF1alpha enhances hematopoietic stem and progenitor homing and engraftment. *Blood* 123, 203–207. doi: 10.1182/blood-2013-07-516336

Trecki, J., and Unterwald, E. M. (2009). Modulation of cocaine-induced activity by intracerebral administration of CXCL12. *Neuroscience* 161, 13–22. doi: 10.1016/j.neuroscience.2009.03.027

Walter, H. L., van der Maten, G., Antunes, A. R., Wieloch, T., and Ruscher, K. (2015). Treatment with AMD3100 attenuates the microglial response and improves outcome after experimental stroke. *J. Neuroinflammation* 12:24. doi: 10.1186/s12974-014-0232-1

Wang, X., Li, C., Chen, Y., Hao, Y., Zhou, W., Chen, C., et al. (2008). Hypoxia enhances CXCR4 expression favoring microglia migration via HIF-1alpha activation. *Biochem. Biophys. Res. Commun.* 371, 283–288. doi: 10.1016/j.bbrc.2008.04.055

Wasserman, J. K., and Koeberle, P. D. (2009). Development and characterization of a hemorrhagic rat model of central post-stroke pain. *Neuroscience* 161, 173–183. doi: 10.1016/j.neuroscience.2009.03.042

Wilson, N. M., Jung, H., Ripsch, M. S., Miller, R. J., and White, F. A. (2011). CXCR4 signaling mediates morphine-induced tactile hyperalgesia. *Brain Behav. Immun.* 25, 565–573. doi: 10.1016/j.bbi.2010.12.014

Xie, F., Wang, Y., Li, X., Chao, Y. C., and Yue, Y. (2016). Early repeated administration of CXCR4 antagonist AMD3100 dose-dependently improves neuropathic pain in rats after L5 spinal nerve ligation. *Neurochem. Res.* 41, 2289–2299. doi: 10.1007/s11064-016-1943-8

Yang, F., Fu, H., Lu, Y. F., Wang, X. L., Yang, Y., Yang, F., et al. (2014). Post-stroke pain hypersensitivity induced by experimental thalamic hemorrhage in rats is region-specific and demonstrates limited efficacy of gabapentin. *Neurosci. Bull.* 30, 887–902. doi: 10.1007/s12264-014-1477-5

Yang, F., Sun, W., Luo, W. J., Yang, Y., Yang, F., Wang, X. L., et al. (2017). SDF1-CXCR4 signaling contributes to the transition from acute to chronic pain state. *Mol. Neurobiol.* 54, 2763–2775. doi: 10.1007/s12035-016-9875-5

Yang, F., Sun, W., Yang, Y., Wang, Y., Li, C. L., Fu, H., et al. (2015). SDF1-CXCR4 signaling contributes to persistent pain and hypersensitivity via regulating excitability of primary nociceptive neurons: involvement of ERK-dependent Nav1.8 up-regulation. *J. Neuroinflammation* 12:219. doi: 10.1186/s12974-015-0441-2

Yang, L., Wang, M., Guo, Y. Y., Sun, T., Li, Y. J., Yang, Q., et al. (2016). Systemic inflammation induces anxiety disorder through CXCL12/CXCR4 pathway. *Brain Behav. Immun.* 56, 352–362. doi: 10.1016/j.bbi.2016.03.001

Yang, Y., Yang, F., Yang, F., Li, C. L., Wang, Y., Li, Z., et al. (2016). Gabapentinoid insensitivity after repeated administration is associated with down-regulation of the alpha$_2$delta-1 subunit in rats with central post-stroke pain hypersensitivity. *Neurosci. Bull.* 32, 41–50. doi: 10.1007/s12264-015-0008-3

Youn, S. W., Lee, S. W., Lee, J., Jeong, H. K., Suh, J. W., Yoon, C. H., et al. (2011). COMP-Ang1 stimulates HIF-1alpha-mediated SDF-1 overexpression and recovers ischemic injury through BM-derived progenitor cell recruitment. *Blood* 117, 4376–4386. doi: 10.1182/blood-2010-07-295964

Zhang, L., Berta, T., Xu, Z. Z., Liu, T., Park, J. Y., and Ji, R. R. (2011). TNF-alpha contributes to spinal cord synaptic plasticity and inflammatory pain: distinct role of TNF receptor subtypes 1 and 2. *Pain* 152, 419–427. doi: 10.1016/j.pain.2010.11.014

Zhang, Z., Yan, J., Taheri, S., Liu, K. J., and Shi, H. (2014). Hypoxia-inducible factor 1 contributes to N-acetylcysteine's protection in stroke. *Free Radic. Biol. Med.* 68, 8–21. doi: 10.1016/j.freeradbiomed.2013.11.007

Zhao, P., Waxman, S. G., and Hains, B. C. (2007). Modulation of thalamic nociceptive processing after spinal cord injury through remote activation of thalamic microglia by cysteine cysteine chemokine ligand 21. *J. Neurosci.* 27, 8893–8902. doi: 10.1523/JNEUROSCI.2209-07.2007

18

Ca^{2+}-Dependent and Ca^{2+}-Independent ATP Release in Astrocytes

Yingfei Xiong [1,2†], Suhua Sun [1†], Sasa Teng [1], Mu Jin [1,3*] and Zhuan Zhou [1*]*

[1] *State Key Laboratory of Biomembrane and Membrane Biotechnology and Peking-Tsinghua Center for Life Sciences and PKU-IDG/McGovern Institute for Brain Research, Institute of Molecular Medicine, Peking University, Beijing, China,* [2] *Department of Neurosurgery, Affiliated Hospital of The Air Force Institute of Aeromedicine, Beijing, China,* [3] *Department of Anesthesiology, Beijing Friendship Hospital, Capital Medical University, Beijing, China*

Keywords: glial transmitter, ATP, astrocyte, exocytosis, P2X7, calcium, mechanical stimulation

***Correspondence:**
Yingfei Xiong
yfxiong@139.com
Mu Jin
jinmu0119@tom.com
Zhuan Zhou
zzhou@pku.edu.cn

† These authors have contributed equally to this work.

Like neurons, astrocytes are abundant in the central nervous system. They contact all types of cells in the brain, communicate with them, and modulate their activity by releasing gliotransmitters, including glutamate, ATP, etc. (Newman, 2003; Halassa and Haydon, 2010; Hamilton and Attwell, 2010; Volterra et al., 2014; Verkhratsky and Nedergaard, 2018), although the role of astrocytes in neurotransmission is still debated (Nedergaard and Verkhratsky, 2012; Fiacco and McCarthy, 2018).

ATP is considered to be a powerful extracellular messenger in both the peripheral and central nervous systems (Edwards et al., 1992; Fields and Stevens, 2000; Burnstock, 2007; Verkhratsky et al., 2009). *Via* activating multiple receptors in glia and neurons, ATP signaling participates in many important functions including cell development and synaptic plasticity (Evans et al., 1992; Newman, 2003; Zhang et al., 2003; Agresti et al., 2005; Abbracchio et al., 2009; Butt, 2011; Wurm et al., 2011; Verkhratsky et al., 2016). However, unlike the mechanism of neurotransmission *via* quantal exocytosis (Katz, 1959, 1969; Augustine and Neher, 1992; Neher, 1998; Sudhof, 2004; Pankratov et al., 2006, 2007; Sudhof and Rothman, 2009), the mechanisms by which ATP is released remain controversial. Because ATP is easily hydrolyzed, monitoring its real-time release is a challenge.

EARLY STUDIES OF NON-QUANTAL ATP RELEASE FROM ASTROCYTES

Early studies usually applied indirect methods such as dye-uptake, ATP analog labeling, and luciferase-luciferin system tests to indirectly detect ATP release (Duan et al., 2003). Based on these studies, ATP was thought to be released through specific channels, such as connexin/pannexin hemichannels, "maxi-anion" channels, and P2X7 receptor channels (**Figure 1A,** Stout et al., 2002; Duan et al., 2003; Bao et al., 2004; Suadicani et al., 2006; Kang et al., 2008; Liu et al., 2008; Iglesias et al., 2009; Bennett et al., 2012). These studies used indirect measurements based on measuring dyes or currents through channels that can pass molecules larger than ATP.

QUANTAL ATP RELEASE UNDER PHYSIOLOGICAL CONDITIONS

Patch-clamp recording *via* sniffer cells (Young and Poo, 1983; Liu et al., 2011) is a powerful tool with which to investigate the ATP release mechanism due to its high spatiotemporal resolution (Hollins and Ikeda, 1997; Pangrsic et al., 2007; Karanauskaite et al., 2009; Lalo et al., 2014; Liu et al., 2014; Lee et al., 2015). By using ATP-sniffer cells, Lalo et al. recorded quantal ATP release in astrocytes freshly isolated from mouse cortex. ATP is released by Ca^{2+}-dependent exocytosis

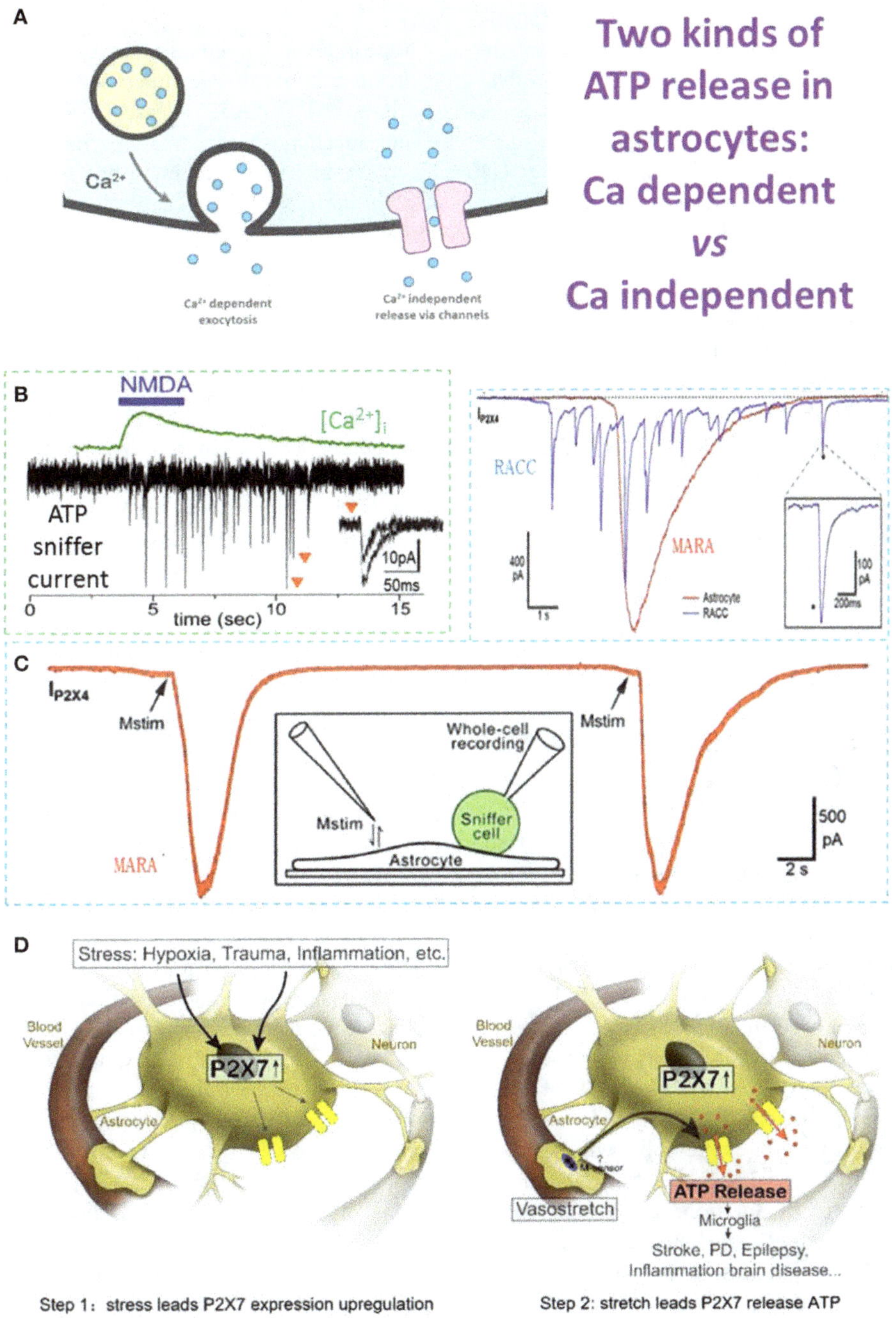

FIGURE 1 | Ca^{2+}-dependent and Ca^{2+}-independent ATP release in astrocytes. **(A)** Two types of ATP release: Ca^{2+}-dependent quantal release and Ca^{2+}-independent non-quantal release. In the Ca^{2+}-dependent pathway, secretory vesicles with packaged ATP are trafficked to the plasma membrane where they dock and fuse with it on arrival of a stimulus; this typically depends on an increase of cytosolic Ca^{2+}. In the Ca^{2+}-independent pathway, ATP is released through channels expressed on the astrocyte plasma membrane, such as the swelling-induced anion channel, connexin hemichannels activated by lower Ca^{2+} concentrations, and ionotropic purinergic receptor channels. **(B)** N-methyl-D-aspartate (NMDA)-induced Ca^{2+}-dependent quantal ATP release from a freshly-isolated astrocyte, recorded by a HEK293-P2X2 sniffer cell. **(C)** Typical MARA current (IP2X4) recorded with an HEK293A sniffer-cell expressing P2X4 (ATP sniffer) on an astrocyte. The astrocyte was stimulated twice and the MARA signal was reproducible. Inset, cartoon of the experimental protocol for MARA recording; Upper right: MARA vs Ca^{2+}-dependent ATP release of burst quantal events from a rat chromaffin cell (RACC). **(D)** Two-step hypothesis of MARA-mediated brain diseases. Left. Step 1: Stress induces upregulation of P2X7 receptor expression in astrocytes. The stressors include hypoxia (i.e., stroke and ischemia), trauma, and CNS diseases associated with inflammation or neurodegeneration. Right. Step 2: Stretch leads to P2X7-mediated ATP release from astrocytes. The abundant ATP is an emergency signal for the neural immune system that recruits and activates microglia in the ATP "hot-spots" and promotes disease. All data adapted from Xiong et al. (2018), except B adapted from Lalo et al. (2014). All data with reproduction permission from the original publishers.

following the activation of metabotropic and ionotropic receptors or direct UV-uncaging. The ATP release is SNARE protein-dependent and is eliminated by pretreatment with bafilomycin, a blocker of vacuolar-type H-ATPase. The kinetics of sniffer-cell responses are consistent with the millisecond time-scale, suggesting that ATP exocytosis is from synaptic-like small vesicles (**Figures 1A,B**). The ATP released from astrocytes (1) activates P2X receptors in neighboring neurons to enhance excitatory signaling, and (2) down-regulates inhibitory synaptic signaling (Lalo et al., 2014).

In addition to the tiny quantal ATP release arising from Ca^{2+}-dependent exocytosis in freshly-isolated astrocytes, an earlier report proposed that Ca^{2+}-dependent lysosome exocytosis is responsible for quantal ATP release in cultured astrocytes (Zhang et al., 2007). They provided three lines of indirect evidence: (1) the uptake of a fluorescent ATP analog, MANT-ATP, into lysosomes; (2) the presence of ATP in biochemically-purified lysosomes; and (3) the real-time visualization of Ca^{2+}-dependent lysosome exocytosis by total internal reflection fluorescence microscopy imaging of a false neurotransmitter (FM2-10). Although these data raised the possibility of quantal ATP release *via* lysosomal exocytosis, direct recording of quantal ATP release from glial lysosomes was absent.

STRETCH-INDUCED CA^{2+}-INDEPENDENT ATP RELEASE THROUGH P2X7 CHANNELS

It was our original goal to record quantal ATP release in cultured hippocampal astrocytes using ATP-sniffer cells. To our surprise, with three independent assays of lysosomal exocytosis, our real-time sniffer recordings clearly denied ATP release by Ca^{2+}-dependent lysosomal exocytosis (Xiong et al., 2018). Instead, following a gentle membrane stretch, Ca^{2+}-independent non-vesicular ATP release occurred in the cultured astrocytes (Xiong et al., 2018). In contrast to the quantal spikes of ATP release in chromaffin cells, the Mechanically-induced ATP Release from Astrocytes (MARA) displays a single spike with distinct kinetic characteristics. MARA is ~300-fold greater in release content and ~50 times longer in duration (**Figure 1C**, upper right inset). Mechanistically, MARA-mediated ATP release is (1) Ca^{2+}-independent, (2) not *via* lysosome exocytosis, and (3) mitochondria-dependent (Xiong et al., 2018). The P2X7 receptor channel is essential for MARA-mediated ATP release, because it is profoundly inhibited by Brilliant Blue G, a selective P2X7 antagonist, as well as by RNA interference-based P2X7 knockdown (Xiong et al., 2018). Considering that the open pore of the P2X7 channel allows the permeation of cytosolic molecules of molecular weight ≤900 Da, P2X7 channels should be able to release ATP (507 Da) (Yan et al., 2008; Nagasawa et al., 2009), and other purines, such as AMP, ADP, or adenosine. Together, MARA occurs when mechanical stretch triggers ATP efflux through P2X7 channel pores in "activated" astrocytes expressing P2X7 receptors in culture/*in vitro* or hypoxia/trauma/disease *in vivo* (Nagasawa et al., 2009). Hypoxia might decrease mitochondrial-dependent cytosolic ATP level and partially compensate the MARA signal for the increase in P2X7 expression.

One important open question left by Xiong (Xiong et al., 2018) is the identity of the mechanical sensor by which MARA initiates mechanical-P2X7-ATP release in astrocytes. Since P2X7 itself is not mechanosensor, we hypothesize that a mechanosensor [such as piezo 1 protein (Zhao et al., 2018)] binds P2X7 and "transactivates" mechanical force to activate P2X7 [One recent example of protein-transactivation is that a voltage-sensor channel activates another binding protein of vesicle fusion-pore (Chai et al., 2017)]. Following the discovery of MARA, identification of this sensor is critical for treating possible MARA-mediated brain diseases as proposed below. In addition to P2X7 (Xiong et al., 2018), hemichannels such as connexins (Stout et al., 2002) and large conductance Cl^- channel (Liu et al., 2008) have been reported to mediate the non-vesicular ATP release from astrocytes. At present it is unclear about the relative contributions to ATP release among these channel types, because the stimulations in these studies were different. These ATP-release pathways through different channels may play condition-dependent roles in astrocytes' functions.

A HYPOTHESIS OF MARA IN BRAIN PROTECTION AND DISEASES

In contrast to the Ca^{2+}-dependent quantal ATP release in freshly-isolated astrocytes, the Ca^{2+}-independent ATP release event (MARA) in cultured astrocytes is ~10,000 times greater (**Figure 1C**). P2X7 is up-regulated under ischemic conditions *in vivo* (Nagasawa et al., 2009), and this could contribute to the release of large amounts of ATP from cellular sources, and in the extracellular space it is quickly hydrolyzed to ADP, AMP, and adenosine, which activate their receptors and play roles in brain protection and damage (Neary et al., 2003; Choo et al., 2013; Rodrigues et al., 2015). On the other hand, the large ATP release *via* MARA would recruit microglia, leading to protective or pathological pathways (Dou et al., 2012). Thus, we propose that MARA could be a mechanism underlying brain diseases such as those associated with hypoxia/ischemia and trauma, as well as other neurological disorders (Parkinson's disease, Alzheimer's disease, and epilepsy) (**Figure 1D**). Our hypothesis is detailed as follows. The first step is MARA genesis, which is dependent on P2X7 receptor expression. The expression level of P2X7 receptors in astrocytes is up-regulated under either of the two conditions: (1) in culture, which is an extremely stressful condition lacking blood and other support for astrocytes (Narcisse et al., 2005; Bartlett et al., 2014; Burnstock, 2017; Xiong et al., 2018); and (2) *in vivo*, when astrocytes suffer stresses, such as hypoxia (i.e., ischemia and stroke), trauma, or some CNS diseases (for example, inflammation and/or neurodegeneration; Ballerini et al., 1996; Narcisse et al., 2005; Nagasawa et al., 2009; Burnstock, 2017). Indeed, few astrocytes express P2X7 receptors in the intact hippocampal slices under physiological conditions (Nagasawa et al., 2009; Xiong et al., 2018). As the P2X7 receptor is a key modulator of aerobic glycolysis, it has an intrinsic ability to reprogram cell metabolism to meet the needs imposed by adverse environmental conditions (Amoroso et al., 2012). The up-regulation of P2X7 receptors in astrocytes is not only a form of adaptation to stress, but is also a necessary preparation for MARA to execute its repair function in its brain region (**Figure 1D**, step 1). The second step is MARA activation following mechanical stimulation, which could be generated by arterioles *in vivo*. Astrocytes regulate cerebral blood flow to match the metabolic requirements of the brain (Gordon et al., 2008) by eliciting the vasoconstriction or vasodilation of arterioles. A tension change due to such vasoconstriction/dilation is an effective physiological mechanical stimulus (membrane stretch) for astrocytes, which are activated by deformation of their surroundings on a timescale of milliseconds (Janmey and Miller, 2011). Physiological arteriole stretch does not trigger MARA because P2X7 receptors are rare in astrocytes under normal conditions. When stresses such as hypoxia/trauma occur, the expression of astrocytic P2X7 receptors is up-regulated, and if the stress persists and exceeds a threshold, the abnormal changes in arteriole tension directly trigger MARA (**Figure 1D**, step 2). So, noxious stimuli up-regulate the P2X7 receptor expression in astrocytes, and this can be considered as an adaptive change in response to stress and the beginning of pathological damage, because up-regulation of P2X7 receptor expression is the prerequisite for non-quantal ATP release/MARA. When the steady-state is disturbed, MARA occurs and triggers a sustained increase in extracellular ATP, which then acts to alert the presence of a "hot spot." This signal recruits and activates microglia, which scavenge and phagocytize injured cells and cellular debris, increase the susceptibility of neurons to damage, promote astrogliosis, and mount neuroinflammatory responses (Parpura et al., 2012; Rivera et al., 2016). In this sense, both astrocytes and microglia comprise the immune system of the CNS: each of a vast number astrocytes performs an immobile surveillance for its specific location in the brain, signaling stress, while each microglia cell acts as a mobile patrol, responding to the astrocytic "alert" signal—an unusually large increase in extracellular ATP concentration. To a certain extent, MARA from a few astrocytes recruits a few microglia to clear the site and protect the brain. However, when the stress is excessive, MARA from a large number of astrocytes may recruit too many microglia, overkill healthy tissue, and cause irreversible brain damage (disease).

AUTHOR CONTRIBUTIONS

All authors listed have made a substantial, direct and intellectual contribution to the work, and approved it for publication.

ACKNOWLEDGMENTS

We thank Prof. I. C. Bruce (Peking University) for helpful comments, and Xi Wu for art of **Figures 1A** and **D**. This work was supported by the National Natural Science Foundation of China (31330024, 31761133016, 21790394, 31171026, 31327901, 31521062, and 21790390), the National Key Research and Development Program of China (2016YFA0500401), and the National Basic Research Program of China (2012C B518006).

REFERENCES

Abbracchio, M. P., Burnstock, G., Verkhratsky, A., and Zimmermann, H. (2009). Purinergic signalling in the nervous system: an overview. *Trends Neurosci.* 32, 19–29. doi: 10.1016/j.tins.2008.10.001

Agresti, C., Meomartini, M. E., Amadio, S., Ambrosini, E., Volonté, C., Aloisi, F., et al. (2005). ATP regulates oligodendrocyte progenitor migration, proliferation, and differentiation: involvement of metabotropic P2 receptors. *Brain Res. Brain Res. Rev.* 48, 157–165. doi: 10.1016/j.brainresrev.2004.12.005

Amoroso, F., Falzoni, S., Adinolfi, E., Ferrari, D., and Di Virgilio, F. (2012). The P2X7 receptor is a key modulator of aerobic glycolysis. *Cell Death Amp. Dis.* 3:e370. doi: 10.1038/cddis.2012.105

Augustine, G. J., and Neher, E. (1992). Calcium requirements for secretion in bovine chromaffin cells. *J. Physiol.* 450, 247–271. doi: 10.1113/jphysiol.1992.sp019126

Ballerini, P., Rathbone, M. P., Di Iorio, P., Renzetti, A., Giuliani, P., D'alimonte, I., et al. (1996). Rat astroglial P2Z (P2X7) receptors regulate intracellular calcium and purine release. *Neuroreport* 7, 2533–2537. doi: 10.1097/00001756-199611040-00026

Bao, L., Locovei, S., and Dahl, G. (2004). Pannexin membrane channels are mechanosensitive conduits for ATP. *FEBS Lett.* 572, 65–68. doi: 10.1016/j.febslet.2004.07.009

Bartlett, R., Stokes, L., and Sluyter, R. (2014). The P2X7 receptor channel: recent developments and the use of P2X7 antagonists in models of disease. *Pharmacol. Rev.* 66, 638–675. doi: 10.1124/pr.113.008003

Bennett, M. V., Garré, J. M., Orellana, J. A., Bukauskas, F. F., Nedergaard, M., and Saez, J. C. (2012). Connexin and pannexin hemichannels in inflammatory responses of glia and neurons. *Brain Res.* 1487, 3–15. doi: 10.1016/j.brainres.2012.08.042

Burnstock, G. (2007). Physiology and pathophysiology of purinergic neurotransmission. *Physiol. Rev.* 87, 659–797. doi: 10.1152/physrev.00043.2006

Burnstock, G. (2017). Purinergic signalling and neurological diseases: an update. *CNS Neurol. Disord. Drug Targets* 16, 257–265. doi: 10.2174/1871527315666160922104848

Butt, A. M. (2011). ATP: a ubiquitous gliotransmitter integrating neuron-glial networks. *Semin. Cell Dev. Biol.* 22, 205–213. doi: 10.1016/j.semcdb.2011.02.023

Chai, Z., Wang, C., Huang, R., Wang, Y., Zhang, X., Wu, Q., et al. (2017). CaV2.2 gates calcium-independent but voltage-dependent secretion in mammalian sensory neurons. *Neuron* 96, 1317–1326.e1314. doi: 10.1016/j.neuron.2017.10.028

Choo, A. M., Miller, W. J., Chen, Y. C., Nibley, P., Patel, T. P., Goletiani, C., et al. (2013). Antagonism of purinergic signalling improves recovery from traumatic brain injury. *Brain* 136, 65–80. doi: 10.1093/brain/aws286

Dou, Y., Wu, H. J., Li, H. Q., Qin, S., Wang, Y. E., Li, J., et al. (2012). Microglial migration mediated by ATP-induced ATP release from lysosomes. *Cell Res.* 22, 1022–1033. doi: 10.1038/cr.2012.10

Duan, S., Anderson, C. M., Keung, E. C., Chen, Y., Chen, Y., and Swanson, R. A. (2003). P2X7 receptor-mediated release of excitatory amino acids from astrocytes. *J. Neurosci.* 23, 1320–1328. doi: 10.1523/JNEUROSCI.23-04-01320.2003

Edwards, F. A., Gibb, A. J., and Colquhoun, D. (1992). ATP receptor-mediated synaptic currents in the central nervous system. *Nature* 359, 144–147. doi: 10.1038/359144a0

Evans, R. J., Derkach, V., and Surprenant, A. (1992). ATP mediates fast synaptic transmission in mammalian neurons. *Nature* 357, 503–505. doi: 10.1038/357503a0

Fiacco, T. A., and McCarthy, K. D. (2018). Multiple lines of evidence indicate that gliotransmission does not occur under physiological conditions. *J. Neurosci.* 38, 3–13. doi: 10.1523/JNEUROSCI.0016-17.2017

Fields, R. D., and Stevens, B. (2000). ATP: an extracellular signaling molecule between neurons and glia. *Trends Neurosci.* 23, 625–633. doi: 10.1016/S0166-2236(00)01674-X

Gordon, G. R., Choi, H. B., Rungta, R. L., Ellis-Davies, G. C., and MacVicar, B. A. (2008). Brain metabolism dictates the polarity of astrocyte control over arterioles. *Nature* 456, 745–749. doi: 10.1038/nature07525

Halassa, M. M., and Haydon, P. G. (2010). Integrated brain circuits: astrocytic networks modulate neuronal activity and behavior. *Annu. Rev. Physiol.* 72, 335–355. doi: 10.1146/annurev-physiol-021909-135843

Hamilton, N. B., and Attwell, D. (2010). Do astrocytes really exocytose neurotransmitters? *Nat. Rev. Neurosci.* 11, 227–238. doi: 10.1038/nrn2803

Hollins, B., and Ikeda, S. R. (1997). Heterologous expression of a P2x-purinoceptor in rat chromaffin cells detects vesicular ATP release. *J. Neurophysiol.* 78, 3069–3076. doi: 10.1152/jn.1997.78.6.3069

Iglesias, R., Dahl, G., Qiu, F., Spray, D. C., and Scemes, E. (2009). Pannexin 1: the molecular substrate of astrocyte "hemichannels". *J. Neurosci.* 29, 7092–7097. doi: 10.1523/JNEUROSCI.6062-08.2009

Janmey, P. A., and Miller, R. T. (2011). Mechanisms of mechanical signaling in development and disease. *J. Cell Sci.* 124, 9–18. doi: 10.1242/jcs.071001

Kang, J., Kang, N., Lovatt, D., Torres, A., Zhao, Z., Lin, J., et al. (2008). Connexin 43 hemichannels are permeable to ATP. *J. Neurosci.* 28, 4702–4711. doi: 10.1523/JNEUROSCI.5048-07.2008

Karanauskaite, J., Hoppa, M. B., Braun, M., Galvanovskis, J., and Rorsman, P. (2009). Quantal ATP release in rat beta-cells by exocytosis of insulin-containing LDCVs. *Pflugers Arch.* 458, 389–401. doi: 10.1007/s00424-008-0610-6

Katz, B. (1959). Nature of the nerve impulse. *Rev. Mod. Phys* 8, 629–638. doi: 10.1103/RevModPhys.31.466

Katz, B. (1969). The release of neural transmitter substances. *J. Neurol. Neurosurg. Psychiatry* 32:638. doi: 10.1136/jnnp.32.6.638

Lalo, U., Palygin, O., Rasooli-Nejad, S., Andrew, J., Haydon, P. G., and Pankratov, Y. (2014). Exocytosis of ATP from astrocytes modulates phasic and tonic inhibition in the neocortex. *PLoS Biol.* 12:e1001747. doi: 10.1371/journal.pbio.1001747

Lee, J., Chun, Y. E., Han, K. S., Lee, J., Woo, D. H., and Lee, C. J. (2015). Ca($^{2+}$) Entry is required for mechanical stimulation-induced ATP release from astrocyte. *Exp. Neurobiol.* 24, 17–23. doi: 10.5607/en.2015.24.1.17

Liu, H. T., Toychiev, A. H., Takahashi, N., Sabirov, R. Z., and Okada, Y. (2008). Maxi-anion channel as a candidate pathway for osmosensitive ATP release from mouse astrocytes in primary culture. *Cell Res.* 18, 558–565. doi: 10.1038/cr.2008.49

Liu, T., Li, H., Gounko, N. V., Zhou, Z., Xu, A., Hong, W., et al. (2014). Detection of insulin granule exocytosis by an electrophysiology method with high temporal resolution reveals enlarged insulin granule pool in BIG3-knockout mice. *Am. J. Physiol. Endocrinol. Metab.* 307, E611–E618. doi: 10.1152/ajpendo.00208.2014

Liu, T., Sun, L., Xiong, Y., Shang, S., Guo, N., Teng, S., et al. (2011). Calcium triggers exocytosis from two types of organelles in a single astrocyte. *J. Neurosci.* 31, 10593–10601. doi: 10.1523/JNEUROSCI.6401-10.2011

Nagasawa, K., Escartin, C., and Swanson, R. A. (2009). Astrocyte cultures exhibit P2X7 receptor channel opening in the absence of exogenous ligands. *Glia* 57, 622–633. doi: 10.1002/glia.20791

Narcisse, L., Scemes, E., Zhao, Y., Lee, S. C., and Brosnan, C. F. (2005). The cytokine IL-1beta transiently enhances P2X7 receptor expression and function in human astrocytes. *Glia* 49, 245–258. doi: 10.1002/glia.20110

Neary, J. T., Kang, Y., Willoughby, K. A., and Ellis, E. F. (2003). Activation of extracellular signal-regulated kinase by stretch-induced injury in astrocytes involves extracellular ATP and P2 purinergic receptors. *J. Neurosci.* 23, 2348–2356. doi: 10.1523/JNEUROSCI.23-06-02348.2003

Nedergaard, M., and Verkhratsky, A. (2012). Artifact versus reality–how astrocytes contribute to synaptic events. *Glia* 60, 1013–1023. doi: 10.1002/glia.22288

Neher, E. (1998). Vesicle pools and Ca^{2+} microdomains: new tools for understanding their roles in neurotransmitter release. *Neuron* 20, 389–399. doi: 10.1016/S0896-6273(00)80983-6

Newman, E. A. (2003). New roles for astrocytes: regulation of synaptic transmission. *Trends Neurosci.* 26, 536–542. doi: 10.1016/S0166-2236(03)00237-6

Pangrsic, T., Potokar, M., Stenovec, M., Kreft, M., Fabbretti, E., Nistri, A., et al. (2007). Exocytotic release of ATP from cultured astrocytes. *J. Biol. Chem.* 282, 28749–28758. doi: 10.1074/jbc.M700290200

Pankratov, Y., Lalo, U., Verkhratsky, A., and North, R. A. (2006). Vesicular release of ATP at central synapses. *Pflugers Arch.* 452, 589–597. doi: 10.1007/s00424-006-0061-x

Pankratov, Y., Lalo, U., Verkhratsky, A., and North, R. A. (2007). Quantal release of ATP in mouse cortex. *J. Gen. Physiol.* 129, 257–265. doi: 10.1085/jgp.200609693

Parpura, V., Heneka, M. T., Montana, V., Oliet, S. H., Schousboe, A., Haydon, P. G., et al. (2012). Glial cells in (patho)physiology. *J. Neurochem.* 121, 4–27. doi: 10.1111/j.1471-4159.2012.07664.x

Rivera, A., Vanzulli, I., and Butt, A. M. (2016). A central role for ATP signalling in glial interactions in the CNS. *Curr. Drug Targets* 17, 1829–1833. doi: 10.2174/1389450117666160711154529

Rodrigues, R. J., Tomé, A. R., and Cunha, R. A. (2015). ATP as a multi-target danger signal in the brain. *Front. Neurosci.* 9:148. doi: 10.3389/fnins.2015.00148

Stout, C. E., Costantin, J. L., Naus, C. C., and Charles, A. C. (2002). Intercellular calcium signaling in astrocytes via ATP release through connexin hemichannels. *J. Biol. Chem.* 277, 10482–10488. doi: 10.1074/jbc.M109902200

Suadicani, S. O., Brosnan, C. F., and Scemes, E. (2006). P2X7 receptors mediate ATP release and amplification of astrocytic intercellular Ca^{2+} signaling. *J. Neurosci.* 26, 1378–1385. doi: 10.1523/JNEUROSCI.3902-05.2006

Sudhof, T. C. (2004). The synaptic vesicle cycle. *Annu. Rev. Neurosci.* 27, 509–547. doi: 10.1146/annurev.neuro.26.041002.131412

Sudhof, T. C., and Rothman, J. E. (2009). Membrane fusion: grappling with SNARE and SM proteins. *Science* 323, 474–477. doi: 10.1126/science.1161748

Verkhratsky, A., Krishtal, O. A., and Burnstock, G. (2009). Purinoceptors on neuroglia. *Mol. Neurobiol.* 39, 190–208. doi: 10.1007/s12035-009-8063-2

Verkhratsky, A., Matteoli, M., Parpura, V., Mothet, J. P., and Zorec, R. (2016). Astrocytes as secretory cells of the central nervous system: idiosyncrasies of vesicular secretion. *EMBO J.* 35, 239–257. doi: 10.15252/embj.201592705

Verkhratsky, A., and Nedergaard, M. (2018). Physiology of Astroglia. *Physiol. Rev.* 98, 239–389. doi: 10.1152/physrev.00042.2016

Volterra, A., Liaudet, N., and Savtchouk, I. (2014). Astrocyte Ca^{2+} signalling: an unexpected complexity. *Nat. Rev. Neurosci.* 15, 327–335. doi: 10.1038/nrn3725

Wurm, A., Pannicke, T., Iandiev, I., Francke, M., Hollborn, M., Wiedemann, P., et al. (2011). Purinergic signaling involved in Muller cell function in the mammalian retina. *Prog. Retin. Eye Res.* 30, 324–342. doi: 10.1016/j.preteyeres.2011.06.001

Xiong, Y., Teng, S., Zheng, L., Sun, S., Li, J., Guo, N., et al. (2018). Stretch-induced Ca($^{2+}$) -independent ATP release in hippocampal astrocytes. *J. Physiol.* 596, 1931–1947. doi: 10.1113/JP275805

Yan, Z., Li, S., Liang, Z., Tomić, M., and Stojilkovic, S. S. (2008). The P2X7 receptor channel pore dilates under physiological ion conditions. *J. Gen. Physiol.* 132, 563–573. doi: 10.1085/jgp.200810059

Young, S. H., and Poo, M. M. (1983). Spontaneous release of transmitter from growth cones of embryonic neurones. *Nature* 305, 634–637. doi: 10.1038/305634a0

Zhang, J. M., Wang, H. K., Ye, C. Q., Ge, W., Chen, Y., Jiang, Z. L., et al. (2003). ATP released by astrocytes mediates glutamatergic activity-dependent heterosynaptic suppression. *Neuron* 40, 971–982. doi: 10.1016/S0896-6273(03)00717-7

Zhang, Z., Chen, G., Zhou, W., Song, A., Xu, T., Luo, Q., et al. (2007). Regulated ATP release from astrocytes through lysosome exocytosis. *Nat. Cell Biol.* 9, 945–953. doi: 10.1038/ncb1620

Zhao, Q., Zhou, H., Chi, S., Wang, Y., Wang, J., Geng, J., et al. (2018). Structure and mechanogating mechanism of the Piezo1 channel. *Nature* 554, 487–492. doi: 10.1038/nature25743

Permissions

All chapters in this book were first published by Frontiers; hereby published with permission under the Creative Commons Attribution License or equivalent. Every chapter published in this book has been scrutinized by our experts. Their significance has been extensively debated. The topics covered herein carry significant findings which will fuel the growth of the discipline. They may even be implemented as practical applications or may be referred to as a beginning point for another development.

The contributors of this book come from diverse backgrounds, making this book a truly international effort. This book will bring forth new frontiers with its revolutionizing research information and detailed analysis of the nascent developments around the world.

We would like to thank all the contributing authors for lending their expertise to make the book truly unique. They have played a crucial role in the development of this book. Without their invaluable contributions this book wouldn't have been possible. They have made vital efforts to compile up to date information on the varied aspects of this subject to make this book a valuable addition to the collection of many professionals and students.

This book was conceptualized with the vision of imparting up-to-date information and advanced data in this field. To ensure the same, a matchless editorial board was set up. Every individual on the board went through rigorous rounds of assessment to prove their worth. After which they invested a large part of their time researching and compiling the most relevant data for our readers.

The editorial board has been involved in producing this book since its inception. They have spent rigorous hours researching and exploring the diverse topics which have resulted in the successful publishing of this book. They have passed on their knowledge of decades through this book. To expedite this challenging task, the publisher supported the team at every step. A small team of assistant editors was also appointed to further simplify the editing procedure and attain best results for the readers.

Apart from the editorial board, the designing team has also invested a significant amount of their time in understanding the subject and creating the most relevant covers. They scrutinized every image to scout for the most suitable representation of the subject and create an appropriate cover for the book.

The publishing team has been an ardent support to the editorial, designing and production team. Their endless efforts to recruit the best for this project, has resulted in the accomplishment of this book. They are a veteran in the field of academics and their pool of knowledge is as vast as their experience in printing. Their expertise and guidance has proved useful at every step. Their uncompromising quality standards have made this book an exceptional effort. Their encouragement from time to time has been an inspiration for everyone.

The publisher and the editorial board hope that this book will prove to be a valuable piece of knowledge for researchers, students, practitioners and scholars across the globe.

List of Contributors

Jian Jing Siew and Yijuang Chern
Taiwan International Graduate Program in Molecular Medicine, National Yang-Ming University and Academia Sinica, Taipei, Taiwan
Institute of Biomedical Sciences, Academia Sinica, Taipei, Taiwan

Lisanne E. Wisse, Carola G. M. van Berkel, Timo J. ter Braak, Emiel Polder, Marjo S. van der Knaap and Truus E. M. Abbink
Pediatrics, VU University Medical Center, Amsterdam, Netherlands

Renske Penning, Esther A. Zaal, Celia R. Berkers and Maarten A. F. Altelaar
Biomolecular Mass Spectrometry and Proteomics Group, Utrecht Institute for Pharmaceutical Sciences, Bijvoet Center for Biomolecular Research, Utrecht University, Utrecht, Netherlands

Justin W. Kenney and Christopher G. Proud
Centre for Biological Sciences, University of Southampton, Southampton, United Kingdom

Dave Speijer
Medical Biochemistry, Academic Medical Center, Amsterdam, Netherlands

Susanna Amadio and Savina Apolloni
IRCCS Santa Lucia Foundation, Experimental Neuroscience, Rome, Italy

Paola Fabbrizio
IRCCS Santa Lucia Foundation, Experimental Neuroscience, Rome, Italy
Department of Systems Medicine, Tor Vergata University, Rome, Italy

Cinzia Volonté
IRCCS Santa Lucia Foundation, Experimental Neuroscience, Rome, Italy
CNR, Institute of Cell Biology and Neurobiology, Rome, Italy

Kacper A. Walentynowicz, Natalia Ochocka, Kamil Wojnicki, Karolina Stepniak, Jakub Mieczkowski, Iwona A. Ciechomska and Bozena Kaminska
Laboratory of Molecular Neurobiology, Neurobiology Center, Nencki Institute of Experimental Biology, Polish Academy of Sciences, Warsaw, Poland

Maria Pasierbinska
Laboratory of Molecular Neurobiology, Neurobiology Center, Nencki Institute of Experimental Biology, Polish Academy of Sciences, Warsaw, Poland
Glia Sp. z o.o, Warsaw, Poland

Meagan Milton and Patrice D. Smith
Department of Neuroscience, Carleton University, Ottawa, ON, Canada

Francesca Gullo, Michela Ceriani, Barbara Costa and Marzia Lecchi
Department of Biotechnology and Biosciences and Milan Center for Neuroscience, University of Milano-Bicocca, Milan, Italy

Alessia D'Aloia and Enzo Wanke
Department of Biotechnology and Biosciences, University of Milano-Bicocca, Milan, Italy

Andrew Constanti
Department of Pharmacology, School of Pharmacy, University College London, London, United Kingdom

Kathleen Jordan, Joseph Murphy and Cassie S. Mitchell
Laboratory for Pathology Dynamics, Department of Biomedical Engineering, Georgia Institute of Technology, Emory University School of Medicine, Atlanta, GA, United States

Anjanya Singh
Laboratory for Pathology Dynamics, Department of Biomedical Engineering, Georgia Institute of Technology, Emory University School of Medicine, Atlanta, GA, United States
School of Medicine, University of Texas Southwestern Medical Center, Dallas, TX, United States

David A. Menassa and Diego Gomez-Nicola
Biological Sciences, Faculty of Natural and Environmental Sciences, Southampton General Hospital, University of Southampton, Southampton, United Kingdom

Kimberley D. Bruce, Sachi Gorkhali and Robert H. Eckel
Division of Endocrinology, Metabolism, & Diabetes, Department of Medicine, University of Colorado Denver Anschutz Medical Campus, Aurora, CO, United States

Katherine Given and Wendy B. Macklin
Department of Cell and Developmental Biology, University of Colorado School of Medicine, Aurora, CO, United States

Alison M. Coates
School of Health Sciences, Sansom Institute for Health Research, University of South Australia, Adelaide, SA, Australia

Kristen E. Boyle
Department of Pediatrics, Section of Nutrition, University of Colorado School of Medicine, Aurora, CO, United States

Zhai Yang, Qing Xie, Cheng-Liang Hu, Qiong Jiang, Hui-Fan Shen and Wei-Jiang Zhao
Center for Neuroscience, Shantou University Medical College, Shantou, China

Melitta Schachner
Center for Neuroscience, Shantou University Medical College, Shantou, China
Keck Center for Collaborative Neuroscience and Department of Cell Biology and Neuroscience, Rutgers University, Piscataway, NJ, United States

Angélique Boerboom, Valérie Dion and Rachelle Franzen
GIGA-Neurosciences, University of Liège, Liège, Belgium

Alain Chariot
GIGA-Molecular Biology of Diseases, University of Liège, Liège, Belgium
Walloon Excellence in Lifesciences and Biotechnology (WELBIO), Wavre, Belgium

Tian Zhou, Zijing Huang, Xiaowei Zhu, Xiaowei Sun, Yan Liu, Bing Cheng, Mei Li, Yizhi Liu, Chang He and Xialin Liu
State Key Laboratory of Ophthalmology, Zhongshan Ophthalmic Center, Sun Yat-sen University, Guangzhou, China

Giampiero Leanza
Department of Drug Sciences, University of Catania, Catania, Italy

Rosario Gulino
Department of Biomedical and Biotechnological Sciences, University of Catania, Catania, Italy

Robert Zorec
Laboratory of Neuroendocrinology-Molecular Cell Physiology, Faculty of Medicine, Institute of Pathophysiology, University of Ljubljana, Ljubljana, Slovenia
Laboratory of Cell Engineering, Celica Biomedical, Ljubljana, Slovenia

Tao Li, Lingyun Wang, Teng Ma, Shouyu Wang, Jianqin Niu, Hongli Li and Lan Xiao
Chongqing Key Laboratory of Neurobiology, Department of Histology and Embryology, Third Military Medical University, Chongqing, China

Alex Plata, Albina Lebedeva, Pavel Denisov, Olga Nosova and Alexey Pimashkin
UNN Institute of Neuroscience, N. I. Lobachevsky State University of Nizhny Novgorod, University of Nizhny Novgorod, Nizhny Novgorod, Russia

Tatiana Y. Postnikova
Laboratory of Molecular Mechanisms of Neural Interactions, Sechenov Institute of Evolutionary Physiology and Biochemistry, Russian Academy of Sciences, St. Petersburg, Russia
Department of Medical Physics, Peter the Great St. Petersburg Polytechnic University, St. Petersburg, Russia

Alexey Brazhe
Department of Biophysics, Faculty of Biology, M. V. Lomonosov Moscow State University, Moscow, Russia

Aleksey V. Zaitsev
Laboratory of Molecular Mechanisms of Neural Interactions, Sechenov Institute of Evolutionary Physiology and Biochemistry, Russian Academy of Sciences, St. Petersburg, Russia
Institute of Experimental Medicine, Almazov National Medical Research Centre, St. Petersburg, Russia

Dmitri A. Rusakov
UNN Institute of Neuroscience, N. I. Lobachevsky State University of Nizhny Novgorod, University of Nizhny Novgorod, Nizhny Novgorod, Russia
UCL Institute of Neurology, University College London, London, United Kingdom

Alexey Semyanov
UNN Institute of Neuroscience, N. I. Lobachevsky State University of Nizhny Novgorod, University of Nizhny Novgorod, Nizhny Novgorod, Russia
Department of Molecular Neurobiology, Shemyakin-Ovchinnikov Institute of Bioorganic Chemistry, Russian Academy of Sciences, Moscow, Russia
All-Russian Research Institute of Medicinal and Aromatic Plants, Moscow, Russia

R. Ayana and Soumya Pati
Department of Life Sciences, School of Natural Sciences, Shiv Nadar University, Lucknow, India

Shailja Singh
Department of Life Sciences, School of Natural Sciences, Shiv Nadar University, Lucknow, India
Special Centre for Molecular Medicine, Jawaharlal Nehru University, New Delhi, India

Fei Yang, Wen-Jun Luo and Jiang-Lin Wang
Institute for Biomedical Sciences of Pain, Tangdu Hospital, The Fourth Military Medical University, Xi'an, China

Wei Sun, Yan Wang, Fan Yang, Chun-Li Li, Na Wei and Xiao-Liang Wang
Institute for Biomedical Sciences of Pain, Tangdu Hospital, The Fourth Military Medical University, Xi'an, China
Key Laboratory of Brain Stress and Behavior, People's Liberation Army (PLA), Xi'an, China

Su-Min Guan
School of Stomatology, The Fourth Military Medical University, Xi'an, China

Jun Chen
Institute for Biomedical Sciences of Pain, Tangdu Hospital, The Fourth Military Medical University, Xi'an, China
Key Laboratory of Brain Stress and Behavior, People's Liberation Army (PLA), Xi'an, China
Beijing Institute for Brain Disorders, Beijing, China

Suhua Sun, Sasa Teng and Zhuan Zhou
State Key Laboratory of Biomembrane and Membrane Biotechnology and Peking-Tsinghua Center for Life Sciences and PKU-IDG/McGovern Institute for Brain Research, Institute of Molecular Medicine, Peking University, Beijing, China

Yingfei Xiong
State Key Laboratory of Biomembrane and Membrane Biotechnology and Peking-Tsinghua Center for Life Sciences and PKU-IDG/McGovern Institute for Brain Research, Institute of Molecular Medicine, Peking University, Beijing, China
Department of Neurosurgery, Affiliated Hospital of The Air Force Institute of Aeromedicine, Beijing, China

Mu Jin
State Key Laboratory of Biomembrane and Membrane Biotechnology and Peking-Tsinghua Center for Life Sciences and PKU-IDG/McGovern Institute for Brain Research, Institute of Molecular Medicine, Peking University, Beijing, China
Department of Anesthesiology, Beijing Friendship Hospital, Capital Medical University, Beijing, China

Index

www.ingramcontent.com/pod-product-compliance
Lightning Source LLC
LaVergne TN
LVHW081029210926
848485LV00004B/981